Texts and
Monographs
in Physics

Texts and Monographs in Physics

R. Bass: **Nuclear Reactions with Heavy Ions** (1980).

A. Böhm: **Quantum Mechanics** (1979).

O. Bratteli and D.W. Robinson: **Operator Algebras and Quantum Statistical Mechanics.** Volume I: C*- and W*-Algebras. Symmetry Groups, Decomposition of States (1979). Volume II: Equilibrium States. Models in Quantum Statistical Mechanics (1981).

K.Chadan and P.C. Sabatier: **Inverse Problems in Quantum Scattering Theory** (1977).

G. Gallavotti: **The Elements of Mechanics I** (1983).

W. Glöckle: **The Quantum Mechanical Few-Body Problem** (1983).

J.M. Jauch and F. Rohrlich: **The Theory of Photons and Electrons: The Relativistic Quantum Field Theory of Charged Particles with Spin One-half,** Second Expanded Edition (1980).

J. Kessler: **Polarized Electrons (1976).**

G. Ludwig: **Foundations of Quantum Mechanics I** (1983).

G. Ludwig: **Foundations of Quantum Mechanics II** (1985).

R.G. Newton: **Scattering Theory of Waves and Particles,** Second Edition (1982).

H. Pilkuhn: **Relativistic Particle Physics** (1979).

R.D. Richtmyer: **Principles of Advanced Mathematical Physics.** Volume I (1978). Volume II (1981)

W. Rindler: **Essential Relativity: Special, General, and Cosmological,** Revised Second Edition (1980).

P.Ring and P. Schuck: **The Nuclear Many-Body Problem** (1980).

R.M. Santilli: **Foundations of Theoretical Mechanics.** Volume I: The Inverse Problem in Newtonian Mechanics (1978). Volume II: Birkhoffian Generalization of Hamiltonian Mechanics (1983).

M.D. Scadron: **Advanced Quantum Theory and Its Applications Through Feynman Diagrams** (1979).

C. Truesdell and S. Bharatha: **The Concepts and Logic of Classical Thermodynamics as a Theory of Heat Engines: Rigourously Constructed upon the Foundation Laid by S. Carnot and F. Reech** (1977).

F.J. Yndurain: **Quantum Chromodynamics: An Introduction to the Theory of Quarks and Gluons** (1983).

G. Ludwig

Foundations of Quantum Mechanics II

Translated by Carl A. Hein

With 54 Illustrations

Springer-Verlag
New York Berlin Heidelberg Tokyo

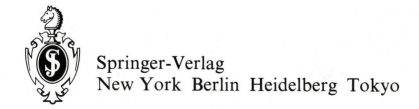

G. Ludwig
Institut für Theoretische Physik
Universität Marburg
Renthof 7
Federal Republic of Germany

Carl A. Hein (*Translator*)
Dunster House
Swanson Road
Boxboro, MA 01719
U.S.A.

Editors

Wolf Beiglböck
Institut für Angewandte Mathematik
Universität Heidelberg
Im Neuenheimer Feld 5
D-6900 Heidelberg 1
Federal Republic of Germany

Joseph L. Birman
Department of Physics
The City College of the
City University of New York
New York, NY 10031
U.S.A.

Elliott H. Lieb
Department of Physics
Joseph Henry Laboratories
Princeton University
Princeton, NJ 08540
U.S.A.

Tullio Regge
Istituto de Fisica Teorica
Universita di Torino
C. so M. d'Azeglio, 46
10125 Torino
Italy

Walter Thirring
Institut für Theoretische Physik
der Universität Wien
Boltzmanngasse 5
A-1090 Wien
Austria

Library of Congress Cataloging in Publication Data
Ludwig, Gunther, 1918–
Foundations of quantum mechanics.
(Texts and monographs in physics)
Translation of: Die Grundlagen der
Quantenmechanik.
Bibliography: p.
Includes index.
1. Quantum theory. I. Title. II. Series.
QC174.12.L8318 1983 530.1′2 82-10437

Original German edition: *Die Grundlagen der Quantenmechanik.* Berlin–Heidelberg–New York: Springer-Verlag, 1954.

Typeset by Composition House Ltd., Salisbury, England.
Printed and bound by R. R. Donnelley & Sons, Harrisonburg, Virginia.
Printed in the United States of America.

9 8 7 6 5 4 3 2 1

ISBN 0-387-13009-8 Springer-Verlag New York Berlin Heidelberg Tokyo
ISBN 3-540-13009-8 Springer-Verlag Berlin Heidelberg New York Tokyo

Dedicated to my wife

Preface

In this second volume on the Foundations of Quantum Mechanics we shall show how it is possible, using the methodology presented in Volume I, to deduce some of the most important applications of quantum mechanics. These deductions are concerned with the structures of the microsystems rather than the technical details of the construction of preparation and registration devices. Accordingly, the only new axioms (relative to Volume I) which are introduced are concerned with the relationship between ensemble operators W, effect operators F, and certain construction principles of the preparation and registration devices. The applications described here are concerned with the measurement of atomic and molecular structure and of collision experiments.

An additional and essential step towards a theoretical description of the preparation and registration procedures is carried out in Chapter XVII. Here we demonstrate how microscopic collision processes (that is, processes which can be described by quantum mechanics) can be used to obtain novel preparation and registration procedures if we take for granted the knowledge of only a few macroscopic preparation and registration procedures. By clever use of collision processes we are often able to obtain very precise results for the operators W and F which describe the *total* procedures from a very imprecise knowledge of the *macroscopic parts* of the preparation and registration processes. In this regard experimental physicists have done brilliant work. In this sense Chapter XVII represents a general theoretical foundation for the procedures used by experimental physicists.

Thus Chapters II to XVII represent a complete foundation of quantum theory as far as experimental practice is concerned. Fundamental questions about the relationships between quantum mechanics and the objective description of macroscopic systems (that is, a "statistical mechanics" description), and the related problem of the completion of the measurement process in the macroscopic domain are not treated in this book. Readers will find a few critical comments and discussion of these problems in Chapter XVIII. A fundamental treatment of these problems can be found in [7].

I hope that this second volume, and Volume I, will contribute to the elimination of false problems in quantum mechanics. The two volumes present a formulation of quantum mechanics which is self-consistent and can describe *all* possible experiments in the application range of the theory. Whether this description can fulfill the conscious or unconscious ideological and aesthetic desires of all readers is another problem.

References in the text are made as follows: For references to other sections of the same chapter, we shall only list the section number of the reference, for example, §3.1. For references to other chapters, the chapter is also given; for example, XVII, §2.3 refers to Chapter XVII, Section 2.3. The formulas are numbered as follows: (2.3.10) refers to the 10th formula in Section 2.3 of the current chapter. References to formulas in other chapters are given, for example, by XIV (2.3.10). References to the Appendix are given by AV, §2, where AV denotes Appendix V.

Again, as in the case of Volume I, I wish to express my deep gratitude to Mr. Carl A. Hein who has undertaken the difficult task of translating these two volumes into English. The translation has been especially difficult because of the wide range of topics discussed in these two volumes, ranging from the coverage of various specialized areas in advanced mathematics, philosophical questions in quantum mechanics, and the discussion of theoretical and experimental aspects of quantum physics. I would also like to thank Mr. Hein for accommodating my wishes to make late alterations to the text which I hope will make the book more understandable to readers.

Marburg, January 1984 G. LUDWIG

Contents (Volume II)

Contents (Volume I)

CHAPTER IX

Representation of Hilbert Spaces by Function Spaces

The representation of Hilbert spaces in terms of function spaces is of great importance in the application of quantum mechanics. We have already encountered such representations in VII, §2 in the form $\mathscr{L}^2(\mathbf{R}^3, dk_1\, dk_2\, dk_3)$ and in VII, §3 in the form $\mathscr{L}^2(\mathbf{R}^3, dx_1\, dx_2\, dx_3)$, where we have considered isomorphic maps of $\mathscr{L}^2(\mathbf{R}^3, dx_1\, dx_2\, dx_3)$ into $\mathscr{L}^2(\mathbf{R}^3, dk_1\, dk_2\, dk_3)$ which are defined by Fourier transforms. Such (and similar) representations are especially suitable for the treatment of physical problems. This chapter does not introduce any new axioms, that is, any new physical laws; instead, we introduce a number of suitably chosen tools for the analysis of the physical laws in the previous chapters. In this chapter we shall consider only a single Hilbert space $\mathscr{H}$, that is, a single-system type.

1 Maximal Decision Observables

The starting point for our discussion of particular representations of a Hilbert space is the discussion of measurement scales for observables presented in IV, §2.5, and, in particular, the discussion of "scale observables" as described in IV, D 2.5.6.

We shall now assume that we are given a complete Boolean ring $\Sigma = \Sigma(y_1, \ldots, y_n)$ of decision effects, the scale for which is uniquely determined by the specification of n scale observables $A_1, \ldots, A_n$ (which mutually commute, see the remarks preceding IV, D 2.5.6). In IV, §2.5 we have seen that, to each Boolean ring Σ of decision effects there exists a finite number of scale observables $A_1, \ldots, A_n$ for which $\Sigma = \Sigma(y_1, \ldots, y_n)$. Indeed, we may choose $n = 1$.

Because we frequently find that certain A_ν (and the corresponding scales) are physically determined on the basis of transformation groups (see, for example, VII and VIII) we will therefore not assume that $n = 1$. Since it is not, however, difficult to extend the structures described for the case $n = 1$ to the case $n > 1$, the following discussion will be more transparent if, in the following, we examine the case in which $n = 1$, and then carry out the extension to $n > 1$ wherever necessary.

A Boolean ring Σ is said to be maximal, as a subset of G, if there does not exist a Boolean ring $\tilde{\Sigma} \subset G$ for which $\Sigma \subsetneqq \tilde{\Sigma}$. A maximal Σ is, on the basis of IV, Th. 1.4.7, complete. According to IV, Th. 1.3.6 Σ is maximal if there exists no $E \in G$, $E \notin \Sigma$ which is commensurable with all elements of Σ.

Th. 1.1. *To each $\Sigma \subset G$ there exists a maximal $\tilde{\Sigma} \subset G$ for which $\Sigma \subset \tilde{\Sigma}$.*

The proof of this result follows directly from Zorn's lemma, because, for a totally ordered subset of Boolean rings $\Sigma_\lambda \subset G$ for which $\Sigma \subset \Sigma_\lambda$, the set $\bigcup_\lambda \Sigma_\lambda$ is a Boolean ring.

Since, according to IV, §2.5 each Σ may be represented as $\Sigma(y_1, \ldots, y_n)$ we shall, for the purpose of discussion of the representations of Hilbert space, always use maximal Boolean rings $\Sigma(y_1, \ldots, y_n) \subset G$.

D 1.1. A maximal Boolean subring of G is said to be a maximal decision observable. A n-tuple of commuting $A_1, \ldots, A_n$ is said to be maximal if the Boolean ring $\Sigma(y_1, \ldots, y_n)$ generated by this n-tuple is maximal.

If $\{A_\lambda\}$ is a set (not necessarily finite) of self-adjoint commuting operators and if Σ is the Boolean ring generated by the spectral families of the A_λ then there exists a maximal $\tilde{\Sigma}$ for which $\Sigma \subset \tilde{\Sigma}$. According to IV, Th. 2.5.6, to $\tilde{\Sigma}$ there exists a y such that $\tilde{\Sigma} = \tilde{\Sigma}(y)$ and, according to IV, Th. 2.5.9 there exists a maximal scale observable A for which the corresponding Boolean ring is equal to $\tilde{\Sigma}$. Since $\tilde{\Sigma}(y_\lambda) \subset \Sigma \subset \tilde{\Sigma}$, A_λ is, according to IV, Th. 2.5.11, a function $f_\lambda(A)$. To each set $\{A_\lambda\}$ of coexistent (and therefore commensurable) scale observables there exists a scale observable A for which all the A_λ are functions of A.

We shall now introduce the following notation: For the Boolean ring $\Sigma \subset G$ and $\varphi \in \mathscr{H}$ let $\mathscr{T}(\Sigma, \varphi)$ denote the (closed) subspace of $\mathscr{H}$ generated by $\{E\varphi \mid E \in \Sigma\}$; let $E(\Sigma, \varphi)$ denote the projection operator corresponding to $\mathscr{T}(\Sigma, \varphi)$. Since $\mathbf{1} \in \Sigma$ we find that $\varphi \in \mathscr{T}(\Sigma, \varphi)$.

Th. 1.2. *$E(\Sigma, \varphi)$ is commensurable to all elements of Σ.*

PROOF. For $\chi \in \mathscr{T}(\Sigma, \varphi)$ it follows that $E\chi \in \mathscr{T}(\Sigma, \varphi)$ for all $E \in \Sigma$, since this result is trivial for all χ of the form $\tilde{E}\varphi$ where $\tilde{E} \in \Sigma$, since $E\tilde{E} \in \Sigma$. For $\psi \in \mathscr{H}$ we find that

$E(\Sigma, \varphi)\psi \in \mathscr{T}(\Sigma, \varphi)$. Therefore, for $E \in \Sigma$ we obtain $EE(\Sigma, \varphi)\psi \in \mathscr{T}(\Sigma, \varphi)$, that is, $E(\Sigma, \varphi)EE(\Sigma, \varphi)\psi = EE(\Sigma, \varphi)\psi$. Since ψ was arbitrary it follows that

$$E(\Sigma, \varphi)EE(\Sigma, \varphi) = EE(\Sigma, \varphi).$$

From the adjoint equation we obtain $E(\Sigma, \varphi)EE(\Sigma, \varphi) = E(\Sigma, \varphi)E$ and we have proven that E and $E(\Sigma, \varphi)$ commute.

Th. 1.3. *For Σ there exists a $\varphi \in \mathscr{H}$ for which $w = P_\varphi$ is effective on Σ, that is, for $E \in \Sigma$ the following relationship is satisfied:*

$$\mathrm{tr}(P_\varphi E) = \langle \varphi, E\varphi \rangle = \langle \varphi, E^2\varphi \rangle = \|E\varphi\|^2 = 0 \Rightarrow E = 0.$$

If Σ is maximal, then there exists a $\varphi \in \mathscr{H}$ for which $\mathscr{T}(\Sigma, \varphi) = \mathscr{H}$.

PROOF. If Σ is not maximal, then, according to Th. 1.1 there exists a maximal $\tilde{\Sigma}$ such that $\Sigma \in \tilde{\Sigma}$. If, according to the second part of Th. 1.3, there exists a φ such that $\mathscr{T}(\Sigma, \varphi) = \mathscr{H}$, then, if for $\tilde{E} \in \Sigma$ we obtain $\|\tilde{E}\varphi\| = 0$, then for ψ from the projection space of $\tilde{E}$ and for arbitrary $E \in \tilde{E}$ we obtain

$$\langle \varphi, E\varphi \rangle = \langle \tilde{E}\psi, E\varphi \rangle = \langle \psi, \tilde{E}E\varphi \rangle = \langle \psi, E\tilde{E}\varphi \rangle = 0,$$

that is, ψ is orthogonal to all $\{E\varphi \mid E \in \tilde{E}\}$ and therefore is orthogonal to $\mathscr{T}(\Sigma, \varphi) = \mathscr{H}$, from which it follows that $\psi = 0$ and $\tilde{E} = 0$.

Now we need only prove the second part of Th. 1.3. We shall assume that Σ is maximal. Let $\{\chi_\nu\}$ denote a dense denumerable subset in $\mathscr{H}$. We define

$$\mathscr{H}_n = \bigvee_{\nu=1}^{n} \mathscr{T}(\Sigma, \chi_\nu).$$

Therefore $\{\mathscr{H}_n\}$ is an increasing sequence of subspaces. Since $\chi_\nu \in \mathscr{T}(\Sigma, \chi_\nu)$ we find that $\chi_\nu \in \mathscr{H}_n$ for $\nu \leq n$. Thus it follows that $\bigvee_n \mathscr{H}_n = \mathscr{H}$.

We shall now show that, to each $\mathscr{H}_n$ there exists a ψ_n such that $\mathscr{H}_n = \mathscr{T}(\Sigma, \psi_n)$. We shall prove this result by induction.

For $\mathscr{H}_1$ we need only choose $\psi_1 = \chi_1$. Therefore, according to the induction hypothesis $\mathscr{H}_{n-1} = \mathscr{T}(\Sigma, \psi_{n-1})$. Therefore we find that $\mathscr{H}_n = \mathscr{T}(\Sigma, \psi_{n-1}) \vee \mathscr{T}(\Sigma, \chi_n)$. Let E_n denote the projection operator on $\mathscr{H}_n$.

We therefore obtain $E_n = E(\Sigma, \psi_{n-1}) \vee E(\Sigma, \chi_n)$. Since, according to Th. 1.2, $E(\Sigma, \psi_{n-1})$ and $E(\Sigma, \chi_n)$ are commensurable with all $E \in \Sigma$, then, according to IV, Th. 1.3.5, E_n is commensurable with all $E \in \Sigma$. Since Σ is maximal, we therefore find that $E_n \in \Sigma$ and $E(\Sigma, \chi_n) \in \Sigma$.

We now choose ψ_n as follows:

$$\psi_n = \psi_{n-1} + (\mathbf{1} - E_{n-1})\chi_n.$$

By multiplication with an $E \in \Sigma$ it follows that

$$E\psi_n = E\psi_{n-1} + E(\mathbf{1} - E_{n-1})\chi_n.$$

Since $E_{n-1} \in \Sigma$ and therefore $\mathbf{1} - E_{n-1} \in \Sigma$ we find that $E(\mathbf{1} - E_{n-1}) \in \Sigma$ and we obtain $E(\mathbf{1} - E_{n-1})\chi_n \in \mathscr{T}(\Sigma, \chi_n)$. Thus it follows that $E\psi_n \in \mathscr{H}_n$ and therefore $\mathscr{T}(\Sigma, \psi_n) \subset \mathscr{H}_n$.

Since $E_{n-1}E\psi_n = E_{n-1}E\psi_{n-1} = EE_{n-1}\psi_{n-1} = E\psi_{n-1}$ we obtain $E\psi_{n-1} \in \mathscr{T}(\Sigma, \psi_n)$ and $\mathscr{H}_{n-1} = \mathscr{T}(\Sigma, \psi_{n-1}) \subset \mathscr{T}(\Sigma, \psi_n)$. Since $(\mathbf{1} - E_{n-1})E\psi_n = E(\mathbf{1} - E_{n-1})\psi_{n-1} + E(\mathbf{1} - E_{n-1})\chi_n = E(\mathbf{1} - E_{n-1})\chi_n = (\mathbf{1} - E_{n-1})E\chi_n$ (because

$E_{n-1}\psi_{n-1} = \psi_{n-1}$!) we obtain $(\mathbf{1} - E_{n-1})\mathscr{T}(\Sigma, \chi_n) = \mathscr{T}(\Sigma, \chi_n) \cap \mathscr{H}_{n-1}^{\perp} \subset \mathscr{T}(\Sigma, \psi_n)$. Since $E(\Sigma, \chi_n)$ and E_{n-1} are commensurable,

$$\mathscr{H}_n = \mathscr{H}_{n-1} \vee \mathscr{T}(\Sigma, \chi_n) = \mathscr{H}_{n-1} \vee [\mathscr{T}(\Sigma, \chi_n) \cap \mathscr{H}_{n-1}^{\perp}]$$

and we therefore obtain $\mathscr{H}_n \subset \mathscr{T}(\Sigma, \psi_n)$. Thus we have proven that $\mathscr{H}_n = \mathscr{T}(\Sigma, \psi_n)$. If $\mathscr{H}_n = \mathscr{H}$ from a given value of n, then the theorem is proven. Since $\bigvee_n \mathscr{H}_n = \mathscr{H}$. $E_n f \to f$ for each $f \in \mathscr{H}$. If we set $\eta_1 = \psi_1, \ldots, \eta_n = (E_n - E_{n-1})\psi_n$ then $\mathscr{T}(\Sigma, \eta_n) = \mathscr{H}_n \cap \mathscr{H}_{n-1}^{\perp}$. For $a_n \neq 0$ and $\sum_n \|a_n\|^2 < \infty$ we find that

$$\varphi = \sum_n a_n \frac{\eta_n}{\|\eta_n\|} \in \mathscr{H}$$

and, since $(E_n - E_{n-1})\varphi = a_n(\eta_n/\|\eta_n\|)$ it follows that $\eta_n \in \mathscr{T}(\Sigma, \varphi)$ and we therefore obtain $\mathscr{H} = \bigvee_n \mathscr{T}(\Sigma, \eta_n) \subset \mathscr{T}(\Sigma, \varphi)$, that is, $\mathscr{T}(\Sigma, \varphi) = \mathscr{H}$.

In the next section we shall obtain the following result as a corollary:

Th. 1.4. *If $\|E\psi\|^2$ is an effective measure over a complete Boolean ring $\Sigma \subset G$ and there exists a $\varphi \in \mathscr{H}$ for which $\mathscr{T}(\Sigma, \varphi) = \mathscr{H}$, then $\mathscr{T}(\Sigma, \psi) = \mathscr{H}$.*

From this result it follows that

Th. 1.5. *If Σ is complete, and if there exists a $\varphi \in \mathscr{H}$ for which $\mathscr{T}(\Sigma, \varphi) = \mathscr{H}$, then to each $\psi \in \mathscr{H}$ there corresponds a unique $E_0 \in \Sigma$ for which $E_0\psi = 0$ and $E\psi \neq 0$ for all $E \in \Sigma$ for which $E < \mathbf{1} - E_0$. In addition, $\mathscr{T}(\Sigma, \psi) = (\mathbf{1} - E_0)\mathscr{H}$, that is, $E(\Sigma, \psi) = \mathbf{1} - E_0 \in \Sigma$.*

Proof. We define E_0 as the union of all $E \in \Sigma$ for which $E\psi = 0$; since Σ is complete $E_0 \in \Sigma$; therefore E_0 is the largest $E \in \Sigma$ for which $E\psi = 0$. In the subspace $\mathscr{H}_0 = (\mathbf{1} - E_0)\mathscr{H}$ we find that $\Sigma_0 = \{E(\mathbf{1} - E_0) | E \in \Sigma\}$ and $\varphi_0 = (\mathbf{1} - E_0)\varphi$ satisfies all assumptions of Th. 1.4. Therefore $\psi = (\mathbf{1} - E_0)\psi$ is a vector from $\mathscr{H}_0$ for which $\mathscr{T}(\Sigma_0, \psi) = \mathscr{H}_0$, and we therefore find that

$$\mathscr{T}(\Sigma, \psi) = \mathscr{H}_0 = (\mathbf{1} - E_0)\mathscr{H},$$

that is, the projection $E(\Sigma, \psi)$ which corresponds to $\mathscr{T}(\Sigma, \psi)$ is an element of Σ.

Th. 1.6. *If Σ is complete, and there exists a $\varphi \in \mathscr{H}$ for which $\mathscr{T}(\Sigma, \varphi) = \mathscr{H}$, then Σ is maximal.*

Proof. Suppose that Σ is not maximal. Then there exists a $\tilde{E} \neq 0$ which is commensurable with all $E \in \Sigma$. For $\psi \in \tilde{E}\mathscr{H}$ it follows that $\mathscr{T}(\Sigma, \psi) \subset \tilde{E}\mathscr{H}$ and $E(\Sigma, \psi) < \tilde{E}$. Since $\psi \in \mathscr{T}(\Sigma, \psi)$ it follows that $\bigvee_{\psi \in \tilde{E}\mathscr{H}} \mathscr{T}(\Sigma, \psi) = \tilde{E}\mathscr{H}$, that is, $\bigvee_{\psi \in \tilde{E}\mathscr{H}} E(\Sigma, \psi) = \tilde{E}$. Since, according to Th. 1.5, $E(\Sigma, \psi) \in \Sigma$ and since Σ is complete, then $\tilde{E} \in \Sigma$ in contradiction to the assumption that $\tilde{E} \notin \Sigma$.

Th. 1.7. *For a complete Σ the following two conditions are equivalent:*

(i) *Σ is maximal.*
(ii) *There exists a $\varphi \in \mathscr{H}$ such that $\mathscr{T}(\Sigma, \varphi) = \mathscr{H}$.*

The proof follows directly from Th. 1.3 and Th. 1.6.

Th. 1.8. *If Σ is complete, but not maximal, then there exists at most a denumerable set of pairwise orthogonal $E_n \in G$ which are commensurable to all elements of Σ; the Boolean ring $\tilde{\Sigma}$ generated by Σ and by the E_n is maximal.*

Proof. Since the E_n are pairwise orthogonal, there can be at most denumerably many E_n. We now consider sets $\{\varphi_\nu\}$ of vectors for which the $\mathscr{T}(\Sigma, \varphi_\nu)$ are pairwise orthogonal. Among these sets there are maximal sets, a result which follows directly from Zorn's lemma. Let $\{\varphi_n\}$ denote such a maximal set. Let $\mathscr{T}_n = \mathscr{T}(\Sigma, \varphi_n)$ and $E_n = E(\Sigma, \varphi_n)$; let $\tilde{\Sigma}$ denote the closed Boolean ring generated by Σ and the E_n.

If $\tilde{\Sigma}$ was not maximal, then there exists a $E_0 \notin \tilde{\Sigma}$ which is commensurable with all $E \in \tilde{\Sigma}$. Thus we obtain

$$E_0 = \sum_n E_0 E_n + E_0\left(\mathbf{1} - \sum_n E_n\right),$$

where $E_0 E_n = E_n E_0$.

Since $\mathscr{T}_n = \mathscr{T}(\Sigma, \varphi_n) = \mathscr{T}(\Sigma, E_n \varphi_n)$, the Boolean ring $\Sigma_n = \{EE_n | E \in \Sigma\}$ is, according to Th. 1.7, maximal with respect to the Hilbert space $\mathscr{T}_n$; therefore $E_0 E_n \in \Sigma_n \subset \tilde{\Sigma}$. Therefore $\sum_n E_0 E_n \in \tilde{\Sigma}$.

If $\mathbf{1} - \sum_n E_n \neq 0$, then there would exist a χ where $\chi = (\mathbf{1} - \sum_n E_n)\chi$. Since $\mathscr{T}(\Sigma, \chi) = \mathscr{T}(\Sigma, (\mathbf{1} - \sum_n E_n)\chi) = (\mathbf{1} - \sum_n E_n)\mathscr{T}(\Sigma, \chi)$ it follows that $\mathscr{T}(\Sigma, \chi)$ is $\perp$ to all $\mathscr{T}_n$, in contradiction to the fact that the $\{\varphi_n\}$ are maximal. Therefore $E_0 = \sum_n E_0 E_n \in \tilde{\Sigma}$ and $\tilde{\Sigma}$ is therefore maximal.

Th. 1.9. *The E_n in Th.* 1.8 *can be so chosen such that from $EE_n = 0$ for an $E \in \Sigma$ it follows that $EE_m = 0$ for all $m \geq n$.*

Proof. We shall begin by using the E_n in Th. 1.8, and defining the $\tilde{E}_n$ for which Th. 1.9 is satisfied. With $\tilde{\Sigma}$ obtained from the proof of Th. 1.8, according to Th. 1.3 there exists a φ for which $\mathscr{T}(\tilde{\Sigma}, \varphi) = \mathscr{H}$. Since Σ is complete, the support (IV, Th. 2.1.5) of the measure $\|EE_n\varphi\|^2 = \mu_n(E)$ is defined in Σ and will be denoted by $E^{(n)}$. We therefore obtain $E^{(n)} \in \Sigma$. We therefore obtain $(\mathbf{1} - E^{(n)})E_n = 0$, that is, $E_n E^{(n)} = E_n$. Furthermore, from $EE_n = 0$ for $E \in \Sigma$ it follows that $E \leq \mathbf{1} - E^{(n)}$, that is, $EE^{(n)} = 0$. We recursively define:

$$E^{(1)'} = E^{(1)},$$

$$E^{(2)'} = E^{(2)} - E^{(2)}(\mathbf{1} - E^{(1)}) = E^{(2)}E^{(1)},$$

$$E^{(3)'} = E^{(3)} - E^{(3)}(\mathbf{1} - E^{(1)} \vee E^{(2)}) = E^{(3)}(E^{(1)} \vee E^{(2)}),$$

$$E^{(4)'} = E^{(4)} - E^{(4)}(\mathbf{1} - E^{(1)} \vee E^{(2)} \vee E^{(3)}) = E^{(4)}(E^{(1)} \vee E^{(2)} \vee E^{(3)}),$$

$$\vdots$$

and we then define

$$E^{(2)''} = E^{(2)'},$$

$$E^{(3)''} = E^{(3)'}E^{(2)'},$$

$$E^{(4)''} = E^{(4)'}(E^{(2)'} \vee E^{(3)'}),$$

$$\vdots$$

and, in the same way, we obtain $E^{(3)'''}$, $E^{(4)'''}$, etc.

The set of all projection operators obtained in the above manner are elements of Σ. With the aid of the above, we recursively define the $\tilde{E}_n$:

$$\begin{aligned}\tilde{E}_1 &= E_1E^{(1)} + E_2[E^{(2)}(\mathbf{1} - E^{(1)})] + E_3[E^{(3)}(\mathbf{1} - E^{(1)} \vee E^{(2)})]\\ &\quad + E_4[E^{(4)}(\mathbf{1} - E^{(1)} \vee E^{(2)} \vee E^{(3)}) + \cdots,\\ \tilde{E}_2 &= E_2E^{(2)'} + E_3[E^{(3)'}(\mathbf{1} - E^{(2)'})] + E_4[E^{(4)}(\mathbf{1} - E^{(2)'} \vee E^{(3)'})] + \cdots,\\ \tilde{E}_3 &= E_3E^{(3)''} + E_4[E^{(4)''}(\mathbf{1} - E^{(3)''})] + \cdots,\\ &\ \vdots\end{aligned}$$

From $E_nE^{(n)} = E_n$ and $E_nE_m = 0$ for $n \neq m$ it follows that $\tilde{E}_n$ are pairwise orthogonal. We must show that the complete Boolean ring generated by the $\tilde{E}_n$ and Σ is maximal and, from $E\tilde{E}_n = 0$ for a $E \in \Sigma$ we also obtain $E\tilde{E}_m = 0$ for all $m \geq n$.

Let $E\tilde{E}_n = 0$, for example, $E\tilde{E}_2 = 0$, then it follows that $E_2EE^{(2)'} = 0$, $E_3E[E^{(3)'}(\mathbf{1} - E^{(2)'})] = 0, \ldots$. Thus it follows that $EE^{(2)'} \leq \mathbf{1} - E^{(2)}$ and that with $E^{(2)'} = E^{(2)}E^{(1)}$ the relation $EE^{(2)'} = 0$ and then we obtain the relation $E_3EE^{(3)'} = 0$, that is, $EE^{(3)'} < \mathbf{1} - E_3$ and we therefore obtain $EE^{(3)'} = 0$.

Similarly, it follows that $EE^{(4)'} = 0$, etc. Thus it follows that $EE^{(2)''} = 0$, $EE^{(3)''} = 0$, $EE^{(4)''} = 0, \ldots$ and we finally obtain $E\tilde{E}_3 = 0$. Similarly, from $E\tilde{E}_n = 0$ it follows that $E\tilde{E}_{n+1} = 0$ and therefore $E\tilde{E}_m = 0$ for all $m \geq n$.

From the definition of $\tilde{E}_n$ it follows that

$$\begin{aligned}\tilde{E}_1E^{(1)} &= E_1E^{(1)} = E_1,\\ \tilde{E}_1E^{(2)} &= E_1E^{(2)} + E_2(\mathbf{1} - E^{(1)}),\\ \tilde{E}_1E^{(3)} &= E_1E^{(3)} + E_2(E^{(3)} - E^{(1)}E^{(3)}) + E_3(\mathbf{1} - E^{(1)} \vee E^{(2)}),\\ &\ \vdots\end{aligned}$$

and we obtain

$$\begin{aligned}E_1 &= \tilde{E}_1E^{(1)},\\ E_2(\mathbf{1} - E^{(1)}) &= \tilde{E}_1E^{(2)} - \tilde{E}_1E^{(1)}E^{(2)} = \tilde{E}_1E^{(2)}(\mathbf{1} - E^{(1)}),\\ E_3(\mathbf{1} - E^{(1)} \vee E^{(2)}) &= \tilde{E}_1E^{(3)} - \tilde{E}_1E^{(1)}E^{(3)} - \tilde{E}_1E^{(2)}E^{(3)}(\mathbf{1} - E^{(1)}),\\ &\ \vdots\end{aligned}$$

Similarly, we proceed with $\tilde{E}_2, \tilde{E}_3, \ldots$ and recognize that the $E_1, E_2, \ldots$ are elements of the Boolean ring generated by Σ and the $\tilde{E}_1, \tilde{E}_2$. Thus the complete Boolean ring generated by Σ and the $\tilde{E}_n$ is equal to that generated by Σ and the E_n, and is therefore also maximal.

2 Representation of $\mathscr{H}$ as $\mathscr{L}^2(\mathrm{Sp}(A), \mu)$ where $\mathrm{Sp}(A)$ is the Spectrum of a Scale Observable A

The fact that every complete Boolean ring $\Sigma \subset G$ can be generated by a scale observable A was explained in IV, §2.5 and emphasized in §1. In §1 we have also stressed the fact that these considerations can also be applied to the case in which Σ is generated by an n-tuple of commuting scale observables $A_1, \dots, A_n$. Since we cannot use Th. 1.4–1.8, we shall, for the present, only assume that there exists a $\varphi \in \mathscr{H}$ for which $\mathscr{T}(\Sigma, \varphi) = \mathscr{H}$. Then, for each $f \in \mathscr{H}$ there exists finitely many $E_i \in \Sigma$ and complex numbers a_i such that, given an $\varepsilon > 0$

$$\left\| f - \sum_{i=1}^{n} a_i E_i \varphi \right\| < \varepsilon$$

is satisfied. We may assume that the E_i are pairwise orthogonal, because we may use, instead of the E_i, the pairwise orthogonal atoms of the finite Boolean subring of Σ generated by the E_i.

According to IV, Th. 2.5.8 we may identify each $E \in \Sigma$ with a measurable subset k of the spectrum $\mathrm{Sp}(A)$ of a scale observable A. Let P denote the set of these measurable subsets k. k is uniquely defined by E up to a set of measure 0. For finitely many orthogonal E_i, since $E_i \wedge E_j = E_i E_j = 0$ for $i \neq j$, we may choose the corresponding subsets k_i such that $k_i \cap k_j = \varnothing$ for $i \neq j$. Instead of E_i we shall write $E(k_i)$ (where $E(k)$ is defined in the sense of $\sigma(k)$ in IV, Th. 2.5.8); therefore, to each $f \in \mathscr{H}$ and to each $\varepsilon > 0$ there are finitely many disjoint measurable subsets k_i of $\mathrm{Sp}(A)$ for which

$$\left\| f - \sum_{i=1}^{n} a_i E(k_i) \right\| < \varepsilon. \tag{2.1}$$

For a decomposition $\mathrm{Sp}(A) = \bigcup_{i=1}^{n} k_i$ ($k_i \cap k_j = \varnothing$ for $i \neq j$) of the spectrum of A into measurable subsets k_i we define a stepwise function g on $\mathrm{Sp}(A)$ as follows

$$g(\alpha) = \gamma_i \quad \text{for } \alpha \in k_i, \tag{2.2}$$

where the γ_i are complex numbers. From the step function $g(\alpha)$ we obtain a vector from $\mathscr{H}$ as follows:

$$g(\alpha) \to \sum_{i=1}^{n} \gamma_i E(k_i)\varphi. \tag{2.3}$$

For a pair of decompositions $\{k_i\}$ and $\{k'_j\}$ it follows that

$$\left\langle \sum_j \gamma'_j E(k'_j)\varphi, \sum_i \gamma_i E(k_i)\varphi \right\rangle = \sum_{i,j} \bar{\gamma}'_j \gamma_i \| E(k'_j) E(k_i)\varphi \|^2.$$

Since $E(k'_j)E(k_i) = E(k'_j \cap k_i)$, and from the σ-additive measure over P given by

$$\mu(k) = \| E(k)\varphi \|^2, \tag{2.4}$$

we obtain

$$\left\langle \sum_j \gamma'_j E(k'_j)\varphi, \sum_i \gamma_i E(k_i)\varphi \right\rangle = \sum_{i,j} \bar{\gamma}'_j \gamma_i \mu(k'_j \cap k_i). \tag{2.5}$$

For μ defined by (2.4) the set of step functions is dense in the Hilbert space $\mathscr{L}^2(\mathrm{Sp}(A), \mu)$. From (2.3) and (2.5) it follows that (2.3) is an isomorphic mapping of this subset into $\mathscr{H}$ which has a unique extension as an isomorphism

$$\mathscr{L}^2(\mathrm{Sp}(A), \mu) \to \mathscr{H}. \tag{2.6}$$

We write this isomorphism (2.6) in the following form (where $f(\lambda)$ is measurable and quadratically integrable):

$$f(\alpha) \to f = \int f(\alpha)\, dE(\alpha)\varphi, \tag{2.7}$$

where the integral on the right-hand side can be considered to be defined by this isomorphism. Thus it follows that, for a sequence of step functions $g_\nu(\alpha)$ for which $g_\nu(\alpha) \to f(\alpha)$ (in the sense of the norm in $\mathscr{L}^2(\mathrm{Sp}(A), \mu)$, $g_\nu = \sum_i \gamma_i^{(\nu)} E(k_i^{(\nu)})$ is a Cauchy sequence in $\mathscr{H}$ (where the $\gamma_i^{(\nu)}$ are the values corresponding to the g_ν according to (2.2)) which satisfies

$$g_\nu \to \int f(\alpha)\, dE(\alpha)\varphi. \tag{2.8}$$

We can also choose to use (2.8) as the definition of the integral.

Here we remark that

$$\int f(\alpha)\, dE(\alpha) \tag{2.9}$$

defines a function of the operator A for measurable $f(\alpha)$; $f(\alpha)$ need not be quadratically integrable. The operator $B = f(A)$ is a "normal" operator; its definition domain is the set of all $\psi \in \mathscr{H}$ for which $\int |f(\alpha)|^2\, d\|E(\alpha)\psi\|^2 < \infty$. If $f(\alpha)$ is bounded, then so is $f(A)$, where the latter is already defined in IV, §2.5 (see IV, Th. 2.5.11).

To each operator B in $\mathscr{H}$ there exists a corresponding operator $\bar{B}$ in $\mathscr{L}^2(\mathrm{Sp}(A), \mu)$ defined by the isomorphism (2.6), (2.7) (see AIV, §13) as follows:

$$f'(\alpha) = \bar{B} f(\alpha) \to Bf = \int f'(\alpha)\, dE(\alpha)\varphi.$$

The mode of action of

$$A = \int \alpha\, dE(\alpha) \tag{2.10}$$

is particularly simple. If we replace the integral in (2.10) by the approximation sum

$$\sum_i \alpha_i (E(\alpha_i) - E(\alpha_{i-1})),$$

then it is easy to show that

$$\tilde{A} f(\alpha) = \alpha f(\alpha). \tag{2.11}$$

Similarly, it is easy to prove that for a measurable set $k \in P$

$$\bar{E}(k)f(\alpha) = \begin{cases} f(\alpha) & \text{for } \alpha \in k, \\ 0 & \text{for } \alpha \notin k. \end{cases} \tag{2.12}$$

Let ψ be a vector in $\mathscr{H}$ and let $\psi(\alpha)$ be the corresponding function in $\mathscr{L}^2(\mathrm{Sp}(A), \mu)$. Then, for a finite sum it follows that

$$\sum_i \gamma_i \bar{E}(k_i)\psi(\alpha) \to \sum_i \gamma_i E(k_i)\psi$$

and for $g(\alpha)$ defined according to (2.2) we obtain

$$g(\alpha)\psi(\alpha) \to \sum_i \gamma_i E(k_i)\psi.$$

The subspace $\mathscr{T}(\Sigma, \psi)$ of $\mathscr{H}$ therefore corresponds the subspace of $\mathscr{L}^2(\mathrm{Sp}(A), \mu)$ spanned by all $g(\alpha)\psi(\alpha)$ where $g(\alpha)$ are arbitrary step functions; let $\bar{\mathscr{T}}(\Sigma, \psi)$ denote this subspace. Thus we find that $\bar{\mathscr{T}}(\Sigma, \psi)$ is spanned by all $h(\alpha)\psi(\alpha)$ where the $h(\alpha)$ are bounded measurable functions.

Let k_ψ denote the measurable set

$$k_\psi = \{\alpha \,|\, \alpha \in \mathrm{Sp}(A) \text{ and } \psi(\alpha) \neq 0\}. \tag{2.13}$$

Then we obtain

$$\bar{\mathscr{T}}(\Sigma, \psi) = \{f(\alpha) \,|\, f(\alpha) \in \mathscr{L}^2(\mathrm{Sp}(A), \mu) \text{ and } f(\alpha) = 0 \text{ for } \alpha \in \mathrm{Sp}(A)\backslash k_\psi\}$$

and we therefore obtain

$$\mathscr{T}(\Sigma, \psi) = E(k_\psi)\mathscr{H}.$$

According to (2.13), $(\mathbf{1} - \bar{E}(k_\psi))\psi(\alpha) = 0$, and we therefore find that

$$(\mathbf{1} - E(k_\psi))\psi = 0. \tag{2.14}$$

If the measure $\|E\psi\|^2$ is effective over Σ then we must have $\mathbf{1} - E(k_\psi) = \mathbf{0}$ and we obtain $E(k_\psi)\mathscr{H} = \mathscr{H}$, that is, $\mathscr{T}(\Sigma, \psi) = \mathscr{H}$, from which we have also proven Th. 1.5 and 1.8. In particular, the operator A we used at the beginning must have been a maximal scale observable.

If ψ is, in addition to φ, a second vector for which $\mathscr{T}(\Sigma, \psi) = \mathscr{H}$ then with the help of ψ we may define an isomorphism in the same way as we have done with φ as follows:

$$\begin{gathered} \mathscr{L}^2(\mathrm{Sp}(A), \tilde{\mu}) \to \mathscr{H}, \\ \tilde{f}(\alpha) \to f = \int \tilde{f}(\alpha)\, dE(\alpha)\psi, \end{gathered} \tag{2.15}$$

where

$$\tilde{\mu}(k) = \|E(k)\psi\|^2 = \int_k |\psi(\alpha)|^2 \, d\mu(\alpha). \tag{2.16}$$

For a step function $\tilde{g}(\alpha)$ of the form (2.2) where we replace γ_i by $\tilde{\gamma}_i$ it follows that

$$\tilde{g}(\alpha) \to \sum_i \tilde{\gamma}_i E(k_i)\psi \leftarrow \sum_i \tilde{\gamma}_i \bar{E}(k_i)\psi(\alpha),$$

where $\bar{E}(k)\psi(\alpha)$ is given in (2.12). Therefore

$$\tilde{g}(\alpha) \to \sum_i \tilde{\gamma}_i E(k_i)\psi \leftarrow \tilde{g}(\alpha)\psi(\alpha).$$

This result can be extended to all $\tilde{f}(\alpha) \in \mathscr{L}^2(\mathrm{Sp}(A), \mu)$, that is, the isomorphisms (2.6), (2.7) and (2.13) lead to an isomorphism

$$\begin{aligned} &\mathscr{L}^2(\mathrm{Sp}(A), \tilde{\mu}) \to \mathscr{L}^2(\mathrm{Sp}(A), \mu), \\ &\qquad \tilde{f}(\alpha) \to f(\alpha) = \tilde{f}(\alpha)\psi(\alpha). \end{aligned} \tag{2.17}$$

We call $\mathscr{L}^2(\mathrm{Sp}(A), \mu)$ together with the isomorphism (2.6) an A-representation of the Hilbert space $\mathscr{H}$ (obtained by means of φ). In an A-representation the special simple relations (2.11) and (2.12) hold. Between the A-representations obtained by means of φ and ψ the relationship (2.17) is satisfied, where $\psi(\alpha)$ is an arbitrary measurable function which is quadratically integrable with respect to μ and is nonzero almost everywhere on $\mathrm{Sp}(A)$. According to (2.7) it is clear that in the representation obtained by means of φ the vector φ corresponds to the function $\varphi(\alpha) = 1$.

In an A-representation the probability $\mathrm{tr}(P_f E)$ may be simply calculated as follows for $E \in \Sigma$: To E there exists a measurable set k for which $E = E(k)$. Then, according to (2.12) we obtain

$$\begin{aligned} \mathrm{tr}(P_f E) = \langle f, E(k)f \rangle &= \int \overline{f(\alpha)} \bar{E}(k) f(\alpha)\, d\mu(\alpha) \\ &= \int_k |f(\alpha)|^2\, d\mu(\alpha), \end{aligned} \tag{2.18}$$

where (2.18) is the probability that in the measurement of A for an ensemble $W = P_f$ the scale value of A lies in the subset k.

Is it possible to extend the formula (2.18) to general W? For $W = \sum_\nu w_\nu P_{\psi_\nu}$, $w_\nu \geq 0$ and $\sum w_\nu = 1$ it follows that $\mathrm{tr}(WE(k)) = \sum_\nu w_\nu \,\mathrm{tr}(P_{\psi_\nu} E(k))$ and we therefore obtain

$$\mathrm{tr}(WE(k)) = \sum_\nu w_\nu \int |\psi_\nu(\alpha)|^2\, d\mu(\alpha). \tag{2.19}$$

What is the form of the operator $\overline{W}$ in $\mathscr{L}^2(\mathrm{Sp}(A), \mu)$? For

$$\bar{P}_\psi f(\alpha) = \psi(\alpha) \int \overline{\psi(\alpha')} f(\alpha')\, d\mu(\alpha')$$

it follows that

$$\overline{W} f(\alpha) = \sum_\nu w_\nu \psi_\nu(\alpha) \int \overline{\psi_\nu(\alpha')} f(\alpha')\, d\mu(\alpha'). \tag{2.20}$$

Since $\sum_\nu |w_\nu| = \sum_\nu w_\nu = 1$ there exists a function $W(\alpha, \alpha')$ in $\mathrm{Sp}(A)^2$ which is, for fixed α, measurable in α' where $W(\alpha', \alpha) = \overline{W(\alpha, \alpha')}$ and $W(\alpha, \alpha) \geq 0$, so that it is possible to write (2.20) in the form

$$f'(\alpha) = \overline{W}f(\alpha) = \int W(\alpha, \alpha')f(\alpha')\, d\mu(\alpha'). \tag{2.21}$$

(2.19) is transformed into

$$\mathrm{tr}(WE(k)) = \int_k W(\alpha, \alpha)\, d\mu(\alpha). \tag{2.22}$$

From $\mathrm{tr}(W) = 1$ it follows from (2.22) that, for the special case in which $k = \mathrm{Sp}(A)$,

$$\int W(\alpha, \alpha)\, d\mu(\alpha) = 1. \tag{2.23}$$

The representation of W by means of an integral kernel $W(\alpha, \alpha')$ cannot be extended to the case of arbitrary self-adjoint operators B, not even, for example, to $\overline{A}$ in (2.11). In order to overcome these computational difficulties we shall present new methods in §§3–5.

The simple form (2.11) of A permits us to define arbitrary measurable functions $\chi(A)$ as follows: If χ is an arbitrary real μ-measurable function on $\mathrm{Sp}(A)$, then we define

$$\overline{\chi(A)}f(\alpha) = \chi(\alpha)f(\alpha). \tag{2.24}$$

$\chi(A)$ is therefore a self-adjoint operator with domain of definition of all the $f(\alpha)$ for which

$$\int |\chi(\alpha)f(\alpha)|^2\, d\mu(\alpha) < \infty.$$

This represents a generalization of the function concept introduced in IV, §2.5 to the case of unbounded functions.

In particular, A itself can be an unbounded self-adjoint operator. Then $\mathrm{Sp}(A)$ will be an unbounded subset on the real axis. The domain of definition of A is then the set of all $f(\alpha) \in \mathscr{L}^2(\mathrm{Sp}(A), \mu)$ for which $\alpha f(\alpha) \in \mathscr{L}^2(\mathrm{Sp}(A), \mu)$.

In the A-representation described above it is customary to define an "renormalization" in which the spectrum of A is decomposed into three disjoint sets k_d, k_{cc}, k_{sc} as described in IV, §2.5 (page 147–148): $\mathrm{Sp}(A) = k_d \cup k_{cc} \cup k_{sc}$. Accordingly μ is decomposed into $\mu = \mu_d + \mu_{cc} + \mu_{sc}$ where $\mu_d(k) = \mu(k_d \cap k)$, $\mu_{cc}(k) = \mu(k_{cc} \cap k)$ and $\mu_{sc}(k) = \mu(k_{sc} \cap k)$. The set k_d is identical with the set of discontinuity points of the spectral family $E(\alpha)$ of A. The set k_d is denumerable; let α_ν denote the points of k_d. Let $P_\nu = (E(\alpha_\nu) - E(\alpha_\nu -)$, then we find that $E(k_d) = \sum_\nu P_\nu$. The projections $E(k_d)$, $E(k_{cc})$ and

$E(k_{sc})$ are pairwise orthogonal, and $\mathbf{1} = E(k_d) + E(k_{cc}) + E(k_{sc})$. For $f_d = E(k_d)f$, $f_{cc} = E(k_{cc})f$, $f_{sc} = E(k_{sc})f$ the right-hand side of (2.7) takes on the form $f = f_d + f_{cc} + f_{sc}$ where

$$f_d = \sum_\nu f(\alpha_\nu)P_\nu\varphi,$$
$$f_{cc} = \int_{k_c} f(\alpha)\, dE(\alpha)\varphi, \tag{2.25}$$
$$f_{sc} = \int_{k_{sc}} f(\alpha)\, dE(\alpha)\varphi.$$

From $\mathscr{H} = \mathscr{T}(\Sigma, \varphi)$ it follows that all the P_ν are one-dimensional projections, because from (2.25) it follows that, for a vector f satisfying $P_\nu f = f$

$$f = f(\alpha_\nu)P_\nu\varphi,$$

that is, all vectors from the projection space of P_ν are multiples of a single vector $P_\nu\varphi$.

The set of solutions of the eigenvalue equation $Af = \alpha_\nu f$ is given by the set of vectors in the projection space of P_ν. $Af = \alpha f$ has a nonzero solution only for the values of $\alpha_\nu \in k_d$. If the eigenspace for the eigenvalue α_ν of A is one-dimensional, then this eigenvalue is said to be nondegenerate. From $\mathscr{H} = \mathscr{T}(\Sigma, \varphi)$ it follows that the eigenvalues α_ν are nondegenerate.

For $\tilde{\varphi}_\nu = P_\nu\varphi$ it follows from (2.25) that

$$f = \sum_\nu f(\alpha_\nu)\tilde{\varphi}_\nu + f_{cc} + f_{sc}. \tag{2.26}$$

Since the $\tilde{\varphi}_\nu$ are pairwise orthogonal and are orthogonal to f_{cc}, f_{sc} from (2.26) it follows that

$$\langle \tilde{\varphi}_\nu, f\rangle = f(\alpha_\nu)\|\tilde{\varphi}_\nu\|^2$$

and that

$$f = \sum_\nu \frac{\tilde{\varphi}_\nu}{\|\tilde{\varphi}_\nu\|}\left\langle \frac{\tilde{\varphi}_\nu}{\|\tilde{\varphi}_\nu\|}, f\right\rangle + f_c + f_{sc}. \tag{2.27}$$

Here $\|\tilde{\varphi}_\nu\|^2$ is the measure $\mu(k_\nu)$ where k_ν consists of the single point α_ν. Equation (2.27) suggests the following renormalization: Instead of using the $\tilde{\varphi}_\nu$ in the expansion (2.27) we use the normalized eigenvectors $\varphi_\nu = \tilde{\varphi}_\nu\|\tilde{\varphi}_\nu\|^{-1}$. Then we obtain

$$f = \sum_\nu \varphi_\nu\langle \varphi_\nu, f\rangle + f_{cc} + f_{sc}. \tag{2.28}$$

We now define

$$\langle \varphi_\nu, f\rangle = \langle \alpha_\nu | f\rangle, \tag{2.29}$$

where we consider $\langle \alpha_\nu | f\rangle$ to be a complex function defined on k_d. We obtain

$$\langle \alpha_\nu | f\rangle = f(\alpha_\nu)\sqrt{\mu(k_\nu)}, \tag{2.30}$$

where k_ν is the set consisting of a single point α_ν. By analogy with (2.30), for $\alpha \in k_{cc}$ we define

$$\langle \alpha | f \rangle = f(\alpha)\sqrt{\rho(\alpha)}, \tag{2.31}$$

where $\rho(\alpha)$ is defined over k_{cc} by the equation

$$\mu(k) = \int_k \rho(\alpha)\, d\alpha \quad \text{for all} \quad k \subset k_{cc}, \tag{2.32}$$

where $d\alpha$ is the Lebesgue measure. For $\alpha \in k_{sc}$ we set $\langle \alpha | f \rangle = f(\alpha)$. In this way we obtain a new isomorphism in addition to that given by (2.6) and (2.7) as follows:

$$\mathscr{L}^2(\mathrm{Sp}(A), \mu_n) \to \mathscr{H}, \tag{2.33}$$
$$\langle \alpha | f \rangle \to f,$$

with the "normed" measure

$$\mu_n(k) = N(k \cap k_d) + \int_{k \cap k_{cc}} d\alpha + \int_{k \cap k_{sc}} d\mu(\alpha), \tag{2.34}$$

where $N(k \cap k_d)$ is the number of elements of $k \cap k_d$. μ_n is not necessarily bounded, that is, $\mu_n(\mathrm{Sp}(A)) = \infty$ is possible!

If $k_{sc} = \varnothing$, then (2.18) takes on the renormalized form:

$$\mathrm{tr}(P_f E(k)) = \sum_{\substack{\nu \\ \alpha_\nu \in k \cap k_d}} |\langle \alpha_\nu | f \rangle|^2 + \int_k |\langle \alpha | f \rangle|^2 \, d\alpha. \tag{2.35}$$

The existence of a set $k_{sc} \neq \varnothing$ appears to be peculiar for a physical measurement scale. From the viewpoint of the mathematical arbitrariness of the choice of a measurement scale as described in IV, §2.5 we cannot produce an argument against the existence of the set k_{sc}. If, in the choice of scales we introduce an additional structure to that described in IV, §2.5, we introduce new questions about the sets k_d, k_{cc}, k_{sc}; for example, with respect to the Hamiltonian operator H as a scale observable, where H describes the time translation of the registration apparatus relative to the preparation apparatus. On the basis of this physical meaning for H the "measurement scale" for H attains a new physical meaning (see, for example, X, §1 or XVI, §4.3).

3 Improper Scalar and Vector Functions Defined on Sp(A)

In practice it is often desirable to use different representations of a Hilbert space for quantum mechanics. In many respects the distinction between the sets k_d and k_{cc} and, if it exists k_{sc}, is inconvenient. For this reason methods have been developed which permit the points of k_{cc} to be treated as "simply" as are those of k_d. In this way we also seek to obtain "solutions" f of the eigenvalue equation $Af = \alpha f$ for points of k_c. These f clearly cannot be elements of $\mathscr{H}$.

One of these methods is that of the Gelfand triple [1]. A second method is that developed by Nikodym [2]. In either case there is no "simple" accessible method which is sufficiently reliable to permit us to attempt to guarantee (on the basis of general theorems) the mathematical validity of the computed results. For the most part these methods are used as heuristic principles, to obtain quick results, the mathematical validity of which can later be verified in individual cases. In concrete examples it is not unusual to find unsatisfactory results which do not, however, place the theory itself in question. Perhaps, in quantum mechanics, a restriction to the mathematics of Hilbert space is inconvenient. Such a restriction is not necessary since operators in Hilbert space are only representations of a special Banach space structure (see III, §3). In this representation there is an additional underlying structure which we have denoted by $\mathscr{D}$ (see the discussions in VII, VIII). It is conceivable that the development of a "theory of $\mathscr{D}$" could lead to new practical methods (see also [3]).

Since we do not yet have a satisfactory method for the reformulation of mathematical problems in $\mathscr{H}$ we shall now present a brief outline of the method of Nikodym in order that we may at least have a general method which underlies the somewhat messy mathematical calculations which will appear later in this book.

In order to obtain a greater similarity between the general expansion of f on the right-hand side of (2.7) and the expansion of f with respect to a complete orthonormal basis φ_ν (the case in which $\mathrm{Sp}(A) = k_d$), $f = \sum_\nu \varphi_\nu \langle \varphi_\nu, f \rangle$, we shall return to (2.1).

According to IV, Th. 2.5.8 (and the remarks preceding the theorem), for $k \in P$ we obtain

$$E(k) = \bigwedge_u \bigvee_{\nu=1}^{\infty} E(I_\nu), \tag{3.1}$$

where $\bigwedge_u$ is to be taken over all coverings u of k. For each $\varphi \in \mathscr{H}$, $\bigvee_{\nu=1}^{\infty} E(I_\nu)\varphi$ may be approximated in the norm to arbitrary accuracy by $\bigvee_{\nu=1}^{N} E(I_\nu)\varphi$. Here we may assume that the intervals I_ν are mutually exclusive because we may easily replace $\bigvee_{\nu=1}^{N} E(I_\nu)$ by $\bigvee_{\mu}^{N'} E(I'_\mu)$ with disjoint I'_μ. From (3.1) it follows that $E(k)\varphi$ can be approximated arbitrarily well by $\bigvee_{\nu=1}^{\infty} E(I_\nu)\varphi$ for a particular covering u. Therefore, $E(k)\varphi$ can be arbitrarily well approximated by $\bigvee_{\nu=1}^{N} E(I_\nu)\varphi$ where the intervals I_ν are disjoint, that is, it can be approximated by the sum $\sum_{\nu=1}^{N} E(I_\nu)\varphi$.

From (2.1) it follows that each f can be approximated by finite sums $\sum_\rho a_\rho E(I_\rho)\varphi$ where the I_ρ are disjoint intervals. Therefore, to each $\varepsilon > 0$ there exists a partition of $\mathbf{R}$ into finitely many disjoint intervals I_ρ and there exists a finite set of complex numbers a_ρ such that

$$\left\| f - \sum_\rho a_\rho E(I_\rho)\varphi \right\| < \varepsilon.$$

Let I_ν^n be, for fixed n, a partition of $\mathbf{R}$ into intervals where δ_n is the maximum length of the intervals and suppose $\delta_n \to 0$. Then there exists for each $\varepsilon > 0$

a δ_n and a set of numbers a_ν^n such that

$$\left\| f - \sum_\nu a_\nu^n E(I_\nu^n)\varphi \right\| < \varepsilon.$$

Since (for fixed n) the $E(I_\nu^n)$ are pairwise orthogonal, $\| f - \sum_\nu a_\nu^n E(I_\nu^n)\varphi\|$ will take on its minimum value (see AIV, §2) when

$$a_\nu^n = \frac{\langle E(I_\nu^n)\varphi, f\rangle}{\|E(I_\nu^n)\varphi\|^2}.$$

Therefore we obtain

$$\lim_{n\to\infty} \sum_\nu E(I_\nu^n)\varphi \frac{\langle E(I_\nu^n)\varphi, f\rangle}{\|E(I_\nu^n)\varphi\|^2} = f \tag{3.2}$$

for each $f \in \mathscr{H}$. For the case in which A has only a discrete spectrum, with α_ν denoting the elements of k_d and $E(\alpha_\nu) - E(\alpha_\nu -) = P_\nu$ (see (2.27)) we obtain

$$\sum_\nu P_\nu\varphi \frac{\langle P_\nu\varphi, f\rangle}{\|P_\nu\varphi\|^2} = f. \tag{3.3}$$

For $\varphi_\nu = P_\nu\varphi\|P_\nu\varphi\|^{-1}$, (3.3) is an expansion of f with respect to the normed eigenvectors φ_ν of A. We may write φ_ν as the limit as $\alpha' < \alpha_\nu$ as follows:

$$\lim_{\alpha'\to\alpha_\nu} \frac{(E(\alpha_\nu) - E(\alpha'))\varphi}{\|(E(\alpha_\nu) - E(\alpha'))\varphi\|} = \frac{P_\nu\varphi}{\|P_\nu\varphi\|} = \varphi_\nu. \tag{3.4}$$

In general, if α_ν is not a point of discontinuity of the spectral family $E(\alpha)$ then the limit of

$$\chi_{\alpha,\alpha'} = \frac{(E(\alpha) - E(\alpha'))\varphi}{\|(E(\alpha) - E(\alpha'))\varphi\|} \tag{3.5}$$

does not exist for $\alpha' \to \alpha$, otherwise $\lim_{\alpha'\to\alpha} \chi_{\alpha\alpha'}$ would be a normalized eigenvector of A with eigenvalue α.

If α' is close to α then $\chi_{\alpha\alpha'}$ will approximately represent an eigenvector of A since

$$\|A\chi_{\alpha\alpha'} - \alpha\chi_{\alpha\alpha'}\| \le \alpha - \alpha'. \tag{3.6}$$

We obtain (3.6) from

$$A\chi_{\alpha\alpha'} - \alpha\chi_{\alpha\alpha'} = \|(E(\alpha) - E(\alpha'))\varphi\|^{-1} \int_{\alpha'}^{\alpha} (\alpha'' - \alpha)\, dE(\alpha'')\varphi$$

and from

$$\begin{aligned} \|A\chi_{\alpha\alpha'} - \alpha\chi_{\alpha\alpha'}\|^2 &= \|(E(\alpha) - E(\alpha'))\varphi\|^{-2} \int_{\alpha'}^{\alpha} (\alpha'' - \alpha)^2\, d\|E(\alpha'')\varphi\|^2 \\ &\le \|(E(\alpha) - E(\alpha'))\varphi\|^{-2}(\alpha' - \alpha)^2\|(E(\alpha) - E(\alpha'))\varphi\|^2 \\ &= (\alpha' - \alpha)^2. \end{aligned}$$

Equation (3.2) says that, for sufficiently large n each f can be expressed to arbitrary accuracy by an expansion in terms of "approximate" eigenvectors $\chi_{\alpha\alpha'}$ of the form (3.5). This result motivates the following definition of an "improper" vector function $\tilde{\chi}(\alpha)$ on Sp(A).

For fixed α, $\tilde{\chi}(\alpha)$ is a mapping of the interval $-\infty < \alpha' < \alpha$ into $\mathscr{H}$, that is, there exists a function $\chi(\alpha, \alpha')$ which is defined for all (α, α') for which $\alpha' < \alpha$ where $\chi(\alpha, \alpha') \in \mathscr{H}$; $\tilde{\chi}(\alpha)$ is defined as the function $\chi(\alpha, \alpha')$ as a function of α' for fixed α.

Similarly, an improper complex function $\tilde{a}(\alpha)$ on Sp(A) is defined with the aid of a complex function $a(\alpha, \alpha')$.

For $f \in \mathscr{H}$, $\langle \chi(\alpha, \alpha'), f \rangle = a(\alpha, \alpha')$ defines, for an improper vector function $\tilde{\chi}(\alpha)$, an improper complex function $\tilde{a}(\alpha)$. Here we write $\tilde{a}(\alpha) = \langle \tilde{\chi}(\alpha), f \rangle$.

We may also define improper functions for the case of several variables. For example, we may define an improper vector function $\tilde{\chi}(\alpha_1, \alpha_2)$ by means of function $\chi(\alpha_1, \alpha_1'; \alpha_2, \alpha_2')$ for $\alpha_1' < \alpha_1, \alpha_2' < \alpha_2$ where $\chi(\alpha_1, \alpha_1'; \alpha_2, \alpha_2') \in \mathscr{H}$.

An improper function $\tilde{\eta}(\alpha_1, \alpha_2)$ is defined by $\eta(\alpha_1, \alpha_1'; \alpha_2, \alpha_2') = a(\alpha_1, \alpha_1')\chi(\alpha_2, \alpha_2')$; here we write $\tilde{\eta}(\alpha_1, \alpha_2) = \tilde{a}(\alpha_1)\tilde{\chi}(\alpha_2)$. Similarly, we define $\tilde{a}(\alpha_1, \alpha_2) = \langle \tilde{\chi}(\alpha_1), \tilde{\chi}(\alpha_2) \rangle$ by $\langle \chi(\alpha_1, \alpha_1'), \chi(\alpha_2, \alpha_2') \rangle$; similarly, we may define $\tilde{\chi}(\alpha_1) + \tilde{\chi}(\alpha_2)$, etc.. Note, however, that we can define an improper function of only one variable $\tilde{a}(\alpha)\tilde{\chi}(\alpha)$ by $a(\alpha, \alpha')\chi(\alpha, \alpha')$. Then

$$\tilde{a}(\alpha)\tilde{\chi}(\alpha) = [\tilde{a}(\alpha_1)\tilde{\chi}(\alpha_2)]_{\alpha_1 = \alpha_2 = \alpha}$$

does not hold! For an improper function $\tilde{\eta}(\alpha_1, \alpha_2)$ of two variables we cannot simply set $\alpha_1 = \alpha_2$ instead, we must write $[\tilde{\eta}(\alpha_1, \alpha_2)]_{\alpha_1 = \alpha_2}$. Here we hope that no misunderstanding will result from the following formulae.

Let k be a measurable set of the spectrum; let us consider, as in the case of IV, §2.5, a covering u of k by means of countably many intervals I_ν: $k \subset \bigcup_\nu I_\nu$. From such a covering we may select a covering consisting of disjoint intervals as follows: Choose $\tilde{I}_{11} = I_1$; $I_2 \dotplus I_1$ is then a set which consists of finitely many disjoint intervals, so that $I_2 \cup I_1$ may be represented as the union of the disjoint intervals $\tilde{I}_{11}$ and those in $I_2 \dotplus I_1$. If $\bigcup_{\mu=1}^m I_\mu$ may be represented as the union of disjoint intervals, then the same is true for $I_{m+1} \dotplus \bigcup_{\mu=1}^m I_\mu$. In this way we recursively prove our assertion. We need only consider the coverings consisting of disjoint intervals instead of all possible coverings. Since k is measurable, there exist sequences of intervals I_ν^n such that $E^n = \sum_\nu E(I_\nu^n)$ is a projection operator and $E^n \to E(k)$.

We could also assume that the maximum length δ_n of the intervals I_ν^n tend towards zero; to do so requires only that we divide the intervals I_ν^n.

An improper vector function $\tilde{\chi}(\alpha)$ is said to be summable over Sp(A) if for every measurable subset k of Sp(A) the limit $\sum_\nu \chi(I_\nu^n)$ exists and is independent of the sequence of coverings $u_n = \bigcup_\nu I_\nu^n$ of disjoint intervals for which $\delta_n \to 0$; for I: $\alpha' < \cdots \le \alpha$ we have written $\chi(I)$ instead of $\chi(\alpha, \alpha')$.

For a summable vector function $\tilde{\chi}(\alpha)$ we write, for the definition

$$\int_k \tilde{\chi}(\alpha) = \lim_{n\to\infty} \sum_\nu \chi(I_\nu^n). \tag{3.7}$$

We note that, according to the definition (3.7), only the points $\alpha \in \mathrm{Sp}(A)$ occur in the sum $\int$, since k must be a subset of Sp(A). If the spectrum is discrete, that is, $\mathrm{Sp}(A) = k_d$, it follows that for a summable function $\tilde{\chi}(\alpha)$ the limit $\lim_{\varepsilon \to 0} \chi(\alpha_\nu, \alpha_\nu - \varepsilon) = \chi_\nu \in \mathscr{H}$ exists for the points $\alpha_\nu \in \mathrm{Sp}(A)$, where the latter follows from the fact that (3.7) is satisfied for $k = \{\alpha_\nu\}$, that is, for a set consisting of a single point α_ν; from (3.7) it follows that for an arbitrary k:

$$\int_k \tilde{\chi}(\alpha) = \sum_{\alpha_\nu \in k} \chi_\nu. \tag{3.8}$$

If $E(I)\chi(I) = \chi(I)$, then, from (3.7) it follows that for

$$g = \int_{\mathrm{Sp}(A)} \tilde{\chi}(\alpha): \int_k \tilde{\chi}(\alpha) = E(k)g. \tag{3.9}$$

The improper vector function $\tilde{\chi}(\alpha)$ defined by $\chi(I) = E(I)f$ for fixed $f \in \mathscr{H}$ is summable, and trivially satisfies the equation $E(I)\chi(I) = \chi(I)$. We obtain:

$$\int_{\mathrm{Sp}(A)} \tilde{\chi}(\alpha) = f, \qquad \int_k \tilde{\chi}(\alpha) = E(k)f.$$

In this case $\int_{\mathrm{Sp}(A)} \tilde{\chi}(\alpha)$ is nothing other than an alternative notation for

$$\int_{-\infty}^{\infty} dE(\alpha)f.$$

Two improper vector functions $\tilde{\chi}(\alpha)$ and $\tilde{\eta}(\alpha)$ are said to be equivalent (written $\tilde{\chi}(\alpha) \sim \tilde{\eta}(\alpha)$) if, for all measurable k, the following equation is satisfied:

$$\int_k \tilde{\chi}(\alpha) = \int \tilde{\eta}(\alpha). \tag{3.10}$$

From the above derivation it follows that, for example, the improper vector function $\tilde{\chi}(\alpha)$, defined in terms of $E(I)f$, is equivalent to the improper vector function $\tilde{\eta}(\alpha)$ defined by means of

$$E(I)\varphi \frac{\langle E(I)\varphi, f\rangle}{\|E(I)\varphi\|^2},$$

$$\tilde{\varphi}(\alpha) := \frac{E(I)\varphi}{\|E(I)\varphi\|}$$

(here := means that the expression on the left is defined by the expression to the right of the equal sign) is an improper vector function, and satisfies

$$\tilde{\eta}(\alpha) = \tilde{\varphi}(\alpha)\langle\tilde{\varphi}(\alpha), f\rangle. \tag{3.11}$$

We therefore obtain the expansion

$$\int_{\mathrm{Sp}(A)} \tilde{\varphi}(\alpha)\langle\tilde{\varphi}(\alpha), f\rangle = f. \tag{3.12}$$

The improper vector function $\tilde{\varphi}(\alpha)$ itself need not be summable; but the improper function $\tilde{\eta}(\alpha)$ given by (3.11) must be.

If Sp(A) is discrete, then (3.12) becomes

$$\sum_{\nu} \psi_\nu \langle \psi_\nu, f \rangle = f,$$

where ψ_ν are the normed(!) eigenvectors of A:

$$\psi_\nu = \lim_{\varepsilon \to 0} \frac{(E(\alpha_\nu) - E(\alpha_\nu - \varepsilon))\varphi}{\|(E(\alpha_\nu) - E(\alpha_\nu - \varepsilon))\varphi\|}. \tag{3.13}$$

Let $\rho(k)$ be a positive σ-additive measure over the set P of measurable subsets of Sp(A). It is not assumed that $E(k)\varphi \neq 0$ implies that $\rho(k) \neq 0$. In addition, $\rho(k)$ need not necessarily be normalized; it is permissible that $\rho(\mathrm{Sp}(A)) = \infty$. However, for all finite intervals I for which $E(I) \neq 0$ we assume that $\rho(I) \neq 0$ and $\rho(I)$ is finite. Thus an improper function is defined by $\rho(I)$, which we denote by $d\rho(\alpha)$.

For a partition of the α-line into intervals we obtain:

$$\sum_{\nu} \frac{E(I_\nu)\varphi}{\|E(I_\nu)\varphi\|} \left\langle \frac{E(I_\nu)\varphi}{\|E(I_\nu)\varphi\|}, f \right\rangle$$
$$= \sum_{\nu} \frac{E(I_\nu)\varphi}{\|E(I_\nu)\varphi\|\sqrt{\rho(I_\nu)}} \left\langle \frac{E(I_\nu)\varphi}{\|E(I_\nu)\varphi\|\sqrt{\rho(I_\nu)}}, f \right\rangle \rho(I_\nu).$$

For the improper vector function

$$\tilde{\psi}(\alpha) := \frac{E(I)\varphi}{\|E(I)\varphi\|\sqrt{\rho(I)}} \tag{3.14}$$

we obtain, in the limit

$$f = \int_{\mathrm{Sp}(A)} \tilde{\psi}(\alpha)\langle \tilde{\psi}(\alpha), f \rangle \, d\rho(\alpha). \tag{3.15}$$

It is customary to separate the discrete spectrum from the continuous spectrum: $\mathrm{Sp}(A) = k_d \cup k_c$ where $k_c = k_{cc} \cup k_{sc}$ ($k_c = \mathrm{Sp}_c(A)$, $k_d = \mathrm{Sp}_d(A)$ in IV, D 2.5.4). For $P_\nu = E(\alpha_\nu) - E(\alpha_\nu -)$ (α_ν are the points of the discrete spectrum) we define

$$\tilde{E}(\alpha) = E(\alpha) - \sum_{\alpha_\nu \leq \alpha} P_\nu \tag{3.16}$$

and

$$A = A_d + \tilde{A}_c, \tag{3.17}$$

where

$$A_d = \sum_{\nu} \alpha_\nu P_\nu \quad \text{and} \quad \tilde{A}_c = \int \alpha \, d\tilde{E}(\alpha). \tag{3.18}$$

We define the improper vector function $\tilde{\psi}_c(\alpha)$ in a similar manner as (3.14) substituting $\tilde{E}(\alpha)$ for $E(\alpha)$. For ψ_c defined according to (3.13) we obtain, instead of (3.15),

$$f = \sum_{\nu} \psi_\nu \langle \psi_\nu, f \rangle + \int_{k_c} \tilde{\psi}_c(\alpha) \langle \tilde{\psi}_c(\alpha), f \rangle \, d\rho(\alpha). \tag{3.19}$$

Both forms (3.15) and (3.19) are, in principle, possible; for "physical" purposes (3.19) is more useful, because in (3.15) the discrete part of the spectrum is difficult to determine. In order to make the formula more transparent, we set $A_d = 0$, that is, $k_d = \varnothing$.

For this case we shall discuss (3.15) further.

We now choose $\rho(k) = \mu(k)$ where μ is defined by (2.4). Then we obtain

$$\langle \tilde{\psi}(\alpha), f \rangle : \left\langle \frac{E(I)\varphi}{\|E(I)\varphi\|^2}, f \right\rangle = \frac{1}{\mu(I)} \int_{\alpha-\varepsilon}^{\alpha} f(\alpha') \, d\mu(\alpha'), \tag{3.20}$$

where $f(\alpha')$ is the μ-measurable function which corresponds to f. Thus it follows that, in the limit $\varepsilon \to 0$,

$$\lim_{\varepsilon \to 0} \left\langle \frac{(E(\alpha) - E(\alpha - \varepsilon))\varphi}{\|(E(\alpha) - E(\alpha - \varepsilon))\varphi\|^2}, f \right\rangle = f(\alpha) \tag{3.21}$$

holds modulo sets of measure zero. For $\rho = \mu$ we may replace the improper vector function $\langle \tilde{\psi}(\alpha), f \rangle$ by the μ-measurable (proper) function $f(\alpha)$ given by (3.21).

If we choose $\rho(k) = l(k)$ where l is the Lebesgue measure (where we prefer to use $d\alpha$ instead of $dl(\alpha)$), we may not, in general, expect that in the limit $\varepsilon \to 0$ $\langle \tilde{\psi}(\alpha), f \rangle$ will be equal to an l-measurable function $f(\alpha)$. This is the case only if the spectrum of A is absolutely continuous. If this is the case, then for a suitably chosen l-measurable function $k(\alpha)$ we obtain

$$\mu(I) = \int_I k(\alpha) \, d\alpha.$$

Then we obtain almost everywhere

$$\lim_{\varepsilon \to 0} \left\langle \frac{E(I)\varphi}{\|E(I)\varphi\|^2}, f \right\rangle \sqrt{\frac{\mu(I)}{l(I)}} = f(\alpha)\sqrt{k(\alpha)} = \mathbf{f}(\alpha), \tag{3.22}$$

where

$$\|f\|^2 = \int |f(\alpha)|^2 \, d\mu(\alpha) = \int |\mathbf{f}(\alpha)|^2 \, d\alpha. \tag{3.23}$$

Equation (3.15) is always correct. In l measure we obtain

$$\begin{aligned} \langle f, g \rangle &= \int \langle f, \tilde{\psi}(\alpha) \rangle \langle \tilde{\psi}(\alpha), g \rangle \, d\alpha \\ &\doteq \int \overline{\mathbf{f}(\alpha)} \mathbf{g}(\alpha) \, d\alpha, \end{aligned} \tag{3.24}$$

where the final equal sign $\stackrel{\vee}{=}$ holds only if the spectrum of A is absolutely continuous. Then, instead of (3.22) we write

$$\langle\tilde{\psi}(\alpha), f\rangle \stackrel{\vee}{=} \mathbf{f}(\alpha). \tag{3.25}$$

An improper vector function $\tilde{\chi}(\alpha)$ is said to belong to the domain of definition of an operator A if the vectors $\chi(\alpha, \alpha - \varepsilon)$ belong to the definition domain of A for all ε which satisfy $0 < \varepsilon < \varepsilon_0$ for some ε_0. The above $\tilde{\varphi}(\alpha)$, $\tilde{\psi}(\alpha)$ belong therefore to the domain of definition of A. If we substitute the vector Af for f in (3.12) and observe that $\langle\tilde{\varphi}(\alpha), Af\rangle = \langle A\tilde{\varphi}(\alpha), f\rangle$, we obtain

$$Af = \int \tilde{\varphi}(\alpha)\langle A\tilde{\varphi}(\alpha), f\rangle.$$

We will now show that

$$\tilde{\varphi}(\alpha)\langle A\tilde{\varphi}(\alpha), f\rangle \sim \tilde{\varphi}(\alpha)\alpha\langle\tilde{\varphi}(\alpha), f\rangle:$$

$$\begin{aligned} |\langle(A - \alpha\mathbf{1})E(I)\varphi, f\rangle| &= |\langle(A - \alpha\mathbf{1})E(I)\varphi, E(I)f\rangle| \\ &\leq \|(A - \alpha\mathbf{1})E(I)\varphi\| \; \|E(I)f\|. \end{aligned}$$

According to (3.6) we obtain

$$\|(A - \alpha\mathbf{1})E(I)\varphi\| \leq (\alpha - \alpha')\|E(I)\varphi\|.$$

Let δ be the maximum length of the intervals I_ν, then

$$\begin{aligned} &\left\| \sum_\nu \frac{E(I_\nu)\varphi}{\|E(I_\nu)\varphi\|} \left\langle A \frac{E(I_\nu)\varphi}{\|E(I_\nu)\varphi\|}, f \right\rangle - \sum_\nu \frac{E(I_\nu)\varphi}{\|E(I_\nu)\varphi\|} \alpha_\nu \left\langle \frac{E(I_\nu)\varphi}{\|E(I_\nu)\varphi\|}, f \right\rangle \right\|^2 \\ &\quad = \sum_\nu \left| \left\langle (A - \alpha_\nu \mathbf{1}) \frac{E(I_\nu)\varphi}{\|E(I_\nu)\varphi\|}, f \right\rangle \right|^2 \leq \delta \sum_\nu \|E(I_\nu)f\|^2 = \delta\|f\|^2. \end{aligned}$$

For $\delta \to 0$ we obtain

$$Af = \int \tilde{\varphi}(\alpha)\alpha\langle\tilde{\varphi}(\alpha), f\rangle = \int \tilde{\psi}(\alpha)\alpha\langle\tilde{\psi}(\alpha), f\rangle \, d\alpha. \tag{3.26}$$

We may therefore replace $A\tilde{\varphi}(\alpha)$ by $\alpha\tilde{\varphi}(\alpha)$ and $A\tilde{\psi}(\alpha)$ by $\alpha\tilde{\psi}(\alpha)$, respectively; here we use the notation $A\tilde{\varphi}(\alpha) \sim \alpha\tilde{\varphi}(\alpha)$, $A\tilde{\psi}(\alpha) \sim \alpha\tilde{\psi}(\alpha)$. We call $\tilde{\varphi}(\alpha)$ or $\tilde{\psi}(\alpha)$ the improper eigenvectors of A.

From the continuity of the inner product and, from (3.15), it follows that, for $\rho = l$:

$$\langle\tilde{\psi}(\alpha'), f\rangle = \int \langle\tilde{\psi}(\alpha'), \tilde{\psi}(\alpha)\rangle\langle\tilde{\psi}(\alpha), f\rangle \, d\alpha. \tag{3.27}$$

Therefore $\langle\tilde{\psi}(\alpha'), \tilde{\psi}(\alpha)\rangle$ is an improper function of the two(!) variables α', α. This function is called the Dirac delta function, which is often written $\delta(\alpha' - \alpha)$ instead of $\delta(\alpha, \alpha')$.

For the construction of the $\tilde{\varphi}(\alpha)$ and $\tilde{\psi}(\alpha)$ we have made use of our knowledge of the spectral family $E(\alpha)$. If the spectral family $E(\alpha)$ is not known, then we may often succeed by proceeding as follows: From

$$E(\alpha)f = \int_{-\infty}^{\alpha} \tilde{\psi}(\alpha')\langle\tilde{\psi}(\alpha'), f\rangle \, d\alpha' \tag{3.28}$$

we seek (often heuristically) to guess $\tilde{\psi}(\alpha)$ as an improper eigenvector for A from the equation $A\tilde{\psi}(\alpha) \sim \alpha\tilde{\psi}(\alpha)$ and then construct the operator $E(\alpha)$ from (3.28). If, for example, $A = (1/i)(d/dx)$ is an operator in the Hilbert space $\mathscr{H}$ defined on quadratically integrable functions over x from $-\infty$ to ∞ then we seek solutions of the equation $(1/i)(d/dx)g(\alpha, x) = \alpha g(\alpha, x)$. We obtain $g(\alpha, x) = ae^{i\alpha x}$. The $g(\alpha, x)$ are not vectors(!) in $\mathscr{H}$ since $\int_{-\infty}^{\infty} |g(\alpha, x)|^2 \, dx$ is not convergent. But the $\int_{\alpha-\varepsilon}^{\alpha} e^{i\alpha' x} \, d\alpha'$ are vectors in $\mathscr{H}$ and define an improper vector function $\tilde{\chi}(\alpha, x)$. With the correct normalization we can set

$$\tilde{\psi}(\alpha, x) := \frac{1}{\sqrt{2\pi}} \int_{\alpha-\varepsilon}^{\alpha} e^{i\alpha' x} \, d\alpha'. \tag{3.29}$$

Thus, from (3.28) we obtain

$$E(\alpha)f(x) = \frac{1}{2\pi} \int_{-\infty}^{\alpha} e^{i\alpha' x} \left(\int_{-\infty}^{\infty} e^{-i\alpha' x'} f(x') \, dx' \right) d\alpha'. \tag{3.30}$$

The representation of the vector f in terms of the improper function $\langle\tilde{\psi}(\alpha), f\rangle$ (or, if it is permissible, in terms of the $\mathbf{f}(\alpha)$) is called the A-representation. It is not uniquely determined, because, instead of $\tilde{\psi}(\alpha)$ we could have chosen $e^{ig(\alpha)}\tilde{\psi}(\alpha)$ as the improper vector function providing $g(\alpha)$ is chosen subject to the requirement that the appropriate integrals exist; instead of $\langle\tilde{\psi}(\alpha), f\rangle$ we get the improper function $e^{-ig(\alpha)}\langle\tilde{\psi}(\alpha), f\rangle$ to represent f.

If B is another maximal self-adjoint operator and $\tilde{\chi}(\beta)$ is the corresponding set of improper eigenvectors of B, then we may obtain an expansion of the $\tilde{\chi}(\beta)$ with respect to the $\tilde{\psi}(\alpha)$ as follows:

$$\tilde{\chi}(\beta) = \int \tilde{\psi}(\alpha)\langle\alpha|\beta\rangle \, d\alpha, \quad \text{where} \quad \langle\alpha|\beta\rangle = \langle\tilde{\psi}(\alpha), \chi(\beta)\rangle. \tag{3.31}$$

$\langle\alpha|\beta\rangle$ is therefore an improper function of the two variables α and β. According to (3.31) and (3.15) (with l instead of ρ) $\langle\alpha|\beta\rangle$ are nothing other than the improper eigenvectors $\tilde{\chi}(\beta)$ of B in the A-representation. In the A-representation we obtain (where we use the same symbols for the operator B):

$$B\langle\alpha|\beta\rangle \sim \beta\langle\alpha|\beta\rangle. \tag{3.32}$$

For $\langle\beta|\alpha\rangle = \overline{\langle\alpha|\beta\rangle} = \langle\tilde{\chi}(\beta), \tilde{\psi}(\alpha)\rangle$ it follows that

$$\tilde{\psi}(\alpha) = \int \tilde{\chi}(\beta)\langle\beta|\alpha\rangle \, d\beta, \tag{3.33}$$

so that, for the operator A in the B-representation we obtain

$$A\langle\beta|\alpha\rangle \sim \alpha\langle\beta|\alpha\rangle, \tag{3.34}$$

that is, the $\langle\beta|\alpha\rangle$ are the improper eigenfunctions of A in the B-representation.

A bounded operator D can be characterized by the matrix

$$\langle \nu | D | \mu \rangle = \langle \varphi_\nu, D\varphi_\mu \rangle$$

with respect to a complete orthonormal basis φ_ν. The $\langle \nu | D | \mu \rangle$ can be considered to be the expansion coefficients of $D\varphi_\mu$:

$$D\varphi_\mu = \sum_\nu \varphi_\nu \langle \nu | D | \mu \rangle.$$

We may obtain a corresponding matrix for D with respect to an A-representation as follows:

$$D\tilde{\psi}(\alpha) = \int \tilde{\psi}(\alpha')\langle \alpha' | D | \alpha \rangle \, d\alpha', \tag{3.35}$$

where

$$\langle \alpha' | D | \alpha \rangle = \langle \tilde{\psi}(\alpha'), D\tilde{\psi}(\alpha) \rangle.$$

We may express the operator D in the A-representation as follows:

$$D\langle \tilde{\psi}(\alpha), f \rangle = \int \langle \alpha | D | \alpha' \rangle \langle \tilde{\psi}(\alpha'), f \rangle \, d\alpha'. \tag{3.36}$$

According to (3.35), $\langle \alpha' | D | \alpha \rangle$ is an improper function of the two variables α', α. For the **1**-operator we obtain the following special case:

$$\begin{aligned} \langle \alpha' | \mathbf{1} | \alpha \rangle &= \langle \tilde{\psi}(\alpha'), \tilde{\psi}(\alpha) \rangle \\ &= \delta(\alpha, \alpha') = \delta(\alpha - \alpha'). \end{aligned} \tag{3.37}$$

4 Transformation of One Representation into Another

At the end of the previous section we have considered two different representations of a Hilbert space, the A- and B-representations. We shall now introduce an elegant notation, due to Dirac, which facilitates the transformation from one representation to another, and simplifies the calculations, providing that it is clear what is meant, in mathematical terms, by the notation.

The simplification arises from the fact that, for the case of the discrete spectrum as well as for the continuous spectrum of A, the same symbol $|\alpha\rangle$ is used for the proper eigenvectors and improper eigenvectors of A, instead of ψ_ν for the eigenvectors and $\tilde{\psi}(\alpha)$ for the improper eigenvectors of the continuous spectrum. The real number α in $|\alpha\rangle$ denotes the corresponding value in Sp(A). If f is an arbitrary vector in $\mathscr{H}$ we then write $|f\rangle$ instead of f. With the new notation the expansion of f with respect to the proper and improper eigenvectors of A is written:

$$|f\rangle = \sum_\nu |\alpha_\nu\rangle\langle\alpha_\nu|f\rangle + \int |\alpha\rangle\langle\alpha|f\rangle \, d\alpha, \tag{4.1}$$

where $\langle\alpha_\nu|f\rangle = \langle\psi_\nu, f\rangle$, $\langle\alpha|f\rangle = \langle\tilde{\psi}(\alpha), f\rangle$. The sum in (4.1) is taken over all discrete eigenvalues, that is, over the discrete spectrum of A.

To each vector $f \in \mathscr{H}$ there exists a linear form $l(g) = \langle f, g\rangle$ on $\mathscr{H}$; we denote this linear form by $\langle f|$. The value of this linear form for a $g \in \mathscr{H}$ is $\langle f|g\rangle = \langle f, g\rangle$. All bounded linear forms over $\mathscr{H}$ are of the form $\langle f|$ for a suitably chosen f (see AIV, §4). The set of $\langle f|$ is therefore identical to the Banach space which is dual to $\mathscr{H}$ (see AIV, §4).

For improper eigenvectors $|\alpha\rangle$ we define an improper linear form $\langle\alpha|$ by $\langle\alpha|g\rangle = \langle\tilde{\psi}(\alpha), g\rangle$ for $g \in \mathscr{H}$.

On the basis of the results of §3 it is not necessary to provide additional explanation of the new notation. We shall only provide a few examples of the relationship between the new and old notations as follows:

$$P_f = |f\rangle\langle f| \quad \text{for} \quad f \in \mathscr{H} \text{ satisfies } \|f\| = 1; \tag{4.2}$$

$$E(\alpha) = \sum_{\substack{\nu \\ \alpha_\nu \leq \alpha}} |\alpha_\nu\rangle\langle\alpha_\nu| + \int_{-\infty}^{\alpha} |\alpha\rangle\langle\alpha|\, d\alpha; \tag{4.3}$$

$$A = \sum_\nu \alpha_\nu |\alpha_\nu\rangle\langle\alpha_\nu| + \int \alpha|\alpha\rangle\langle\alpha|\, d\alpha. \tag{4.4}$$

In the A-representation the vectors $f \in \mathscr{H}$ are represented by "functions" $\langle\alpha|f\rangle$. We then find that

$$\langle f, g\rangle = \sum_\nu \langle f|\alpha_\nu\rangle\langle\alpha_\nu|g\rangle + \int \langle f|\alpha\rangle\langle\alpha|g\rangle\, d\alpha. \tag{4.5}$$

Clearly $\langle f|\alpha\rangle = \overline{\langle\alpha|f\rangle}$. In the A-representation, from (3.26) we obtain

$$A\langle\alpha|f\rangle = \alpha\langle\alpha|f\rangle, \tag{4.6}$$

that is, A is the operator which multiplies the function $\langle\alpha|f\rangle$ by α. If D is an operator which admits a matrix representation (for example, if D is bounded) then in the A-representation we find that:

$$D\langle\alpha|f\rangle = \sum_\nu \langle\alpha|D|\alpha_\nu\rangle\langle\alpha_\nu|f\rangle + \int \langle\alpha|D|\alpha'\rangle\langle\alpha'|f\rangle\, d\alpha'. \tag{4.7}$$

If B is a second maximal self-adjoint operator, then, by analogy with (4.1) we obtain

$$|f\rangle = \sum_\mu |\beta_\mu\rangle\langle\beta_\mu|f\rangle + \int |\beta\rangle\langle\beta|f\rangle\, d\beta. \tag{4.8}$$

For

$$\begin{aligned} |\beta\rangle &= \sum_\nu |\alpha_\nu\rangle\langle\alpha_\nu|\beta\rangle + \int |\alpha\rangle\langle\alpha|\beta\rangle\, d\alpha, \\ |\alpha\rangle &= \sum_\mu |\beta_\mu\rangle\langle\beta_\mu|\alpha\rangle + \int |\beta\rangle\langle\beta|\alpha\rangle\, d\beta, \end{aligned} \tag{4.9}$$

it follows that $\langle\beta|\alpha\rangle = \overline{\langle\alpha|\beta\rangle}$ and

$$\langle\beta|f\rangle = \sum_{\nu}\langle\beta|\alpha_\nu\rangle\langle\alpha_\nu|f\rangle + \int\langle\beta|\alpha\rangle\langle\alpha|f\rangle\,d\alpha, \tag{4.10}$$

$$\langle\alpha|f\rangle = \sum_{\mu}\langle\alpha|\beta_\mu\rangle\langle\beta_\mu|f\rangle + \int\langle\alpha|\beta\rangle\langle\beta|f\rangle\,d\beta. \tag{4.11}$$

Equations (4.10) and (4.11) are the transformation equations between the A- and B-representations and vice versa.

The $\langle\beta|\alpha\rangle$ are the (proper and improper) eigenvectors of B in the A-representation

$$B\langle\alpha|\beta\rangle = \beta\langle\alpha|\beta\rangle \tag{4.12}$$

and $\langle\beta|\alpha\rangle$ are the eigenvectors of A in the B-representation

$$A\langle\beta|\alpha\rangle = \alpha\langle\beta|\alpha\rangle. \tag{4.13}$$

Here we use the equal sign $=$ instead of the equivalence symbol $\sim$ (in the sense of §3 for improper functions).

For the matrix of the operator D we obtain

$$\begin{aligned}\langle\alpha|D|\alpha'\rangle = {} & \sum_{\mu,\rho}\langle\alpha|\beta_\mu\rangle\langle\beta_\mu|D|\beta_\rho\rangle\langle\beta_\rho|\alpha'\rangle \\ & + \sum_{\mu}\int\langle\alpha|\beta_\mu\rangle\langle\beta_\mu|D|\beta\rangle\langle\beta|\alpha'\rangle\,d\beta \\ & + \sum_{\rho}\int\langle\alpha|\beta\rangle\langle\beta|D|\beta_\rho\rangle\langle\beta_\rho|\alpha'\rangle\,d\beta \\ & + \iint\langle\alpha|\beta\rangle\langle\beta|D|\beta'\rangle\langle\beta'|\alpha'\rangle\,d\beta\,d\beta'. \end{aligned} \tag{4.14}$$

The ensemble operator $W \in K$ (because it is bounded) corresponds to a matrix $\langle\alpha|W|\alpha'\rangle$ in the A-representation. For an arbitrary complete orthonormal basis η_ν, noting that

$$\sum_{\nu}\langle\alpha|\eta_\nu\rangle\langle\eta_\nu|\alpha'\rangle = \delta(\alpha, \alpha'),$$

the physically important quantity $\operatorname{tr}(WF)$ is given by

$$\begin{aligned}\operatorname{tr}(WF) &= \iint\sum_{\nu}\langle\eta_\nu|\alpha\rangle\langle\alpha|WF|\alpha'\rangle\langle\alpha'|\eta_\nu\rangle\,d\alpha\,d\alpha' \\ &= \iint\delta(\alpha, \alpha')\langle\alpha|WF|\alpha'\rangle\,d\alpha\,d\alpha',\end{aligned}$$

where $\iint$ means that we take the sum over discrete values and the integral over continuous eigenvalues α, α'. We therefore obtain

$$\operatorname{tr}(WF) = \sum_{\nu}\langle\alpha_\nu|WF|\alpha_\nu\rangle + \iint\delta(\alpha, \alpha')\langle\alpha|WF|\alpha'\rangle\,d\alpha\,d\alpha'. \tag{4.15}$$

We will now use (4.15) to illustrate the computation of probabilities.

Let $E(\alpha)$ denote the spectral family of the operator A, that is, in the A-representation we have

$$E(\alpha')\langle\alpha|f\rangle = \begin{cases} \langle\alpha|f\rangle & \text{if } \alpha \leq \alpha', \\ 0 & \text{if } \alpha > \alpha'. \end{cases}$$

For $\alpha_2 < \alpha_1$ we shall denote $E(\alpha_1) - E(\alpha_2)$ by $E(I)$. Suppose that an ensemble W is prepared for which $\operatorname{tr}(WE(I)) = 1$, that is, for the ensemble W a measurement value $\alpha \in I$ for the scale observable A is obtained with certainty. Therefore we find that $W = E(I)WE(I)$, and (4.15) is transformed into

$$\operatorname{tr}(WF) = \sum_{\substack{\nu \\ \alpha_\nu \in I}} \langle\alpha_\nu|WF|\alpha_\nu\rangle + \int_{\alpha_2}^{\alpha_1}\int_{\alpha_2}^{\alpha_1} \delta(\alpha, \alpha')\langle\alpha|WF|\alpha'\rangle \, d\alpha \, d\alpha'. \tag{4.16}$$

If B is a maximal scale observable and F is the corresponding projection such that the measurement value of B falls in the interval $J\colon \beta_2 < \cdots \leq \beta_1$, then it follows that

$$\begin{aligned}\langle\alpha''|F|\alpha'\rangle = &\sum_{\substack{\mu \\ \beta_\mu \in J}} \langle\alpha''|\beta_\mu\rangle\langle\beta_\mu|\alpha'\rangle \\ &+ \int_{\beta_2}^{\beta_1} \langle\alpha''|\beta\rangle\langle\beta|\alpha'\rangle \, d\beta.\end{aligned} \tag{4.17}$$

Using (4.17) we may calculate (4.16) in the general case. Since the latter formula is more encompassing, we will explicitly state only certain special cases which illustrate the general procedure.

Let us consider the special case W in which the measurement value of A does not have any dispersion. Then $W = P_{\varphi_\nu}$ for some φ_ν where φ_ν is the proper eigenvector for A with eigenvalue α_ν. Thus it follows that $\langle\alpha_\nu|W|\alpha_\nu\rangle = 1$, and the remaining matrix elements of W are equal to zero. Therefore it follows that

$$\begin{aligned}\operatorname{tr}(P_{\varphi_\nu}F) &= \langle\alpha_\nu|WF|\alpha_\nu\rangle \\ &= \langle\alpha_\nu|W|\alpha_\nu\rangle\langle\alpha_\nu|F|\alpha_\nu\rangle = \langle\alpha_\nu|F|\alpha_\nu\rangle \\ &= \sum_{\substack{\mu \\ \beta_\mu \in J}} \langle\alpha_\nu|\beta_\mu\rangle\langle\beta_\mu|\alpha_\nu\rangle + \int_{\beta_2}^{\beta_1} \langle\alpha_\nu|\beta\rangle\langle\beta|\alpha_\nu\rangle \, d\beta, \\ \operatorname{tr}(P_{\varphi_\nu}F) &= \sum_{\substack{\mu \\ \beta_\mu \in J}} |\langle\alpha_\nu|\beta_\mu\rangle|^2 + \int_{\beta_2}^{\beta_1} |\langle\alpha_\nu|\beta\rangle|^2 \, d\beta.\end{aligned} \tag{4.18}$$

$|\langle\alpha_\nu|\beta_\mu\rangle|^2$ is therefore the probability for the discrete value β_μ and $|\langle\alpha_\nu|\beta\rangle|^2$ is the "probability density" for the continuous values of the spectrum of B.

Suppose that the interval I for which $E(I)WE(I) = W$ is such that I does not contain any discrete eigenvalues of A. Then from (4.17) and (4.18) it follows that

$$\operatorname{tr}(WF) = \sum_{\substack{\mu \\ \beta_\mu \in J}} \int_{\alpha_2}^{\alpha_1} d\alpha \int_{\alpha_2}^{\alpha_1} d\alpha' \langle \beta_\mu | \alpha \rangle \langle \alpha | W | \alpha' \rangle \langle \alpha' | \beta_\mu \rangle$$
$$+ \int_{\beta_2}^{\beta_1} d\beta \int_{\alpha_2}^{\alpha_1} d\alpha \int_{\alpha_2}^{\alpha_1} d\alpha' \langle \beta | \alpha \rangle \langle \alpha | W | \alpha' \rangle \langle \alpha' | \beta \rangle. \tag{4.19}$$

Of particular importance for the evaluation of (4.19) are the "kernels"

$$\begin{aligned} K_\mu(\alpha', \alpha) &= \langle \alpha' | \beta_\mu \rangle \langle \beta_\mu | \alpha \rangle, \\ K_\beta(\alpha', \alpha) &= \langle \alpha' | \beta \rangle \langle \beta | \alpha \rangle. \end{aligned} \tag{4.20}$$

For the measurement values β_μ of the discrete spectrum we obtain the probability

$$w_\mu = \int_{\alpha_2}^{\alpha_1} d\alpha \int_{\alpha_2}^{\alpha_1} d\alpha' K_\mu(\alpha', \alpha) \langle \alpha | W | \alpha' \rangle \tag{4.21}$$

and for the continuous spectrum we obtain the probability density

$$w(\beta) = \int_{\alpha_2}^{\alpha_1} d\alpha \int_{\alpha_2}^{\alpha_1} d\alpha' K_\beta(\alpha', \alpha) \langle \alpha | W | \alpha' \rangle. \tag{4.22}$$

In general little can be said about (4.21) and (4.22). If the interval $I: \alpha_2 < \cdots \leq \alpha_1$ is small, then there can exist values of β_μ and β for which $K_\mu(\alpha', \alpha)$ and $K_\beta(\alpha', \alpha)$ are, for all practical purposes, constant for $\alpha', \alpha \in I$, that is, for an α_0 in I we obtain:

$$K_\mu(\alpha', \alpha) \approx |\langle \beta_\mu | \alpha_0 \rangle|^2, \qquad K_\beta(\alpha', \alpha) \approx |\langle \beta | \alpha_0 \rangle|^2. \tag{4.23}$$

For such β_μ or β (4.21), (4.22) are transformed into

$$w_\mu = |\langle \beta_\mu | \alpha_0 \rangle|^2 c_w, \tag{4.24}$$

$$w(\beta) = |\langle \beta | \alpha_0 \rangle|^2 c_w \tag{4.25}$$

with constant

$$c_w = \int_{\alpha_2}^{\alpha_1} d\alpha \int_{\alpha_2}^{\alpha_1} d\alpha' \langle \alpha | W | \alpha' \rangle. \tag{4.26}$$

Equations (4.24), (4.25) then show the dependence of the probabilities w_μ, or probability density $w(\beta)$, upon the values of the spectrum of B. Equations (4.24), (4.25) hold only for those β_μ and β for which (4.23) is a good approximation for $\alpha', \alpha \in I$. This cannot be the case for all β_μ, β because the following equation must be satisfied:

$$\sum_\mu \int_{\alpha_2}^{\alpha_1} d\alpha \int_{\alpha_2}^{\alpha_1} d\alpha' K_\mu(\alpha', \alpha) \langle \alpha | W | \alpha' \rangle$$
$$+ \int_{-\infty}^{\infty} d\beta \int_{\alpha_2}^{\alpha_1} d\alpha \int_{\alpha_2}^{\alpha_1} d\alpha' K_\beta(\alpha', \alpha) \langle \alpha | W | \alpha' \rangle = 1$$

and (4.23) would lead to the result

$$c_w\left[\sum_\mu |\langle\beta_\mu|\alpha_0\rangle|^2 + \int_{-\infty}^{\infty} |\langle\beta|\alpha_0\rangle|^2 \, d\beta\right] = 1$$

in contradiction to the "normalization" equation

$$\sum_\mu \langle\alpha|\beta_\mu\rangle\langle\beta_\mu|\alpha'\rangle + \int_{-\infty}^{\infty} \langle\alpha|\beta\rangle\langle\beta|\alpha'\rangle \, d\beta = \delta(\alpha, \alpha')$$

(where α and α' lie in the continuous spectrum of A) from which it follows that

$$\sum_\mu |\langle\beta_\mu|\alpha\rangle|^2 + \int_{-\infty}^{\infty} |\langle\beta|\alpha\rangle|^2 \, d\beta = \infty.$$

Nevertheless, the physical interpretation of $\langle\beta|\alpha\rangle$ is most intuitively evident in (4.18), (4.24), (4.25) providing that (4.24) and (4.25) are used with suitable caution.

5 Position and Momentum Representation

This section will serve to illustrate the discussion in the previous section by means of special examples. The standard example, the so-called "harmonic oscillator," is particularly suited to illustrate those aspects of the structure of quantum mechanics which will be considered in X, and is of considerable importance also for "realistic" problems. As an approximation, the harmonic oscillator plays a large role for molecular spectra (see XV) and provides the basis for quantum field theory (Fock representation, see, for example, [4]).

The position and momentum representation was originally obtained from the representation of the Galileo transformation, as we have outlined in VII, §2. The spaces $\mathscr{L}^2(\mathbf{R}^3, dx_1\, dx_2\, dx_3)$ and $\mathscr{L}^2(\mathbf{R}^3, dk_1\, dk_2\, dk_3)$ describe the position and momentum representations, and are isomorphically connected by the Fourier transformation. Here we shall treat the same problem from the other side in order to illustrate the methods of previous sections. Here we shall consider the "one-dimensional" case of a position operator Q and a momentum operator P which satisfies the commutation relation (see VII (4.22))

$$PQ - QP = \frac{1}{i}\mathbf{1}. \tag{5.1}$$

An "harmonic oscillator" is defined by the Hamiltonian operator

$$H = \frac{1}{2m} P^2 + \frac{m\omega^2}{2} Q^2, \tag{5.2}$$

which corresponds to an elementary (one-dimensional) system of mass m in an external field of the (one-dimensional) potential $(m\omega^2/2)Q^2$ (for external fields see VIII, §6).

We now introduce new operators

$$Q' = \sqrt{m\omega}\,Q, \qquad P' = \frac{1}{\sqrt{m\omega}}\,P. \tag{5.3}$$

We obtain

$$P'Q' - Q'P' = \frac{1}{i}\,\mathbf{1} \tag{5.4}$$

and

$$H = \omega H', \quad \text{where} \quad H' = \tfrac{1}{2}(P'^2 + Q'^2). \tag{5.5}$$

It suffices, therefore, to find the solution of the problem for Q', P', H', and the solution for Q, P, H is obtained by simple computation. In order to simplify the notation, we shall omit the $'$ in Q', P' and H'.

We shall now make the following assumptions:

(1) The linear manifold $\mathscr{L} = \bigcap_{n=1,m=1}^{\infty} (\mathscr{D}_{P^n} \cap \mathscr{D}_{Q^n})$ (where $\mathscr{D}$ is the domain of definition of the operators) is dense in $\mathscr{H}$.
(2) $PQ - QP = -i\mathbf{1}$ in $\mathscr{L}$.
(3) $H = \frac{1}{2}(P^2 + Q^2)$ is a self-adjoint operator with spectral decomposition $E(\varepsilon)$ and $E(\varepsilon)f \in \mathscr{L}$ for all $f \in \mathscr{H}$.

The assumptions (1), (2), (3) are satisfied for the position and momentum operators introduced in VII, §4. Conversely, if (1)–(3) (where (3) can be weakened somewhat) are satisfied for the position and momentum operators, then the infinitesimal transformations corresponding to P and Q correspond to a representation of the Galileo group. We refer readers who are interested in these relationships to [5].

Assumption (1) permits us to construct arbitrary products of the form $P^{\alpha_1}Q^{\beta_1}P^{\alpha_2}Q^{\beta_2}\ldots$. Therefore we find that $\mathscr{L} \subset \mathscr{D}_H$.

We define the operators:

$$A = \frac{1}{\sqrt{2}}(Q + iP) \quad \text{and} \quad B = \frac{1}{\sqrt{2}}(Q - iP). \tag{5.6}$$

In $\mathscr{L}$ we therefore find that $B = A^+$, and that the following relationships are satisfied:

$$AA^+ - A^+A = \mathbf{1} \quad \text{and} \quad H = A^+A + \tfrac{1}{2}\mathbf{1}. \tag{5.7}$$

Therefore, in $\mathscr{L}$ instead of H we may consider the operator $N = A^+A$. From assumption (3) we obtain:

$$N = \int_{-\infty}^{\infty} \lambda\, d\tilde{E}(\lambda), \tag{5.8}$$

where $\tilde{E}(\lambda)$ is the spectral family for N; from the latter we may easily obtain the spectral family for H.

The spectrum of N cannot be negative, since

$$\langle f, Nf\rangle = \|Af\|^2 \geq 0$$

for all f in $\mathscr{L}$.

According to Th. 1.3 (even in the case in which N is not maximal) there exists a φ such that

$$(\tilde{E}(\alpha) - \tilde{E}(\beta))\varphi \neq 0 \quad \text{for} \quad \tilde{E}(\alpha) - \tilde{E}(\beta) \neq 0.$$

If λ is a value from the spectrum of N, then for the vectors

$$\chi(\lambda, \varepsilon) = \frac{(\tilde{E}(\lambda) - \tilde{E}(\lambda - \varepsilon))\varphi}{\|(\bar{E}(\lambda) - \tilde{E}(\lambda - \varepsilon))\varphi\|}$$

we obtain

$$N\chi(\lambda, \varepsilon) - \lambda\chi(\lambda, \varepsilon) \to 0.$$

Conversely, if we are given a set of vectors $\chi(\varepsilon)$ for which $\|\chi_\varepsilon\| > \delta$ and $N\chi_\varepsilon - \lambda'\chi_\varepsilon \to 0$ then λ' is a point of the spectrum of N; on the other hand, if λ' is not a point of the spectrum, then there exists an interval $\lambda' - \eta \cdots \lambda' + \eta$ for which $\tilde{E}(\lambda + \eta) - \tilde{E}(\lambda - \eta) = 0$, and for every vector f we would find that $\|(N - \lambda'\mathbf{1})f\| \geq \eta\|f\|$. From $(N - \lambda'\mathbf{1})\chi(\varepsilon) \to 0$ for $f = \chi(\varepsilon)$ it would follow that $\chi(\varepsilon) \to 0$ in contradiction to the assumption that $\|\chi(\varepsilon)\| > \delta$.

From (5.7) it follows that in $\mathscr{L}$ the following relationships are satisfied: $AN - NA = A$ and $A^+N - NA^+ = -A^+$. Since the $\chi(\lambda, \varepsilon)$ lie in $\mathscr{L}$ we find that

$$NA\chi(\lambda, \varepsilon) = AN\chi(\lambda, \varepsilon) - A\chi(\lambda, \varepsilon)$$

and

$$NA\chi(\lambda, \varepsilon) - (\lambda - 1)A\chi(\lambda, \varepsilon) = A(N - \lambda\mathbf{1})\chi(\lambda, \varepsilon).$$

For $(N - \lambda\mathbf{1})\chi(\lambda, \varepsilon) = h$ we obtain

$$\begin{aligned}\|Ah\|^2 &= \langle h, Nh\rangle = \langle h, (N^2 - \lambda N)\chi(\lambda, \varepsilon)\rangle \\ &\leq \|h\|\,\|(N^2 - \lambda N)\chi(\lambda, \varepsilon)\| \leq \|h\|\lambda\varepsilon\end{aligned}$$

(the last relation is obtained in a manner similar to (3.6)). Therefore we find that $Ah \to 0$, that is,

$$(N - (\lambda - 1)\mathbf{1})A\chi(\lambda, \varepsilon) \to 0. \tag{5.9}$$

In a similar way it is possible to show that

$$(N - (\lambda + 1)\mathbf{1})A^+\chi(\lambda, \varepsilon) \to 0. \tag{5.10}$$

Since $\|A^+\chi(\lambda, \varepsilon)\|^2 = \langle\chi(\lambda, \varepsilon), \quad (N + \mathbf{1})\chi(\lambda, \varepsilon)\rangle = 1 + \|A\chi(\lambda, \varepsilon)\|^2$ it therefore follows that if λ is a value of the spectrum of N then so is $\lambda + 1$.

If $\|A\chi(\lambda, \varepsilon)\| > \delta$ for all ε, then $\lambda - 1$ will also be a value of the spectrum of N. Since the spectrum of N is nonnegative it follows that there must be an integer n for which there exists a sequence $\varepsilon_\nu \to 0$ such that $A^{n+1}\chi(\lambda, \varepsilon_\nu) \to 0$

and $\|A^n\chi(\lambda, \varepsilon_\nu)\| > \delta$. In addition $(N - (\lambda - n)\mathbf{1})A^n\chi(\lambda, \varepsilon_\nu) \to 0$ must be satisfied, that is, $\lambda - n$ must belong to the spectrum of N; in addition $(N - (\lambda - n - 1)\mathbf{1})A^{n+1}\chi(\lambda, \varepsilon_\nu) \to 0$ and therefore $NA^{n+1}\chi(\lambda, \varepsilon_\nu) \to 0$. From $NA^n\chi(\lambda, \varepsilon_\nu) = A^+A^{n+1}\chi(\lambda, \varepsilon_\nu)$ and $\|A^+A^{n+1}\chi(\lambda, \varepsilon_\nu)\|^2 = \|A^{n+1}\chi(\lambda, \varepsilon_\nu)\|^2 + \langle A^{n+1}\chi(\lambda, \varepsilon_\nu), \; NA^{n+1}\chi(\lambda, \varepsilon_\nu)\rangle$ it follows that $A^+A^{n+1}\chi(\lambda, \varepsilon_\nu) \to 0$, that is, $NA^n\chi(\lambda, \varepsilon_\nu) \to 0$. From $(N - (\lambda - n)\mathbf{1})A^n\chi(\lambda, \varepsilon_\nu) \to 0$ it follows that $(\lambda - n)A^n\chi(\lambda, \varepsilon_\nu) \to 0$. Since $\|A^n\chi(\lambda, \varepsilon_\nu)\| > \delta$ we must have $\lambda = n$, that is, λ is a positive integer $n \geq 0$. Therefore the spectrum N is discrete, and can only contain integers $n \geq 0$.

Since $NA^n\chi(\lambda, \varepsilon_\nu) \to 0$ and $\|\chi(\lambda, \varepsilon_\nu)\| > \delta$ the value 0 is a value in the spectrum of N. Therefore there exists an eigenvector ϕ_0 of N with eigenvalue $n = 0$. We may assume that $\|\phi_0\| = 1$. From $N\phi_0 = 0$ and from $\|N\phi_0\|^2 = \|A^+A\phi_0\|^2 = \|A\phi_0\|^2 + \|A^2\phi_0\|^2$ it follows that

$$A\phi_0 = 0. \tag{5.11}$$

Since $N\phi_0 = 0\phi_0$, it follows that, for $\chi(\lambda, \varepsilon) = \phi_0$ in (5.10), by recursion $N(A^+)^n\phi_0 = n(A^+)^n\phi_0$. The spectrum of N consists of all integers $n \geq 0$ and $(A^+)^n\phi_0$ is an eigenvector of N with eigenvalue n.

We set

$$\phi_n = a_n(A^+)^n\phi_0, \tag{5.12}$$

where a_n is the normalization factor. From the normalization requirement $\|\phi_n\| = 1$ it follows that

$$1 = \|\phi_n\|^2 = \left|\frac{a_n}{a_{n-1}}\right|^2 \langle A^+\phi_{n-1}, A^+\phi_{n-1}\rangle = \left|\frac{a_n}{a_{n-1}}\right|^2 \langle \phi_{n-1}, AA^+\phi_{n-1}\rangle$$
$$= \left|\frac{a_n}{a_{n-1}}\right|^2 \langle \phi_{n-1}, (N + \mathbf{1})\phi_{n-1}\rangle = n\left|\frac{a_n}{a_{n-1}}\right|^2.$$

Therefore the ϕ_n will be normalized if we set $a_n = 1/\sqrt{n!}$; we obtain

$$\phi_n = \frac{1}{\sqrt{n!}}(A^+)^n\phi_0, \tag{5.13}$$

and we obtain

$$N\phi_n = n\phi_n. \tag{5.14}$$

Since the ϕ_n are eigenvectors which correspond to different eigenvalues, they are pairwise orthogonal.

If the eigenspace of N for the eigenvalue 0 is multi-dimensional, then we may introduce a complete orthonormal basis $\phi_0^{(k)}$ in this eigenspace. We then define

$$\phi_n^{(k)} = \frac{1}{\sqrt{n!}}(A^+)^n\phi_0^{(k)}. \tag{5.15}$$

The $\phi_n^{(k)}$ are normalized and are pairwise orthogonal, that is, they are an orthonormal set in $\mathscr{H}$. We will now show that they are complete. If this is not the case, then there exists a smallest $n' > 0$ for which the $\phi_{n'}^{(k)}$ do not span the entire eigenspace of N with eigenvalue n', that is, there exists an eigenvector $\psi_{n'}$ (which is orthogonal to all the $\phi_{n'}^{(k)}$) of N with eigenvalue n', which, according to (3), lies in $\mathscr{L}$. From $n' \neq 0$ it follows that $A\psi_{n'} \neq 0$ and $N(A\psi_{n'}) = (n' - 1)(A\psi_{n'})$. Since $A\psi_{n'}$ is also orthogonal to all $\phi_{n'-1}^{(k)}$, this result contradicts the condition that n' is the smallest eigenvalue for which the $\phi_{n'}^{(k)}$ do not span the entire eigenspace.

Let $\mathscr{H}^{(k)}$ denote the subspace spanned by $\phi_n^{(k)}$ for fixed k; clearly

$$\mathscr{H} = \sum_k \oplus \mathscr{H}^{(k)}. \tag{5.16}$$

The operators A, A^+, H, P, Q all act in the various $\mathscr{H}^{(k)}$ in the same way; thus $\mathscr{H}$ can be expressed as a direct product as follows:

$$\mathscr{H} = \mathscr{H}_1 \times \mathscr{H}_2 \tag{5.17}$$

with the correspondence

$$\phi_n^{(k)} \leftrightarrow \phi_n u^{(k)}, \tag{5.18}$$

where the ϕ_n form a complete orthonormal basis for $\mathscr{H}_1$ and the $u^{(k)}$ form a complete orthonormal basis for $\mathscr{H}_2$, and the operators A, A^+, H, P, Q all have the form

$$A \times \mathbf{1}, \quad A^+ \times \mathbf{1}, \quad \text{etc.} \tag{5.19}$$

and the $A, A^+ \ldots$, as operators in $\mathscr{H}_1$ obey (5.11), (5.13) and (5.14). It therefore suffices to consider the above operators in $\mathscr{H}_1$ where N is maximal (according to Th. 1.6 we need only choose $\varphi = \sum_n \alpha_n \phi_n$ where $\alpha_n \neq 0$ and $\sum_n |\alpha_n|^2 < \infty$). For $\mathscr{H}_1$ there exists an N-representation given by $f \leftrightarrow \langle n|f\rangle$. From

$$P = \frac{1}{i}\frac{1}{\sqrt{2}}(A - A^+), \qquad Q = \frac{1}{\sqrt{2}}(A + A^+) \tag{5.20}$$

it follows that

$$\begin{aligned} P\phi_n &= \frac{1}{i}\frac{1}{\sqrt{2}}(\sqrt{n}\,\phi_{n-1} - \sqrt{n+1}\,\phi_{n+1}), \\ Q\phi_n &= \frac{1}{\sqrt{2}}(\sqrt{n}\,\phi_{n-1} + \sqrt{n+1}\,\phi_{n+1}). \end{aligned} \tag{5.21}$$

For $f = \sum_n \phi_n\langle n|f\rangle$ it therefore follows that, in the N-representation that

$$N\langle n|f\rangle = n\langle n|f\rangle,$$

$$P\langle n|f\rangle = \langle n|Pf\rangle = \frac{i}{\sqrt{2}}(\sqrt{n}\langle n-1|f\rangle - \sqrt{n+1}\langle n+1|f\rangle),$$

$$Q\langle n|f\rangle = \langle n|Qf\rangle = \frac{1}{\sqrt{2}}(\sqrt{n}\langle n-1|f\rangle + \sqrt{n+1}\,\langle n+1|f\rangle),$$

$$\langle n|P|m\rangle = \frac{1}{i}\frac{1}{\sqrt{2}}(\sqrt{m}\,\delta_{n,m-1} - \sqrt{m+1}\,\delta_{n,m+1}),$$

$$\langle n|Q|m\rangle = \frac{1}{\sqrt{2}}(\sqrt{m}\,\delta_{n,m-1} + \sqrt{m+1}\,\delta_{n,m+1}). \tag{5.22}$$

In order to find the Q-representation of $\mathscr{H}_1$ it is necessary to solve the eigenvalue equation for Q:

$$Q\langle n|x\rangle = x\langle n|x\rangle. \tag{5.23}$$

Using (5.22), (5.23) is transformed into

$$\sqrt{\frac{n+1}{2}}\langle n+1|x\rangle + \sqrt{\frac{n}{2}}\langle n-1|x\rangle = x\langle n|x\rangle$$

if we substitute

$$\langle n|x\rangle = \frac{1}{\sqrt{2^n n!}}\langle 0|x\rangle H_n(x) \tag{5.24}$$

from (5.23) we obtain the recursion formula

$$H_n = 2xH_{n-1} - 2(n-1)H_{n-2} \tag{5.25}$$

with the initial value $H_0 = 1$. The functions defined by this recursion formula are called the Hermite polynomials. They are identical to the expansion coefficients of the "generating function"

$$f(x, t) = e^{-t^2+2tx} = e^{x^2}e^{-(t-x)^2} = \sum_{n=0}^{\infty} H_n(x)\frac{t^n}{n!} \tag{5.26}$$

from which it follows that

$$H_n(x) = \frac{\partial^2}{\partial t^2} f(x, t)|_{t=0} = (-1)^n e^{x^2}\frac{d^n e^{-x^2}}{dx^n}. \tag{5.27}$$

Since

$$\frac{\partial}{\partial x} f(x, t) = 2tf(x, t)$$

from (5.26) we obtain the relation

$$H_n'(x) = 2nH_{n-1}(x) \tag{5.28}$$

and, from

$$\frac{\partial}{\partial t} f(x, t) + 2(t - x)f(x, t) = 0$$

we obtain the relation (5.25). Since from (5.26) it follows that $H_0(x) = 1$, we have proven the equivalence of (5.25) and (5.26).

The $\langle n|x\rangle$ in (5.24) are improper eigenvectors since $\sum_{n=0}^{\infty} |\langle n|x\rangle|^2 = \infty$. We note, however, that

$$\langle n|\varphi\rangle = \int_{x-\varepsilon}^{x} \langle n|x'\rangle \, dx'$$

are approximate eigenvectors of Q since (as we will shortly find) $\sum_{n=0}^{\infty} |\langle n|\varphi\rangle|^2 < \infty$. We will now determine $\langle n|0\rangle$ from the normalization condition

$$\sum_{n=0}^{\infty} \langle x'|n\rangle\langle n|x\rangle = \delta(x - x'). \tag{5.29}$$

For this purpose, using (5.27), we obtain

$$g(x, x') = \sum_n \frac{1}{2^n n!} H_n(x)H_n(x') = e^{x'^2} \sum_n \frac{1}{2^n n!} H_n(x)\left(-\frac{d}{dx'}\right)^n e^{-x'^2}.$$

From the Fourier expansion

$$e^{-x'^2} = \frac{1}{2\sqrt{\pi}} \int_{-\infty}^{\infty} e^{-i\omega x'} e^{-\omega^2/4} \, d\omega$$

we obtain

$$\left(-\frac{d}{dx'}\right)^n e^{-x'^2} = \frac{1}{2\sqrt{\pi}} \int_{-\infty}^{\infty} e^{-i\omega x'} (i\omega)^n e^{-\omega^2/4} \, d\omega$$

and, using the generating function in (5.26), it follows that

$$\begin{aligned} g(x, x') &= e^{x'^2} \frac{1}{2\sqrt{\pi}} \int_{-\infty}^{\infty} e^{-i\omega x'} e^{-\omega^2/4} \sum_n \frac{1}{2^n n!} H_n(x)(i\omega)^n \, d\omega \\ &= e^{x'^2} \frac{1}{2\sqrt{\pi}} \int_{-\infty}^{\infty} e^{-i\omega x'} e^{-\omega^2/4} e^{-(i\omega/2)^2 + 2(i\omega/2)x} \, d\omega \\ &= \sqrt{\pi} e^{x'^2} \frac{1}{2\pi} \int_{-\infty}^{\infty} e^{i\omega(x - x')} \, d\omega = \sqrt{\pi} e^{x'^2} \delta(x - x'), \end{aligned}$$

where we have used the following identity from the Fourier integral theorem

$$\frac{1}{2\pi} \int_{-\infty}^{\infty} e^{-i\omega(x - x')} \, d\omega = \delta(x - x').$$

Therefore, instead of (5.24) we can write,

$$\langle n|x\rangle = \frac{1}{\sqrt{2^n n!\sqrt{\pi}}} H_n(x)e^{-x^2/2}. \tag{5.30}$$

Each vector $f \in \mathscr{H}_1$ can therefore be represented in the Q-representation by a function $\langle x|f\rangle$ as follows:

$$\langle x|f\rangle = \sum_n \langle x|n\rangle\langle n|f\rangle.$$

In particular, the operator P can be expressed in the Q-representation as follows:

$$P\langle x|f\rangle = \sum_{n,m} \langle x|n\rangle\langle n|P|m\rangle\langle m|f\rangle.$$

From (5.28) and (5.25) it follows that:

$$\sum_n \langle x|n\rangle\langle n|P|m\rangle = \frac{1}{i}\frac{d}{dx}\langle x|m\rangle$$

and we therefore obtain

$$P\langle x|f\rangle = \frac{1}{i}\frac{d}{dx}\langle x|f\rangle. \tag{5.31}$$

The Q-representation can therefore be chosen such that (5.31). In the Q-representation we must, of course, have $Q\langle x|f\rangle = x\langle x|f\rangle$, as can easily be shown.

In the Q-representation the Hamiltonian operator H is given by

$$H = \frac{1}{2}(P^2 + Q^2) = \frac{1}{2}\left[\left(\frac{1}{i}\frac{d}{dx}\right)^2 + x^2\right].$$

In particular, we therefore obtain

$$H\langle x|n\rangle = \frac{1}{2}\left[\left(\frac{1}{i}\frac{d}{dx}\right)^2 + x^2\right]\langle x|n\rangle = \left(n + \frac{1}{2}\right)\langle x|n\rangle. \tag{5.32}$$

If, instead, we begin with the Q-representation which was given in VII, §4, then we could have used (5.32) in order to calculate $\langle x|n\rangle$ and then transform to the N-representation. We have taken the opposite route in order to demonstrate the methods of §§1–4.

We will now briefly outline how we may make the transition from the Q- to the P-representation. In order to determine the transformation coefficients $\langle x|p\rangle$ we proceed from the fact that in the Q-representation the $\langle x|p\rangle$ must be (improper) eigenvectors of P:

$$\frac{1}{i}\frac{d}{dx}\langle x|p\rangle = p\langle x|p\rangle \tag{5.33}$$

from which it follows that

$$\langle x|p\rangle = \frac{1}{\sqrt{2\pi}} e^{ipx}. \tag{5.34}$$

The factor already satisfies the normalization condition

$$\int_{-\infty}^{\infty} \langle p'|x\rangle\langle x|p\rangle = \delta(p - p').$$

The relation (5.34) may be formulated precisely in the sense of §4. $\langle x|p\rangle$ is an improper function of two variables x and p:

$$\langle x|p\rangle\colon \int_{p-\varepsilon}^{p} dp' \int_{x-\delta}^{x} dx'\langle x'|p\rangle.$$

The reader may visualize the concepts of §4 by means of this example.

We shall now use the $\langle x|n\rangle$ and $\langle x|p\rangle$ in order to explicitly compute the probabilities which were discussed at the end of §4.

Given an ensemble W for which the energy has zero dispersion, that is, $W = P_{\phi_n}$ for some n, then the probability of a position measurement is, according to (4.18) given by the probability density

$$w(x) = |\langle x|n\rangle|^2.$$

In Figure 1 $\langle x|n\rangle$ is displayed; the behavior of $|\langle x|n\rangle|^2$ as a function of x can readily be ascertained.

Conversely, given an ensemble W for which the position is in the interval $x - \varepsilon, x$ with certainty, it follows that the probability that a measurement of the energy will be $(n + \frac{1}{2})$ is, according to (4.21), given by

$$w_n = \int_{x-\varepsilon}^{x} dx' \int_{x-\varepsilon}^{x} dx'' K_n(x'', x')\langle x'|W|x''\rangle, \tag{5.35}$$

where

$$K_n(x', x'') = \langle x'|n\rangle\langle n|x''\rangle.$$

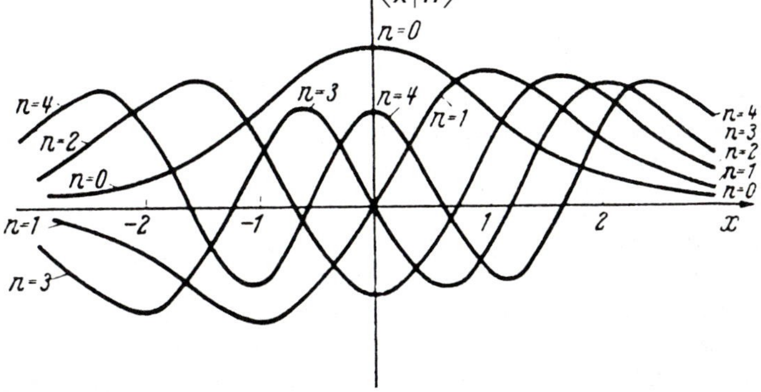

Figure 1 Eigenfunctions $\langle x|n\rangle$ of harmonic oscillators.

From Figure 1 it is clear that, for small ε, the function $K_n(x', x'')$ is, for all practical purposes, constant in the interval $x - \varepsilon \cdots x$ providing that n is not too large! If n is not too large, then, according to (4.24) we obtain

$$w_n \approx |\langle x|n\rangle|^2 c_w, \quad \text{where} \quad c_w = \int_{x-\varepsilon}^{x} dx' \int_{x-\varepsilon}^{x} dx'' \langle x'|W|x''\rangle. \tag{5.36}$$

From this example it is clear why (5.36) cannot be a good approximation to (5.35) for all n.

The above derivation makes it possible to determine those ensembles P_φ for which equality holds in the Heisenberg uncertainty relation IV (8.3.18). According to the discussion immediately following IV (8.3.17) equality can hold if there exists a real number α for which the relation

$$(Q' + i\alpha P') = 0 \tag{5.37}$$

holds, where

$$P' = P - \operatorname{tr}(P_\varphi P)\mathbf{1} = P - \bar{p}\mathbf{1},$$
$$Q' = Q - \operatorname{tr}(P_\varphi Q)\mathbf{1} = Q - \bar{q}\mathbf{1}.$$

$Q' + i\alpha P'$ is, however, except for a factor and an origin displacement, identical with the operator A or $B = A^+$ in (5.6). Therefore (5.37) can only be satisfied for $\alpha > 0$ and must correspond to the "ground state" of the harmonic oscillator. We may express (5.37) in the Q-representation as follows:

$$\left[x - \bar{q} + \alpha \frac{d}{dx} - i\alpha\bar{p}\right]\langle x|\varphi\rangle = 0.$$

Thus it follows that

$$\langle x|\varphi\rangle = \frac{1}{\sqrt{2\alpha\sqrt{\pi}}} e^{-[(x-\bar{q})^2/2\alpha] - i\bar{p}(x-\bar{q})} \tag{5.38}$$

which is also normalized, is a Hilbert space vector only if $\alpha > 0$. For the position probability density in the ensemble $W = P_\varphi$ we therefore obtain

$$|\langle x|\varphi\rangle|^2 = \frac{1}{2\alpha\sqrt{\pi}} e^{-[(x-\bar{q})^2/\alpha]}. \tag{5.39}$$

This form is known as the "Gauss distribution (Gaussian)."

6 Degenerate Spectra

The A-representation of a Hilbert space is possible if A is maximal, that is, if the complete Boolean ring Σ generated by the spectral representation of A is maximal (D 1.1). Σ is maximal if there is no decision effect $E \notin \Sigma$ which is commensurable with all elements of Σ. If α_ν is an eigenvalue of A and P_ν

is the projection operator onto the eigenspace for the eigenvalue α_ν then each $P_\nu \in \Sigma$ and each $E \leq P_\nu$ are commensurable with all elements of Σ. If A is maximal, then from $E \leq P_\nu$ it follows that either $E = P_\nu$ or $E = 0$, that is, P_ν must be one-dimensional. Therefore, to α_ν there exists a uniquely determined eigenvector (up to a factor) φ_ν. In this case we say that the eigenvalue is nondegenerate. If Sp(A) has only discrete values and is nondegenerate then Σ is maximal (see §5) and $P_{\varphi_\nu} \in \Sigma$ for the complete orthonormal basis of the eigenvectors of A. If $E \neq 0$ is commensurable with all P_{φ_ν} (and therefore commutes with all P_{φ_ν}) then it follows that $EP_{\varphi_\nu} = P_{\varphi_\nu}E \leq P_{\varphi_\nu}$, that is, $E\varphi_\nu = \varphi_\nu$ or $E\varphi_\nu = 0$ and $E = \sum'_\nu P_{\varphi_\nu}$ where Σ' is the sum over all ν for which $E\varphi_\nu = \varphi_\nu$. Thus it follows that $E \in \Sigma$.

D 6.1. The dimension of a projection P_ν onto an eigenspace of A corresponding to the eigenvalue α_ν is called the degree (order) of the degeneracy for α_ν. Note that the statements "α_ν is nondegenerate" and "α_ν is degenerate of order 1" have the same meaning.

We shall now outline how we may obtain an A-representation for a Hilbert space for an A which is not maximal, and how the notion of degree of degeneracy may be extended to the continuous spectrum.

As we have already mentioned at the beginning of §1, if we are initially given a nonmaximal A, then there exists a finite set of commensurable decision observables $A, B, \ldots$, the combination of which is maximal (D 1.1). The discussion of previous sections is applicable to this case. If, for example, A and B are commensurable, then instead of an A-representation we will obtain an (A, B)-representation by means of proper or improper functions in the (α, β)-plane as follows: $f \leftrightarrow \langle \alpha, \beta | f \rangle$. Since the extension of the preceding formulas is trivial, we shall not discuss them here. In the following we shall present the case of position representation for several Q_ν.

In order to define the notion of the degree of degeneracy more generally we shall begin with Th. 1.9. Suppose that E_n ($n = 1, 2, \ldots$) is a denumerable set of pairwise orthogonal decision effects which are commensurable with A; so that the complete Boolean ring generated by the spectral family of A and by the E_n is maximal and that from $EE_n = 0$ for an E in the Boolean ring generated by the spectral family it follows that $EE_m = 0$ for all $m > n$. For the operator $B = \sum_{n=0} nE_n$ and $E_0 = \mathbf{1} - \sum_{n=1} E_n$ we find that A and B are a maximal pair (D 1.1) of scale observables, that is, the complete Boolean ring Σ generated by the spectral family of A and by the E_n ($n = 0, 1, 2, \ldots$) is maximal. According to Th. 1.3 there exists a φ for which $\mathscr{T}(\Sigma, \varphi) = \mathscr{H}$. Therefore the discussions in §2 can be directly applied to A, B.

For an interval L in the (α, β)-plane of the form $L = (\alpha_1, \alpha_2] \times (\beta_1, \beta_2]$ we define the measure μ as follows:

$$\mu(L) = \|E^{(A)}(I)E^{(B)}(J)\varphi\|^2, \tag{6.1}$$

where $I = (\alpha_1, \alpha_2]$, $J = (\beta_1, \beta_2]$ and $E^{(A)}(\lambda)$, $E^{(B)}(\lambda)$ are the spectral families of A and B, respectively. $\mathscr{H}$ can therefore be represented in terms of $\mathscr{L}^2(\mathbf{R}^2, \mu)$.

Because of the special character of B we may simplify the representation using functions $f(\alpha, n)$ for $n = 1, 2, \ldots$ for which

$$\sum_{n=0} \int |f(\alpha, n)|^2 \, d\mu_n(\alpha) < \infty, \tag{6.2}$$

where

$$\mu_n(I) = \|E^{(A)}(I)E_n\varphi\|^2. \tag{6.3}$$

The inner product is given by

$$\langle f, g \rangle = \sum_n \int \overline{f(\alpha, n)} g(\alpha, n) \, d\mu_n(\alpha). \tag{6.4}$$

We denote the space of quadratically integrable functions $f(\alpha, n)$ (in the sense of (6.2)) by $\mathscr{L}^2(\mathbf{R} \times N_+, \mu)$. Let $\mathrm{Sp}_n(A)$ denote the set $\{\alpha |$ for all α_1, α_2 where $\alpha_1 < \alpha < \alpha_2$, it follows $\|(E^{(A)}(\alpha_2) - E^{(A)}(\alpha_1))E_n\varphi\|^2 \neq 0\}$.

We shall now show that $\mathrm{Sp}_n(A) \supset \mathrm{Sp}_m(A)$ for $m > n$: If $\alpha \notin \mathrm{Sp}_n(A)$ then there exists an interval $\alpha_1 < \alpha < \alpha_2$ such that $(E^{(A)}(\alpha_2) - E^{(A)}(\alpha_1))E_n\varphi = 0$ and therefore we obtain $(E^{(A)}(\alpha_2) - E^{(A)}(\alpha_1))E_n = 0$. Thus, from Th. 1.9 it also follows that $(E^{(A)}(\alpha_2) - E^{(A)}(\alpha_1))E_m = 0$ for $m > n$ from which it follows that $\mathrm{Sp}_n(A) \supset \mathrm{Sp}_m(A)$.

We shall say that n is the degree of degeneracy of the value α in the continuous spectrum of A if n is the largest number for which $\alpha \in \mathrm{Sp}_n(A)$. Here we permit the value $n = \infty$.

It is easy to generalize the special case of operators Q and P which satisfy the commutation relations (5.1) onto a set of operators Q_i, P_i where $i = 1, 2, \ldots, f$ which satisfy the commutation relations:

$$P_i Q_k - Q_k P_i = \frac{1}{i} \delta_{ik} \mathbf{1}, \qquad Q_i Q_k - Q_k Q_i = 0, \qquad P_i P_k - P_k P_i = 0.$$

We introduce the operators

$$A_i = \frac{1}{\sqrt{2}} (Q_i + iP_i), \qquad A_i^+ = \frac{1}{\sqrt{2}} (Q_i - iP_i)$$

and the operator $N = \sum_{i=1}^{f} A_i^+ A_i$. In the same manner as in §5 it is easy to show that N has eigenvalues $n = 0, 1, 2, \ldots$ and there exists a vector ϕ_0 for which $A_i \phi_0 = 0$ for all A_i. For the complete orthonormal basis $\phi_0^{(k)}$ for the eigenspace of N with eigenvalue 0 the set of vectors

$$\phi^{(k)}_{n_1 n_2 \cdots n_f} = \frac{1}{\sqrt{\prod_i n_i!}} \prod_j (A_0^+)^{n_j} \phi_0^{(k)}$$

forms a complete orthonormal basis in $\mathscr{H}$.

We may, as we have done in §5, express the Hilbert space $\mathscr{H}$ in the form $\mathscr{H} = \mathscr{H}_b \times \mathscr{H}_2$ where the Q_i, P_i are expressed in the form $Q_i \times \mathbf{1}$, $P_i \times \mathbf{1}$; then, in the position representation $\mathscr{H}_b$ takes on the following form

$$\begin{aligned} \varphi \in \mathscr{H}_b: \qquad \varphi &\leftrightarrow \langle x_1, x_2, \ldots, x_f | \varphi \rangle, \\ Q_j \varphi &\leftrightarrow x_j \langle x_1, x_2, \ldots, x_f | \varphi \rangle, \\ P_j \varphi &\leftrightarrow \frac{1}{i} \frac{\partial}{\partial x_j} \langle x_1, x_2, \ldots, x_f | \varphi \rangle. \end{aligned}$$

$\mathscr{H}_b$ is often called the orbit space (see VII and VIII).

CHAPTER X

Equations of Motion

In VIII we have established that VIII (6.26) and (6.27) are, in a formal sense, the most general form for the description of the time translation of a registration apparatus relative to a preparation apparatus. In VIII, §6 we indicated that we would like to call VIII (6.2.7) the "equations of motion for quantum mechanics." In this chapter we shall only be concerned with formal transformations of the above time translations. For that reason we shall take VIII (6.27) and (6.16) as our starting point. It will then be possible to repeat all the formal transformations for VIII (6.26) instead of VIII (6.27).

1 The Heisenberg Picture

We shall call the description of time translation based upon VIII (6.27), that is,

$$\frac{dB_t}{dt} = i(H_t B_t - B_t H_t) + \frac{\partial B_t}{\partial t} \tag{1.1}$$

the so-called Heisenberg picture. In the computation of statistical quantities and expectation values of the form $\operatorname{tr}(WB_t)$, W is taken to be constant.

For the special case in which B_t does not "explicitly" depend on t, that is, $\partial B_t/\partial t = 0$, it follows that

$$B_t = U_t B_0 U_t^+, \tag{1.2}$$

where, according to VIII (6.16), the unitary operator U satisfies the differential equation

$$\frac{dU_t}{dt} = iH_t U_t, \tag{1.3}$$

where $U_0 = \mathbf{1}$. The solution of the differential equation (1.1) for the case $\partial B_t/\partial t = 0$ follows, therefore, from the solution of (1.3) given by (1.2). For the special case in which H is independent of time, a formal solution of (1.3) is given by

$$U_t = e^{iHt}. \tag{1.4}$$

Let A and H be a complete set of commensurable scale observables; let α and ε denote the values of the spectrum of the self-adjoint operators A and H, respectively. Then U_t is the operator which multiplies the vectors $\langle \varepsilon, \alpha | \psi \rangle$ (the vectors ψ in the HA-representation) by $e^{i\varepsilon t}$. If we represent B_0 by a matrix, then we may express B_t by the following matrix:

$$\langle \varepsilon, \alpha | B_t | \varepsilon', \alpha' \rangle = e^{i(\varepsilon - \varepsilon')t} \langle \varepsilon, \alpha | B_0 | \varepsilon', \alpha' \rangle. \tag{1.5}$$

The fact that B_t is obtained from B_0 by means of a unitary transformation U_t implies that B_t has the same spectrum as B_0 because the spectral representation $E_t(\lambda)$ of B_t is obtained from the $E_0(\lambda)$ of B_0 as follows:

$$E_t(\lambda) = U_t E_0(\lambda) U_t^+. \tag{1.6}$$

The solution of the eigenvalue equation $B_t \varphi_t = \beta \varphi_t$ follows simply from $B_0 \varphi_0 = \beta \varphi_0$ and is given by $\varphi_t = U_t \varphi_0$. The physical meaning of this situation is as follows: The possible measurement values for a scale observable cannot change with time if there is no time variable external field.

From (1.6) it follows that the statistics of the measurement values do change with time. The statistics of the measurement values will remain constant with time for every ensemble only if B_0 commutes with all H_t for all t—then we will obtain $B_t = B_0$ and $E_t(\lambda) = E_0(\lambda)$. In the case in which $B_t = B_0$ we say that $B = B_0 = B_t$ is a constant or integral of motion.

For our first example of the application of the Heisenberg picture we shall consider the case of a one-dimensional harmonic oscillator under the influence of an external field. This example is more than an academic one; it plays an important role in many realistic physical problems—for example, as an approximation in molecular physics (see XV, §8) and as a problem in quantum field theory.

For the external field we shall choose the simple potential field given by $\phi(x, t) = -xk(t)$. Then the Hamiltonian operator H is given by

$$H = \frac{1}{2m} P^2(t) + \frac{m\omega^2}{2} Q^2(t) - Qk(t), \tag{1.7}$$

where $P(t)$ and $Q(t)$ are the momentum and position operators, respectively.

In addition to scale observables $P(t)$, $Q(t)$, we shall also discuss the time behavior of the "energy"

$$H_0(t) = \frac{1}{2m} P^2(t) + \frac{m\omega^2}{2} Q^2(t).$$

For each value of t the operators $P(t)$, $Q(t)$ satisfy the Heisenberg commutation relation (see IX (5.1))

$$P(t)Q(t) - Q(t)P(t) = \frac{1}{i} \mathbf{1}. \tag{1.8}$$

Since much of the mathematics development for this problem is completely analogous to that presented in IX, §5 we shall only present an outline.

We introduce the operator $A(t)$ and its adjoint $A^+(t)$ as follows:

$$\begin{aligned} A(t) &= \frac{1}{\sqrt{2}}\left(\sqrt{m\omega}\, Q(t) + \frac{i}{\sqrt{m\omega}} P(t)\right), \\ A^+(t) &= \frac{1}{\sqrt{2}}\left(\sqrt{m\omega}\, Q(t) - \frac{i}{\sqrt{m\omega}} P(t)\right). \end{aligned} \tag{1.9}$$

From which it follows that:

$$\begin{aligned} A(t)A^+(t) - A^+(t)A(t) &= \mathbf{1}, \\ H_0(t) &= \omega A^+(t)A(t) + \frac{\omega}{2} \mathbf{1}, \\ H_t = H_0(t) &- \frac{1}{\sqrt{2m\omega}} (A(t) + A^+(t))k(t). \end{aligned} \tag{1.10}$$

For each value of t there exists a (normalized) vector $\Psi_0(t) \in \mathscr{H}$ which satisfies $A(t)\Psi_0(t) = 0$. The vectors

$$\Psi_n(t) = \frac{1}{\sqrt{n!}} A^+(t)^n \Psi_0(t) \tag{1.11}$$

form a complete orthonormal basis in $\mathscr{H}$. They are also eigenvectors of $H_0(t)$:

$$H_0(t)\Psi_n(t) = (n + \tfrac{1}{2})\omega\Psi_n(t). \tag{1.12}$$

According to the Heisenberg picture we obtain the time evolution of the observables of interest $P(t)$, $Q(t)$, $H_0(t)$ if we obtain $A(t)$ from the differential equation (1.1):

$$\frac{1}{i} \dot{A}(t) = H_t A(t) - A(t) H_t = -\omega A(t) + \frac{k(t)}{\sqrt{2m\omega}} \mathbf{1}.$$

The solution is very simple:

$$A(t) = e^{-i\omega t}(A(0) + f(t)\mathbf{1}), \tag{1.13}$$

where

$$f(t) = \frac{i}{\sqrt{2m\omega}} \int_0^t k(\tau)e^{i\omega\tau}\, d\tau. \tag{1.14}$$

We shall write $\Psi_n(0) = \phi_n$, so that from (1.10) we therefore obtain:

$$\phi_n = \frac{1}{\sqrt{n!}} A^+(0)^n \phi_0. \tag{1.15}$$

According to the general considerations about the eigenvectors and eigenvalues of B_t we may obtain the eigenvectors of $H_0(t)$ from those of $H_0(0)$ by means of the operator U_t. Equation (1.13) gives us the value of the right-hand side of the following equation:

$$A(t) = U_t A(0) U_t^+.$$

Then from (1.11) we obtain

$$\Psi_n(t) = U_t A^+(0)^n U_t^+ \; \Psi_0(t). \tag{1.16}$$

The eigenvector $\Psi_0(t)$ of $H_0(t)$ corresponding to the eigenvalue $\omega/2$ must have the form

$$\Psi_0(t) = \alpha(t) U_t \phi_0, \tag{1.17}$$

where we may choose $\alpha(t) = 1$. For computational purposes, however, it is desirable to refrain from this choice. From (1.16), (1.17) it therefore follows that, in general,

$$\Psi_n(t) = \alpha(t) U_t \phi_n \tag{1.18}$$

U_t is therefore determined if $\Psi_n(t)$ is expressed in terms of the orthonormal basis of the ϕ_n and if $\alpha(t)$ is given.

The expansion

$$\Psi_n(t) = \sum_m \phi_m \langle \phi_m, \Psi_n(t) \rangle \tag{1.19}$$

is determined when we have computed $\langle \phi_m, \Psi_n \rangle$. From (1.11) it follows that

$$\langle \phi_m, \Psi_n(t) \rangle = \frac{1}{\sqrt{n!}} \langle A(t)^n \phi_m, \Psi_0(t) \rangle. \tag{1.20}$$

According to (1.13) we obtain

$$A(t)^n = e^{-in\omega t} \sum_{\nu=0}^{n} \binom{n}{\nu} f^\nu A(0)^{n-\nu}. \tag{1.21}$$

Furthermore, from (1.15) and the first equation in (1.10) it follows that

$$A(0)^p \phi_m = \sqrt{m(m-1) \cdots (m-p+1)}\, \phi_{m-p}. \tag{1.22}$$

In order to find $\langle\phi_m, \Psi_n\rangle$ it therefore suffices to compute $\langle\phi_\rho, \Psi_0\rangle$:

$$\begin{aligned}\langle\phi_\rho, \Psi_0(t)\rangle &= \frac{1}{\sqrt{\rho!}}\langle A^+(0)^\rho\phi_0, \Psi_0(t)\rangle = \frac{1}{\sqrt{\rho!}}(\phi_0, A(0)^\rho\Psi_0(t)\rangle \\ &= \frac{1}{\sqrt{\rho!}}\langle\phi_0, [A(t)e^{i\omega t} - f(t)]^\rho\Psi_0(t)\rangle \\ &= \frac{1}{\sqrt{\rho!}}[-f(t)]^\rho\langle\phi_0, \Psi_0(t)\rangle.\end{aligned}$$

Since $\|\Psi_0(t)\|^2 = \sum_\rho |\langle\phi_\rho, \Psi_0(t)\rangle|^2 = 1$, we must have $|\langle\phi_0, \Psi_0(t)\rangle|^2 = e^{-|f|^2}$. Since we have not yet chosen the factor $\alpha(t)$ in (1.17) we may choose $\langle\phi_0, \Psi_0(t)\rangle = e^{-(1/2)|f|^2}$. Then we obtain

$$\langle\phi_\rho, \Psi_0(t)\rangle = \frac{1}{\sqrt{\rho!}}[-f(t)]^\rho e^{-(1/2)|f(t)|^2}. \tag{1.23}$$

Then, from (1.20) to (1.23) we obtain:

$$\begin{aligned}\langle\phi_m, \Psi_n(t)\rangle = \sqrt{\frac{m!}{n!}}\, e^{in\omega t}e^{-(1/2)|f(t)|^2}[-f(t)]^{m-n}\cdot \\ \cdot\sum_{\nu=\sigma}^{n}\binom{n}{\nu}[-|f(t)|^2]^\nu\frac{1}{(m-n+\nu)!},\end{aligned} \tag{1.24}$$

where $\sigma = 0$ for $m \geq n$ and $\sigma = n - m$ for $m \leq n$.

The ρth derivative of the Laguerre polynomial

$$L_n(x) = \sum_{\nu=0}^{n}\binom{n}{\nu}\frac{n!}{\nu!}(-x)^\nu$$

is given by

$$L_n^\rho(x) = \sum_{\nu=0}^{n}\binom{n}{\nu}\frac{n!}{(\nu-\rho)!}(-x)^{\nu-\rho}(-1)^\rho.$$

Therefore, for $m \leq n$ we obtain

$$\langle\phi_m, \Psi_n(t)\rangle = \sqrt{\frac{m!}{n!}}\frac{1}{n!}\, e^{in\omega t}e^{-(1/2)|f(t)|^2}(\overline{f(t)})^{n-m}L_n^{n-m}(|f(t)|^2) \tag{1.25a}$$

for $m \geq n$ we obtain

$$\langle\phi_m, \Psi_n(t)\rangle = \sqrt{\frac{n!}{m!}}\frac{1}{m!}\, e^{in\omega t}e^{-(1/2)|f(t)|^2}(-f(t))^{m-n}L_n^{m-n}(|f(t)|^2). \tag{1.25b}$$

Thus the expansion (1.19) and the operator U_t defined by (1.18) is determined up to the factor $\alpha(t)$.

With the aid of the $\langle\phi_m, \Psi_n(t)\rangle$ given in (1.25a) and (1.25b) we may now answer the following physical question: Suppose we are given a preparation apparatus which prepares an ensemble W for which the measurement of

$H_0(0)$ results in the value $\omega(m + \frac{1}{2})$ with certainty. Therefore we must have $\mathrm{tr}(WP_{\phi_m}) = 1$, from which it follows that $W = P_{\phi_m}$. What is the probability for the same ensemble that a measurement of $H(t)$ will result in a meaurement value of $\omega(n + \frac{1}{2})$ (that is, for a time displacement t of the registration apparatus for $H_0(0)$ relative to the preparation apparatus)? It is given by

$$\mathrm{tr}(WP_{\Psi_n(t)}) = \mathrm{tr}(P_{\phi_m} P_{\Psi_n(t)}) = |\langle \phi_m, \Psi_n(t)\rangle|^2. \tag{1.26}$$

It remains to compute the factor $\alpha(t)$ in (1.18). From (1.17) and (1.3) it follows that:

$$\begin{aligned}\dot{\Psi}_0(t) &= \dot{\alpha} U_t \phi_0 + \alpha \dot{U}_t \phi_0 = \frac{\dot{\alpha}}{\alpha} \alpha U_t \phi_0 + i\alpha H_t U_t \phi_0 \\ &= \frac{\dot{\alpha}}{\alpha} \Psi_0(t) + iH_t \Psi_0(t)\end{aligned}$$

and we obtain

$$\langle \Psi_0(t), \dot{\Psi}_0(t)\rangle = \frac{\dot{\alpha}}{\alpha} + i\langle \Psi_0(t), H_t \Psi_0(t)\rangle.$$

From H_t in (1.10) it follows that

$$\frac{\dot{\alpha}}{\alpha} = -\frac{i\omega}{2} + \langle \Psi_0(t), \dot{\Psi}_0(t)\rangle.$$

In order to compute the inner product $\langle \Psi_0, \dot{\Psi}_0\rangle$ we begin with $\Psi_0 = \sum_n \phi_n \langle \phi_n, \Psi_0\rangle$ and use (1.23); we obtain:

$$\Psi_0(t) = \sum_n \phi_n (-f(t))^n e^{-(1/2)|f(t)|^2} \frac{1}{\sqrt{n!}}.$$

By differentiating with respect to time it follows that:

$$\dot{\Psi}_0(t) = \sum_n \phi_n \frac{1}{\sqrt{n!}} e^{-(1/2)|f|^2} \{n(-f)^{n-1}(-\dot{f}) - \tfrac{1}{2}(-f)^n(\bar{f}\dot{f} + \dot{\bar{f}} f)\}.$$

Therefore we obtain

$$\langle \Psi_0, \dot{\Psi}_0(t)\rangle = \tfrac{1}{2}(\bar{f}\dot{f} - \dot{\bar{f}} f) = \frac{1}{2} \frac{k(t)}{\sqrt{2m\omega}} i(\overline{f(t)} e^{i\omega t} + f(t) e^{-i\omega t})$$

from which we obtain the following equation for α:

$$\frac{\dot{\alpha}}{\alpha} = -\frac{i\omega}{2} + \frac{i}{2} \frac{k(t)}{\sqrt{2m\omega}} (\bar{f}(t) e^{i\omega t} + f(t) e^{-i\omega t}).$$

With $\alpha(0) = 1$ we finally obtain

$$\alpha(t) = \exp\left\{-\frac{i\omega t}{2} + \frac{i}{2} \frac{1}{\sqrt{2m\omega}} \int_0^t k(\tau)(\overline{f(\tau)} e^{i\omega\tau} + f(\tau) e^{-i\omega\tau})\, d\tau\right\}. \tag{1.27}$$

Thus we have completely determined U_t.

For our second example illustrating the Heisenberg picture we shall consider an elementary system of mass m in the absence of an external field. Here we have $H = (1/2m) \sum_{i=1}^{3} P_i^2$. For $P_i(t)$ and $Q_i(t)$ it follows from (1.1) that

$$\dot{P}_i(t) = 0, \qquad \dot{Q}_i(t) = \frac{1}{m} P_i(t),$$

P_i is therefore constant with time, and we obtain

$$Q_i(t) = Q_i(0) + \frac{t}{m} P_i. \tag{1.28}$$

Thus it follows that the expectation value for the position for the ensemble W is given by

$$M(Q_i(t)) = \operatorname{tr}(WQ_i(t)) = M(Q_i(0)) + \frac{t}{m} M(P_i), \tag{1.29}$$

that is, the expectation value of the position moves with constant velocity $(1/m)M(P_i)$. If we set

$$Q_i'(t) = Q_i(t) - M(Q_i(t))\mathbf{1},$$
$$P_i' = P_i - M(P_i)\mathbf{1},$$

then, from (1.28), (1.29) it follows that:

$$Q_i'(t) = Q_i'(0) + \frac{t}{m} P_i'. \tag{1.30}$$

For the dispersion

$$\operatorname{disp}(Q_i(t)) = \operatorname{tr}(WQ_i'(t)^2) \tag{1.31}$$

we obtain the relation:

$$\operatorname{disp}(Q_i(t)) = \operatorname{disp}(Q_i(0)) + \frac{t}{m} M(Q_i'(0)P_i' + P_i'Q_i'(0))$$
$$+ \frac{t^2}{m^2} \operatorname{disp}(P_i).$$

It is easy to compute the time t_m for which $\operatorname{disp}(Q_i(t))$ is a minimum. For t_m we obtain

$$M(Q_i'(t_m)P_i' + P_i'Q_i'(t_m)) = 0$$

from which it follows that

$$\operatorname{disp}(Q_i(t)) = \operatorname{disp}(Q_i(t_m)) + \frac{(t - t_m)^2}{m^2} \operatorname{disp}(P_i). \tag{1.32}$$

Equation (1.32) shows that the dispersion of Q_i is quadratic in its time dependence relative to its minimum value. In addition to (1.32) the Heisenberg uncertainty relation IV (8.3.18) must be satisfied

$$[\operatorname{disp}(P_i)]^{1/2}[\operatorname{disp}(Q_i(t)]^{1/2} \geq \tfrac{1}{2}. \tag{1.33}$$

If (1.33) is an equality for $t = t_m$ then at other times the Heisenberg uncertainty relation can only be satisfied as an inequality.

In addition to the expectation values, it is also possible to obtain answers to questions concerning the probabilities for different position and momentum values. In order to make computations, we shall consider the special ensemble $W = P_\varphi$.

The probability for a momentum value in a region π of momentum space is constant with time and is given by

$$\int_\pi |\langle p_1, p_2, p_3 | \varphi\rangle|^2 \, dp_1 \, dp_2 \, dp_3, \tag{1.34}$$

where $\langle p_1, p_2, p_3 | \varphi\rangle$ is the representation of φ in the momentum representation. We will now compute the probability for a measurement of the position $Q_i(t)$. For this purpose we shall transform from the momentum representation to the $(Q_1(t), Q_2(t), Q_3(t))$ representation. In order to obtain this representation, we seek the solution of the equations

$$Q_j(t) q_{x_1, x_2, x_3} = x_j g_{x_1, x_2, x_3} \tag{1.35}$$

in the momentum representation, where $Q_j(0)$ is the operator $-(1/i)(\partial/\partial p_j)$. Here (1.35) becomes

$$\left(-\frac{1}{i}\frac{\partial}{\partial p_j} + \frac{t}{m} p_j\right)\langle p_1, p_2, p_3 | x_1, x_2, x_3\rangle = x_j \langle p_1, p_2, p_3 | x_1, x_2, x_3\rangle$$

from which it follows that

$$\langle p_1, p_2, p_3 | x_1, x_2, x_3\rangle = \frac{1}{(2\pi)^{3/2}} \exp\left[-i \sum_{j=1}^{3} \left(x_j p_j - \frac{t^2}{2m} p_j^2\right)\right].$$

From

$$\langle x_1, x_2, x_3 | \varphi\rangle = \int \langle x_1, x_2, x_3 | p_1, p_2, p_3\rangle\langle p_1, p_2, p_3 | \varphi\rangle \, dp_1 \, dp_2 \, dp_3$$

we find that the probability that a measurement of $Q_j(t)$ will result in a measurement value in the domain $\mathscr{V}$ is given by:

$$\int_{\mathscr{V}} |\langle x_1, x_2, x_3 | \varphi\rangle|^2 \, dx_1 \, dx_2 \, dx_3.$$

If φ is so chosen that, at time $t = 0$, it satisfies the Heisenberg uncertainty relations with an equal sign (where we may obtain an evaluation of IX (5.37) which is analogous to IX (5.38) in the momentum representation) we then obtain

$$\langle p_1, p_2, p_3 | \varphi\rangle = \left(\frac{2}{\pi a^2}\right)^{3/4} \exp\left[-i \sum_{i=1}^{3} (p_i - \pi_i) y_i\right] \exp\left[-\frac{1}{a^2} \sum_{i=1}^{3} (p_i - \pi_i)^2\right]$$

and it follows that

$$\begin{aligned}
&\langle x_1, x_2, x_3 | \varphi\rangle \\
&= \frac{1}{(2\pi)^{3/2}} \left(\frac{2}{\pi a^2}\right)^{3/4} \int \exp\left[i \sum_{j=1}^{3} \left(x_j p_j - \frac{t}{2m} p_j^2\right) - i \sum_{j=1}^{3} (p_j - \pi_j) y_j\right] \\
&\quad \cdot \exp\left[-\left(\frac{1}{a^2}\right) \sum_{j=1}^{3} (p_j - \pi_j)^2\right] dp_1\, dp_2\, dp_3 \\
&= \frac{1}{(2\pi a^2)^{3/4}} b^{-3/2} \exp\left[-\frac{1}{4a^2 |b|^2} \sum_{j=1}^{3} \left[x_j - \left(y_j + \frac{\pi_j t}{m}\right)\right]^2\right] \\
&\quad \cdot \exp\left[-\frac{i}{|b|^2} \sum_{j=1}^{3} \left[\frac{1}{a^4} (x_j - y_j)\pi_j - \frac{t}{8m} (x_j - y_j)^2 - \frac{t^2}{2ma^4} \pi_j^2\right]\right] \\
&\quad \cdot \exp\left[i \sum_{j=1}^{3} \pi_j y_j\right],
\end{aligned}$$

where $b = 1/a^2 + it/m$.

Here it again follows that the probability distribution $|\langle x_1, \ldots, |\varphi\rangle|^2$ propagates with velocity $(1/m)\pi_j$ where $M(p_j) = \pi_j$. In this way we obtain a more precise description of the behavior of this probability distribution (not only the dispersion given by (1.32)).

2 The Schrödinger Picture

The expression $\operatorname{tr}(WB_t)$ is used to compute all probabilities and expectation values. If B_t does not explicitly depend upon the time t, then from (1.2) we obtain the expression $\operatorname{tr}(WU_t B_0 U_t^+)$. From the property of the trace functional we may exchange U_t^+ with $WU_t B_0$; we obtain

$$\operatorname{tr}(W_0 U_t B_0 U_t^+) = \operatorname{tr}(U_t^+ W_0 U_t B_0), \tag{2.1}$$

where we have written $W = W_0$ in order to emphasize the fact that $W = W_0$ is constant in time. We now define a time dependent W_t as follows:

$$W_t = U_t^+ W_0 U_t \tag{2.2}$$

we then obtain

$$\operatorname{tr}(W_0 B_t) = \operatorname{tr}(W_t B_0). \tag{2.3}$$

In §1 we have called the expression on the left side of the equation (2.3) the Heisenberg picture. We call the expression on the right-hand side the Schrödinger picture. For many purposes it is often mathematically more convenient to work with the right-hand side of (2.3), the Schrödinger picture. In particular, the product representation of the Hilbert space introduced in VIII (2.1) for $t = 0$ does not change with time (see VIII, §2). The position representation of the Hilbert space (in which the momentum operators have the form $(1/i)(\partial/\partial x)$) can be taken to be fixed with time.

If B_t depends explicitly upon the time, that is, if $\partial B_t/\partial t \neq 0$ then, in the Schrödinger picture we may define an operator

$$\tilde{B}_t = U_t^+ B_t U_t \tag{2.4}$$

which, in the case that $\partial B_t/\partial t = 0$ is (according to (1.2)) equal to B_0. Instead of (2.3) we find that, in general

$$\operatorname{tr}(W_0 B_t) = \operatorname{tr}(W_t \tilde{B}_t). \tag{2.5}$$

For the special case in which $W_0 = P_{\varphi_0}$ we obtain

$$\begin{aligned} U_t^+ P_{\varphi_0} U_t f &= U_t^+ \varphi_0 \langle \varphi_0, U_t f\rangle = U_t^+ \varphi_0 \langle U_t^+ \varphi_0, f\rangle \\ &= \varphi_t \langle \varphi_t, f\rangle = P_{\varphi_t} f, \end{aligned}$$

where

$$\varphi_t = U_t^+ \varphi_0. \tag{2.6}$$

Taking the adjoint of equation (1.3), we obtain the following differential equation for φ_t:

$$\dot{\varphi}_t = \dot{U}_t^+ U_t \varphi_0 = \dot{U}_t^+ U_t \varphi_t = -iU_t^+ H_t U_t \varphi_t.$$

Using definition (2.4) it follows that

$$-\frac{1}{i}\dot{\varphi}_t = \tilde{H}_t \varphi_t. \tag{2.7a}$$

In the same way we obtain the following differential equation for W_t

$$\frac{dW_t}{dt} = i(W_t \tilde{H}_t - \tilde{H}_t W_t). \tag{2.7b}$$

(2.7a) and (2.7b) are called the Schrödinger equations.

What is the form of the operator $\tilde{H}_t$?

As we already mentioned in VIII, §6, in the expression VIII (5.8) for H_t we are to use the operators $\vec{P}_i(t)$, $\vec{Q}_i(t)$, $\vec{S}_i(t)$. Since $\vec{P}_i(t) = U_t \vec{P}_i(0) U_t^+$, etc., we find that we obtain the operator $\tilde{H}_t$ if we replace all $\vec{P}_i$, $\vec{Q}_i$, $\vec{S}_i$ in VIII (5.8) by $\vec{P}_i(0)$, $\vec{Q}_i(0)$, $\vec{S}_i(0)$, that is, if all $\vec{P}_i$, $\vec{Q}_i$, $\vec{S}_i$ are chosen to be constant with time. Nevertheless there can be a remaining time dependence in $\tilde{H}_t$ which may be due to the time dependence of the external field $\vec{A}$, φ, that is, through an "explicit" time dependence of H_t.

In order to clarify this situation, let us consider an elementary system of charge e in an external time dependent field. Then we would have

$$\begin{aligned} H_t = \frac{1}{2m_i} \sum_{i=1}^{3} [P_i(t) - eA_i(Q_1(t), Q_2(t), Q_3(t), t)]^2 \\ + e\varphi(Q_1(t), Q_2(t), Q_3(t), t). \end{aligned} \tag{2.8}$$

From (2.8) it follows that

$$\tilde{H}_t = \frac{1}{2m} \sum_{i=1}^{3} [P_i(0) - eA_i(Q_1(0), Q_2(0), Q_3(0), t)]^2 + e\varphi(Q_1(0), Q_2(0), Q_3(0), t). \tag{2.9}$$

If we know that we are in the Schrödinger picture, we may omit the symbol (0) in (2.9) and write

$$\tilde{H}_t = \frac{1}{2m} \sum_{i=1}^{3} [P_i - eA_i(Q_1, Q_2, Q_3, t)]^2 + e\varphi(Q_1, Q_2, Q_3, t), \tag{2.10}$$

where we know that the P_i, Q_i are operators which are constant with time.

With (2.10) the Schrödinger equation (2.7a), as expressed in the position representation corresponding to Q_1, Q_2, Q_3 is given by

$$-\frac{1}{i}\frac{\partial}{\partial t}\langle x_1, x_2, x_3 | \varphi_t\rangle = \frac{1}{2m} \sum_{k=1}^{3} \left[\frac{1}{i}\frac{\partial}{\partial x_k} - eA_k(x_1, x_2, x_3, t)\right]^2 \langle x_1, x_2, x_3 | \varphi_t\rangle + e\varphi(x_1, x_2, x_3, t)\langle x_1, x_2, x_3 | \varphi_t\rangle. \tag{2.11}$$

In Schrödinger's heuristic development of quantum mechanics equation (2.11) is "guessed" with remarkable intuition. From (2.11) it follows that a wave packet in a "weakly" varying electromagnetic field behave much like mass points of classical mechanics with the Hamiltonian

$$H = \frac{1}{2m} \sum_{k=1}^{3} (P_k - eA_k(q_1, q_2, q_3, t))^2 + e\varphi(q_1, q_2, q_3, t)$$

(see, for example, [6], XI, §1.2).

If H_t does not explicitly depend upon time, then in both the Heisenberg and Schrödinger pictures H_t is a time-constant operator: $H_t = \tilde{H}_t = H_0$. The Schrödinger equation (2.7a) is then given by:

$$-\frac{1}{i}\dot{\varphi}_t = H\varphi_t, \quad \text{where } H = H_0. \tag{2.12}$$

Let ε denote the values of the spectrum of H and let α denote the degeneracy index for H. In the H-representation (2.12) becomes

$$-\frac{1}{i}\frac{\partial}{\partial t}\langle \varepsilon, \alpha | \varphi_t\rangle = \varepsilon\langle \varepsilon, \alpha | \varphi_t\rangle \tag{2.13}$$

the solution of which is given by

$$\langle \varepsilon, \alpha | \varphi_t\rangle = e^{-i\varepsilon t}\langle \varepsilon, \alpha | \varphi_0\rangle. \tag{2.14}$$

For examples we shall again consider the case of a harmonic oscillator and a free elementary system.

Let U_t denote the unitary operator for the harmonic oscillator which was computed in §1. The solution of the Schrödinger equation (2.7a) is given by

$$\varphi_t = U_t^+ \varphi_0$$

and, in the $N = A^+(0)A(0)$-representation we obtain

$$\begin{aligned}\langle n|\varphi_t\rangle &= \sum_m \langle n|U_t^+|m\rangle\langle m|\varphi_0\rangle \\ &= \sum_m \overline{\langle m|U_t|n\rangle}\langle m|\varphi_0\rangle.\end{aligned}$$

According to (1.18)

$$\langle \phi_m, \Psi_n\rangle = \alpha(t)\langle \phi_m, U_t\phi_n\rangle = \alpha(t)\langle m|U_t|n\rangle$$

and we find that

$$\langle m|U_t|n\rangle = \overline{\alpha(t)}\langle \phi_m, \Psi_n\rangle.$$

Therefore we obtain

$$\begin{aligned}\langle n|\varphi_t\rangle &= \alpha(t)\sum_m \langle \Psi_n, \phi_m\rangle\langle m|\varphi_0\rangle \\ &= \alpha(t)\sum_m \overline{\langle \phi_m \Psi_n\rangle}\langle m|\varphi_0\rangle, \end{aligned} \tag{2.15}$$

where $\langle \phi_m, \Psi_n\rangle$ is given by (1.25a), (1.25b) and $\alpha(t)$ is given by (1.27).

Instead of taking a somewhat circuitous route using the Heisenberg picture, we shall instead directly solve the Schrödinger equation (2.7a) in the position representation where the latter is given by

$$\begin{aligned}-\frac{1}{i}\frac{\partial}{\partial t}\langle x|\varphi_t\rangle &= \frac{1}{2m}\left(\frac{1}{i}\frac{\partial}{\partial x}\right)^2\langle x|\varphi_t\rangle \\ &\quad + \frac{m\omega^2}{2}x^2\langle x|\varphi_t\rangle - k(t)x\langle x|\varphi_t\rangle. \end{aligned} \tag{2.16}$$

For $f(t)$ as defined in (1.14) we shall set

$$\begin{aligned} y(t) &= \frac{1}{\sqrt{2m\omega}}(e^{-i\omega t}f(t) + e^{i\omega t}\overline{f(t)}), \\ \pi(t) &= \frac{1}{i}\sqrt{\frac{m\omega}{2}}(e^{-i\omega t}f(t) - e^{i\omega t}\overline{f(t)}). \end{aligned} \tag{2.17}$$

$y(t)$ and $\pi(t)$ are solutions for the motion $x(t)$, $p(t)$ for a mass point according to classical mechanics! Thus we find that solutions of (2.16) are given by

$$\begin{aligned} g_n(x, t) &= \frac{1}{\sqrt{2^n n!\sqrt{\pi}}}(m\omega)^{1/2}H_n((m\omega)^{1/2}(x - y(t))e^{-(m\omega/2)(x-y(t))^2} \\ &\quad \cdot e^{i\pi(t)x}e^{-i\int_0^t (\pi(\tau)/2m)\,d\tau}e^{-i\omega(n+1/2)t}, \end{aligned} \tag{2.18}$$

where

$$\frac{\sqrt{m\omega}}{\sqrt{2^2 n!\sqrt{\pi}}} H_n((m\omega)^{1/2}x)e^{-(m\omega/2)x^2}$$

are the normalized eigenfunctions of

$$\frac{1}{2m} P^2 + \frac{m\omega^2}{2} Q^2$$

which correspond to the eigenvalues $(n + \frac{1}{2})\omega$ (see IX (5.30)), that is,

$$\sqrt{\frac{m\omega}{2^2 n!\sqrt{\pi}}} H_n((m\omega)^{1/2}x)e^{-(m\omega^2/2)x^2} = \langle x|n\rangle. \tag{2.19}$$

From (2.18) it therefore follows that $g_n(x, 0) = \langle x|n\rangle$. The most general solution of the Schrodinger equation (2.16) is therefore given by

$$\langle x|\varphi_t\rangle = \sum_{n=0}^{\infty} g_n(x, t)\langle n|\varphi_0\rangle. \tag{2.20}$$

Clearly (2.15) and (2.20) must be the same solution. Since

$$\langle x|\varphi_t\rangle = \sum_n \langle x|n\rangle\langle n|\varphi_t\rangle$$

and

$$\langle n|\varphi_t\rangle = \int_{-\infty}^{\infty} \langle n|x\rangle\langle x|\varphi_t\rangle\, dx$$

from (2.20) it follows that

$$\langle n|\varphi_t\rangle = \sum_{m=0}^{\infty} \left[\int_{-\infty}^{\infty} \langle n|x\rangle g_m(x, t)\, dx\right]\langle m|\varphi_0\rangle. \tag{2.21}$$

Since the $\langle m|\varphi_0\rangle$ were arbitrary, by comparison of (2.21) with (2.15) we obtain

$$\alpha(t)\langle\Psi_m(t), \phi_m\rangle = \int_{-\infty}^{\infty} \langle n|x\rangle g_m(x, t)\, dx. \tag{2.22}$$

We now seek to answer the same physical question as we have in §1. Suppose that we prepare an ensemble for which the measurement of the observable H_0 (which is constant in time in the Schrödinger picture!) has, at time 0, the eigenvalue $(m + \frac{1}{2})\omega$ with certainty. Therefore we must have $W_0 = P_{\varphi_0}$ and $\langle n|\varphi_0\rangle = \delta_{nm}$. If, for the same ensemble we measure the observable H_0 "at time t" we must use $W_t = P_{\varphi_t}$. For the initial value $\langle n|\varphi_0\rangle = \delta_{nm}$ from (2.21) it follows that

$$\langle n|\varphi_t\rangle = \int_{-\infty}^{\infty} \langle n|x\rangle g_m(x, t)\, dx. \tag{2.23}$$

The probability that the measurement value will be $\omega\,(n + \frac{1}{2})$ is given by $|\langle n|\varphi_t\rangle|^2$ in complete agreement with (1.26).

For a free elementary system the Schrödinger equation, as expressed in the momentum representation, is given by

$$-\frac{1}{i}\frac{\partial}{\partial t}\langle p_1, p_2, p_3|\varphi_t\rangle = \frac{1}{2m}\sum_{i=1}^{3} p_i^2\langle p_1, p_2, p_3|\varphi_t\rangle.$$

The general solution is given by

$$\langle p_1, p_2, p_3|\varphi_t\rangle = \exp\left[-\frac{i}{2m}\left(\sum_{i=1}^{3} p_i^2\right)t\right]\langle p_1, p_2, p_3|\varphi_0\rangle.$$

The solution of the Schrödinger equation (2.12) in the position representation

$$-\frac{1}{i}\frac{\partial}{\partial t}\langle x_1, x_2, x_3|\varphi_t\rangle = \frac{1}{2m}\sum_{j=1}^{3}\left(\frac{1}{i}\frac{\partial}{\partial x_j}\right)^2\langle x_1, x_2, x_3|\varphi_t\rangle$$

is therefore given by

$$\begin{aligned}\langle x_1, x_2, x_3|\varphi_t\rangle &= \int\langle x_1, x_2, x_3|p_1, p_2, p_3\rangle\langle p_1, p_2, p_3|\varphi_\tau\rangle\, dp_1, dp_2, dp_3\\ &= \frac{1}{(2\pi)^{3/2}}\int\exp\left[i\left[\sum_{j=1}^{3}\left(x_j p_j - \frac{t}{2m}p_j^2\right)\right]\right]\\ &\quad\cdot\langle p_1, p_2, p_3|\varphi_0\rangle\, dp_1, dp_2, dp_3,\end{aligned}$$

where

$$\langle p_1, p_2, p_3|\varphi_0\rangle = \int\langle p_1, p_2, p_3|x_1, x_2, x_3\rangle\langle x_1, x_2, x_3|\varphi_0\rangle\, dx_1\, dx_2\, dx_3.$$

If we let

$$\langle x_1, x_2, x_3|\varphi_0\rangle = \left(\frac{a^2}{2\pi}\right)^{3/4}\exp\left[-a^2\sum_{j=1}^{3}\left[\frac{(x_j - y_j)^2}{4} - \frac{i}{a^2}\pi_j(x_j - y_j)\right]\right],$$

then from $\langle x_1, x_2, x_3|\varphi_t\rangle$ we obtain the same physical results for the time behavior for the position probability distribution as we obtained for the Heisenberg picture at the end of §1.

3 The Interaction Picture

Frequently we find that it is possible to separate the Hamiltonian operator into two components $H = H^0 + H^{12}$, where the spectral representation of H^0 is easy to determine. In applications H^0 frequently has the following form: $H^0 = H_1^0 \times \mathbf{1} + \mathbf{1} \times H_2^0$, which refers to the product decomposition $\mathscr{H} = \mathscr{H}_1 \times \mathscr{H}_2$ "at time $t = 0$" for a composite system containing a system of type 1 and of type 2. H_1^0, H_2^0 are the Hamiltonians for systems of type 1 and type 2, respectively; see VIII (2.1). H^{12} is often called the interaction

Hamiltonian, from which we obtain the expression "Interaction Picture." Here we assume that H^0 does not explicitly depend upon the time. Thus, (according to §2) $\tilde{H}^0$ is a time-independent operator. Thus it follows that

$$e^{i\tilde{H}^0 t} = e^{i\tilde{H}_1^0 t} \times e^{i\tilde{H}_2^0 t}. \tag{3.1}$$

Thus it follows that

$$e^{i\tilde{H}^0 t}(B_1 \times B_2)e^{-i\tilde{H}^0 t} = (e^{i\tilde{H}_1^0 t}B_1 e^{-i\tilde{H}_1^0 t}) \times (e^{i\tilde{H}_2^0 t}B_2 e^{-i\tilde{H}_2^0 t}). \tag{3.2}$$

Equation (3.2) means that the product representation $\mathscr{H} = \mathscr{H}_1 \times \mathscr{H}_2$ remains invariant under unitary transformations of the form (3.1), a result which is very practical (see XVI).

With W_t defined by (2.2) it follows that, for the expectation value of an observable in (2.5)

$$\begin{aligned} \operatorname{tr}(W_t \tilde{B}_t) &= \operatorname{tr}(e^{i\tilde{H}^0 t} W_t e^{-i\tilde{H}^0 t} e^{i\tilde{H}^0 t} \tilde{B}_t e^{-i\tilde{H}^0 t}) \\ &= \operatorname{tr}(V_t^+ W_0 V_t e^{i\tilde{H}^0 t} \tilde{B}_t e^{-i\tilde{H}^0 t}), \end{aligned} \tag{3.3}$$

where

$$V_t = U_t e^{-i\tilde{H}^0 t}. \tag{3.4}$$

If we simplify our notation as follows:

$$\mathbf{B}_t = e^{i\tilde{H}_0 t}\tilde{B}_t e^{-i\tilde{H}_0 t} \quad \text{and} \quad \mathbf{W}_t = V_t^+ W_0 V_t \tag{3.5}$$

then the expectation values can be written in the form

$$\operatorname{tr}(\mathbf{W}_t \mathbf{B}_t). \tag{3.6}$$

In (3.6) the operator $\mathbf{W}_t$ which describes the ensemble and the operator $\mathbf{B}_t$ are time dependent. The time dependence in (3.6) is therefore divided among both factors as follows: $\mathbf{B}_t$ depends upon the time t as if H^0 was the Hamiltonian operator, that is, in the case (3.1) as if there was no interaction between the subsystems. $\mathbf{B}_t$ therefore corresponds to the Heisenberg picture without interaction.

The time dependence of $\mathbf{W}_t$ arises therefore from the interaction. Similarly, as in the Schrödinger picture, for the special case $W = P_{\psi_0}$ it follows from (3.5) that:

$$\mathbf{W}_t = P_{\psi_t}, \quad \text{where} \quad \psi_t = V_t^+ \psi_0 = e^{i\tilde{H}^0 t} U_t^+ \psi_0. \tag{3.7}$$

From differentiating with respect to t and using (1.3) it follows that

$$\begin{aligned} -\frac{1}{i}\dot{\psi}_t &= -\tilde{H}^0 \psi_t - e^{i\tilde{H}_0 t}\left(\frac{1}{i}\dot{U}_t^+ \psi_0\right) \\ &= -\tilde{H}^0 \psi_t + e^{i\tilde{H}_0 t} U_t^+ H_t \psi_0 \\ &= -\tilde{H}^0 \psi_t + e^{i\tilde{H}^0 t} U_t^+ H_t U_t e^{-i\tilde{H}^0 t} \psi_t. \end{aligned}$$

We find that

$$U_t^+ H_t U_t = \tilde{H}_t = \tilde{H}^0 + \tilde{H}_t^{12}.$$

Since $\tilde{H}^0$ commutes with $e^{i\tilde{H}_0 t}$, we therefore finally obtain

$$-\frac{1}{i}\dot{\psi}_t = \mathbf{H}_t^{12}\psi_t, \tag{3.8}$$

where

$$\mathbf{H}_t^{12} = e^{i\tilde{H}^0 t}\tilde{H}_t^{12}e^{-i\tilde{H}^0 t} \tag{3.9}$$

represents the operator H^{12} in the time dependence of the Heisenberg picture without interaction (see (3.5)).

As in §§1 and 2 we shall again consider the example of forced oscillation of a harmonic oscillator with

$$H^0 = \frac{1}{2m}P^2 + \frac{m\omega^2}{2}Q^2, \qquad H^{12} = -Qk(t). \tag{3.10}$$

For this case equation (3.8) becomes

$$-\frac{1}{i}\dot{\psi}_t = -k(t)\mathbf{Q}(t)\psi_t, \tag{3.11}$$

where $\mathbf{Q}(t)$ is to be computed according to the equation of motion of a free oscillator

$$\dot{\mathbf{Q}}(t) = \frac{1}{m}\mathbf{P}(t), \qquad \dot{\mathbf{P}}(t) = -m\omega^2\mathbf{Q}(t),$$

that is,

$$\mathbf{Q}(t) = Q(0)\cos\omega t + \frac{1}{m\omega}P(0)\sin\omega t. \tag{3.12}$$

For the $Q(0)$-representation of the Hilbert space, in which $P(0) = (1/i)(d/dx)$, (3.11) is transformed into

$$\frac{1}{i}\frac{\partial}{\partial t}\langle x|\psi_t\rangle = k(t)\left[x\cos\omega t + \frac{1}{im\omega}\sin\omega t\frac{\partial}{\partial x}\right]\langle x|\psi_t\rangle. \tag{3.13}$$

For a special solution for this equation we choose the form

$$g(x, t) = e^{i[x\kappa(t) + u(t)]}. \tag{3.14}$$

For (3.13), with this choice, we obtain

$$(x\dot{\kappa} + \dot{u}) = k(t)x\cos\omega t + \frac{1}{m\omega}k(t)\kappa(t)\sin\omega t,$$

that is,

$$\dot{\kappa} = k(t)\cos\omega t, \qquad \dot{u} = \frac{1}{m\omega}k(t)\kappa(t)\sin\omega t$$

and we therefore obtain

$$\kappa(t) = \frac{\sqrt{2m\omega}}{2i}[f(t) - \overline{f(t)}],$$
$$u(t) = \int_0^t \frac{1}{m\omega} k(\tau)\kappa(\tau) \sin \omega\tau \, d\tau, \tag{3.15}$$

where $f(t)$ is defined according to (1.14). In order to find a general solution of (3.13) we observe that

$$\langle x | \psi_t \rangle = \chi(x - w(t))g(x, t),$$

where

$$w(t) = \sqrt{\frac{1}{2m\omega}}(f(t) + \overline{f(t)})$$

is, for arbitrary $\chi(z)$, a solution of (3.13) providing that $g(x, t)$ is a solution. Thus, for the most general solution we obtain:

$$\langle x | \psi_t \rangle = \chi(x - w(t))e^{i[x\kappa(t) + u(t)]}. \tag{3.16}$$

For $t = 0$ we obtain $w(0) = 0$, $\kappa(0) = 0$, $u(0) = 0$. $\chi(x)$ is therefore equal to $\langle x | \psi_0 \rangle$ so that we obtain:

$$\langle x | \psi_t \rangle = \langle x - w(t) | \psi_0 \rangle e^{i[x\kappa(t) + u(t)]}. \tag{3.17}$$

Since the energy observable $\tilde{H}^0 = (1/2m)P^2 + (m\omega^2/2)Q^2$ is constant with time in the interaction picture, then so are the eigenvectors $\langle x | n \rangle$ (in the $Q(0)$-representation). The probability that the energy value $\omega(n + \frac{1}{2})$ is measured at time t, assuming that the ensemble is, at time $t = 0$, dispersion free with respect to the measurement value $\omega(m + \frac{1}{2})$ is given by

$$\left| \int_{-\infty}^{\infty} \langle n | x \rangle \langle x - w(t) | m \rangle e^{i[x\kappa(t) + u(t)]} \, dx \right|^2, \tag{3.18}$$

which, according to (1.25a, b), (1.26) agrees with $|\langle \phi_m, \Psi_n \rangle|^2$ and, according to (2.23) agrees with $|\langle n | \varphi_t \rangle|^2$.

4 Time Reversal Transformations

In VII, §7 we have considered parity transformations and have discussed the problem that it is nontrivial to realize such a transformation by means of r-automorphisms. The problem of the realization of time reversal transformations is even more difficult to comprehend.

The time reversal transformations correspond to the time "reflection" $x'_\nu = x_\nu$, $t' = -t$. What is the physical meaning of such a transformation? The transformation r_1 in VII (7.1) can be intuitively visualized with the aid of a mirror, which gives us some idea on how we may build a "mirror image" of a registration apparatus. What, however, do we mean by a registration apparatus for which the time is reversed? Here we may consider the idea of making a motion picture of experiments with microsystems, and running the film backwards. By analogy with a mirror image, we may seek to realize such an experiment by running a motion picture backwards, but it is clear that this is impossible because the experiments are "irreversible." We cannot realize the following process in reverse: microsystems are emitted by a preparation apparatus, and are detected by a counter (as a registration apparatus) which, in turn, responds. The response of the counter is not reversible, and cannot be realized in reverse. In general, the time direction of the action of the preparation apparatus on the registration apparatus by means of microsystems is that direction in which entropy increases in the macroprocesses connected with the preparation and registration. This directedness is an important fundamental structure underlying quantum mechanics which cannot be avoided—see, for example, [6], XVI and [7].

Therefore we shall not, at first, discuss the question whether there exists a realization of time reversal by means of registration procedures. Instead, we shall define a time reversal transformation T as a B-continuous effect automorphism for which we obtain a representation of the "extended Galileo group" by means of effect automorphisms, where the Galileo group is extended by the addition of the time reversal transformation

$$s\colon x'_\nu = x_\nu,\ t' = -t.$$

With the help of VII (1.1) it follows that

$$(A, \vec{\delta}. \vec{\eta}, \gamma)s = s(A, -\vec{\delta}, \vec{\eta}, -\gamma). \tag{4.1}$$

With the reflection operator r in VII, §7 it follows that

$$rs = sr. \tag{4.2}$$

The operator T can, in principle, transform one system type into another, a fact of great importance in elementary particle physics. Here we shall assume (in analogy to the case of the spatial reflection (parity) operator) that T leaves system types invariant. According to V, §5, in the Hilbert space $\mathscr{H}$ of every system type there exists a unitary or anti-unitary operator V which satisfies

$$Ty = VyV^{-1} \tag{4.3}$$

for $y \in \mathscr{B}'(\mathscr{H})$. From $s^2 = e$ it follows that $T^2 y = y$ and that $V^2 = e^{i\alpha}\mathbf{1}$.

From (4.1) and (4.2) it follows that

$$U(A, \vec{\delta}, \vec{\eta}, \gamma)VU(A, -\vec{\delta}, \vec{\eta}, -\gamma) = \lambda(A, \vec{\delta}, \vec{\eta}, \gamma)V, \tag{4.4}$$

$$U(r)VU(r)^{-1} = e^{i\beta}V. \tag{4.5}$$

In Hilbert space $\mathscr{H}^n$ where $\mathscr{H} = \mathscr{H}_b \times \imath_s$ we define the following anti-unitary operator C as follows: Let u_+, u_- denote the basis vectors of $\imath_s$

which were defined in VII, §3. The vectors of $\mathscr{H}^n$ may be written in the following form (where $\vec{Q}_i$ is defined as multiplication by $\vec{r}_i$, P_i is the operator $1/i \operatorname{grad}_{\vec{r}_i}$):

$$\sum_{\alpha_1, \ldots, \alpha_n} \langle \vec{r}_1, \vec{r}_2, \ldots, \vec{r}_n | \varphi; \alpha_1, \ldots, \alpha_n \rangle u_{\alpha_1} u_{\alpha_2} \cdots u_{\alpha_n},$$

where each α_i runs through the indices $+$ and $-$. We set

$$\begin{aligned} C &\sum_{\alpha_1, \ldots, \alpha_n} \langle \vec{r}_1, \vec{r}_2, \ldots, \vec{r}_n | \varphi; \alpha_1, \ldots, \alpha_n \rangle\, u_{\alpha_1} \cdots u_{\alpha_n} \\ &= \sum_{\alpha_1, \ldots, \alpha_n} \overline{\langle \vec{r}_1, \vec{r}_2, \ldots, \vec{r}_n | \varphi; \alpha_1, \ldots, \alpha_n \rangle} (\alpha_1, \alpha_2, \ldots, \alpha_n) u_{-\alpha_1} u_{-\alpha_2} \cdots u_{-\alpha_n}, \end{aligned} \tag{4.6}$$

where $(\alpha_1, \alpha_2, \ldots, \alpha_n) = 1$ for an even number of $\alpha_i = -1$ and $= -1$ for an odd number. C then transforms $\{\mathscr{H}^n\}_+$ and $\{\mathscr{H}^n\}_-$ into themselves. It is easy to extend the definition of C to the more general case of a composite system consisting of n electrons and several atomic nuclei; here it would be necessary to introduce additional position coordinates and additional spin vectors.

It follows that $C^2 = (-1)^n \mathbf{1}$ and $C^{-1} = (-1)^n C$. Here it is easy to prove that

$$\begin{aligned} C\vec{Q}_i &= \vec{Q}_i C, \\ C\vec{P}_i &= -\vec{P}_i C, \\ C\vec{S}_i &= -\vec{S}_i C, \\ CH &= HC, \end{aligned} \tag{4.7}$$

where H is the Hamiltonian operator (without external fields) given by VIII (5.8).

For the reflection operator defined in VII (7.4) for $\mathscr{H} = \mathscr{H}_b \times \imath_s$, $U(r)$, defined in (4.5), as an operator in $\mathscr{H}$ is given by $U(r) = (R \times \mathbf{1}) \times (R \times \mathbf{1}) \times \cdots \times (R \times \mathbf{1})$. From (4.6) it follows that

$$CU(r) = U(r)C. \tag{4.8}$$

From (4.7) it follows that

$$CU(A, \vec{\delta}, \vec{\eta}, \gamma) = U(A, -\vec{\delta}, \vec{\eta}, -\gamma)C. \tag{4.9}$$

If we multiply (4.4) and (4.5) on the right by C^{-1}, we obtain

$$U(A, \vec{\delta}, \vec{\eta}, \gamma)(VC)^{-1})U(A, \vec{\delta}, \vec{\eta}, \gamma)^{-1} = \lambda(A, \delta, \vec{\eta}, \gamma)VC^{-1}, \tag{4.10}$$

$$U(r)(VC^{-1})U(r)^{-1} = e^{i\beta}VC^{-1}. \tag{4.11}$$

From (4.10) it follows that the $\lambda(A, \delta, \vec{\eta}, \gamma)$ form a one-dimensional representation of the Galileo group; therefore $\lambda(\cdots) = 1$. For an elementary system it therefore follows that VC^{-1} must be a multiple of the unit operator, and since a number in V is arbitrary, we may set

$$V = C. \tag{4.12}$$

We shall require that (4.12) also holds for the case of a composite system. We therefore find that, from (4.3)

$$Ty = CyC^{-1}. \tag{4.13}$$

For the special case in which y is an element of L, that is, y is an effect, then there exists a B-continuous effect automorphism defined by (4.13). Note that this result does not, in any way, explain how the transformation (4.13) is to be realized in terms of registration procedures. The following question we cannot answer: Let (b_0, b) be an effect procedure; for which other effect procedures (b'_0, b') is the equation $C\psi(b_0, b)C^{-1} = \psi(b'_0, b')$ satisfied?

By analogy with the transformation from the Heisenberg to the Schrödinger picture, we may write

$$\mathrm{tr}(WCBC^{-1}) = \mathrm{tr}(C^{-1}WCB). \tag{4.14}$$

We may therefore apply the time reversal transformation to the ensemble W; we obtain

$$W \to C^{-1}WC = CWC^{-1}, \tag{4.15}$$

where we note that $C^{-1} = (-1)^n C$. Again, we have not explained how, given a preparation procedure a, we may obtain another a' such that $\varphi(a') = C\varphi(a)C^{-1}$.

According to (4.9) we obtain the special case

$$Ce^{iH\gamma} = e^{-iH\gamma}C. \tag{4.16}$$

In the Schrödinger picture we find that

$$W_{t'} = e^{-iH(t-\tau)}W_\tau e^{iH(t-\tau)}. \tag{4.17}$$

Let us define

$$W'_{t'} = CW_{2\tau - t'}C^{-1} \tag{4.18}$$

then, from (4.17) it follows that

$$\begin{aligned} W'_{t'} &= Ce^{iH(t'-\tau)}W_\tau e^{-iH(t-\tau)}C^{-1} \\ &= e^{-iH(t'-\tau)}CW_\tau C^{-1}e^{iH(t'-\tau)} \\ &= e^{-iH(t'-\tau)}W'_\tau e^{iH(t'-\tau)}. \end{aligned} \tag{4.19}$$

Therefore $W'_{t'}$ is also an ensemble which satisfies the Schrödinger picture.

If we subject W_t to normal time development from time t_1 to $t_2 = \tau > t_1$ then at the time τ, $W_{\tau'}$ is transformed into $CW_\tau C^{-1}$; if we then subject $CW_\tau C^{-1}$ to normal time development of τ to the same interval $t_2 - t_1 = \tau - t_1$ until $\tau + (\tau - t_1) = 2\tau - t_1$, then, according to (4.19) we obtain at time $2\tau - t_1$ the ensemble $CW_{t_1}C^{-1}$, that is, the ensemble which is time reversed with respect to the "initial ensemble" W_{t_1}. This means that we can run the "quantum mechanical Schrödinger motion picture" W_t backwards in the form $CW_{2\tau - t}C^{-1}$, that is, this "backwards" version is a physical possibility if the formulation of quantum mechanics described here is a g–G-closed theory in the sense of [8], §§8 and 10.

It is therefore an important question whether for a given preparation procedure a, it is possible to produce an a' (at least, in principle) such that,

$$\varphi(a') \approx C\varphi(a)C^{-1}$$

is satisfied (at least approximately). Clearly, for large systems, in particular, macrosystems, this is not the case because of their irreversible nature; see, for example, [6], XV, [7], X and the remarks in Chapter XVIII. A macroscopic extrapolation of quantum mechanics cannot therefore be a g–G-closed theory for macrosystems than such an extrapolation of quantum mechanics (see [7], X).

In XVI, §7 we shall return to a discussion of the question whether the "Quantum Mechanics of Microsystems" can be a g–G-closed theory.

CHAPTER XI

The Spectrum of One-Electron Systems

In IV, §4 we suggested that, in order to obtain a more concrete and encompassing development of the theory, we need to establish the connection between well-defined (with respect to their technical structure) registration methods b_0, the corresponding Boolean ring $\mathscr{R}(b_0)$ and the operators $g = \psi(b_0, b)$ by introducing axioms which specify the mathematical form for the operators $g = \psi(b_0, b)$.

We shall not number these axioms; instead we only note that they are additional axioms in the sense of the mathematical picture. We also cannot give a comprehensive description of the technical details of construction in order to formulate these new mapping principles. We are only able to obtain an overview of the preparation and registration procedures by using the concise language of the physicist in which previous knowledge of the physical phenomena associated with the structure of the apparatus is already assumed. In accord with this mode of expression we shall, for example, speak of a "spectral apparatus" without making reference to textbooks of experimental physics concerning the structure of such an apparatus, or its resolution capability. In XVII we shall present the first step in the development of an extension of the theory which will provide a more precise description of the structure of the preparation and registration apparatus and we shall ascertain the form of the maps $\mathscr{Q}' \xrightarrow{\varphi} K$, $\mathscr{F} \xrightarrow{\psi} L$ for some examples. In XVIII we shall examine the extension problem in more detail without attempting to obtain a solution to the problem. A solution of this problem can be found in [7], XI. For that reason we must, at least for the present, be satisfied with the prospect of having to guess or discover individual axioms for particular preparation and registration methods.

It is advisable to pursue this route by beginning with special cases (one-electron spectra) rather than the general case (for example, the problem of atomic and molecular spectra). It is easier to understand how the special case leads to the general case rather than to understand an overview of the general case. For this reason we urge readers to be patient in their consideration of the systematics when they encounter a previously introduced axiom at a later point in the development as a special case of a later axiom. In this sense the previous comments may serve as a justification for the following discussion of the light emission for the case of a single electron. This type of presentation permits us to show the essentials more concisely and the reader may eventually be able to carry out the necessary generalizations.

1 The Effect of the Emission of a Photon

For the present we shall not attempt to develop a systematic theory in order to establish the correspondence described earlier between an operator g and the construction rules for the registration apparatus. Here we shall often use partially known theories, and other theories (obtained by guesswork) in order to obtain an expression for the desired operator. We will then introduce the latter as an "axiom" in our theory.

As is well known (see, for example, [6], XI, §1), the description of the emission of photons has been very successful. The so-called "spectral apparatus" is well suited to determine the wavelength of a photon emitted by an atom, and to register the photon on a "detector" (for example, a photographic plate, photomultiplier, etc.). In terms of the discussions presented in II an atom produced by a preparation apparatus a produces a detection response b (with the aid of a photon) in the registration apparatus b_0, where b_0 is a special spectral apparatus. Normally, a spectral apparatus is an integrating apparatus. It does not register each individual photon, but instead records the intensity of the spectral lines, that is, it approximately measures the number of photons which are confined to the region of a spectral line.

Which operator g should correspond to the following effect: Within a time interval Δt a light quantum (photon) of frequency ω is registered by the spectral apparatus? Here we shall not inquire about the details of the spectral apparatus—such as the width of the slit, the spatial orientation of the slit, etc.; instead, we will only inquire about the effect, that "an atom emits a photon of frequency ω in the direction of solid angle $d\Omega$ during the time interval Δt." In this way we may obtain, with the help of numerical factors, the "intensity" of a spectral line measured by a concrete apparatus.

This example demonstrates typical aspects of the theory of many registration apparatuses: First a process is treated in a purely quantum mechanical way (for example, the emission of a photon by an atom). The next step in the process is then treated classically (for example, the propagation of the photon in the spectral apparatus is treated in terms of wave optics). In the third step we encounter the problem of registration (here, let us consider a photographic plate) where we need to know either from the theory or from experiment whether all or only a known fraction of systems subjected to the preceding processes are actually registered. Here we shall consider only the first part—the emission of a photon by an atom; we shall assume that the remaining theory of the spectral apparatus is already known.

At present we are unable to describe the emission process for a photon purely from quantum mechanics, because it would first be necessary to develop a quantum mechanical theory for the electromagnetic field. Because there is no mathematically correct formulation of such a quantum field theory at present, we have to rely on a very extensive workable and practical system of computation schemes (the so-called quantum electrodynamics) which functions well for most problems. We will not use these methods to compute the desired effect of the emission of a photon; instead, we shall give a somewhat simple "rule of computation" for this problem, from which we may obtain the desired operator.

Our starting point will be the classical description of the emission of electromagnetic radiation by means of a charge-current distribution, as computed using Maxwell's equations. We will first describe the charge-current distributions classically by means of two functions $\rho(x, t)$ and $\vec{j}(x, t)$. We shall then introduce operators for the classical expressions for the charge-current distribution for a single electron around an atomic nucleus. These operators will be inferred from the use of the wave-picture for the electron. Since, in the following computations ρ and $\vec{j}$ will occur linearly, the computations for the corresponding operators ρ, $\vec{j}$ will be completely analogous to those obtained for the classical functions ρ and $\vec{j}$.

The fields generated by ρ and $\vec{j}$ are computed from the scalar potential φ and vector potential $\vec{A}$ by means of the equations

$$\vec{E} = -\dot{\vec{A}} - \operatorname{grad} \varphi, \qquad \vec{B} = \operatorname{curl} \vec{A}.$$

For φ and $\vec{A}$ we use the formulae for retarded potentials:

$$\varphi(\vec{x}, t) = \int \frac{\rho(\vec{x}', t - |\vec{x} - \vec{x}'|)}{|\vec{x} - \vec{x}'|} d^3\vec{x}', \tag{1.1}$$

$$\vec{A}(\vec{x}, t) = \int \frac{\vec{j}(\vec{x}', t - |\vec{x} - \vec{x}'|}{|\vec{x} - \vec{x}'|} d^3\vec{x}'. \tag{1.2}$$

We shall consider these retarded potentials for distances which are much greater than the region of the charge and current distribution, where the latter is characterized by a distance d which roughly corresponds to the "diameter" of the atom. In this section we shall obtain a quantitative value for d from quantum mechanics. For the origin of our coordinate system we will choose the center of mass of the atom; for all practical purposes, since the mass of the electron can almost be ignored compared to the mass of the nucleus, the center of mass coincides with the nucleus (see VIII, §§2, 4, 5). In the transition to quantum mechanics we retain the convention that the position operators (after separating out the motion of the center of mass) describe the position relative to center of mass (see, for example, VIII, §2). Here, for simplicity, we assume that the velocity of the center of mass is so small that the Doppler frequency shift of the spectral line—the so-called Doppler broadening—can be neglected. Our discussions can easily be extended in order to take into account the effect of the motion of the center of mass, that is, for the case in which an ensemble W has an appreciable probability for large center of mass motion.

We will now consider an expansion of φ and $\vec{A}$ according to the following two viewpoints:

(1) The point $\vec{x}$ lies far outside the sources of the field, that is, $|\vec{x}| \gg d$.
(2) The wavelength λ of the emitted radiation is much greater than the size of the atom, that is, $\lambda \gg d$.

We shall now introduce an expression of φ and $\vec{A}$ in powers of $d/|\vec{x}|$ and d/λ as follows:

First, it follows that (since $|\vec{x}'|/|\vec{x}| < d/|\vec{x}|$)

$$\frac{1}{|\vec{x} - \vec{x}'|} = \frac{1}{|\vec{x}|}\left(1 + \frac{\vec{x}' \cdot \vec{x}}{|\vec{x}|^2} + \cdots\right). \tag{1.3}$$

We shall obtain an expansion of ρ and $\vec{j}$ in terms of the retardation in $t - |\vec{x} - \vec{x}'|$. A time derivative of ρ corresponds to (if we consider ρ expanded in a time Fourier series) a power of the frequency ω, which is proportional to $1/\lambda$. Thus, in the following, we can read, from which power in d/λ the individual terms of the expansion arise. We then obtain (where the position dependence of ρ is supressed):

$$\rho(t - |\vec{x} - \vec{x}'|) = \rho(t - |\vec{x}|) + \dot{\rho}(t - |\vec{x}|)\varepsilon + \tfrac{1}{2}\ddot{\rho}(t - |\vec{x}|)\varepsilon^2 + \cdots,$$

where $\varepsilon = |\vec{x}| - |\vec{x} - \vec{x}'| = \vec{x}' \cdot \vec{x}/|\vec{x}| + \cdots$. If we carry out the expansion only to terms of first order in $\vec{x}'/|\vec{x}|$ (that is, $d/|\vec{x}|$ and d/λ) we obtain

$$\frac{\rho}{|\vec{x} - \vec{x}'|} = \frac{\rho(t - |\vec{x}|)}{|\vec{x}|} + \frac{\vec{x}}{|\vec{x}|^2} \cdot \frac{\vec{x}'}{|\vec{x}|}\rho(t - |\vec{x}|) + \frac{\vec{x}}{|\vec{x}|} \cdot \frac{\vec{x}'}{|\vec{x}|}\dot{\rho}(t - |\vec{x}|). \tag{1.4}$$

The so-called dipole moment is given by

$$\vec{d}(t - |\vec{x}|) \stackrel{\text{def}}{=} \int \vec{x}' \rho(\vec{x}', t - |\vec{x}|)\, d^3\vec{x}' \tag{1.5}$$

and the charge (which is time independent) is given by

$$Q = \int \rho(\vec{x}', t - |\vec{x}|)\, d^3\vec{x}'.$$

We therefore obtain

$$\varphi(\vec{x}, t) = \frac{Q}{|\vec{x}|} + \frac{\vec{x}}{|\vec{x}|^3} \cdot \vec{d}(t - |\vec{x}|) + \frac{\vec{x}}{|\vec{x}|^2} \cdot \dot{\vec{d}}(t - |\vec{x}|). \tag{1.6}$$

For the vector potential $\vec{A}$ we obtain an analogous expression, where ρ is replaced by $\vec{j}$. In the analogous expression for $\vec{A}$ the individual terms can be rewritten in such a way that their physical interpretation will be more clearly evident. Using the continuity equation $\dot{\rho} + \operatorname{div}\vec{j} = 0$ and, integrating by parts, (here the surface integrals vanish) we obtain:

$$\int \vec{j}(\vec{x}', t - |\vec{x}|) d^3\vec{x}' = \int \vec{j}\cdot\nabla'\vec{x}'\, d^3\vec{x}' = -\int \vec{x}' \operatorname{div}\vec{j}\, d^3\vec{x}'$$

$$= \int \dot{\rho}\vec{x}'\, d^3\vec{x}' = \dot{\vec{d}}(t - |\vec{x}|),$$

where ∇' is the vector operator $(\partial/\partial x_1', \partial/\partial x_2', \partial/\partial x_3')$.

For an arbitrary constant vector $\vec{a}$ we obtain

$$\int \vec{a}\cdot\vec{x}\vec{j}\, d^3\vec{x}' = \int \vec{a}\cdot\vec{x}'\vec{j}\cdot\nabla'\vec{x}'\, d^3\vec{x}' = -\int \vec{x}\nabla'\cdot(\vec{j}\vec{a}\cdot\vec{x}')\, d^3\vec{x}'$$

$$= -\int \vec{x}'\vec{a}\cdot\vec{x}' \operatorname{div}\vec{j}\, d^3x' - \int \vec{x}'\vec{j}\cdot\nabla'(\vec{a}\cdot\vec{x}') d^3\vec{x}'$$

$$= \int \vec{x}'\vec{a}\cdot\vec{x}'\dot{\rho}\, d^3x' - \int \vec{x}'\vec{j}\cdot\vec{a}\, d^3\vec{x}'.$$

We may neglect the term $\int \vec{x}'\vec{a}\cdot\vec{x}'\dot{\rho}\, d^3\vec{x}'$ because it is of order of magnitude $d/|\vec{x}|\, d/\lambda$ in the expansion. We may therefore set

$$\int \vec{a}\cdot\vec{x}'\vec{j}\, d^3\vec{x}' = \tfrac{1}{2}\int \vec{a}\cdot\vec{x}'\vec{j}\, d^3\vec{x}' - \tfrac{1}{2}\int \vec{x}'\vec{j}\cdot\vec{a}\, d^3\vec{x}'$$

$$= -\tfrac{1}{2}\int \vec{a}\times(\vec{x}'\times\vec{j})\, d^3\vec{x}'.$$

Similarly, we obtain

$$\int \vec{a}\cdot\vec{x}\frac{\partial\vec{j}}{\partial t} d^3\vec{x}' = \frac{\partial}{\partial t}\int \vec{a}\cdot\vec{x}\vec{j}\, d^3\vec{x}' = -\frac{1}{2}\frac{\partial}{\partial t}\int \vec{a}\times(\vec{x}'\times\vec{j})\, d^3\vec{x}'.$$

We define the magnetic dipole moment

$$\vec{m} = \tfrac{1}{2}\int \vec{x}'\times\vec{j}\, d^3\vec{x}'. \tag{1.7}$$

Thus we obtain

$$\vec{A}(\vec{x}, t) = \frac{1}{|\vec{x}|}\dot{\vec{d}} - \frac{\vec{x}}{|\vec{x}|^3}\times\vec{m} - \frac{\vec{x}}{|\vec{x}|^2}\times\dot{\vec{m}}. \tag{1.8}$$

From (1.6) and (1.8) we may compute the fields $\vec{E}$ and $\vec{B}$.

When we make the transition to quantum mechanics we will replace ρ and $\vec{j}$ by operators. In order to discover these operators we shall make use of well-known expressions for ρ and $\vec{j}$ from the wave picture (see, for example, [6], XI (1.3.4), (1.3.5))(without external fields!).

$$\rho = e|\psi|^2$$

$$\vec{j} = \frac{e}{m}\left[\bar{\psi}\left(\frac{1}{i}\operatorname{grad}\psi\right) + \left(-\frac{1}{i}\operatorname{grad}\bar{\psi}\right)\psi\right],$$

where we assume that

$$\int|\psi|^2\, d^3\vec{x}' = 1.$$

For the "position representation at time $=0$" (that is, for the representation corresponding to $Q_1(0)$, $Q_2(0)$, $Q_3(0)$) we choose the following operators:

$$\rho(\vec{x}', 0) = e\delta(\vec{x} - \vec{x}'),$$

$$j(\vec{x}', 0) = \frac{e}{2m}\left[\delta(\vec{x} - \vec{x}')\left(\frac{1}{i}\operatorname{grad}_{\vec{x}}\right) + \frac{1}{i}\operatorname{grad}_{\vec{x}}\delta(\vec{x} - \vec{x}')\right],$$

where these are to be applied to functions of $\vec{x}$. These expressions correspond to the description of the electron as a point charge. These operators should only be considered in a symbolic sense, because only the integrals of these expressions over $\vec{x}'$ using test functions have any meaning. The operators $\rho(\vec{x}', t)$ and $\vec{j}(\vec{x}', t)$ are obtained from the time displacement operator e^{iHt} where H is the Hamiltonian operator for the "inner motion," that is, after separating out the motion of the center of mass. In symbolic form we may write

$$\rho(\vec{x}, t) = e\delta(Q(t) - \vec{x}'),$$

$$\vec{j}(\vec{x}, t) = \frac{e}{2m}[\delta(\vec{Q}(t) - \vec{x}')\vec{P}(t) + \vec{P}(t)\delta(\vec{Q}(t) - \vec{x}')],$$

where $\vec{Q}(t)$ is the position operator at time t and $\vec{P}(t)$ is the momentum operator at time t.

From (1.5) it follows that

$$\vec{d}(t - |\vec{x}|) = e\vec{Q}(t - |\vec{x}|) \tag{1.9}$$

a result that we could easily have guessed.

From (1.7) it follows that

$$\vec{m} = (e/4m)(\vec{Q} \times \vec{P} - \vec{P} \times \vec{Q}) = (e/2m)(\vec{Q} \times \vec{P})$$

because, in the cross product the noncommuting components of $\vec{Q}$ and $\vec{P}$ do not appear. Since $\vec{L} = \vec{Q} \times \vec{P}$ is the angular momentum about the nucleus (see VIII, §2) we obtain

$$\vec{m}(t - |\vec{x}|) = (e/2m)\vec{L}(t - |\vec{x}|). \tag{1.10}$$

To (1.10) we must add the eigenmoment of the electron which depends on the spin as an additional magnetic moment; this topic will be discussed later (see XIV, §6) and has already been discussed in the context of interactions with an external field $\vec{B}$ in VIII (5.8).

With the operators (1.9) and (1.10) $\vec{E}$ and $\vec{B}$ then become operators, that is, we have undergone a transition to a quantum electrodynamic description of emission of radiation from an atom. As in the classical case we obtain only an approximation if we substitute the "free motion" of the atom for $\vec{d}(t)$ and $\vec{m}(t)$ and neglect the radiation reaction of the electromagnetic field upon the atom (for this approximation, see VIII, §5). The small width of a spectral line is evidence that this approximation is very good, otherwise, the emission would strongly affect the time dependence and broaden the spectral line.

First we shall consider electric dipole radiation. The corresponding quantitities will be identified with the index (1). In the fields $\vec{E}$ and $\vec{B}$ we shall neglect all terms which decrease more rapidly than $1/|\vec{x}|$; in the so-called radiation zone we obtain:

$$\vec{E}^{(1)} = \frac{1}{|\vec{x}|}\frac{\vec{x}}{|\vec{x}|} \times \left(\frac{\vec{x}}{|\vec{x}|} \times \ddot{\vec{d}}\right),$$

$$\vec{B}^{(1)} = -\frac{\vec{x}}{|\vec{x}|^2} \times \ddot{\vec{d}}. \tag{1.11}$$

$\vec{B}$ and $\vec{E}$ are therefore orthogonal, and both are orthogonal to the direction $\vec{x}/|\vec{x}|$.

We may now consider the use of a spectral apparatus to decompose the fields into individual frequencies. The frequencies with which a given operator changes are given by

$$A(t) = e^{iHt}A(0)e^{-iHt} = \sum_{n,m} E_n A(0) E_m e^{i(\varepsilon_n - \varepsilon_m)t} + \cdots,$$

where the E_m are projections onto the eigenspaces of H with eigenvalues ε_n and the remaining terms $(+\cdots)$ are terms from the continuous spectrum.

We therefore obtain the component of $\vec{A}(t)$ which oscillates with frequency $\omega_{nm} = \varepsilon_n - \varepsilon_m$ as follows:

$$A_{nm}(t) = E_n A(0) E_m e^{i\omega_{nm}t} + E_m A(0) E_n e^{-i\omega_{nm}t}. \tag{1.12}$$

If we substitute the corresponding term $\vec{d}_{nm}$ for $\vec{d}$ into (1.11), we then obtain the corresponding field intensities $\vec{E}^{(1)}_{nm}$ and $\vec{B}^{(1)}_{nm}$.

The Poynting energy-flux vector $\vec{S}$ is given by $(1/4\pi)\,\vec{E} \times \vec{B}$ and we therefore obtain

$$\vec{S}^{(1)}_{nm} = \frac{1}{4\pi}\frac{1}{|\vec{x}|^2}\left|\frac{\vec{x}}{|\vec{x}|} \times \ddot{\vec{d}}_{nm}\right|^2 \frac{\vec{x}}{|\vec{x}|}. \tag{1.13}$$

In order to avoid misunderstanding, we emphasize the fact that $\vec{x}$ is a parameter which has nothing to do with the operator $\vec{d}$: $\vec{x}$ is, for example, the position of the slit in the spectral apparatus, through which, with the help of (1.13) the energy stream is computed, while $\vec{d}$ is the operator which describes the motion of the electron relative to the nucleus!. The direction $\vec{e} = \vec{x}/|\vec{x}|$ from the radiating atoms to the slit of the spectral apparatus is therefore associated with the apparatus.

Let $\Delta\Omega$ denote an element of solid angle; the energy passing through a surface area $|\vec{x}|^2\Delta\Omega$ per unit time in a beam of radiation is given by

$$\frac{1}{4\pi}|\vec{e} \times \ddot{\vec{d}}_{nm}|^2 \, \Delta\Omega. \tag{1.14}$$

If, for $\vec{d}_{nm}$ we substitute an expression of the form (1.12), then, since $E_n E_m = E_m E_n = 0$ for $m \neq n$, we obtain:

$$\frac{\omega_{nm}^4}{4\pi}(\vec{e} \times E_n\vec{d}(0)E_m) \cdot (\vec{e} \times E_m\vec{d}(0)E_n)\,\Delta\Omega$$

$$+ \frac{\omega_{nm}^4}{4\pi}(\vec{e} \times \vec{E}_m\vec{d}(0)E_n) \cdot (\vec{e} \times E_n\vec{d}(0)E_m)\,\Delta\Omega.$$

The second term differs from the first only by the series of factors. For the translation of the classical formula according to the correspondence principle there is an uncertainty. Here the physical consideration is that energy can only be emitted when there is an accompanying energy loss, that is, if the atom (for $\varepsilon_n > \varepsilon_m$) makes a transition from state n into state m. That is, there is no radiation with frequency ω_{nm} if the atom is in state m. This is, however, the case only if the second term is written in the same order(!) as the first. Thus, for $\varepsilon_n > \varepsilon_m$ the energy emitted in the direction $\vec{e}$ into the solid angle $\Delta\Omega$ per unit time is given by:

$$\Delta\Omega\,\frac{\omega_{nm}^4}{2\pi}(\vec{e} \times E_n\,\vec{d}(0)E_m) \cdot (\vec{e} \times E_m\vec{d}(0)E_n). \tag{1.15}$$

If we divide this expression by the energy ω_{nm} of a photon, we find that the "effect" that, in the time interval Δt a photon of frequency ω_{nm} is emitted in direction $\vec{e}$ into the solid angle $\Delta\Omega$ is given by

$$\Delta^{(2)}F_{n\to m} = \Delta t\,\Delta\Omega\,\frac{\omega_{nm}^3}{2\pi}(\vec{e} \times E_n\vec{d}(0)E_m)(\vec{e} \times E_m\vec{d}(0)E_n). \tag{1.16}$$

By integrating over solid angle, we then find that the effect that, in the time interval Δt, a photon of frequency ω_{nm} is emitted is given by

$$\Delta^{(1)}F_{n\to m} = \Delta t\,\frac{4\omega_{nm}^3}{3}E_n\vec{d}(0)E_m \cdot \vec{d}(0)E_n, \tag{1.17}$$

where we note that $\int \sin^2\theta\, d\Omega = 8\pi/3$.

Given an ensemble corresponding to the operator $W \in K$, the probability for the "effect" $\Delta_{n\to m}^{(2)}F$ or $\Delta_{n\to m}^{(1)}F$ is given by $\operatorname{tr}(W(\Delta F_{n\to m}))$. Effects for other frequencies ω which are not transition frequencies do not occur, that is, are zero!

The formulas (1.16), (1.17) therefore describe the fact that the frequencies of the spectral lines are to be computed according to the "Bohr frequency condition" $\omega_{nm} = \varepsilon_n - \varepsilon_m$. Furthermore, with the help of (1.17) we may confirm the fact that the effect $\Delta^{(1)}F_{n\to m}$ (and therefore also $\Delta^{(2)}F_{n\to m}$) are, for different m and n, coexistent (IV, §1.2), a fact which is necessary for the "combined registration" of different spectral lines by the same spectral apparatus (registration method). In order to show that this is the case we shall now estimate

$$\Delta^{(1)}F_{n\to m} \leq \Delta t \tfrac{4}{3}\omega_{nm}^3 E_n \eta_{nm},$$

where η_{nm} is the largest eigenvalue of the operator

$$E_n \vec{d}(0) E_m \cdot \vec{d}(0) E_n.$$

By means of the expression

$$\tau_n^{-1} = \frac{4}{3} \sum_{\substack{m \\ (\varepsilon_n < \varepsilon_m)}} \omega_{nm}^3 \eta_{nm}$$

a time τ_n is determined for which τ_n^{-1} is an order of magnitude estimate for the width of the spectral line. The time τ_n is called the lifetime of the nth state. The formula (1.17) that $\Delta^{(1)}F_{n\to m}$ is proportional to Δt is only valid for $\Delta t \ll \tau_n$. It is, of course, possible to extend the above considerations for the "effect of the emission photons" to much longer times t, where we take into account the fact that several photons can be emitted in sequence; here, we would obtain more complicated expressions which cannot be expressed in closed form, since we already know that from the transition probabilities obtained from (1.17) a "master equation" of the form VIII (5.4) must be solved. For this reason we shall only consider the case in which $\Delta t \ll \tau_n$! We therefore obtain

$$\sum_{\substack{m \\ (\varepsilon_m < \varepsilon_n)}} \Delta^{(1)}F_{nm} \leq E_n$$

from which we obtain

$$\sum_{\substack{n, m \\ (\varepsilon_m < \varepsilon_n)}} \Delta^{(1)}F_{n\to m} \leq \sum_n E_n \leq 1;$$

therefore we find that the effects $\Delta^{(1)}F_{n\to m}$ are coexistent.

As we have previously mentioned, we have only introduced the "derivation" of (1.16), to obtain the expression (1.16). From here on we will consider (1.16) as an axiom for the selection of the operator $\Delta^{(2)}F_{n\to m}$ together with a corresponding (not given explicitly) mapping rule between the characteristics of the construction of a "spectral apparatus" b_0 with a registration b for which $\Psi(b_0, b) = \Delta F_{n\to m}$ where $\Delta F_{n\to m}$ are obtained from $\Delta^{(2)}F_{n\to m}$ by integration over solid angle (as measured from the slit of the apparatus) and from other factors which are also determined by the apparatus.

The "spectral apparatus" b_0 therefore corresponds to an observable (in the sense of VI, §1)

$$\mathscr{R}(b_0) \overset{\psi_{b_0}}{\longrightarrow} L,$$

where the atoms of the Boolean ring are mapped onto the $\Delta F_{n\to m}$. This observable is, however, not a decision observable! The frequency ω may be used as a measurement scale in the sense of IV, §2.5. This scale has a discrete spectrum with discrete values ω_{nm}. Here the scale used has a physically well-defined sense in terms of the construction of the spectral apparatus. The observable measured by the spectral apparatus cannot be confused with the decision observable H (where H is the Hamiltonian operator) although the point spectrum of the frequencies measured by the spectral apparatus contains the eigenvalues of H as differences $\omega_{nm} = \varepsilon_n - \varepsilon_m$!

The expression for magnetic dipole radiation is similar to that obtained above for electric dipole radiation; here we obtain

$$\vec{E}^{(2)} = \frac{\vec{x}}{|\vec{x}|^2} \times \ddot{\vec{m}}, \qquad \vec{B}^{(2)} = \frac{\vec{x}}{|\vec{x}|^2} \times \left(\frac{\vec{x}}{|\vec{x}|} \times \ddot{\vec{m}}\right),$$

the Poynting vector is given by

$$\vec{S}^{(2)} = \frac{1}{4\pi} \frac{1}{|\vec{x}|^2} \left| \frac{\vec{x}}{|\vec{x}|} \times \ddot{\vec{m}} \right|^2 \frac{\vec{x}}{|\vec{x}|}.$$

Because of this similarity, we shall not list the corresponding formulas.

As we have noted in VIII, §5, the effect of the emission of light is such that the operator for the time translation by γ is not precisely given by $e^{iH\gamma}$. We must, in addition, describe the "slow" changes in W by an equation of the form VIII (5.4).

According to (1.17) it is possible to choose, for A_{nm} in VIII (5.4) (axiomatically!)

$$\begin{aligned} A_{nm} &= \sqrt{\frac{4\omega_{nm}^3}{3}} E_m \vec{d}(0) E_n \quad \text{for } \varepsilon_m < \varepsilon_n, \\ A_{nm} &= 0 \qquad\qquad\qquad\qquad \text{otherwise.} \end{aligned} \tag{1.18}$$

On the basis of equation VIII (5.4) the meaning of τ_n introduced above becomes clearer, since VIII (5.4) describes how an "excited" state decays. If W_γ is the solution of VIII (5.4) then for the probability for the effect consisting of the emission of a photon of frequency ω_{nm} in the time interval t to $t + \Delta t$ we would obtain

$$\mathrm{tr}(W_t \Delta^{(1)} F_{n\to m}). \tag{1.19}$$

2 Ensembles Consisting of Bound States

In many cases it is relatively easy to "guess" the operator W which corresponds to an ensemble produced by a given preparation procedure from the structure of the preparation apparatus. We will discuss an example which will enable us to describe many experiments. Again the derivation will be of a heuristic nature. The result for W will be asserted as an axiom (together with a corresponding set of mapping rules for the preparation apparatuses where the latter are constructed according to definite principles) in close analogy to what we have earlier described for the case of effect (spectral) apparatuses.

For the purpose of discussion we shall consider an apparatus which produces an extremely attenuated gas—a so-called ideal gas—composed of the desired material. Since, in an attenuated state it is permissible to neglect the mutual interaction of the individual atoms or molecules, we may consider the individual atoms or molecules to be prepared "in the state of an ideal gas." What operator W are we to assign to the ensemble prepared by this preparation procedure?

According to VIII (1.2) let (we shall neglect the index in H_i)$H_S + H$ denote the Hamiltonian for the atom or molecule where $H_S = \vec{P}^2/2M$ is the Hamiltonian operator of the center of mass and where H is the Hamiltonian operator of the inner structure, where earlier we have called the latter the rest-energy observable (see VIII, §1). Let θ denote the absolute temperature of the gas. Then we set (where $\beta = 1/\theta$)

$$W = CB_S e^{-\beta H}, \tag{2.1}$$

where B_S is an operator which refers only to the center of mass coordinates, and C is a normalization constant determined by the equation $\operatorname{tr}(W) = 1$. (For the basis of this well-known "Boltzmann-ensemble" (2.1) see, for example [2], XV, §6.7.)

The operator B_S is, for the most part, not easily determined theoretically. The following assumption is often found to be useful for the description of an ensemble of an idealized gas: Let us consider the following vectors from the "Hilbert space of the center of mass" in the position representation (see the vectors of minimal $\Delta P \Delta q$ uncertainty in X, §1 and IX, §5):

$$\left(\frac{2}{a^2\pi}\right)^{3/4} e^{-i\vec{p}\cdot(\vec{x}-\vec{x}')} e^{-(|\vec{x}-\vec{x}'|^2)/a^2}. \tag{2.2}$$

We shall consider (2.2) to be a function of $\vec{x}$ in the position representation while $\vec{p}, \vec{x}'$ are fitted quantities. For a we choose a quantity which corresponds to the so-called "mean free path" in the idealized gas present in the preparation apparatus. The projection onto the vector (2.2) will be denoted by $P_{\vec{p},\vec{x}'}$.

We then set

$$B_S = \int e^{-(\beta/2M)\vec{p}^2} P_{\vec{p},\vec{x}'}\, d^3\vec{p}\, d^3\vec{x}', \tag{2.3}$$

where M is the total mass of the atom (or molecule) and $\vec{x}'$ is to be integrated over all space in which the ideal gas is present in the preparation apparatus.

Earlier, in §1 we said that we would, for simplicity, neglect Doppler broadening of spectral lines. By this we mean that for W in (2.1), B_S does not contain any "large" velocities which would result in a large Doppler broadening relative to the resolution capability of the spectral apparatus. Naturally the considerations of §1 can be quantitatively improved with the help of (2.1), (2.3). We shall not, however, do so because we would only obtain a more refined mode of description, which will not, in principle, lead to any new viewpoints.

The formula (2.1) is usable as an expression for the ensemble of the atoms (or molecules) prepared by the preparation procedure providing that θ is sufficiently small and (consequently) β is so large, that is, in

$$e^{-\rho H} = \sum_n e^{-\beta\varepsilon_n} E_n + \cdots$$

only the sum over discrete eigenvalues ε_n plays a role. The E_n are, as in §1, the projections onto the eigenspaces of H. We therefore assume that the spectrum of H which describes the inner structure of the atom (molecule) consists of discrete (and bounded from below) eigenvalues ε_n, and a continuous spectrum. The values for the continuous spectrum do not lead to a convergent expression for $\operatorname{tr}(W)$; if β is so small that the continuous spectrum of H plays an important role in an ideal gas state, then it is necessary to take into account the fact that the gas is actually confined to a finite volume, and it is necessary to replace H by an H' which takes into account the effect of the walls of the enclosure. The continuous spectrum of H describes the so-called unbounded states, in which some of the components of the atoms (molecules) are scattered through the entire space. We then can no longer speak of atoms or molecules as the microsystems prepared by using an idealized gas. The discrete spectrum of H is usually referred to as the set of "bound states." Therefore (2.1) is usable only if we may replace $e^{-\beta H}$ by means of a finite sum $\sum_n e^{-\beta\varepsilon_n} E_n$ over the lowest eigenvalues ε_n of H. If the distance from the smallest eigenvalue ε_0 to the next higher ε_1 is sufficiently large and θ is sufficiently small (that is, β is sufficiently large), then we may replace $e^{-\beta H}$ (because a numerical factor can be taken into C from (2.1)) simply by E_0. In this way we can recognize the great role played by the ground state of an atom or molecule: An ensemble $W \simeq E_0$ is easy to produce. In the theory of chemical bonding we are primarily interested only in this E_0 and the energy eigenvalue ε_0, whose structure we shall investigate extensively in XV.

3 The Spectrum of Hydrogen-like Atoms

A hydrogen-like atom consists of a nucleus of charge Z and a single electron (see VIII, §§2 and 5). Here we shall only be concerned with the so-called inner degrees of freedom, that is, we shall ignore the motion of the center of mass.

We shall also ignore the spin of the electron. We will assume that the Hamiltonian operator is given by:

$$H = \frac{1}{2m}\vec{P}^2 + V(r), \quad \text{where} \quad V(r) = -\frac{Ze^2}{r}. \tag{3.1}$$

In this expression m is the reduced mass, which, because of the relative masses of electron and nucleus, can be taken to be that of the electron; r is the distance from the nucleus to the electron; P is the operator of (relative) momentum, which is practically equal to the momentum of the electron.

Since the operator (3.1) has spherical symmetry, we may apply the results concerning rotations and angular momentum obtained in VII, §3 and VIII, §1. Since H commutes with all rotations about the center of mass, the subspace $\mathscr{T}^l_m$ spanned by $\varphi_\nu(r)Y^l_m(\theta, \varphi)$ for fixed m and l is invariant under H, that is, H only acts upon the variable r in $\varphi_\nu(r)$. We may therefore choose the $\varphi_\nu(r)$ to be (proper or improper) eigenvectors of H.

Since H transforms the different $\mathscr{T}^l_m$ (l fixed; $m = -l, -l-1, \ldots, l-1, l$) isomorphically, we may choose the same $\varphi_\nu(r)$ as the eigenvector of H for all of these values of m. We will characterize these $\varphi_\nu(r)$ by two indices n and l: $\varphi_{nl}(r)$ instead of $\varphi_\nu(r)$. All the $\varphi_{nl} Y^l_m$ where $m = -l, -l+1, \ldots, l-1, l$ belong to the same eigenvalue ε_{nl} of H. ε_{nl} is therefore degenerate of order $(2l+1)$. The eigenspace of H corresponding to the eigenvalue ε_{nl} is invariant under the action of the representation $\mathscr{D}_l$ of the rotation group. The basis vectors $\varphi_{nl} Y^l_m$ of this eigenspace are eigenvectors of $\vec{L}^2$ and L_3 with eigenvalues $l(l+1)$ and m, respectively.

The above assertions are direct consequences of general results; we will now confirm these results directly by computation. For this purpose we shall divide the momentum $\vec{P}$ into two components, one in the direction $\vec{x}$ and the other orthogonal to $\vec{x}$. Only the component which is orthogonal to $\vec{x}$ contributes to the angular momentum $\vec{L} = \vec{x} \times \vec{P}$. It follows that

$$\vec{L}^2 = (\vec{x} \times \vec{P})\cdot(\vec{x} \times \vec{P}) = \vec{x}\cdot[\vec{P} \times (\vec{x} \times \vec{P})]$$

$$= \sum_{i,k=1}^{3} (x_i P_k x_i P_k - x_k P_i x_i P_k).$$

By taking into account the commutation relations and the relation

$$\sum_{i=1}^{3} x_i P_i = r\frac{1}{i}\frac{\partial}{\partial r}$$

we obtain

$$\vec{P}^2 = -\frac{1}{r^2}\frac{\partial}{\partial r}\left(r^2\frac{\partial}{\partial r}\right) + \frac{1}{r^2}\vec{L}^2. \tag{3.2}$$

In spherical coordinates we obtain

$$\vec{L}^2 = -\left[\frac{1}{\sin\theta}\frac{\partial}{\partial\theta}\left(\sin\theta\frac{\partial}{\partial\theta} + \frac{1}{\sin^2\theta}\frac{\partial^2}{\partial\varphi^2}\right)\right]. \tag{3.3}$$

Since

$$\vec{L}^2 Y_m^l = l(l+1) Y_m^l$$

we therefore obtain

$$\vec{P}^2 \varphi_{nl} Y_m^l = \left[-\frac{1}{r^2} \frac{\partial}{\partial r} \left(r^2 \frac{\partial \varphi_{nl}}{\partial r} + \frac{l(l+1)}{r^2} \varphi_{nl} \right) \right] Y_m^l.$$

Thus for φ_{nl} we obtain the equation

$$\frac{1}{2m} \left[-\frac{1}{r^2} \frac{d}{dr} \left(r^2 \frac{d\varphi_{nl}}{dr} \right) + \frac{l(l+1)}{r^2} \varphi_{nl} \right] + V(r)\varphi_{nl} = \varepsilon_{nl} \varphi_{nl}. \tag{3.4}$$

For $V(r) = -Ze^2/r$ (3.4) becomes

$$\varphi_{nl}'' + \frac{2}{r} \varphi_{nl}' + \left(a + 2\frac{b}{r} + \frac{c}{r^2} \right) \varphi_{nl} = 0, \tag{3.5}$$

where

$$a = 2m\varepsilon_{nl}, \qquad b = mZe^2, \qquad c = -l(l+1).$$

We must distinguish between two cases: $\varepsilon_{nl} < 0$ and $\varepsilon_{nl} \geq 0$. In the first case we may, according to the dimension of a, set

$$r_0^2 = -\frac{1}{a}. \tag{3.6}$$

From (3.5) we obtain the asymptotic behavior for large values of r as follows:

$$\varphi_{nl}'' - \frac{1}{r_0^2} \varphi_{nl} \sim 0, \qquad \varphi_{nl} \sim e^{\pm r/r_0}. \tag{3.7}$$

Since φ_{nl} is an element of Hilbert space, for large r we must have

$$\varphi_{nl} \sim e^{-r/r_0}. \tag{3.8}$$

For $\rho = 2r/r_0$ we make the substitution

$$\varphi_{nl} = e^{-\rho/2} u_{nl}(\rho) \tag{3.9}$$

and for $u_{nl}(\rho)$ we obtain

$$u_{nl}'' + \left(\frac{2}{\rho} - 1 \right) u_{nl}' + \left[\left(\frac{b}{\sqrt{-a}} - 1 \right) \frac{1}{\rho} - \frac{l(l+1)}{\rho^2} \right] u_{nl} = 0. \tag{3.10}$$

For a trial solution we set

$$u_{nl}(\rho) = \rho^\lambda \sum_{\mu=0}^{\infty} a_\mu \rho^\mu,$$

where $a_0 \neq 0$. By substituting the trial solution in (3.10), and equating powers of x we obtain

$$\lambda(\lambda - 1) + 2\lambda - l(l+1) = \lambda(\lambda + 1) - l(l+1) = 0$$

from which we conclude that either $\lambda = l$ or $\lambda = -l - 1$. The second solution of (3.9) has a singularity at $r = 0$ and does not belong to the domain of definition of H (VIII, §5). Therefore, only the first solution $\lambda = l$ is usable. We therefore set

$$u_{nl} = \rho^l \sum_{\mu=0}^{\infty} a_\mu \rho^\mu. \tag{3.11}$$

By substituting (3.11) into (3.10) we obtain the following recursion relation for the a_μ:

$$\begin{aligned}[(\mu + l + 1)(\mu + l) + 2(\mu + l + 1) - l(l + 1)]a_{\mu+1} \\ = \left(\mu + l + 1 - \frac{b}{\sqrt{-a}}\right)a_\mu.\end{aligned} \tag{3.12}$$

If the series $\sum_\mu a_\mu \rho^\mu$ does not have a finite number of terms then, for large r, u_{nl} will behave like $e^{+\rho}$ because, for large μ

$$a_\mu \sim \frac{1}{\mu!}, \qquad \varphi_{nl} = u_{nl} e^{-\rho/2}$$

behaves like (for large ρ) $e^{+\rho/2}$ and therefore cannot be an element of Hilbert space. Therefore, the series must terminate; this may be the case for $\mu = n_v$. Thus $a_{n_v+1} = 0$, and, according to (3.12) we must have

$$\frac{b}{\sqrt{-a}} = n_v + l + 1 = n \tag{3.13}$$

from which it follows that

$$\varepsilon_n = -\frac{me^4 Z^2}{2n^2}. \tag{3.14}$$

From (3.13) we obtain the following condition for n:

$$n \geq l + 1. \tag{3.15}$$

In the domain $\varepsilon < 0$, H only has discrete eigenvalues. The eigenvalues ε_n are graphically represented in Figure 2. They determine the spectrum of the hydrogen atom by the equation

$$\omega_{nm} = R\left(\frac{1}{n^2} - \frac{1}{m^2}\right)$$

with the Rydberg-constant

$$R = \frac{me^4}{2}.$$

In the "term-scheme" Figure 2 the perpendicular lines identify the transitions which lead to spectral lines ω_{nm}. For $m = 1$ and $n = 2, 3, \ldots$ we obtain a series of spectral lines, the Lyman series, which lies in the ultraviolet. For $m = 2$ and $n = 3, 4, \ldots$ we obtain the Balmer series, for $m = 3, n = 4, 5, \ldots$ the Paschen series, etc. The entire spectrum for hydrogen and its decomposition into series is represented in Figure 3.

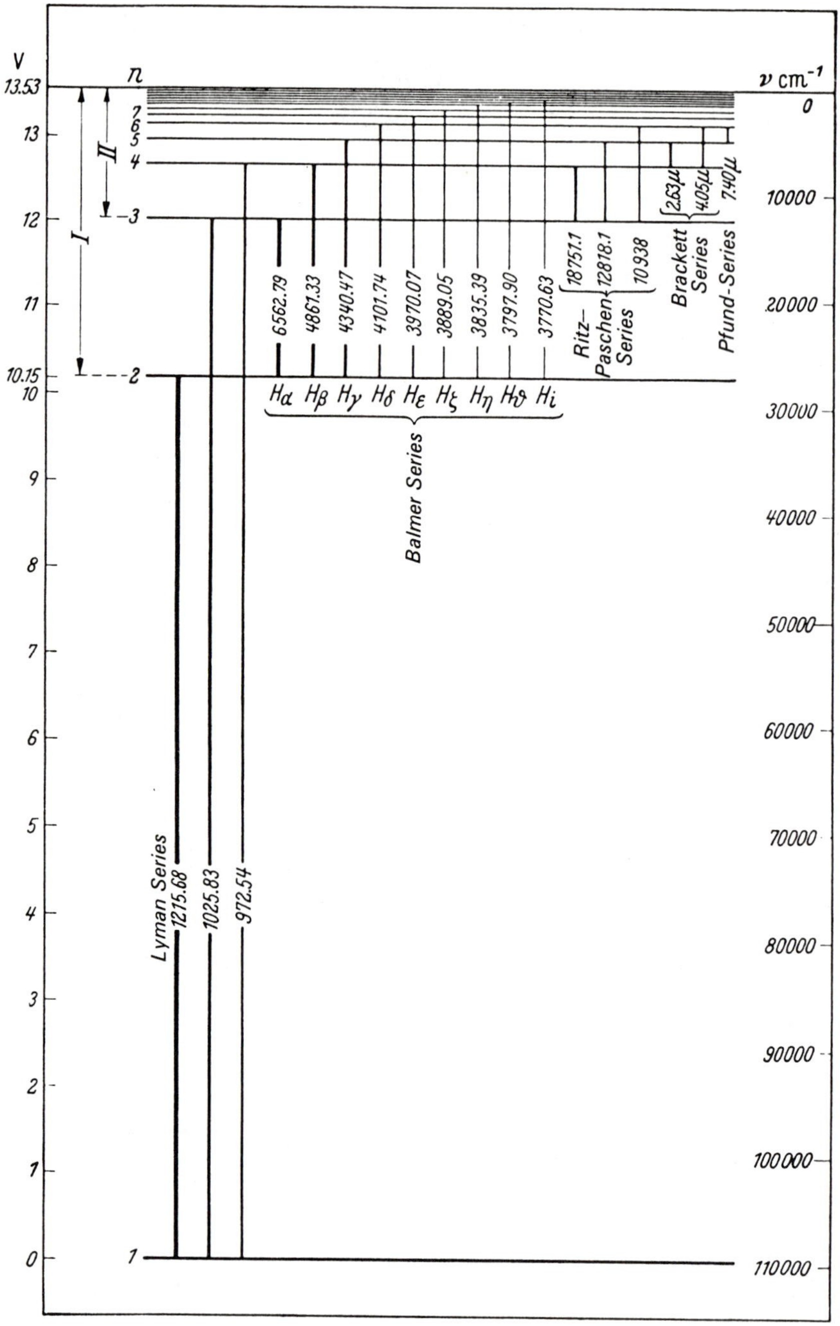

Figure 2 Term scheme of hydrogen.

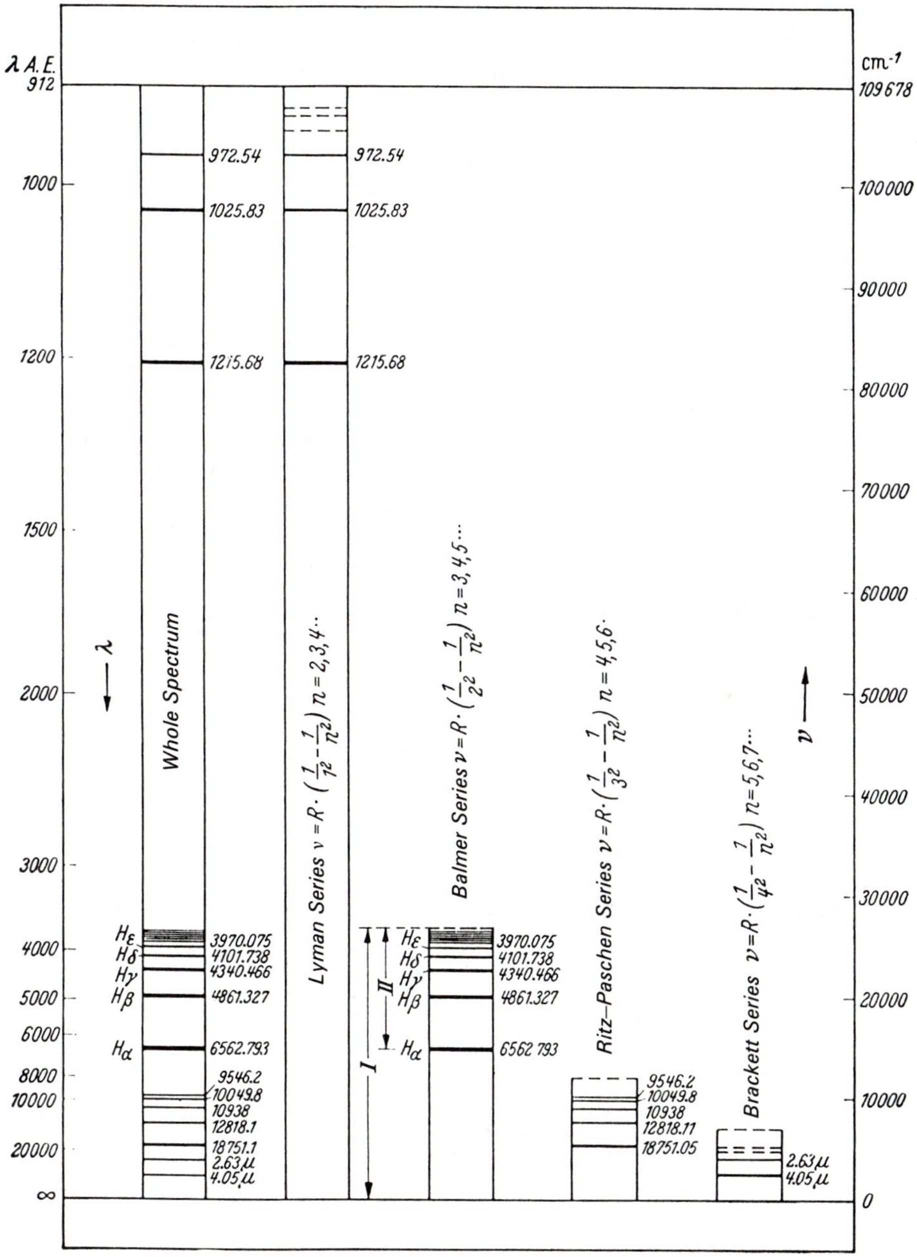

Figure 3 Spectrum for hydrogen.

The eigenvalues ε_n of H are multiply degenerate. To each pair of values n, l there are $2l + 1$ independent eigenvectors of H and of $\vec{L}$. In addition, for fixed n the "angular momentum quantum number" l can take on values $0, 1, \ldots, n - 1$. Therefore the degree of degeneracy of ε_n is equal to

$$\sum_{l=0}^{n-1} (2l + 1) = n^2.$$

From symmetry arguments, that is, since H commutes with L_3 and $\vec{L}$ each eigenvalue ε_{nl} must be $(2l + 1)$-fold degenerate. For the case in which $V(r) = -Ze^2/r$ the eigenvalues ε_{nl} for $l = 0, 1, \ldots, n - 1$ "accidentally" coincide. If $V(r)$ is not the Coulomb potential, then we would expect a greater manifold ε_{nl} of energy eigenvalues.

From experience with hydrogen, and the corresponding ions of the elements helium, lithium, etc. the Balmer terms (3.14) have been exceptionally well confirmed. Only under extremely high resolution is it possible to show that each term is not a single term but consists of multiple adjacent terms. We shall return to this topic later in the book.

4 The Eigenfunctions for the Discrete Spectrum

If we write $u_{nl} = \rho^l w_{nl}(\rho)$ then w_{nl} will be a polynomial of n_νth degree ($n_\nu = n - l - 1$). From the recursion formula (3.12) it follows that

$$w_{nl}(\rho) = c_{nl} F(-n + l + 1, 2l + 2, \rho), \tag{4.1}$$

where c_{nl} is a constant factor and $F(\alpha, \beta, \rho)$ is the confluent hypergeometric function defined by

$$F(\alpha, \beta, x) = 1 + \frac{\alpha}{\beta}\frac{x}{1!} + \frac{\alpha(\alpha + 1)}{\beta(\beta + 1)}\frac{x^2}{2!} + \cdots. \tag{4.2}$$

The $w_{nl}(\rho)$ may also be written in terms of Laguerre polynomials (see X, §1). The latter are defined by

$$L_\lambda^\mu(x) = \frac{d^\mu}{dx^\mu} L_\lambda(x); \qquad L_\lambda(x) = e^x \frac{d^\lambda}{dx^\lambda}(e^{-x}x^\lambda). \tag{4.3}$$

We obtain

$$L_\lambda^\mu(x) = \sum_{\nu=0}^{\lambda} (-1)^\nu \binom{\lambda}{\nu} \frac{\lambda!}{\nu!} x^\nu \tag{4.4}$$

and

$$L^{\mu}(x) = (-1)^{\mu}\lambda!\binom{\lambda}{\mu}F(\lambda - \mu, \mu + 1, x). \tag{4.5}$$

Thus we obtain

$$w_{nl}(\rho) = c_{nl}\frac{(-1)^{2l+1}}{(n+l)!\binom{n+l}{2l+1}}L_{n+l}^{2l+1}(\rho) = D_{nl}L_{n+l}^{2l+1}(\rho). \tag{4.6}$$

The factor c_{nl} or D_{nl} is to be chosen such that the entire eigenfunction

$$D_{nl}\rho^{l}L_{n+l}^{2l+1}(\rho)e^{-\rho/2}Y_m^l(\theta, \varphi)$$

is normalized. Thus we obtain

$$D_{nl}^2\left(\frac{r_0}{2}\right)^3\int\rho^{2(l+1)}e^{-\rho}\left[L_{n+l}^{2l+1}(\rho)\right]^2 d\rho = 1,$$

where r_0 is given by (3.6) and (3.13) since the Y_m^l are normalized on the surface of the sphere. We obtain (see [9], Band II, p. 84):

$$D_{nl}^2 = \left(\frac{2Zme^2}{n}\right)^3\frac{(n-l-1)!}{2n[(n+l)!]^3} \quad \text{and} \quad r_0 = \frac{n}{Zme^2}.$$

Clearly $1/me^2$ is the approximate size of the hydrogen atom ($Z = 1$) in the ground state ($n = 1$).

Finally, in terms of the notation introduced in IX we obtain

$$\langle x_1, x_2, x_3 | \varepsilon_n, l, m\rangle = \left(\frac{2Zme^2}{n}\right)^{3/2}\sqrt{\frac{(n-l-1)!}{2n[(n+l)!]^3}} \cdot\left(2\frac{r}{r_0}\right)^l L_{n+l}^{2l+1}\left(2\frac{r}{r_0}\right)e^{-r/r_0}Y_m^l(\theta, \varphi), \tag{4.7}$$

where r, θ, φ and x_1, x_2, x_3 are related by the equations $x_1 = r\sin\theta\cos\varphi$, $x_2 = r\sin\theta\sin\varphi$, $x_3 = r\cos\theta$. The quantities $\langle x_1, x_2, x_3 | \varepsilon_n, l, m\rangle$ are also the coefficients of the position representation in the $(H, \vec{L}^2, L_3)$-representation. The set of these coefficients is not complete, because all values $\varepsilon \geq 0$ belong to the continuous spectrum of H, as we shall see in the next section.

In Figures 4 and 5 we exhibit the behavior of the radial functions $\varphi_{nl}(r)$. In Figure 4 the function itself is given. In Figure 5 the charge density within a spherical shell of radius r to $r + dr$ is given.

A very intuitive picture of the probability density $|\langle x_1, x_2, x_3 | \varepsilon_n, l, m\rangle|^2$ is given by the White models (see Figure 6).

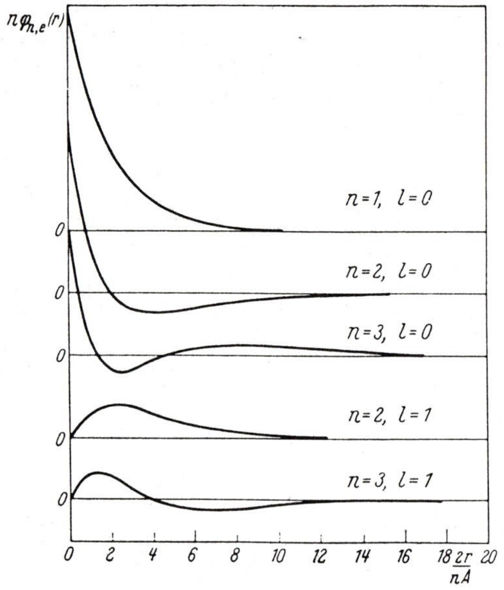

Figure 4 Radial functions for hydrogen. $A = (me^2)^{-1}$.

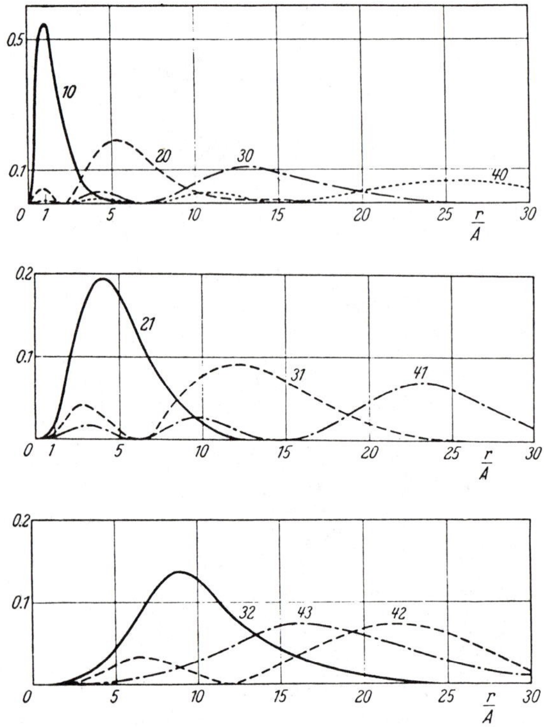

Figure 5 Charge density within a spherical shell. (42 means $n = 4, l = 2$.) $A = (me^2)^{-1}$.

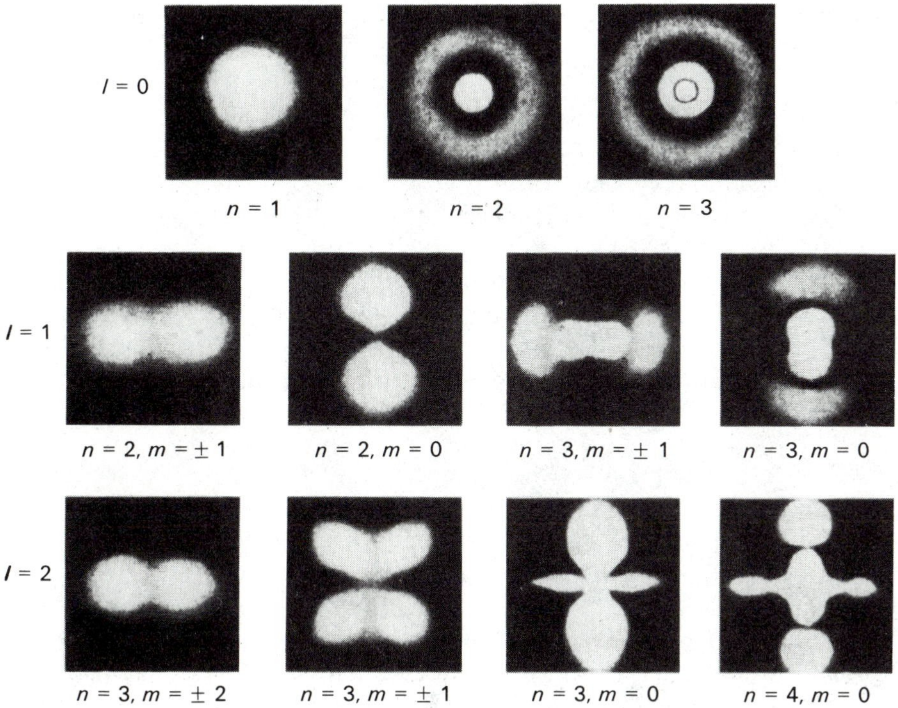

Figure 6 Probability density $|\langle x_1, x_2, x_3, |\varepsilon_n, l, m\rangle|^2$.

5 The Continuous Spectrum

In §3 we have assumed that $\varepsilon < 0$ and therefore $a < 0$, so that we set (according to (3.6)) $r_0^2 = -1/a$. We will now consider the case in which $a > 0$. For this case we set $r_0^2 = 1/a$. The asymptotic behavior of φ_{nl} is given by $\varphi_{nl} \sim e^{\pm i(r/r_0)}$ instead of by (3.7), from which we conclude that for all values $\varepsilon > 0$ the eigenfunctions are improper.

The computations of §3 can be carried out if we set $\rho = 2i(r/r_0)$. For $\varphi_{nl} = e^{-\rho/2}u_{nl}(\rho)$ we again obtain equation (3.10). The power series expansion (3.11) is also valid; the only difference is that the power series does not terminate. We shall now replace ε by the quantity k where k is defined by the equation $k^2/2m = \varepsilon$; k is the wave number of a free electron with energy ε. We then may write

$$n = \frac{b}{\sqrt{-a}} = \frac{Zme^2}{ik}, \tag{5.1}$$

where n is, of course, no longer an integer. In order to obtain a closed analytic expression for the above power series we shall begin with the formula (4.3)

$$L_n(\rho) = e^\rho \frac{d^n}{d\rho^n}(\rho^n e^{-\rho})$$

for the Laguerre polynomial.

If $f(\rho)$ is an analytic function, then the following integral expression holds in general

$$\frac{1}{n!}\frac{d^n}{d\rho^n} f(\rho) = \frac{1}{2\pi i}\oint \frac{f(z)}{(z-\rho)^{n+1}}\,dz,$$

where the integral is taken over a closed path which contains the point $z = \rho$. Thus it follows that:

$$L_n(\rho) = \frac{n!}{2\pi i} e^\rho \oint z^n e^{-z}(z-\rho)^{-n-1}\,dz. \tag{5.2}$$

This expression is, however, a solution of the differential equation

$$\rho L'' + (1-\rho)L' + nL = 0 \tag{5.3}$$

even in the case in which n is not a positive integer. Here we only assume that the integration path is closed. The fact that equation (5.2) is a solution of the differential equation (5.3) can be verified by substitution and integration by parts. The path is closed for arbitrary n (not necessarily only integer or real values of n) if we choose an integration path such as that illustrated in Figure 7 in which the changes of the integrand which result from the branch points at $z = 0$ and $z = \rho$ are compensated.

According to (3.9) we can write the eigenfunctions as follows $\varphi_{nl}(r)Y^l_m(\theta, \varphi)$ where

$$\varphi_{nl} = D' e^{-\rho/2}(i\rho)^l L^{2l+1}_{n+l}(\rho), \tag{5.4}$$

that is,

$$\varphi_{nl} = D e^{-\rho/2}(-i\rho)^l \oint (z+\rho)^{n-l-1} e^{-z} z^{-n-l-1}\,dz, \tag{5.5}$$

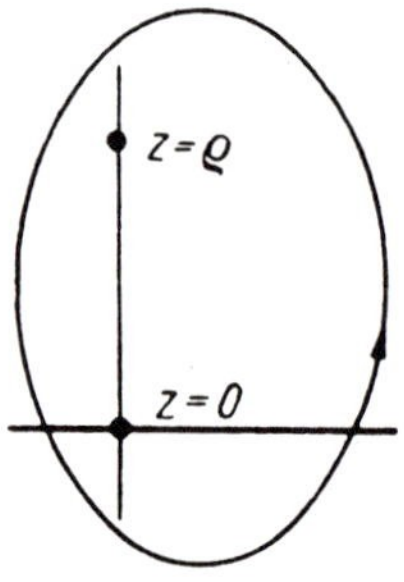

Figure 7

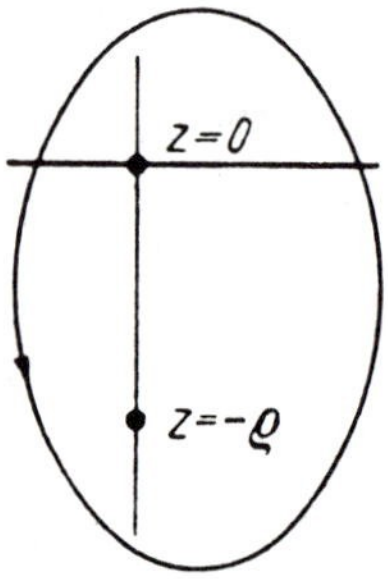

Figure 8

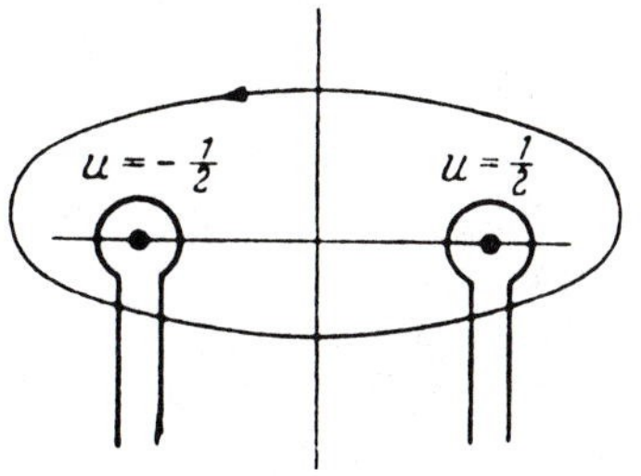

Figure 9

where ρ and n are purely imaginary, and the integration path [obtained by substituting $z \to z + \rho$] is given in Figure 8. The integral can easily be expanded into a power series, since the integration path can be enlarged such that the expansion of $(z + \rho)^{n-l-1}$ in terms of ρ lies in the convergence domain. In this way we obtain the power series expansion (5.2) which was introduced earlier.

We are particularly interested in the bahvior of φ_{nl} for large values of r. For this purpose, let us make the substitution $z = \rho(u - \frac{1}{2})$ in (5.5). We obtain

$$\varphi_{nl} = D(i\rho)^{-l-1} \frac{1}{2\pi} \oint (u + \tfrac{1}{2})^{n-l-1}(u - \tfrac{1}{2})^{-n-l-1} e^{-\rho u}\, du,$$

where the integration path is the closed path in Figure 9 which can be replaced by both loops in Figure 9. We may write $\varphi_{nl} = \frac{1}{2}(\chi_{nl} + \eta_{nl})$ where χ and η are the integrals over each of the loops. If we expand the integrands in powers of $1/\rho$, then the asymptotic behavior for large r is given by

$$\chi_{nl} \sim e^{-\rho/2}\rho^{n-1} \frac{e^{-i\pi(n-l/2)}}{\Gamma(n + l + 1)},$$

$$\eta_{nl} \sim e^{\rho/2}(-\rho)^{-n-1} \frac{e^{-i\pi(n+l/2)}}{\Gamma(-n + l + 1)}$$

and for φ_{nl} we obtain

$$\varphi_{nl} \sim c \frac{\sin(kr + \gamma)}{kr}, \tag{5.6}$$

where

$$c = \frac{e^{-(\pi/2)|n|}}{|\Gamma(n + l + 1)|}; \qquad \gamma = |n| \log 2kr + \alpha - \frac{\pi}{2} l.$$

Here α is determined by $\Gamma(n + l + 1) = |\Gamma(n + l + 1)|e^{i\alpha}$. Thus, up to a normalization factor N, for the continuous spectrum $\langle \vec{x} | \varepsilon, l, m \rangle$ is given by

$$\langle \vec{x} | \varepsilon, l, m \rangle = N\varphi_{nl}(r) Y^l_m(\theta, \varphi),$$

where n is the value given by (5.1) and φ_{nl} is given by (5.4).

6 Perturbation Theory

For spherically symmetric potentials other than $V(r) = Ze^2/r$ it is, for the most part, not possible to obtain an exact calculation for the spectrum of H. Often, for problems which cannot be solved exactly, it is possible to obtain a satisfactory "approximate solution" by using the "method of perturbations." We will now describe a frequently used general perturbation procedure (in §§6 and 7 we will discuss this problem in a more general way than is required for the case of one-electron systems). The method consists of replacing the Hamiltonian operator H by an approximate operator H' which we assume (in the sense of the discussion in [8], §8) is less precise than H, but describes experience with a certain more or less well-known accuracy. The investigation of perturbation theory presented in [10] is nothing other than a mathematical attempt to determine the corresponding "imprecision sets" (see [11] and [8], §§5 and 6), and therefore to solve the problem outlined in [8], §8 concerning the relationships between two theories, one more precise than the other. Here we shall only outline a procedure by which we may obtain an approximate H' from an exact H (which permits us to mathematically treat the problem) in the hope that this H' will be useful (that is, the corresponding theory with a given imprecision set in the sense of [8], §§5 and 6 and [11] will be useful).

This procedure is tailored to the discrete spectrum for H. It can be shown that in the physics of atoms and molecules the Hamiltonian operators can only have discrete eigenvalues which are at most finitely degenerate. This fact will be used in the following procedures.

First Step. We seek to decompose the Hamiltonian operator H into two parts as follows: $H = H_0 + H_1$, such that H_0 contains as much as possible of H, and for which the eigenvalue problem of H_0 can be solved. We may, for example, instead of including $V(r)$ in H, seek to obtain an approximate potential $V_0(r)$ such that $V_0(r)$ contains the "essential" part of $V(r)$, but for which the eigenvalue problem of H_0 (which contains $V_0(r)$) can be solved. H_1 then has the form $H_1 = V(r) - V_0(r)$. We shall call H_0 the zeroth approximation and H_1 the perturbation.

The discrete part of H_0 may be written in the form $\sum_n \varepsilon_n^{(0)} E_n^{(0)}$. According to the previous remarks, each of the projections $E_n^{(0)}$ is finite dimensional.

Second Step. We consider the discrete spectrum $\varepsilon_n^{(0)}$ and determine whether several eigenvalues are "densely spaced" in groups. Two $\varepsilon_{n_1}^{(0)}$ and $\varepsilon_{n_2}^{(0)}$ are to be considered "densely spaced" if the perturbation H_1 displaces the eigenvalues more strongly or about a magnitude $|\varepsilon_{n_1}^{(0)} - \varepsilon_{n_2}^{(0)}|$. If all the eigenvalues are densely spaced in this sense, then we can select arbitrary intervals from the energy scale and collect all $\varepsilon_n^{(0)}$ which lie in the interval into a group; we must then take into account that the computation of the displacement of the eigenvalue $\varepsilon_n^{(0)}$ under the perturbation is incorrect in the neighborhood of the interval boundaries. This can be managed to some extent by the use of overlapping intervals. We will not consider these more complicated cases.

We therefore assume that the $\varepsilon_n^{(0)}$ (or at least those which are of interest with respect to the perturbation) are either widely separated or are divided into groups $\{\varepsilon_{n_i,\nu}^{(0)}\}$ (where ν is the index of the group) which are widely separated and contain only a finite number of indices n_i in a given group. For $\tilde{E}_\nu = \sum_{n_i} E_{n_i}^{(0)}$ (where the sum is taken over the elements n_i of the νth group) we construct the operator

$$H^{(1)} = \sum_n \varepsilon_n^{(0)} E_n^{(0)} + \sum_\nu \tilde{E}_\nu H_1 \tilde{E}_\nu = \sum_\nu \tilde{E}_\nu H \tilde{E}_\nu. \tag{6.1}$$

Here we call $H^{(1)}$ the operator of the first approximation, $H = H^{(1)} + H_2$ where we call the remainder H_2 the perturbation of second order.

The solution of the eigenvalue problem for $H^{(1)}$ consists of the clearly finite-dimensional eigenvalue problems for the various $H_\nu = \tilde{E}_\nu H \tilde{E}_\nu$. For the special case in which $\tilde{E}_\nu$ is identical to $E_n^{(0)}$ (that is, $\varepsilon_n^{(0)}$ is not a neighboring eigenvalue to other eigenvalues of H_0), then

$$H^{(1)} = E_n^{(0)} H E_n^{(0)} = \varepsilon_n^{(0)} E_n^{(0)} + E_n^{(0)} H_1 E_n^{(0)}.$$

Since, in the finite-dimensional subspace determined by $E_n^{(0)}$ the operator $\varepsilon_n^{(0)} E_n^{(0)}$ is a multiple of the unit matrix, we need only determine the eigenvalues of $E_n^{(0)} H_1 E_n^{(0)}$. The eigenvalue equation for $\tilde{E}_\nu H \tilde{E}_\nu$, or for $E_n^{(0)} H_1 E_n^{(0)}$, is often called the secular equation. If we denote the eigenvalues of $E_n^{(0)} H_1 E_n^{(0)}$ by $\varepsilon_{n_\alpha}^{(1)}$, then, as a first approximation, that is, for the eigenspace of $H^{(1)}$ we obtain the values $\varepsilon_n^{(0)} + \varepsilon_{n_\alpha}^{(1)}$; the eigenvalue for the zeroth approximation $\varepsilon_n^{(0)}$ is, for different α, split into the eigenvalues $\varepsilon_n^{(0)} + \varepsilon_{n_\alpha}^{(1)}$ of the first approximation. If $E_n^{(0)}$ is one-dimensional, then we naturally do not find any splitting, and we obtain $\varepsilon_n^{(1)} = \langle \varepsilon_n^{(0)} | H_1 | \varepsilon_n^{(0)} \rangle$, that is, the expectation value of H_1 in the eigenstate of H_0 for the (nondegenerate) eigenvalue $\varepsilon_n^{(0)}$.

We have characterized these two important steps of the zeroth and first approximations without making any assumption concerning the series expansion in order not to unnecessarily burden the approximation procedure with the problem of the expansion of the eigenvalues and eigenvectors in power series of a parameter. For such questions and for questions concerning the validity of such approximations we refer the reader to [10]. Here we shall instead seek to formulate an additional approximation step.

We shall begin by considering the decomposition $H = H^{(1)} + H_2$. This decomposition has the following properties: Let $E_\mu^{(1)}$ denote the projections onto the eigenspace of $H^{(1)}$ with the corresponding eigenvalues $\varepsilon_\mu^{(1)}$ (not to be confused with the above $\varepsilon_{n,\alpha}^{(1)}$: on the contrary, in the above case it is necessary to set $\varepsilon_\mu^{(1)} = \varepsilon_n^{(0)} + \varepsilon_{n,\alpha}^{(1)}$). Then from the definition of H and from the condition that no two eigenvalues $\varepsilon_\mu^{(1)}$ from different groups ν may coincide it follows that all expressions of the form $E_\mu^{(1)} H_2 E_\mu^{(1)}$ must be equal to 0 (otherwise, the groups were incorrectly chosen). Since all the $E_\mu^{(1)} H_2 E_\mu^{(1)} = 0$, we must seek the effect of H_2 in another way than that described in the "second step" above.

In order to make an additional step (a third step with respect to the previous two) we must transform the eigenvectors of $H^{(1)}$ by a unitary transformation at least approximately) into the eigenvectors of $H^{(1)} + H_2$.

We therefore seek a unitary transformation U such that for

$$E'_\mu = U^+ E^{(1)}_\mu U \tag{6.2}$$

the nondiagonal terms $E'_\mu(H^{(1)} + H_2)E'_\rho$ for $\mu \neq \rho$ are as small as possible. In this way the spectrum of $H^{(1)} + H_2$ can be approximated by the solution of the finite-dimensional eigenvalue problem $E'_\mu(H^{(1)} + H_2)E'_\mu$.

From (6.2) it follows that

$$E'_\mu(H^{(1)} + H_2)E'_\rho = U^+ E^{(1)}_\mu U(H^{(1)} + H_2)U^+ E^{(1)}_\rho U.$$

In order to establish a reference point to compare the magnitudes of different terms we shall apply a factor λ to H_2 and set $U = e^{i\lambda S}$. In order to obtain minimal nondiagonal elements we shall choose S such that the expansion of

$$e^{i\lambda S}(H^{(1)} + \lambda H^{(2)}_2)e^{-i\lambda S}$$

has no terms of first order in λ. This will be the case if

$$H_2 = i[H^{(1)}, S]. \tag{6.3}$$

This condition is satisfied for

$$S = \sum_{\substack{\nu,\mu \\ \nu \neq \mu}} \frac{E^{(1)}_\nu H_2 E^{(1)}_\mu}{i(\varepsilon^{(1)}_\nu - \varepsilon^{(1)}_\mu)}. \tag{6.4}$$

From which we obtain

$$e^{i\lambda S}(H^{(1)} + \lambda H_2)e^{-i\lambda S} = H^{(1)} + \frac{\lambda^2}{2} i[S, H_2] + \cdots.$$

For the approximation operator for the third step we select

$$H' = \sum_\mu e^{iS} E^{(1)}_\mu \left(H^{(1)} + \frac{i}{2}[S, H_2]\right) E^{(1)}_\mu e^{iS}. \tag{6.5}$$

In order to find the eigenvalues of H' in (6.5) it suffices to solve the finite-dimensional eigenvalue problems of

$$E^{(1)}_\mu \left(H^{(1)} + \frac{i}{2}[S, H_2]\right) E^{(1)}_\mu \tag{6.6}$$

because the multiplication by e^{-iS} on the left and by e^{iS} on the right does not affect the eigenvalues, but only the eigenvectors. The eigenvectors of (6.6) need only be multiplied by e^{iS} in order to obtain the eigenvectors of (6.5).

The eigenvalue problem (6.6) again has the form of a secular equation. Using (6.4) the operator (6.6) is given by

$$\varepsilon^{(1)}_\mu E^{(1)}_\mu + \sum_{\rho(\neq \mu)} \frac{E^{(1)}_\mu H_2 E^{(1)}_\rho H_2 E^{(1)}_\mu}{\varepsilon^{(1)}_\mu - \varepsilon^{(1)}_\rho}. \tag{6.7a}$$

If the eigenvalue $\varepsilon_\mu^{(1)}$ of $H^{(1)}$ is nondegenerate, that is, if $E_\mu^{(1)}$ is one-dimensional, then from (6.7a) we obtain the eigenvalue

$$\varepsilon_\mu^{(1)} + \sum_{\rho(\neq\mu)} \frac{\langle \varepsilon_\mu^{(1)} | H_2 E_\rho^{(1)} H_2 | \varepsilon_\mu^{(1)} \rangle}{\varepsilon_\mu^{(1)} - \varepsilon_\rho^{(1)}}. \tag{6.7b}$$

In (6.7a) or (6.7b) the sum over ρ can be extended as an integral over the continuous spectrum of $H^{(1)}$ as follows: Let $E^{(1)}(\varepsilon)$ be the spectral family of $H^{(1)}$ (thus $E_\mu^{(1)}$ will be the discontinuities of $E^{(1)}(\varepsilon)$ at locations $\varepsilon_\mu^{(1)}$). Then, in general, we may write:

$$\varepsilon_\mu^{(1)} E_\mu^{(1)} + \lim_{\delta\to 0} \left(\int_{-\infty}^{\varepsilon_\mu^{(1)}-\delta} + \int_{\varepsilon_\mu^{(1)}+\delta}^{\infty} \right) \frac{1}{\varepsilon_\mu^{(1)} - \varepsilon}. \tag{6.7c}$$

$$E_\mu^{(1)} H_2 \, dE^{(1)}(\varepsilon) H_2 E_\mu^{(1)}.$$

7 Perturbation Computations and Symmetry

We will now continue our discussion of the problem described in §6. We shall assume a decomposition of the Hamiltonian H of the form $H = H_0 + H_1$ as in §6 where we assume that the spectrum of H_0 is known. Instead of beginning with the perturbation computation described in §6, we shall consider an operator $H(\lambda) = H_0 + \lambda H_1$ which depends on λ (in the interval $0 \leq \lambda \leq 1$) and seek to determine what we can about the λ dependence of eigenvalues and eigenspaces for the operator $H(\lambda)$, in order to obtain information concerning the spectrum of $H(1) = H_0 + H_1$.

Later we shall encounter Hamiltonian operators which depend on one or more different parameters (see, for example, the discussion of molecular spectra in XV, §§3 and 5). In our discussion of this problem we will permit the parameter λ to take on values which are points in a complete metric space Λ in order to obtain general statements concerning the dependence of the spectrum of the Hamiltonian operator $H(\lambda)$ upon λ.

We will now assume that the lower bound of the continuous spectrum of $H(\lambda)$ is a continuous function $\eta(\lambda)$; often $\eta(\lambda)$ will be constant. We are only interested in the discrete portion of the spectrum where the latter is below $\eta(\lambda)$. If, for a particular value of λ there are infinitely many discrete eigenvalues of $H(\lambda)$ below $\eta(\lambda)$, then let $\eta(\lambda)$ be the one and only one accumulation point of these discrete eigenvalues.

Furthermore, it may be possible to number the eigenvalues of this spectrum below the continuum in such a way that we obtain continuous functions $\varepsilon_n(\lambda)$ for which not all the $\varepsilon_n(\lambda)$ need be defined for all values of λ, namely when the discrete eigenvalue $\varepsilon_n(\lambda)$ exists only for an open subset Λ_n of Λ and on the boundary of Λ_n approach the value of the continuum limit. We then say that the eigenvalue $\varepsilon_n(\lambda)$ emerges from the continuum in the transition from points outside Λ_n to points within Λ_n.

The mathematical problem, to find conditions for $H(\lambda)$ under which the above assumptions and other assumptions to be made later are satisfied is investigated in [10].

Some of the eigenvalues $\varepsilon_n(\lambda)$ can, for definite points λ, coincide. We will therefore not assume that the different functions $\varepsilon_n(\lambda)$ are different for all values of λ. On the contrary, it is precisely those points λ and their immediate neighborhoods where several $\varepsilon_n(\lambda)$ are equal which are of special interest.

For each point λ the projection operators $E_\varepsilon(\lambda)$ onto the eigenspaces corresponding to the different(!) eigenvalues ε are defined where ε may only run through the different values of the $\varepsilon_n(\lambda)$ (where λ is fixed). Therefore, in general, it is not possible to define $E_{\varepsilon_n(\lambda)}$! Therefore we cannot make any assumptions concerning the continuity of the λ dependence of $E_\varepsilon(\lambda)$. A meaningful continuity assumption for the $E_\varepsilon(\lambda)$ can only be formulated as follows (here we assume that all the $E_\varepsilon(\lambda)$ are finite dimensional):

For fixed λ and ε we define the subset $N(\lambda, \varepsilon)$ of all n for which $\varepsilon_n(\lambda) = \varepsilon$. For each point λ' we set $\sigma(\lambda', \lambda) = \{\varepsilon' \mid \varepsilon' = \varepsilon_n(\lambda') \text{ and } n \in N(\lambda, \varepsilon)\}$ and $E_\varepsilon(\lambda', \lambda) = \sum_{\varepsilon' \in \sigma(\lambda', \lambda)} E_{\varepsilon'}(\lambda')$. We therefore obtain $E_\varepsilon(\lambda, \lambda) = E_\varepsilon(\lambda)$. We say that $E_\varepsilon(\lambda)$ is continuous at the point λ if there exists a neighborhood $\mathscr{V}$ of λ for which $E_\varepsilon(\lambda', \lambda)$, as a function of λ' (that is, for $\lambda' \in \mathscr{V}$) has the same dimension as $E_\varepsilon(\lambda)$ and $E_\varepsilon(\lambda', \lambda)$ (as an element of $\mathscr{B}'$) is σ continuous at the location $\lambda' = \lambda$.

We now assume that $E_\varepsilon(\lambda)$ is, in this sense, continuous in all of Λ. The above neighborhood $\mathscr{V}$ need then only be chosen so small that in $\mathscr{V}$ the $\varepsilon_n(\lambda')$ with $n \in N(\lambda, \varepsilon)$ does not coincide with another $\varepsilon_{n'}(\lambda)$ for $n' \notin N(\lambda, \varepsilon)$.

Since the functions $\varepsilon_n(\lambda)$ are continuous, there exists a neighborhood $\mathscr{V}$ of λ, for which the number of different values taken on by $\varepsilon_n(\lambda')$ for $\lambda' \neq \lambda$, $\lambda' \in \mathscr{V}$ and for the $n \in N(\lambda, \varepsilon)$ is constant ($N(\lambda, \varepsilon)$ is finite, because the $\varepsilon_n(\lambda)$ have at most one $\eta(\lambda)$ as an accumulation point!). We say that the eigenvalue ε splits in the neighborhood of λ into the eigenvalues $\varepsilon'(\lambda')$ where $\varepsilon'(\lambda')$ runs through the different values of $\varepsilon_n(\lambda')$ for $n \in N(\lambda, \varepsilon)$. We may make statements concerning this description of splitting if $H(\lambda)$ commutes with the unitary operators of a group representation for all λ.

If U is such a representation operator, then the commutivity of $H(\lambda)$ with U (see AIV, §8) means that all $E_\varepsilon(\lambda)$ commute with U, that is, the finite-dimensional subspace $\imath_\varepsilon(\lambda)$ of the Hilbert space $\mathscr{H}$ determined by $E_\varepsilon(\lambda)$ is left invariant by U and is therefore a representation space for the group. Here we shall only consider compact or finite groups. We will show that the reduction of the representation into irreducible parts in $\imath_\varepsilon(\lambda)$ has implications for the possibilities for the splitting of the eigenvalues ε in the neighborhood of λ.

As we assumed above, there exists a neighborhood $\mathscr{V}$ of λ for which $E_\varepsilon(\lambda', \lambda)$ has the same dimension as $E_\varepsilon(\lambda)$ and $E_\varepsilon(\lambda', \lambda) \to E(\lambda)$ in the σ-topology if $\lambda' \to \lambda$. We will now show that $E_\varepsilon(\lambda', \lambda)$ converges to $E_\varepsilon(\lambda)$ even in the trace-norm topology.

Let the dimension of $E_\varepsilon(\lambda)$ be m and let φ_ν ($\nu = 1, \ldots, m$) be a complete orthonormal basis for the corresponding subspace $\imath_\varepsilon(\lambda)$. Because of the weak continuity we obtain $(\varphi_\nu, E_\varepsilon(\lambda', \lambda)\varphi_\nu) = \|E_\varepsilon(\lambda', \lambda)\varphi_\nu\| \to \langle\varphi_\nu, E_\varepsilon(\lambda)\varphi_\nu\rangle = 1$ and therefore $E_\varepsilon(\lambda', \lambda)\varphi_\nu \to \varphi_\nu$. Thus it follows that there exists a neighborhood of λ such that we may choose a complete system of orthonormal vectors $\psi_\nu(\lambda')$ ($\nu = 1, \ldots, m$; $E_\varepsilon(\lambda', \lambda)$ has the same dimension as $E_\varepsilon(\lambda)$) for $E_\varepsilon(\lambda', \lambda)\mathscr{H}$ such that $\|\psi_\nu(\lambda') - \varphi_\nu\| < \delta$ for all ν. From this result we again find that in the trace norm $\|E_\varepsilon(\lambda', \lambda) - E_\varepsilon(\lambda)\|_s \to 0$ for $\lambda' \to \lambda$.

Since, for each unitary operator

$$\|\mathrm{tr}(E_\varepsilon(\lambda', \lambda)U) - \mathrm{tr}(E_\varepsilon(\lambda)U)\| \leq \|E_\varepsilon(\lambda', \lambda) - E_\varepsilon(\lambda)\|_s$$

the character $\chi_{\lambda'}(g)$ of the representation of a group into the subspace $\imath_\varepsilon(\lambda', \lambda)$ corresponding to $E_\varepsilon(\lambda', \lambda)$ is continuously mapped into the character $\chi_\lambda(g)$ of the representation in $\imath_\varepsilon(\lambda)$, from which we obtain the following estimate (which is independent of the group element g)

$$|\chi_{\lambda'}(g) - \chi_\lambda(g)| \leq \|E_\varepsilon(\lambda', \lambda) - E_\varepsilon(\lambda)\|_s.$$

Since the number $n_\mu(\lambda')$ with which the μth irreducible representation occurring in $\imath_\varepsilon(\lambda', \lambda)$ can be calculated either by means of an integral of $\chi_{\lambda'}(g)$ over the compact group space (finite volume!) or by means of a finite sum over the group elements (AV, §§8 and 10.5) $n_\mu(\lambda')$ is continuously mapped into $n_\mu(\lambda)$, that is, $n_\mu(\lambda')$ is constant in the neighborhood of λ.

The splitting of the eigenvalue ε into eigenvalues ε' in the neighborhood of λ can only occur in such a manner that the sum of the multiplicities of the eigenvalues ε' remains constant. Therefore ε can be split into as many ε' as there exist irreducible representations (counted according to their multiplicities) in the eigenspace corresponding to ε.

Of particular interest is the case in which $H(\lambda_0)$, for a particular value λ_0 of λ, commutes with the representation operators of a group $\mathscr{G}$ but, for $\lambda \neq \lambda_0$ only commutes with the representation operators of a subgroup $\tilde{\mathscr{G}}$ of $\mathscr{G}$. The irreducible representation of $\mathscr{G}$ in $\imath_\varepsilon(\lambda_0)$ splits into several irreducible representations of $\tilde{\mathscr{G}}$. This splitting of an irreducible representation of $\mathscr{G}$ into several irreducible representations of the subgroup $\tilde{\mathscr{G}}$ determines by pure group-theoretical methods the splitting of an eigenvalue ε in the neighborhood of λ_0.

The previous result is one of the most important tools for an understanding which permits us to make at least qualitative statements concerning the spectrum of a Hamiltonian operator when a quantitative computation appears to be impossible. Here the consequences of representations in the eigenspaces becomes more important because the latter determines the structure of the operators for the effect of the emission of a photon as described by (1.16) and (1.17). We will consider this topic in more detail in Chapter XIII.

We shall now extend our discussion of the splitting possibilities for an eigenvalue by considering the representation of a group in the case of the perturbation calculation described in §6.

The decisive step in the perturbation computation is the secular equation $\tilde{E}_\nu H \tilde{E}_\nu$ in (6.1) or $E_\mu^{(1)}(H^{(1)} + (i/2)[S, H_2])E_\mu^{(1)}$ in (6.6) or (6.7a).

If H_0 and H_1 commute with the representation operators of the group, then the following operators H, $\tilde{E}_\nu$, $E_\mu^{(1)}$, S, H_2 also commute with the representation operators. The secular equation than has the following form: In a finite-dimensional subspace $\imath$ of a Hilbert space (the subspace $\imath$ corresponding to $\tilde{E}_\nu$ or to $E_\mu^{(1)}$) we are given a unitary representation of a group and an operator H'' $(=\tilde{E}_\nu H \tilde{E}_\nu$ or $E_\mu^{(1)}(H^{(1)} + (i/2)[S, H_2])E_\mu^{(1)})$ which commutes with the representation operator. We seek the eigenvalues for the operator H''.

In AIV, §14 we have solved the problem of obtaining the most general form of an operator H'' which commutes with a representation. If M is the set of the representation operators for the group then from AIV (14.9) it follows that for H''

$$H'' = \sum_\alpha Q_\alpha(\mathbf{1} \times H^{(\alpha)})Q_\alpha,$$

where the space $\imath$ can, according to AIV (14.7) be written in the form

$$\imath = \sum_\alpha \oplus \mathscr{H}_1^{(\alpha)} \times \mathscr{H}_2^{(\alpha)}$$

and, according to AIV (14.8) the representation operators of the group have the form

$$A = \sum_\alpha Q_\alpha(A^{(\alpha)} \times \mathbf{1})Q_\alpha$$

The secular equation problem for H'' is then reduced to the individual secular equation problems for the individual $H^{(\alpha)}$ in $\mathscr{H}_2^{(\alpha)}$.

The dimension of $\mathscr{H}_2^{(\alpha)}$ is, according to AIV, §14, equal to the multiplicity with which the same irreducible representation (characterized by the index α) occurs in $\imath$. If an irreducible representation occurs only once, then $\mathscr{H}_2^{(\alpha)}$ is one-dimensional and $H^{(\alpha)}$ is the operator, the effect of which is multiplication by an eigenvalue of H'', that is, the solution of the secular equation is trivial. For such an α the $x_{\nu\rho}$ in AIV (14.5) (where ρ takes on only a single index value) are all eigenvectors of H'' corresponding to the same eigenvalue. The $x_{\nu\rho}$ obtained from group-theoretical considerations are often said to be the "correct" linear combinations.

The procedure described here to simplify the secular problem is applicable to innumerable examples. In order to illustrate this procedure we shall give examples; in the following section we will consider the case of the alkali spectrum.

8 The Spectrum of Alkali Atoms

The spectrum of alkali atoms is very similar to the spectrum of hydrogen atoms. In XVII, §3 we will find that the atoms of the alkali metals series of the periodic table (lithium, sodium, potassium, rubidium, cesium) can be treated as one-electron systems, where the electron (the "optical" electron) is considered to move in the Coulomb field of the nucleus and the average field of the remaining electrons. The potential $V(r)$ for the optical electron has, for large values of r, the form $-e^2/r$ corresponding to the fact that the nuclear charge Ze is shielded by the remaining $Z - 1$ electrons. For very small values of r the charge of the nucleus will be completely effective, and we must have $V(r) \sim -Ze^2/r$. We may therefore write the potential as follows:

$$V(r) = -\frac{e^2}{r} + V_1(r), \tag{8.1}$$

where $V_1(r)$ is negative, and nonzero only in the vicinity of the nucleus. Let H_0 denote the Hamiltonian of the hydrogen atom and let $H_1 = V_1$; from the discussion of the perturbation calculation in the previous section we obtain the following result: The eigenspace $\mathscr{T}_n$ corresponding to the energy eigenvalue $\varepsilon_n = -R/n^2$ of H_0 may be decomposed by irreducible representations of the rotation group into

$$\mathscr{T}_n = \sum_{l=0}^{n-1} \oplus \imath^l, \tag{8.2}$$

where $\imath^l$ is the subspace generated by the representation $\mathscr{D}_l$ of the rotation group which corresponds to the angular momentum quantum number l. In general the eigenvalue ε_n is split into n eigenvalues ε_{nl} $(l = 0, 1, \ldots, n - 1)$.

In the first approximation the "correct" linear combinations in the position representation are given by the eigenfunctions in (4.7). The eigenvalues in the first approximation are therefore given by:

$$\varepsilon_{nl} = -\frac{R}{n^2} + \langle \varepsilon_n, l, m | V_1 | \varepsilon_n, l, m \rangle, \tag{8.3}$$

where $\langle \varepsilon_n, l, m | V_1 | \varepsilon_n, l, m \rangle$ cannot depend on m, and is given by

$$\begin{aligned} \langle \varepsilon_n, l, m | V_1 | \varepsilon_n, l, m \rangle &= \left(\frac{2Zme^2}{n}\right)^3 \frac{(n - l - 1)!}{2n[(n + l)!]^3} \\ &\quad \times \int_0^\infty V_1(r)\left(2\frac{r}{r_0}\right)^{2l} \left| L_{n+l}^{2l+1}\left(2\frac{r}{r_0}\right)\right|^2 e^{-2(r/r_0)} r^2 \, dr. \end{aligned} \tag{8.4}$$

Since, for fixed n the distribution

$$\left(2\frac{r}{r_0}\right)^{2l} \left| L_{n+l}^{2l+1}\left(2\frac{r}{r_0}\right)\right|^2 e^{-2(r/r_0)} r^2$$

is, for increasing l, displaced towards larger values of r (see Figure 5) the term ε_{n0} is the lowest, and, for increasing l, the values of ε_{nl} approach $-R/n^2$. This is precisely the behavior of the alkali metal spectra. In Figure 10 the spectrum of the lithium atom is compared to that of the hydrogen atom. The term $n = 1$ does not exist for lithium; we shall explain this fact in XIV, §3. For the other terms $n = 2, 3, \ldots$ we observe the behavior described above.

In Figure 10 we give the conventional term notation: The terms $\varepsilon_{n0}, \varepsilon_{n1}, \varepsilon_{n2}, \varepsilon_{n3}$ are called S, P, D, F, respectively; the remaining terms proceed in alphabetical order.

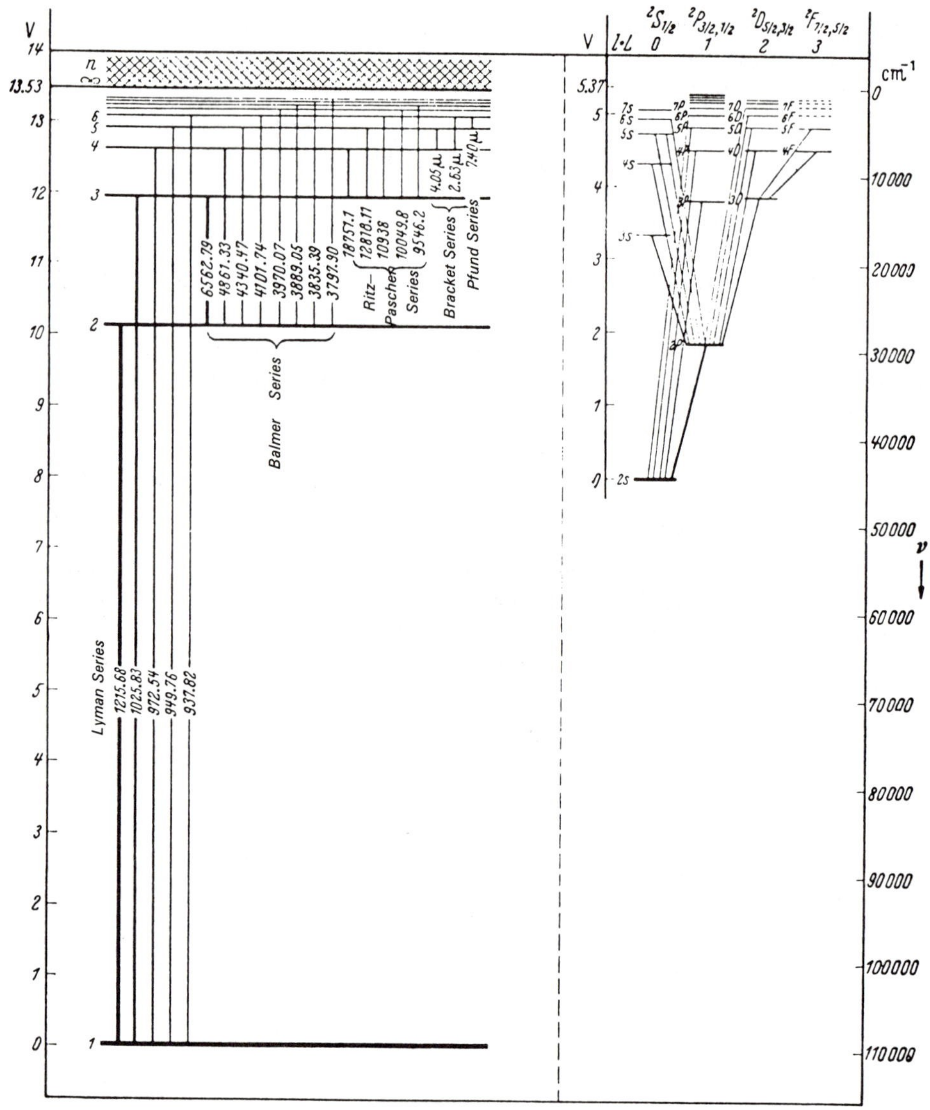

Figure 10 Term scheme of hydrogen (left) and lithium (right).

9 Electron Spin

Precise observations of the spectrum of hydrogen atoms and alkali metal atoms demonstrate that the terms ε_{nl} which we have computed are actually split into additional terms. For the alkali metals this splitting is particularly transparent: The terms for which $l = 0$ remain simple. The terms for which $l \neq 0$ consist of pairs of closely spaced terms.

The Stern–Gerlach experiment, which is described in considerable detail in [6], XI, §7.2, demonstrates that the ground state ε_{10} of hydrogen and the

ε_{20} of the lithium atom are split into two components. This means that we cannot consider the electron to be an elementary object of spin 0. The Stern–Gerlach experiment and the splitting of the terms ε_{nl} for $l \neq 0$ suggests that the electron should be described as a spin $\frac{1}{2}$ particle, as we have already described in VIII, §2.

The Hilbert space $\mathscr{H}$ for an electron can be described (VII, §2) as the product space of a position or momentum Hilbert space $\mathscr{H}_b$ with a two-dimensional spin space $\mathscr{h}_s$: $\mathscr{H} = \mathscr{H}_b \times \mathscr{h}_s$. This is done in the same way in the case of the relative motion of the electron about the atomic nucleus (VIII, §2).

Our previous assumption for the Hamiltonian operator (3.1) can be considered to be an approximation H_0 where

$$H_0 = H_{0b} \times \mathbf{1}, \tag{9.1}$$

where H_{0b} is the operator given by (3.1) which is defined in $\mathscr{H}_b$. H_0 is therefore defined in $\mathscr{H}$ and has the same spectrum as does H_{0b}. The degeneracy must, however, be doubled, since every eigenspace of H_{0b} multiplied by the two-dimensional $\mathscr{h}_s$ yields the corresponding eigenspace of H_0.

The "improved" Hamiltonian operator described in VIII, §5 may be written in the form

$$H = H_0 + H_1. \tag{9.2}$$

10 Addition of Angular Momentum

According to VII, §3 in $\mathscr{H} = \mathscr{H}_b \times \mathscr{h}_s$ the rotation group (as a subgroup of the Galileo group) about the origin (for the relative coordinate system used here, about the nucleus) may be represented by unitary operators of the form $U_D = V_D \times R_D$ (D is an element of the rotation group) where R_D produces the representation $\mathscr{D}_{1/2}$ in $\mathscr{h}_s$. We call the representation in $\mathscr{H}$ the product representation of the two representations in $\mathscr{H}_b$ and $\mathscr{h}_s$. The Hamiltonian operator H given by (9.1), (9.2) will therefore not commute with $V_D \times \mathbf{1}$ as was the case with $H_0 = H_{0b} \times \mathbf{1}$. In the absence of external fields we may, according to VIII, §1 expect that H is spherically symmetric, that is

$$U_D H = H U_D. \tag{10.1}$$

Thus, for the angular momentum operators given in VII (5.3), it follows that

$$J_\nu H = H J_\nu, \quad \text{where} \quad J_\nu = L_\nu \times \mathbf{1} + \mathbf{1} \times S_\nu. \tag{10.2}$$

From $[J_\nu, H] = 0$ it follows that angular momentum is conserved (see X, §1), that is, $\dot{J}_\nu = 0$. Since $L_\nu \times \mathbf{1}$ and $\mathbf{1} \times S_\nu$ do not commute with H there is no exact conservation law for these observables. $L_\nu \times \mathbf{1}$ and $\mathbf{1} \times S_\nu$ remain constant with time only in the approximation $H = H_0$ since

$$[L_\nu \times \mathbf{1}, H_0] = 0; \qquad [\mathbf{1} \times S_\nu, H_0] = 0. \tag{10.3}$$

We note that in VII, §§2 and 7 that we have already called $L_\nu \times \mathbf{1}$ the orbital angular momentum and $\mathbf{1} \times S_\nu$ the spin angular momentum, respectively.

The problem of determining the eigenvalues and eigenfunctions for the angular momentum operator J_v is equivalent to the problem of reducing the product representation given by V_D in $\mathscr{H} = \mathscr{H}_b \times \imath_s$ into irreducible components. We will generalize this problem somewhat by seeking the reduction of $\mathscr{T}_1 \times \mathscr{T}_2$ in the case in which $\mathscr{T}_1$ and $\mathscr{T}_2$ are a pair of arbitrary representation spaces. If $\mathscr{T}_1 = \sum_k \oplus \imath_k$ and $\mathscr{T}_2 = \sum_l \oplus \jmath_l$ where $\imath_k$ and $\jmath_l$ are irreducible then it easily follows that $\mathscr{T}_1 \times \mathscr{T}_2 = \sum_{l,k} \oplus (\imath_k \times \jmath_l)$ where $\imath_k \times \jmath_l$ are invariant subspaces which are not necessarily irreducible. In this way our problem is reduced to the problem of the reduction of $\mathscr{T}_j \times \mathscr{T}_{j'}$ where $\mathscr{T}_j$, $\mathscr{T}_{j'}$ belong to the irreducible representations $\mathscr{D}_j$ and $\mathscr{D}_{j'}$.

It is easy to show that the character of the product representation in $\mathscr{T}_1 \times \mathscr{T}_2$ is equal to the product of the characters of the individual representations in $\mathscr{T}_1$ and $\mathscr{T}_2$. Let $\varphi = 2\alpha$; then, according to VII (3.39) (if we write $\chi(\varphi)$ instead of $\chi(D)$) we obtain

$$\chi_j(\varphi) = \sum_{m=-j}^{j} e^{im\varphi} \tag{10.4}$$

and for the character $\chi(\varphi)$ for the representation $\mathscr{D}_j \times \mathscr{D}_{j'}$ we obtain

$$\begin{aligned}\chi(\varphi) &= \sum_{m=-j}^{j} e^{im\varphi} \sum_{m'=-j'}^{j'} e^{im'\varphi} = \sum_{m,m'} e^{i(m+m')\varphi} \\ &= \sum_{J=|j-j'|}^{j+j'} \sum_{k=-J}^{J} e^{ik\varphi}.\end{aligned} \tag{10.5}$$

Therefore we obtain $\mathscr{D}_j \times \mathscr{D}_{j'} = \mathscr{D}_{j+j'} + \mathscr{D}_{j+j'-1} + \cdots + \mathscr{D}_{|j-j'|}$.

For the case in which the angular momentum of the system described by $\mathscr{T}_j$ is equal to j, and that of the second system described by $\mathscr{T}_{j'}$ equal to j', then the total angular momentum J can assume the values

$$j + j', j + j' - 1, \ldots, |j - j'|.$$

For the case $\mathscr{H} = \mathscr{H}_b \times \imath_s$ we have to reduce the spaces $\mathscr{T}_l \times \imath_s$, that is, we need to add the orbital angular momentum l to the spin angular momentum ($s = \frac{1}{2}$) to obtain a total angular momentum of either $j = l + \frac{1}{2}$ or $j = l - \frac{1}{2}$ if $l \neq 0$; for $l = 0$ only $j = \frac{1}{2}$ is possible. Here we say that the spin can either be added to or subtracted from the orbital angular momentum.

We shall now consider the reduction of the vector space $\mathscr{T}_j \times \mathscr{T}_{j'}$, that is, we shall explicitly identify the subspaces and eigenvectors of $\mathscr{T}_j \times \mathscr{T}_{j'}$ which correspond to the irreducible parts. According to VII (3.35) the basis vectors of $\mathscr{T}_j$ and $\mathscr{T}_{j'}$ may be chosen as follows:

$$U_m = \frac{u_+^{j+m} u_-^{j-m}}{\sqrt{(j+m)!\,(j-m)!}} \quad \text{and} \quad V_{m'} = \frac{v_+^{j'+m'} v_-^{j'-m'}}{\sqrt{(j'+m')!\,(j'-m')!}}, \tag{10.6}$$

where v_+, v_- and u_+, u_- correspond to the same representation $\mathscr{D}_{1/2}$. We then expand the expression

$$K_J = (u_+ v_- - u_- v_+)^{j+j'-J} (u_+ x + u_- y)^{j-j'+J} (v_+ x + v_- y)^{j'-j+J} \tag{10.7}$$

in powers of x and y. The coefficients of

$$X_M^J = \frac{x^{J+M}y^{J-M}}{\sqrt{(J+M)!\,(J-M)!}} \tag{10.8}$$

will, as we shall shortly prove, describe the desired linear combinations W_M^J of the U_m and $V_{m'}$ (up to a normalization factor) which transform according to $\mathscr{D}_J$. K_J may be evaluated in the same way for all $J = j + j', j + j' - 1, |j - j'|$.

In order to prove this transformation property of the coefficients of X_M^J in K_J, we note that K_J is invariant under the action of the group SU_2 if the x, y transform contragradiently (that is, according to the inverse and transposed transformation; A'^{-1} is the contragradient transformation corresponding to A) relative to the u_+, u_-. Therefore the coefficients of X_M^J transform contragradiently with respect to the X_M^J. For the invariant $(u_+ x + u_- y)^{2J}$ we obtain

$$\begin{aligned}(u_+ x + u_- y)^{2J} &= \sum_{\nu=0}^{2J} \frac{(2J)!}{\nu!\,(2J-\nu)!} u_+^\nu x^\nu u_-^{2J-\nu} y^{2J-\nu} \\ &= (2J)! \sum_{M=-J}^{J} \frac{1}{(J+M)!\,(J-M)!} u_+^{J+M} x^{J+M} u_-^{J-M} y^{J-M} \\ &= (2J)! \sum_{M=-J}^{J} \frac{u_+^{J+M} u_-^{J-M}}{\sqrt{(J+M)!\,(J-M)!}} X_M^J. \end{aligned} \tag{10.9}$$

Therefore we find that the $(u_+^{J+M}u_-^{J+M})/\sqrt{(J+M)!\,(J-M)!}$ transform contragradiently with respect to the X_M^J. Therefore the coefficients of X_M^J in K_J transform like

$$\frac{u_+^{J+M}u_-^{J-M}}{\sqrt{(J+M)!\,(J-M)!}}, \tag{10.10}$$

that is, according to $\mathscr{D}_J$.

In order to determine the coefficients of X_M^J in K_J we need to compute

$$(u_+ v_- - u_- v_+)^{j+j'-J} = \sum_{\nu=0}^{j+j'-J} (-1)^\nu \binom{j+j'-J}{\nu} (u_+ v_-)^{j+j'-J-\nu} (u_- v_+)^\nu,$$

$$(u_+ x + u_- y)^{j-j'+J} = \sum_{\mu=0}^{j-j'+J} \binom{j-j'+J}{\mu} (u_+ x)^{j-j'+J-\mu} (u_- y)^\mu,$$

$$(v_+ x + v_- y)^{j'-j+J} = \sum_{\rho=0}^{j'-j+J} \binom{j'-j+J}{\rho} (v_+ x)^\rho (v_- y)^{j'-j+J-\rho}.$$

Instead of μ and ρ we shall use the summation variables $m = j - \nu - \mu$ and $m' = \rho + \nu - j'$. We then obtain

$$K_J = \sum_{mm'} \sum_{\nu} (-1)^\nu \binom{j + j' - J}{\nu} \binom{j - j' + J}{j - \nu - m} \binom{j' - j + J}{j' - \nu + m'}$$
$$\cdot u_+^{j+m} u_-^{j-m} v_+^{j'+m'} v_-^{j'-m'} x^{J+(m+m')} y^{J-(m+m')} \tag{10.11}$$
$$= (j + j' - J)!\,(j - j' + J)!\,(j' - j + J)! \sum_{mm'} c^J_{m,m'} u_m v_{m'} X^J_{m+m'}.$$

With $m + m' = M$ we obtain

$$c^J_{m,m'} = \sum_\nu (-1)^\nu$$
$$\times \frac{\sqrt{(j+m)!(j-m)!(j'+m')!(j'-m')!(J+M)!(J-M)!}}{\nu!(j+j'-J-\nu)!(j-\nu-m)!(J-j'+\nu+m)!(j'-\nu+m')!(J-j+\nu-m')!} \tag{10.12}$$

Thus, using the normalization factor ρ_J we obtain

$$W^J_M = \rho_J \sum_{\substack{mm' \\ (m+m'=M)}} c^J_{m,m'} u_m v_{m'}. \tag{10.13}$$

ρ_J is computed from the normalization condition

$$\|W^J_M\|^2 = 1 = |\rho_J|^2 \sum_{mm'} |c^J_{mm'}|^2.$$

The sum on the right-hand side must be independent of M because $|\rho_J|^2$ cannot depend on M. We will now consider the case in which $M = J$. Then for $m + m' = J$ (noting that in $(j - m - \nu)!$ $\nu \leq j - m$ must be satisfied; in $(J - j - m' + \nu)! = (-j + m + \nu)!$ we must have $j \leq m + \nu$) the only remaining term in the sum is given by

$$c^J_{m,m} = (-1)^{j-m} \frac{\sqrt{(2J)!}}{(J - j' + j)!\,(J + j' - j)!} \sqrt{\frac{(j + m)!\,(j' + m')!}{(j - m)!\,(j' - m')!}}$$
$$= (-1)^{j-m} \sqrt{\frac{(2J)!}{(J - j' + j)!\,(J + j' - j)!} \binom{j + m}{j - m'} \binom{j' + m'}{j - m}}.$$

Therefore we obtain

$$\sum_{\substack{m,m' \\ (m+m'=J)}} |c^J_{m,m'}|^2$$
$$= \frac{(2J)!\,(-1)^{j'+j-J}}{(J - j' + j)!\,(J + j' - j)!} \sum_{\substack{m,m' \\ (m+m'=J)}} \binom{j' - j - J - 1}{j' - m'} \binom{j - j' - J - 1}{j - m},$$

where we have used the relation

$$\binom{u}{v} = (-1)^v \binom{v - u - 1}{v}.$$

From

$$(1+z)^r = \sum_{\nu} \binom{r}{\nu} z^{\nu} \quad \text{and} \quad (1+z)^s = \sum_{\mu} \binom{s}{\mu} z^{\mu}$$

it follows that

$$(1+z)^{r+s} = \sum_{\tau} \binom{r+s}{\tau} z^{\tau} = (1+z)^r (1+z)^s = \sum_{\nu,\mu} \binom{r}{\nu}\binom{s}{\mu} z^{\nu+\mu}.$$

Therefore we obtain

$$\binom{r+s}{\tau} = \sum_{\substack{v,\mu \\ (\nu+\mu=\tau)}} \binom{r}{\nu}\binom{s}{\mu}.$$

Using this fact we obtain

$$\sum_{\substack{m,m' \\ (m+m'=J)}} |c^J_{m,m'}|^2 = \frac{(2J)!\,(-1)^{j'+j-J}}{(J-j'+j)!\,(J+j'-j)!} \binom{-2J-2}{j+j'-J}.$$

From

$$\binom{-2J-2}{j+j'-J} = (-1)^{j+j-J} \binom{j+j'+J+1}{j+j'-J}$$

it follows that

$$\sum_{\substack{m,m' \\ (m+m'=J)}} |c^J_{m,m'}|^2 = \frac{(j+j'+J+1)!}{(J-j'+j)!\,(J+j'-j)!\,(j+j'-J)!\,(2J+1)}$$

and

$$\rho_J = \sqrt{\frac{(2J+1)(J+j'-j)!\,(J-j'+j)!\,(j'+j-J)!}{(j+j'+J+1)!}}.$$

Thus we finally obtain

$$W^J_M = \sum_{\substack{m,m' \\ (m+m'=M)}} d^J_{m,m'} u_m v_{m'} \tag{10.14}$$

with (for the case in which $m + m' = M$)

$$\begin{aligned} d^J_{m,m'} = {} & \frac{1}{\sqrt{(j+j'+J+1)!}} \\ & \cdot \sqrt{(2J+1)(J+j'-j)!\,(J+j'+j)!\,(j+j'-J)!\,(j+m)! \over (j-m)!\,(j'+m')!\,(j'-m')!\,(J+M)!\,(J-M)!} \\ & \cdot \sum_{\nu}^{j+j'-J} (-1)^{\nu} \frac{1}{\nu!\,(j+j'-J-\nu)!\,(j-\nu-m)!\,(J-j'+\nu+m)! \atop (j'-\nu+m)!\,(J-j'+\nu-m)!} \end{aligned} \tag{10.15}$$

The coefficients $d^J_{m,m'}$ are called the *Clebsch–Gordon coefficients*. The $d^J_{m,m'}$ determine a unitary transformation for the transition between the basis consisting of the products $u_m v_{m'}$ to the W^J_M as follows:

$$W^J_M = \sum_{m,m'} u_m v_{m'} \alpha_{m,m;MJ}, \tag{10.16}$$

where the unitary matrix is given by

$$\alpha_{mm';MJ} = \delta_{M,m+m'} d^J_{m,m'}. \tag{10.17}$$

Since the $\alpha_{mm';MJ}$ are real, it easily follows that the inverse is given by

$$u_m v_{m'} = \sum_{M,J} \alpha_{m,m';MJ} W^J_M \tag{10.18}$$

and we therefore obtain

$$u_m v_{m'} = \sum_{J=m+m'}^{j+j'} d^J_{m,m'} W^J_{m+m'}. \tag{10.19}$$

For $j = 1$ and $j' = \frac{1}{2}$ we obtain the special case

$$\begin{aligned} d^{l+1/2}_{m,1/2} &= \sqrt{\frac{l+m+1}{2l+1}}; \qquad & d^{l+1/2}_{m,-1/2} &= \sqrt{\frac{l-m+1}{2l+1}}; \\ d^{l-1/2}_{m,1/2} &= -\sqrt{\frac{l-m}{2l+1}}; \qquad & d^{l-1/2}_{m,-1/2} &= \sqrt{\frac{l+m}{2l+1}}. \end{aligned} \tag{10.20}$$

If $\mathcal{T}_j, \mathcal{T}_{j'}$ are subspaces of the Hilbert spaces $\mathcal{H}_1, \mathcal{H}_2$ corresponding to physical systems 1 and 2 with bases u_m, $v_{m'}$, respectively, then the W^J_M are not only eigenfunctions of the square of the total angular momentum with eigenvalue $J(J+1)$ and are eigenfunctions of the third component of the total angular momentum with eigenvalue M but are also eigenfunctions of the square of the angular momentum of both system 1 and 2 with eigenvalues $j(j+1)$ and $j'(j'+1)$, respectively. They are not, however, eigenfunctions of the third component of angular momentum for systems 1 and 2. This results from the fact that the square of the angular momentum of systems 1 and 2 transform like a scalar quantity under rotations in $\mathcal{H}_1 \times \mathcal{H}_2$ and therefore commute with the operators of the total angular momentum; this is not the case for the individual components of angular momentum for systems 1 and 2.

We shall denote the total angular momentum quantum number for one electron by j. Thus we find that

$$\begin{aligned} &f(r)\left(\sqrt{\frac{l+m+1}{2l+1}}\, Y^l_m u_+ + \sqrt{\frac{l-m}{2l+1}}\, Y^l_{m+1} u_-\right) \\ &= f(r)\left(\sqrt{\frac{j+M}{2j}}\, Y^{j-1/2}_{M-1/2} u_+ + \sqrt{\frac{j-M}{2j}}\, Y^{j-1/2}_{M-1/2} u_-\right) \end{aligned} \tag{10.21}$$

is a vector in $\mathscr{H}_b \times \imath_s$ which is an eigenvector of both $\vec{J}^2$ and J_3 with eigenvalues $j(j+1)$ where $j = l + \frac{1}{2}$ and $M = m + \frac{1}{2}$, respectively, and of $\vec{L}^2$ with eigenvalue $l(l+1)$ and of $\vec{S}^2$ with eigenvalue $\frac{1}{2}(\frac{1}{2}+1)$. The vector

$$g(r)\left(-\sqrt{\frac{l-m+1}{2l+3}}\,Y_m^{l+1}u_+ + \sqrt{\frac{l+m+2}{2l+3}}\,Y_{m+1}^{l+1}u_-\right) \tag{10.22}$$

also corresponds to the same eigenvalues $j(j+1)$ and M of the operators $\vec{J}^2$ and J_3. All vectors of eigenvalue $j(j+1)$ of $\vec{J}^2$ and M of J_3 are obtained by taking the sum of (10.21) and (10.22) with arbitrary $f(r)$ and $g(r)$.

11 Fine Structure of Hydrogen and Alkali Metals

The fine structure of the spectrum of alkali metal atoms is easier to visualize as that of hydrogen. The Hamiltonian operator has the form

$$H = H_{0b} \times \mathbf{1} + H_1.$$

We shall write $H(\lambda) = H_{0b} \times \mathbf{1} + \lambda H_1$ and if we let λ increase from 0 to 1, we will obtain the following behavior for the eigenvalues:

For $\lambda = 0$, H commutes not only with U_D but also with $V_D \times \mathbf{1}$ (and $\mathbf{1} \times R_D$). For the eigenvalue ε_{nl} the corresponding eigenspace of $\mathscr{H}_b \times \imath_s$ is given by $\imath_{nl} \times \imath_s$ where $\imath_{nl}$ is spanned by the eigenvectors $\varphi_{nl}(r)Y_m^l(\theta, \varphi)$, $m = -l, -l+1, \ldots, l$. The representation corresponding to rotations described by $V_D \times \mathbf{1}$ in $\imath_{nl} \times \imath_s$ is given by $\mathscr{D}_l$; the representation described by U_D in $\imath_{nl} \times \imath_s$ is $\mathscr{D}_{l+1/2} + \mathscr{D}_{l-1/2}$ for $l \neq 0$ and $\mathscr{D}_{1/2}$ for $l = 0$. Since, for $0 \leq \lambda \leq 1$ H commutes with U_D the perturbation H_1 splits each term ε_{nl} into two terms (for $l \neq 0$) corresponding to the total angular momentum quantum numbers $j = l + \frac{1}{2}$ and $j = l - \frac{1}{2}$. The terms ε_{n0} cannot be split. According to §7 in the first approximation oi perturbation calculation the correct linear combination for $j = l + \frac{1}{2}$ is obtained from (10.21) by making the substitution $f(r) = \varphi_{nl}(r)$. We obtain

$$\varphi_{nl}(r)\left(\sqrt{\frac{l+m+1}{2l+1}}\,Y_m^l u_+ + \sqrt{\frac{l-m}{2l+1}}\,Y_{m-1}^l u_-\right). \tag{11.1}$$

For $j = l - \frac{1}{2}$ we obtain

$$\varphi_{nl}(r)\left(-\sqrt{\frac{l-m}{2l+1}}\,Y_m^l u_+ + \sqrt{\frac{l+m+1}{2l+1}}\,Y_m^l u_-\right). \tag{11.2}$$

In order to estimate the magnitude of the splitting we must know what H_1 is. Each vector ϕ from $\mathscr{H}_b \times \imath_s$ may be written in the form $\phi = \varphi_+ u_+ + \varphi_- u_-$ where φ_+ and φ_- are vectors in $\mathscr{H}_b$. We therefore obtain $H_1 \varphi u_+ = \psi_{++}u_+ + \psi_{-+}u_-$ and $H_1 \varphi u_- = \psi_{+-}u_+ + \psi_{--}u_-$. The maps $\varphi \to \psi_{++}$, $\varphi \to \psi_{-+}$, $\varphi \to \psi_{+-}$, $\varphi \to \psi_{--}$ represent linear operators in $\mathscr{H}_b$ which we shall denote by $K_{++}\varphi = \psi_{++}$, $K_{-+}\varphi = \psi_{-+}$, etc. Thus we find that $H_1 = \sum_{\nu,\mu} K_{\nu\mu} \times E_{\nu\mu}$ (ν, μ are to be summed over + and − as indices) where $E_{\nu\mu}u_\rho = \delta_{\rho\mu}u_\nu$.

The $E_{\nu\mu}$ may therefore be represented by the four linearly independent operators (matrices) 1, S_1, S_2, S_3. Therefore we obtain

$$H_1 = A \times 1 + \sum_{k=1}^{3} B_k \times S_k. \tag{11.3}$$

Since, under rotations the S_k transform like an (axial) vector, and H_1 must be a scalar, it follows that A must be a scalar and that the B_k must be components of a vector. Since H_1 and the S_k are self-adjoint, A and the B_k, as operators in $\mathscr{H}_b$ must also be self-adjoint.

According to VIII (5.8) H_1 has the following form:

$$H_1 = -\frac{1}{8m^3}\vec{P}^4 + \frac{e}{2m^2}\vec{S}\cdot(\vec{E}\times\vec{P}) + \frac{e^2}{8m^2}\operatorname{div}\vec{E}. \tag{11.4}$$

Since $e\vec{E} = \operatorname{grad} V(r)$ where e is the (positive) elementary charge and $V(r)$ is the potential energy of the electron it follows that (for atoms with no external field):

$$\begin{aligned} A &= -\frac{1}{8m^3}\vec{P}^4 + \frac{1}{8m^2}\Delta V(r), \\ B_k &= \frac{1}{2m^2}(\operatorname{grad} V(r)\times\vec{P})_k. \end{aligned} \tag{11.5}$$

Since $\operatorname{grad} V(r) = V'(r)(\vec{r}/r)$ and $\vec{r}\times\vec{P} = \vec{L}$, it follows that

$$B_k = \frac{1}{2m^2}\frac{V'(r)}{r}L_k. \tag{11.6}$$

According to previous results it follows that $\mathscr{r}_{nl} \times \mathscr{r}_s = \mathscr{o}_{l+1/2} \oplus \mathscr{o}_{l-1/2}$ where $\mathscr{o}_{l+1/2}$ and $\mathscr{o}_{l-1/2}$ are spanned by the vectors (11.1) and (11.2), respectively. For the vectors χ in $\mathscr{o}_j$ ($j = l + \frac{1}{2}$ and $j = l - \frac{1}{2}$) from $J^2 = (\vec{L}+\vec{S})^2 = \vec{L}^2 + \vec{S}^2 + 2\vec{L}\cdot\vec{S}$, $\vec{J}^2\chi = j(j+1)\chi$, $\vec{L}^2\chi = l(l+1)\chi$ and $\vec{S}^2\chi = s(s+1)\chi$ (with $s = \frac{1}{2}$ as the spin angular momentum) we obtain

$$2\vec{L}\cdot\vec{S}\chi = [j(j+1) - l(l+1) - s(s+1)]\chi. \tag{11.7}$$

From (11.3) to (11.6), for a first approximation, we therefore obtain:

$$\varepsilon_{nlj} = \varepsilon_{nl} + C_{nl} + \tfrac{1}{2}D_{nl}[j(j+1) - l(l+1) - s(s+1)], \tag{11.8}$$

where C_{nl} is obtained from A and the D_{nl} are obtained from B_k. D_{nl} is given by the expression:

$$D_{nl} = \frac{1}{2m^2}\int_0^\infty V'(r)r\,|\varphi_{nl}(r)|^2\,dr. \tag{11.9}$$

The separation of the two terms $j = l + \frac{1}{2}$, $j = l - \frac{1}{2}$ is therefore given by

$$\varepsilon_{nl\,l+1/2} - \varepsilon_{nl\,l-1/2} = D_{nl}(l + \tfrac{1}{2}). \tag{11.10}$$

Since $V'(r) > 0$, the energy eigenvalues increase with increasing j. This is intuitively evident since the term $\sum_k B_k \times S_k$ has the form $(1/2m^2)(V'(r)/r)\vec{L}\cdot\vec{S}$.

According to the perturbation computation procedure for ε_{nl} we replace $V'(r)/r$ by its mean value (expectation value)

$$\frac{\overline{V'(r)}}{r} = \int_0^\infty V'(r) r\, |\varphi_{nl}(r)|^2\, dr$$

and we obtain

$$\frac{1}{2m^2} \frac{\overline{V'(r)}}{r} \vec{L} \cdot \vec{S} = D_{nl} \vec{L} \cdot \vec{S}. \tag{11.11}$$

The angular momenta $\vec{L}$ and $\vec{S}$ are therefore coupled with the energy (11.11) where the latter is a maximum when $\vec{L}$ and $\vec{S}$ have the same direction.

The splitting of an alkali metal spectrum by the spin is exhibited by Figure 11. The terms for which $l = 0$ are not split. According to convention, the terms are denoted by S_j, P_j, ..., etc., where S, P correspond to the orbital angular momentum l.

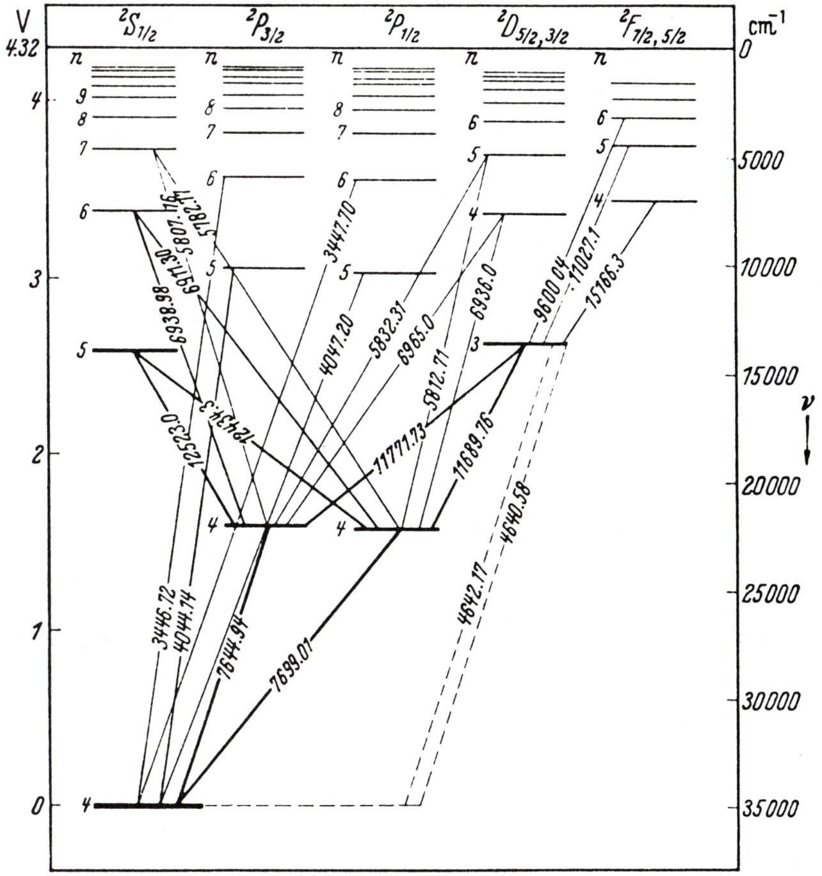

Figure 11 Term scheme of potassium.

For the hydrogen atoms (and corresponding ions) the perturbation energy is given by $H = A \times 1 + C\vec{L} \cdot \vec{S}$ where

$$A = -\frac{1}{8m^3}\vec{P}^4 + \frac{Ze^2}{8m^2}4\pi\delta(Q),$$
$$C = \frac{Ze^2}{2m^2r^3}. \tag{11.12}$$

The eigenfunctions corresponding to the eigenvalue $\varepsilon_n = -me^4Z^2/2n^2$ are given by

$$\chi_{nljM} = \varphi_{nl}(r)W_M^{l+1/2} \quad \text{and} \quad \chi_{nljM} = \varphi_{nl}(r)W_M^{l-1/2}, \tag{11.13}$$

where $j = l + \frac{1}{2}$ or $l - \frac{1}{2}$ and the W_M^j are given by the expressions in the brackets in (11.1) and (11.2). We obtain

$$\begin{aligned}\langle nljM|H_1|n'l'j'M'\rangle = &-\frac{1}{8m^3}\langle nljM|\vec{P}^4|nl'j'M'\rangle \\ &+\frac{Ze^2}{8m^2}\varphi_{n0}^2(0)\delta_{l0}\delta_{ll'}\delta_{j1/2}\delta_{MM'}\delta_{jj'} \\ &+\frac{Ze^2}{2m^2c^2}\left\langle nljM\left|\frac{1}{r^3}\right|nl'j'M'\right\rangle \\ &\tfrac{1}{2}[j(j+1) - l(l+1) - s(s+1)],\end{aligned}$$

where the last term contributes only for the case $l \neq 0$. Since

$$\left\langle nljM\left|\frac{1}{2m}\vec{P}^2 - \frac{Ze^2}{r}\right|n'l'j'M'\right\rangle = \varepsilon_n\delta_{nn'}\delta_{ll'}\delta_{jj'}\delta_{MM'}$$

we therefore obtain

$$\langle nljM|\vec{P}^2|nl'j'M'\rangle = 2m\varepsilon_n\delta_{nn'}\delta_{ll'}\delta_{jj'}\delta_{MM'} + 2mZe^2\left\langle nljM\left|\frac{1}{r}\right|n'l'j'M'\right\rangle.$$

Using the following notation

$$\overline{r^\nu}^{nl} = \int_0^\infty r^\nu|\varphi_{nl}(r)|^2 r^2\,dr$$

from

$$\langle nljM|\vec{P}^4|nl'j'M'\rangle = \sum_{n''l''j''M''}\langle nljM|\vec{P}^2|n''l''j''M''\rangle\langle n''l''j''M''|\vec{P}^2|nl'j'M'\rangle$$

we finally obtain (since $s = \frac{1}{2}$)

$$\begin{aligned}&\langle nljM|H_1|nl'j'M'\rangle \\ &\quad= \delta_{ll'}\delta_{jj'}\delta_{MM'}\left[-\frac{1}{2m}\left(\varepsilon_n^2 + 2Ze^2\varepsilon_n\overline{\frac{1}{r}}^{nl} + Z^2e^4\overline{\frac{1}{r^2}}^{nl}\right)\right. \\ &\qquad\left.+\frac{Ze^2}{8m^2}\varphi_{n0}^2(0)\delta_{l0}\delta_{j1/2} + \frac{Ze^2}{4m^2}\overline{\frac{1}{r^3}}^{nl}\left(j(j+1) - l(l+1) - \tfrac{3}{4}\right)\right].\end{aligned}$$

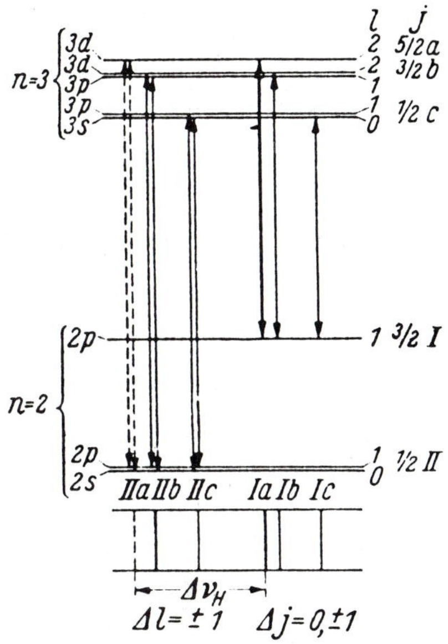

Figure 12 Fine structure of the line $n = 3 \to n = 2$ of the hydrogen atom.

From

$$\overline{\frac{1}{r}}^{nl} = \frac{mZe^2}{n^2}; \qquad \overline{\frac{1}{r^2}}^{nl} = \frac{m^2Z^2e^4}{n^3(l+\frac{1}{2})};$$

$$\overline{\frac{1}{r^3}}^{nl} = \frac{m^3Z^3e^6}{n^4(l+1)(l+\frac{1}{2})l}$$

we obtain

$$\langle nljM|H_1|n'l'j'M'\rangle = \delta_{ll'}\delta_{jj'}\delta_{MM'} \cdot\left[-\frac{1}{2}\frac{mZ^4e^8}{n^3}\left(\frac{1}{l+\frac{1}{2}} - \frac{3}{4n} - \frac{(j(j+1)+l(l+1)+\frac{3}{4})}{l(l+1)(l+\frac{1}{2})}\right) + \frac{1}{8}\frac{Ze^2}{m^2}\varphi_{n0}^2(0)\delta_{l0}\delta_{j1/2}\right].$$

For $j = l \pm \frac{1}{2}$ and for $\varphi_{n0}(0) = (1/\sqrt{2})(2mZe^2/n)^{3/2}$ we obtain, in first approximation, the fine structure formula

$$\varepsilon_{nlj} = -\tfrac{1}{2}me^4Z^2\left[\frac{1}{n^2} + \frac{Z^2e^4}{n^3}\left(\frac{1}{j+\frac{1}{2}} - \frac{3}{4n}\right)\right]. \tag{11.14}$$

In Figure 12 we exhibit the theoretical values for the line $n = 3 \to n = 2$ of the hydrogen atom.

CHAPTER XII

Spectrum of Two-Electron Systems

1 The Hilbert Space and the Hamiltonian Operator for the Internal Motion of Atoms with n Electrons

According to VIII, §4 the Hilbert space for an atom having n electrons has the form $\mathscr{H}_b \times \{\mathscr{H}^n\}_-$ where $\mathscr{H}_b$ is the Hilbert space of the atomic nucleus (considered as an elementary system) and $\mathscr{H} = \mathscr{H}_b \times \imath_s$ is the Hilbert space for a single electron.

In direct analogy with VIII, §2 we will now introduce center of mass and relative coordinates. Let $\vec{P}$ and the $\vec{p}^{(k)}$ denote the momentum operator of the nucleus and the kth electron, respectively. Similarly, let $\vec{Q}$ and $\vec{q}^{(k)}$ denote the position vector of the nucleus and the kth electron. Let M and m denote the mass of the nucleus and the electron, respectively. We define

$$\vec{Q}_s = \frac{M\vec{Q} + m\sum_k \vec{q}^{(k)}}{M + nm} \approx \vec{Q}, \qquad \vec{q}_r^{(k)} = \vec{q}^{(k)} - \vec{Q}.$$

Similarly, for the momentum $\vec{P}_s$ and $\vec{p}_r^{(k)}$ we introduce

$$\vec{P} = \frac{M}{M + nm}\vec{P}_s - \sum_k \vec{p}_r^{(k)}, \qquad \vec{p}^{(k)} = \vec{p}_r^{(k)} + \frac{1}{M + m}\vec{P}_s.$$

Then the $\vec{P}_s$, $\vec{Q}_s$ and $\vec{p}_r^{(k)}$, $\vec{q}_r^{(k)}$ satisfy the same commutation relations as the $\vec{P}$, $\vec{Q}$ and the $\vec{p}^{(k)}$, $\vec{q}^{(k)}$. Furthermore, we find that

$$\frac{1}{2M}\vec{P}^2 + \frac{1}{2m}\sum_k \vec{p}^{(k)2} = \frac{1}{2(M + nm)}\vec{P}_s^2 + \frac{1}{2}\left(\frac{1}{M} + \frac{1}{m}\right)\sum_k \vec{p}_r^{(k)2}. \tag{1.1}$$

Since M is large compared to m, we can set

$$\frac{1}{M} + \frac{1}{m} \sim \frac{1}{m}. \tag{1.2}$$

Thus, from (1.1) it follows that the Hilbert space of an atom with n electrons is of the form $\mathscr{H}_S \times \mathscr{H}_i$ where $\mathscr{H}_S$ describes the center of mass motion and $\mathscr{H}_i = \{\mathscr{H}_r^n\}_-$ where $\mathscr{H}_r = \mathscr{H}_{br} \times \hbar_s$. Therefore $\mathscr{H}_i$ has the same structure as the Hilbert space for n electrons, except that $\mathscr{H}_b$ is replaced by $\mathscr{H}_{br}$, that is, by the orbit space in relative coordinates. From (1.2) it follows that we obtain a good approximation if we do not modify the mass of the electron in the Hamiltonian H_i for the inner motion of the electrons.

We may therefore treat an atom with n electrons in the same way as an n electron system, at least with respect to the inner motion.

We obtain the Hamiltonian operator for the inner motion from VIII (5.8); here Z denotes the charge number of the nucleus, r_i denotes the distance of the ith electron from the nucleus and r_{ik} denotes the distance from the ith to kth electron. The Hamiltonian operator is given by:

$$H = \frac{1}{2m} \sum_{k=1}^{n} \vec{p}^{(k)2} - \sum_{k=1}^{n} \frac{Ze^2}{r_k} + \frac{1}{2} \sum_{\substack{i,k \\ (i \neq k)}} \frac{e^2}{r_{ik}} + H_1, \tag{1.3}$$

where H_1 is a correction term obtained from spin and relativistic effects which we shall examine in more detail in XIV, §2. Instead of (1.3) we shall consider a more general operator

$$H = \frac{1}{2m} \sum_{k=1}^{n} \vec{p}^{(k)2} + \sum_{k=1}^{n} V(r_k) + \frac{1}{2} \sum_{\substack{i,k \\ (i \neq k)}} \frac{e^2}{r_{ik}} + H_1 \tag{1.4}$$

in order to leave the possibility open to describe the alkali metal spectra by using a "mean" field (see XIV, §§3 and 9).

If we identify $\mathscr{H}_i$ from VIII, §1 with the Hilbert space $\mathscr{H}$ of n electrons and we identify H_i from VIII, §1 with H in (1.3) or (1.4) then we may make use of the general results of VIII, §1. Then we have to identify the representation operators $R(A)$ from VIII, §1 with the representation operators V_D in $\mathscr{H}$ for rotations about the nucleus.

According to VIII (1.7) $\mathscr{H}$ may be decomposed as follows:

$$\mathscr{H} = \sum_j \oplus \mathscr{H}^j, \tag{1.5}$$

where $\mathscr{H}^j$ is the eigenspace of $\vec{J}^2$ (where $\vec{J}$ is the total angular momentum about the nucleus) corresponding to the eigenvalue $j(j+1)$. Each $\mathscr{H}^j$ may again, according to VIII (1.8) be written as follows:

$$\mathscr{H}^j = \hbar^j \times k^j, \tag{1.6}$$

where the representation operators V_D in $\mathscr{H}^j$ may be written in the form $V_D \times \mathbf{1}$ and the V_D in $\hbar^j$ corresponds to the irreducible representation $\mathscr{D}_j$. According to VIII, §1 the $\mathscr{H}^j$ are invariant subspaces of the Hamiltonian operator H; according to (1.6) H has the form $\mathbf{1} \times H^j$. Therefore the spectrum of H will be known if the spectrum of the H^j are known. Chapters XII–XIV are, for the most part, concerned with finding the H^j and examining their structure.

2 The Spectrum of Two-Electron Atoms

In the helium atom two electrons move in the field of a doubly charged nucleus. In order to take into account the corresponding ionized atoms of subsequent atoms in the periodic table we shall represent the nuclear charge by Ze. The Hamiltonian operator is given by:

$$H = \frac{1}{2m}(\vec{P}_1^2 + \vec{P}_2^2) - \frac{Ze^2}{r_1} - \frac{Ze^2}{r_2} + \frac{e^2}{r_{12}} + H_1 = H_0 + H_1. \tag{2.1}$$

Since, from the case of the hydrogen atom and the alkali metal atoms, we already know that the influence of the spin is small, we shall first neglect the effect of H_1 and investigate the eigenvalues of H_0.

The Hilbert space for the two electrons is therefore $\{\mathscr{H}^2\}_-$. Clearly $\mathscr{H}^2 = \mathscr{H}_b^2 \times \imath_s^2$. In order to decompose the representation of the symmetric group $\mathbf{S}_2$ in $\mathscr{H}^2$ into irreducible parts, that is, in order to obtain $\{\mathscr{H}^2\}_-$ we shall proceed in two steps as follows:

First we reduce the representation of $\mathbf{S}_2$ in $\mathscr{H}_b^2$ and in $\imath_s^2$ separately (where the permutation operators $\mathbf{P}$ in $\mathscr{H}_b^2$ and $\imath_s^2$ correspond to that defined in $\mathscr{H}^2$). Clearly $\mathbf{S}_2$ has two representations—the symmetric and the antisymmetric. Thus we obtain $\mathscr{H}_b^2 = \{\mathscr{H}_b\}_+ \oplus \{\mathscr{H}_b^2\}_-$ and $\imath_s^2 = \{\imath_s^2\}_+ \oplus \{\imath_s^2\}$. Thus it follows that:

$$\begin{aligned}\mathscr{H}^2 = \{\mathscr{H}_b^2\}_+ \times \{\imath_s^2\}_+ \oplus \{\mathscr{H}_b^2\}_+ \times \{\imath_s^2\}_- \oplus \{\mathscr{H}_b^2\}_- \\ \times \{\imath_s^2\}_+ \oplus \{\mathscr{H}_b^2\}_- \times \{\imath_s^2\}_-.\end{aligned} \tag{2.2}$$

It is easy to show that

$$\begin{aligned}\{\mathscr{H}^2\}_+ &= \{\mathscr{H}_b^2\}_+ \times \{\imath_s^2\}_+ \oplus \{\mathscr{H}_b^2\}_- \times \{\imath_s^2\}_-, \\ \{\mathscr{H}^2\}_- &= \{\mathscr{H}_b^2\}_+ \times \{\imath_s^2\}_- \oplus \{\mathscr{H}_b^2\}_- \times \{\imath_s^2\}_+.\end{aligned} \tag{2.3}$$

According to §1 the Hilbert space $\{\mathscr{H}^2\}_-$ is to be used for the two electrons. Since H_0 does not act upon the vectors of $\imath_s^2$ and commutes with permutations in $\mathscr{H}_b^2$, H_0 transforms the vectors of $\{\mathscr{H}_b^2\}_+$ and $\{\mathscr{H}_b^2\}_-$ into the subspaces $\{\mathscr{H}_b^2\}_+$ and $\{\mathscr{H}_b^2\}_-$, respectively; therefore the same is true for vectors of $\{\mathscr{H}_b^2\}_+ \times \{\imath_s^2\}_-$ and $\{\mathscr{H}_b^2\}_- \times \{\imath_s^2\}_+$. We may therefore consider the spectrum of H_0 separately in $\{\mathscr{H}_b^2\}_+$ and $\{\mathscr{H}_b^2\}_-$.

A precise computation of the eigenvalues of H_0 is not possible. For this reason we shall investigate H_0 using the method of perturbations as follows:

$$H_0 = H_0(1) + H_0(2) + \lambda W, \tag{2.4}$$

where

$$H_0(i) = \frac{1}{2m}\vec{P}_i^2 - \frac{Ze^2}{r_i} \quad \text{and} \quad W = \frac{e^2}{r_{12}}$$

(where we may think of λ as increasing from 0 to 1; see XI, §7).

The eigenvalues of $H_0(1) + H_0(2)$ in $\mathscr{H}_b^2$ are easy to determine. If $\phi_\nu(1)$ is the eigenvector of $H_0(1)$ in $\mathscr{H}_b$ with eigenvalue ε_ν where ν replaces the three indices n, l, m and ϕ_ν is the hydrogen eigenfunction XI (4.7) corresponding to the eigenvalues $\varepsilon_\nu = \varepsilon_{nlm} = \varepsilon_n = -R/n^2$ then the $\phi_\nu(1)\phi_\mu(2)$ are eigenvectors of $H_0(1) + H_0(2)$ in $\mathscr{H}_b^2$ with eigenvalue $\varepsilon_{\nu\mu} = \varepsilon_\nu + \varepsilon_\mu$; $\phi_\mu(1)\phi_\nu(2)$ correspond to the same eigenvalue. In $\{\mathscr{H}_b^2\}_+$ the vectors

$$\phi_\nu(1)\phi_\nu(2) \quad \text{and} \quad \frac{1}{\sqrt{2}}[\phi_\nu(1)\phi_\mu(2) + \phi_\mu(1)\phi_\nu(2)] \tag{2.5}$$

form a complete system of eigenfunctions; similarly

$$\frac{1}{\sqrt{2}}[\phi_\nu(1)\phi_\mu(2) - \phi_\mu(1)\phi_\nu(2)] \tag{2.6}$$

is also such for $\{\mathscr{H}_b^2\}_-$. The eigenvalues $\varepsilon_\nu + \varepsilon_\mu = \varepsilon_{nl} + \varepsilon_{n'l'}$ are multiply degenerate. The lowest eigenvalue is $2\varepsilon_1$, twice the ground state term for the helium ion. The values $\varepsilon_1 + \varepsilon_{n'}$ increase until the continuous spectrum begins at ε_1 (that is, $\varepsilon_{n'} = 0$). The discrete eigenvalue $\varepsilon_1 + \varepsilon_2 = -2R/4 = -R/2$ is larger than the starting value for the continuous spectrum. Under the perturbation W these discrete eigenvalues $\varepsilon_1 + \varepsilon_n$ ($n \geq 2$) become lost in the continuous spectrum; nevertheless the physical significance of these discrete eigenvalues will become evident in other cases which we shall encounter later (anomalous terms, see XIV, §4). If we are interested only in the structure of the discrete spectrum of H we may then expect that the eigenvalues of H for λ continuously increasing from 0 to 1 arise from the eigenvalues $\varepsilon_1 + \varepsilon_n$ of $H_0(1) + H_0(2)$. [For the question whether the eigenvalues change continuously with λ see XI, §§6 and 7 and [10].] The lowest eigenvalue $2\varepsilon_1$ occurs only in $\{\mathscr{H}_b^2\}_+$ for the eigenvector $\phi_1(1)\phi_1(2)$. The remaining $\varepsilon_1 + \varepsilon_n$ ($n \neq 1$) are found in both $\{\mathscr{H}_b^2\}_+$ and $\{\mathscr{H}_b^2\}_-$ and correspond to the eigenvectors

$$\psi_{nlm} = \frac{1}{\sqrt{2}}[\phi_1(1)\phi_{nlm}(2) + \phi_{nlm}(1)\phi_1(2)] \tag{2.7}$$

and

$$\gamma_{nlm} = \frac{1}{\sqrt{2}}[\phi_1(1)\phi_{nlm}(2) - \phi_{nlm}(1)\phi_1(2)], \tag{2.8}$$

respectively. The ψ_{nlm}, as well as the γ_{nlm}, belong to the representation $\mathscr{D}_l$ of the rotation group if we define U_D by means of $U_D(1) \times U_D(2)$, where $U_D(1)$ and $U_D(2)$ act only on vectors in $\mathscr{H}_b(1)$ and $\mathscr{H}_b(2)$, respectively. The total angular momentum is therefore given by the quantum number l for one of the electrons. For increasing λ we therefore expect a splitting of the eigenvalue $\varepsilon_1 + \varepsilon_n$, the effect of which will be different in $\{\mathscr{H}_b^2\}_+$ and $\{\mathscr{H}_b^2\}_-$. The ψ_{nlm} and the γ_{nlm} must be the correct linear combinations, first because W transforms only the vectors of the subspaces $\{\mathscr{H}_b^2\}_+$ and $\{\mathscr{H}_b^2\}_-$ into themselves and second, W commutes with U_D. Thus $F_n W F_n$ (where F_n is a projection operator in the subspace of ψ_{nlm} or γ_{nlm} for fixed n ($l = 0, 1, \ldots, n-1$)) transforms the vectors ψ_{nlm} and γ_{nlm} into themselves up to a factor (for a proof of this result see XI, §7). Because H_0 commutes with U_D the energy eigenvalues do not depend upon m and we obtain

$$\varepsilon_1 + \varepsilon_n + \varepsilon_{nl}^+ \quad \text{and} \quad \varepsilon_1 + \varepsilon_n + \varepsilon_{nl}^-, \tag{2.9}$$

where $\varepsilon_{nl}^+ = \langle \psi_{nlm}, W\psi_{nlm} \rangle$ and $\varepsilon_{nl}^- = \langle \gamma_{nlm}, W\gamma_{nlm} \rangle$. For $\phi_{nlm} = \varphi_{nl}(r) Y_m^l(\theta, \varphi)$ and for the special case in which $\phi_1 = \varphi_1(r) Y^0$ we therefore obtain

$$\varepsilon_{nl}^+ = C_{nl} + A_{nl} \quad \text{and} \quad \varepsilon_{nl}^- = C_{nl} - A_{nl}, \tag{2.10}$$

where

$$C_{nl} = e^2 \int \frac{1}{r_{12}} |\phi_1(\vec{r}_1)|^2 \, |\phi_{nlm}(\vec{r}_2)|^2 \, d^3\vec{r}_1 \, d^3\vec{r}_2$$

and

$$A_{nl} = e^2 \int \frac{1}{r_{12}} \phi_1(\vec{r}_1)\phi_{nlm}(\vec{r}_2)\overline{\phi_1(\vec{r}_2)}\overline{\phi_{nlm}(\vec{r}_1)} \, d^3\vec{r}_1 \, d^3\vec{r}_2 .$$

Intuitively C_{nl} represents the Coulomb interaction energy of two charge clouds of charge density $\rho_1 = e|\phi_1|^2$ and $\rho_2 = e|\phi_2|^2$. A_{nl} is called the exchange integral. The energy eigenvalues so obtained are quantitatively very poor. Essentially better results can be obtained with the aid of the Ritz variation principle, where the latter is described in the next section.

The qualitative results of the previous discussion are as follows: The term schemes are decomposed into two classes; the class to which a given eigenvector is assigned depends on whether it belongs to either $\{\mathscr{H}_b^2\}_+$ or $\{\mathscr{H}_b^2\}_-$ (see Figure 13). The lowest energy value lies in $\{\mathscr{H}_b^2\}_+$; there is no corresponding equivalent eigenvalue in $\{\mathscr{H}_b^2\}_-$ because we cannot construct an antisymmetric function from two identical one-electron functions. For all other terms there exists two equivalent eigenvalues in $\{\mathscr{H}_b^2\}_+$ and $\{\mathscr{H}_b^2\}_-$ which differ by the exchange integral. Since $A_{nl} > 0$ (for more about the exchange integral see §3) the terms of $\{\mathscr{H}_b^2\}_-$ lie somewhat lower than those in $\{\mathscr{H}_b^2\}_+$.

In summary, we may easily obtain the following overview of the terms by means of the following construction principle: Without the mutual interaction of the electrons we could consider a term to be the sum of two-electron terms. Here it is customary to introduce the following notation: We shall denote a state consisting of an electron with principle quantum number n, angular momentum quantum number $l = 0$ (that is, s), and a second electron with principle quantum number n' and angular momentum quantum number $l = 1$ (that is, p) by $ns\, n'p$. By "switching" the interaction we may therefore

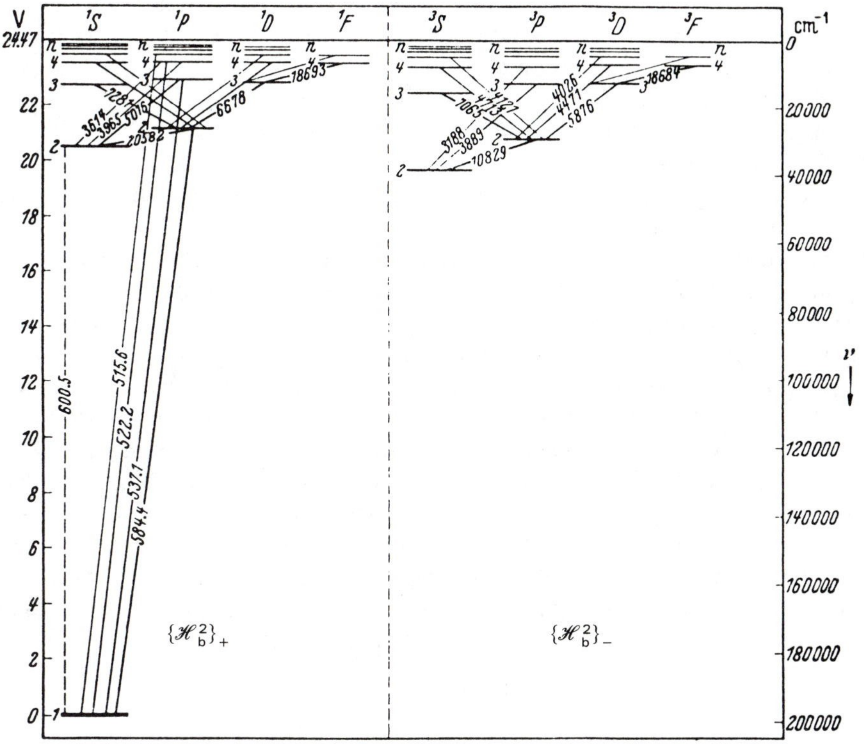

Figure 13 Term scheme of helium.

think of the discrete terms of helium as developing from the "electron configurations" $1s\,ns$, $1s\,np$, $1s\,nd, \ldots$. Each of these electron configuration terms (up to the ground state terms $1s\;1s$) is split by the interaction into two terms, a "symmetric" and an "antisymmetric." For the electron configuration $1s\;1s$ of the ground state we also write $(1s)^2$. This term is symmetric and does not split any further. We will denote the total angular momentum of a term by L. For the helium term L is equal to the angular momentum number l of the "optical electron" which is not in the state $1s$. For the terms $L = 0, 1, 2, \ldots$ we write $S, P, D, \ldots$. For $(1s)^2$ we only obtain a symmetric S term, from $1s\,np$ we obtain two—a symmetric and an antisymmetric P term.

3 Ritz Variational Principle

We may use the Ritz variational principle in order to obtain better numerical estimates for the eigenvalues. Solving the equation $H\varphi = \lambda\varphi$ is equivalent to finding extreme values of $\langle\varphi, H\varphi\rangle$ subject to the constraint $\langle\varphi, \varphi\rangle = 1$. If, for example, $\lambda_1 \le \lambda_2 \le \cdots \le \lambda_n$ are the n lowest eigenvalues of $H(H\varphi_\nu = \lambda_\nu\varphi_\nu)$, each of which is counted according to its degree of degeneracy, then $H' = H - \sum_{\nu=1}^{n}\lambda_\nu P_{\varphi_\nu}$ satisfies the property $\langle\varphi, H'\varphi\rangle \ge$

$\lambda_n\langle\varphi, \varphi\rangle$ for $\langle\varphi, \varphi_\nu\rangle = 0$ and $H'\varphi_n = 0$, $\nu = 1, \ldots, n$, a result which immediately follows from the complete spectral representation of H. If $\langle\varphi, \varphi\rangle = 1$ we therefore obtain $\langle\varphi, H\varphi\rangle = \sum_{\nu=1}^{n}\lambda_\nu|\langle\varphi_\nu, \varphi\rangle|^2 + \langle\varphi, H\varphi\rangle \geq \lambda_1$. The minimum λ_1 of $\langle\varphi, H\varphi\rangle$ will be attained for $\varphi = \varphi_1$. From the additional condition $\langle\varphi, \varphi_1\rangle = 0$ we obtain $\langle\varphi, H\varphi\rangle \geq \lambda_2$, λ_2 will be attained for $\varphi = \varphi_2$; etc.

The method of the Ritz variation problem permits another basis for the perturbation computation described in XI, §7. Let $H = H_0 + H_1$. The eigenvalues and the eigenvectors of H_0 may be known. Two eigenvalues ε_α and ε_β of H_0 may be relatively closely spaced. We apply the variation principle as follows: $\langle\varphi, H\varphi\rangle$ is to be minimized subject to the constraint that $\langle\varphi, \varphi\rangle = 1$ and φ is orthogonal to all eigenvectors of H with lower eigenvalues. Instead of seeking φ in the general case, that is, finding an exact solution of the problem, we shall attempt an approximate solution as follows: $\varphi = \sum_{\nu=1}^{n}\psi_\nu x_\nu$ where the ψ_ν is a complete orthonormal basis in the space $\mathcal{z}_\alpha + \mathcal{z}_\beta$ ($\mathcal{z}_\alpha$ is the eigenspace corresponding to the eigenvalue ε_α of H with the basis $\psi_1, \psi_2, \ldots, \psi_{n_\alpha}$; similarly, $\mathcal{z}_\beta$ is the eigenspace corresponding to the basis $\psi_{n_\alpha+1}, \ldots, \psi_n$). The x_ν are free parameters which are chosen by making $\langle\varphi, H\varphi\rangle$ an extreme value. The chosen φ are orthogonal to the eigenvectors of zeroth approximation for the other eigenvalues of H_0. On the basis of this assumption the φ are clearly only approximate solutions. Setting

$$\langle\psi_\nu, H_1\psi_\mu\rangle = \langle\nu|H_1|\mu\rangle$$

we seek the extremum of

$$\varepsilon_\alpha \sum_{\nu=1}^{n_\alpha} |x_\nu|^2 + \varepsilon_\beta \sum_{\nu=n_\alpha+1}^{n} |x_\nu|^2 + \sum_{\nu,\mu=1}^{n} \bar{x}_\nu\langle\nu|H_1|\mu\rangle x_\mu \tag{3.1}$$

subject to the constraint $\sum_{\nu=1}^{n}|x_\nu|^2 = 1$. If we require that each of the eigenvectors $(x_1, \ldots, x_n)$ is orhtogonal to the preceding eigenvectors, then the solution of the above problem is a series of extreme values $\varepsilon_1^{(1)}, \varepsilon_2^{(1)}, \ldots, \varepsilon_n^{(1)}$ which are the eigenvalues for the eigenvalue problem:

$$(\varepsilon_{\alpha,\beta} - \varepsilon_\nu^{(1)})x_\nu + \sum_{\mu=1}^{n} \langle\nu|H_1|\mu\rangle x_\mu = 0, \tag{3.2}$$

where ε_α or ε_β are chosen depending on whether $\nu \leq n_\alpha$ or $\nu > n_\alpha$. The $\varepsilon_1^{(1)}, \varepsilon_2^{(1)}, \ldots, \varepsilon_n^{(1)}$ are approximate eigenvalues into which the eigenvalues $\varepsilon_\alpha, \varepsilon_\beta$ pass by the influence of the perturbation. Equation (3.2) is identical with the secular equation described in XI, 6 which is used to find the eigenvalues of an operator $\tilde{E}_\nu H \tilde{E}_\nu$ (from the sum in XI (6.1)).

In order to obtain a more precise computation for the ground state of helium we may seek to obtain the minimum of

$$M = \left\langle \varphi, \left[H_0(1) + H_0(2) + \frac{e^2}{r_{12}}\right]\varphi \right\rangle$$

subject to the constraint $\langle \varphi, \varphi \rangle = 1$ by assuming that $\varphi = \phi_1(1)\phi_2(2)$. Using the method of Lagrange multipliers λ we may take into account the constraint as follows:

$$\begin{aligned} &\langle \phi_1(1)\phi_2(2), H_0(1)\phi_1(1)\phi_2(2)\rangle \\ &\quad + \langle \phi_1(1)\phi_2(2), H_0(2)\phi_1(1)\phi_2(2)\rangle \\ &\quad + \left\langle \phi_1(1)\phi_2(2), \frac{e^2}{r_{12}} \phi_1(1)\phi_2(2)\right\rangle \\ &\quad - \lambda\langle \phi_1(1)\phi_2(2), \phi_1(1)\phi_2(2)\rangle = \text{extremum}. \end{aligned} \tag{3.3}$$

If we replace ϕ_k by $\phi_k + \delta\phi_k$ and then replace $\delta\phi_k$ by $i\delta\phi_k$, then, from the condition that the variation of (3.3) must vanish for arbitrary $\delta\phi_k$, we obtain the following pair of equations:

$$\begin{aligned} &H_0(1)\phi_1(1) + \langle \phi_2(2), H_0(2)\phi_2(2)\rangle\phi_1(1) \\ &\quad + \left\langle \phi_2(2), \frac{e^2}{r_{12}} \phi_2(2)\right\rangle\phi_1(1) = \lambda\phi_1(1), \\ &\langle \phi_1(1), H_0(1)\phi_1(1)\rangle\phi_2(2) + H_0(2)\phi_2(2) \\ &\quad + \left\langle \phi_1(1), \frac{e^2}{r_{12}} \phi_1(1)\right\rangle\phi_2(2) = \lambda\phi_2(2) \end{aligned} \tag{3.4}$$

with the extreme value λ of $\langle \phi_1(1)\phi_2(2), H\phi_1(1)\phi_2(2)\rangle$.

If we set $\phi_1 = \phi_2 = \phi$ for the ground state of helium (since the latter is symmetric) we obtain

$$H_0(1)\phi(1) + \left\langle \phi(2), \frac{e^2}{r_{12}} \phi(2)\right\rangle\phi(1) = \varepsilon\phi(1), \tag{3.5}$$

where $\varepsilon = \lambda - \langle \phi(2), H_0(2)\phi(2)\rangle$. This is, however, the eigenvalue equation of an electron with the potential energy $-Ze^2/r_1 - eV(\vec{r}_1)$ where

$$V(\vec{r}_1) = -\left\langle \phi(2), \frac{e}{r_{11}} \phi(2)\right\rangle \tag{3.6}$$

is equal to the potential of the charge density $-e|\varphi(2)|^2$. For the ground state we can assume that ϕ has spherical symmetry; then (3.6) will depend only on r_1. Above all, in order to compute $V(r)$ the quantity ϕ must already be known. It is possible to solve (3.5) by successive approximation where the $V(r)$ from the previous approximation is used to compute the current approximation using (3.4). This method, known as the Hartree method of "selfconsistent" fields is applicable to cases involving more than two electrons.

We will now briefly exhibit the above procedure for the case of the ground state of helium. For the variational problem we shall seek the extremum of $\langle \phi, [H_0(0) + H_0(2) + e^2/r_{12}]\varphi\rangle$. We select the special form

$$\varphi = \frac{\alpha^3}{\pi} e^{-\alpha(r_1 + r_2)}. \tag{3.7}$$

Here only α remains to be chosen. We therefore obtain the variational problem

$$2\left(\frac{1}{2m}\alpha^2 - Ze^2\alpha\right) + \tfrac{5}{8}e^2\alpha = \text{extremum}.$$

The derivative with respect to α must vanish; thus we obtain $\alpha = me^2(Z - \frac{5}{16})$ and a minimum value $\varepsilon = -me^4(Z - \frac{5}{16})^2$.

From this result we may easily compute the ionization energy of helium. The latter is the difference in energy between the helium atom and singly ionized helium. According to XI (3.14) the energy of the helium ion is equal to $-(me^4/2)Z^2$. The iodization energy is equal to

$$me^4[(2 - \tfrac{5}{16})^2 - \tfrac{1}{2}2^2] = 1.695 \text{ Rydberg}.$$

The corresponding experimental value is 1.810 Rydberg = 24.46 eV. In spite of the simple assumption the agreement is reasonably good. If we completely neglect the interaction of the electrons then we would obtain a value of $E = -me^4Z^2$. We shall now give the following intuitive interpretation of the above results: the motion of each electron is such that the charge of the nucleus Ze appears to be shielded by the other electron such that its effective value is $(Z - \frac{5}{16})e$.

For excited states we must require that ϕ_1 and ϕ_2 are different, in order that both equations (3.4) may be simultaneously solved. The fact that one electron must be at a greater distance from the nucleus than the other suggests that, for all practical purposes, we may expect that the eigenfunction $\phi_1(1)$ will be equal to that of the ground state of the helium ion. $|\phi_2(2)|^2$ will then be essentially nonzero where $\phi_1(1) \approx 0$. For this reason we may set $\langle\phi_2(2), (e^2/r_{12})\phi_2(2)\rangle$ in the first equation of (3.4) equal to a constant $\langle\phi_2(2), (e^2/r_2)\phi_2(2)\rangle$. We therefore obtain

$$H_0(1)\phi_1(1) = \left[\lambda - \langle\phi_2(2), H_0(2)\phi_2(2)\rangle - \left\langle\phi_2(2), \frac{e^2}{r_2}\phi_2(2)\right\rangle\right]\phi_1(1). \quad (3.8)$$

The solution is given by

$$\phi_1(1) = (mZe^2)^{3/2}e^{-mZe^2r_1} \quad (3.9)$$

where

$$\left[\lambda - \langle\phi_2(2), H_0(2)\phi_2(2)\rangle - \left\langle\phi_2(2), \frac{e^2}{r_2}\phi_2(2)\right\rangle\right] = \varepsilon_1 = -\frac{mZ^2e^4}{2}.$$

Applying equations (3.4) to $\phi_2(2)$ we obtain

$$\left[\frac{1}{2m}\vec{P}_2^2 + V(r_2)\right]\phi_2(2) = \left(\lambda + \frac{mZ^2e^4}{2}\right)\phi_2(2), \quad (3.10)$$

which is an eigenvalue equation corresponding to an electron in a potential field

$$V(r_2) = -\frac{Ze^2}{r_2} + \left\langle\phi_1(1), \frac{e^2}{r_{12}}\phi_1(1)\right\rangle = \frac{(Z-1)e^2}{r_2} + R(r_2). \quad (3.11)$$

The expression within the bracket depends only on r_2 because $|\phi_1(1)|^2$ is a spherically symmetric charge distribution. $R(r_2)$ is, for all practical purposes, equal to zero for distances greater than $(mZe^2)^{-1}$. For the $(\lambda + mZ^2e^4/2)$ and $\phi_2(2)$ we therefore obtain eigenvalues ε_{nl} and eigenvectors ϕ_{nlm} in the same manner as we have already discussed for the case of the alkali metals spectrum. We may omit the term $n = 1$ because, in the orthogonality condition formulated above we have found that $\phi_1(1)\phi_2(2)$ is orthogonal to the ground state (3.7); as an approximation we may replace this condition by requiring that $\phi_2(2)$ must be orthogonal to (3.9). The energy eigenvalue obtained for the excited state is approximately given by $\lambda = \varepsilon_{nl} - mZ^2e^4/2 = \varepsilon_{nl} + \varepsilon_1$. Therefore we find that the states corresponding to different values of l no longer coincide.

For fixed n and l a particular term can be found in both $\{\mathscr{H}_b^2\}_+$ and $\{\mathscr{H}_b^2\}_-$ because both $\Omega_{nl}^+ = \phi_1(1)\phi_{nl}(2) + \phi_{nl}(1)\phi_1(2)$ as well as $\Omega_{nl}^- = \phi_1(1)\phi_{nl}(2) - \phi_{nl}(1)\phi_1(2)$ belong to the same eigenvalue.

We must therefore obtain better values for the energy if we solve the extremal problem for $\langle\varphi, H\varphi\rangle\|\varphi\|^{-2}$ assuming that $\varphi = a\Omega_{nl}^+ + b\Omega_{nl}^-$. Since H commutes with the permutation of the electrons the selections $\varphi = \Omega_{nl}^+$ and $\varphi = \Omega_{nl}^-$ must yield extremal values. Therefore the energy level $\varepsilon_{nl} + \varepsilon_1$ are therefore split into the energy levels:

$$\langle\Omega_{nl}^+, H\Omega_{nl}^+\rangle\|\Omega_{nl}^+\|^{-2} = \varepsilon_1 + \varepsilon_{nl} + A_{nl}^+ \quad \text{in } \{\mathscr{H}_b^2\}_+ \tag{3.12a}$$

and

$$\langle\Omega_{nl}^-, H\Omega_{nl}^-\rangle\|\Omega_{nl}\|^{-2} = \varepsilon_n + \varepsilon_{nl} - A_{nl}^- \quad \text{in } \{\mathscr{H}_b^2\}_-. \tag{3.12b}$$

Since the ϕ_{nl} are not exactly orthogonal $\|\Omega_{n,l}\|^2$ is not exactly equal to 2. If we neglect this discrepancy then $A_{nl}^+ = A_{nl}^-$ and is equal to the same expression as (2.10) except that we need only replace e^2/r_{12} by $e^2/r_{12} - \langle\phi_1(3), (e^2/r_{32})\phi_1(3)\rangle$. We obtain

$$A_{nl} = e^2 \int\left[\frac{1}{r_{12}} - \int|\langle\vec{r}_3|\varphi_1\rangle|^2 \frac{1}{r_{32}}\, d^3\vec{r}_3\right] \overline{\langle\vec{r}_2|\phi_1\rangle\langle\vec{r}_1|\varphi_{nlm}\rangle}\langle\vec{r}_1|\phi_1\rangle\langle\vec{r}_2|\phi_{nlm}\rangle\, d^3\vec{r}_1\, d^3\vec{r}_2. \tag{3.13}$$

The exchange integral A_{nl} is responsible for the different values of the energy levels in $\{\mathscr{H}_b^2\}_+$ and $\{\mathscr{H}_b^2\}_-$.

A_{nl} and $-A_{nl}$ arise from the Coulomb interaction of electrons in the different states in $\{\mathscr{H}_b^2\}_+$ and $\{\mathscr{H}_b^2\}_-$, respectively, a result which is clearly evident from the first equation in (2.10). How does the energy difference between these states arise? In (2.10), C_{nl} is the interaction energy corresponding to the description of the two electrons as two interacting continuous charge clouds. If we consider a state $\phi_1(1)\phi_2(2)$ then the probability that electron 1 lies in a volume $d^3\vec{r}_1$ about $\vec{r}_1$ and electron 2 lies in a volume $d^3\vec{r}_2$ about $\vec{r}_2$ is, in fact, equal to

$$|\langle r_1|\phi_1\rangle|^2|\langle\vec{r}_2|\phi_2\rangle|^2\, d^3r_1\, d^3r_2.$$

For the state $(1/\sqrt{2})(\phi_1(1)\phi_2(2) \pm \phi_1(2)\phi_2(1))$ the above expression is given by

$$\tfrac{1}{2}[|\langle\vec{r}_1|\phi_1\rangle|^2|\langle\vec{r}_2|\phi_2\rangle|^2 + |\langle\vec{r}_2|\phi_1\rangle|^2|\langle\vec{r}_1|\phi_2\rangle|^2$$
$$\pm \{\overline{\langle\vec{r}_1|\phi_1\rangle\langle\vec{r}_2|\phi_2\rangle}\langle\vec{r}_2|\phi_1\rangle\langle\vec{r}_1|\phi_2\rangle$$
$$+ \overline{\langle\vec{r}_2|\phi_1\rangle\langle\vec{r}_1|\phi_2\rangle}\langle\vec{r}_1|\phi_1\rangle\langle\vec{r}_2|\phi_2\rangle\}]\, d^3\vec{r}_1\, d^3\vec{r}_2.$$

For $\vec{r}_1 = \vec{r}_2$, for example, for the state $\phi_1(1)\phi_2(2)$ the probability is given by $|\langle\vec{r}_1|\phi_1\rangle|^2|\langle r_1|\phi_2\rangle|^2\, d^3\vec{r}_1\, d^3\vec{r}_2$; in the second case, however, we obtain

$$[|\langle\vec{r}_1|\phi_1\rangle|^2|\langle\vec{r}_1|\phi_2\rangle|^2 \pm |\langle\vec{r}_1|\phi_1\rangle|^2|\langle\vec{r}_1|\phi_2\rangle|^2]\, d^3\vec{r}_1\, d^3\vec{r}_2.$$

Therefore either we obtain twice the preceding result or zero. Thus we find the position probability of the electrons are correlated, that is, they are mutually dependent. Intuitively, in the case $\{\mathscr{H}_b^2\}_-$ the electrons "avoid each other's company"; for the case $\{\mathscr{H}_b^2\}_+$ the electrons "prefer each other's company." Thus it is understandable that the Coulomb repulsion for the energy levels in $\{\mathscr{H}_b^2\}_+$ is higher than that in $\{\mathscr{H}_b^2\}_-$. For this reason the exchange integral is often referred to as the Coulomb correlation energy.

4 The Fine Structure of the Helium Spectrum

According to (2.3) the Hilbert space for two electrons is given by

$$\{\mathscr{H}^2\}_- = \{\mathscr{H}_b^2\}_+ \times \{\imath_s\}_- \oplus \{\mathscr{H}_b^2\}_- \times \{\imath_s^2\}_+.$$

In order to exhibit the complete set of eigenfunctions (with spin) for the operator

$$H_0 = \frac{1}{2m}(P_1^2 + P_2^2) - \frac{Ze^2}{r_1} - \frac{Ze^2}{r_2} + \frac{e^2}{r_{12}}$$

we must multiply the eigenvectors in $\{\mathscr{H}_b^2\}_+$ by antisymmetric vectors of $\{\imath_s^2\}_-$ and we must multiply the eigenvectors in $\{\mathscr{H}_b^2\}_-$ by symmetric vectors of $\{\imath_s^2\}_+$.

The space $\imath_s^2$ which is spanned by the vectors $u_+(1)u_+(2)$, $u_+(1)u_-(2)$, $u_-(1)u_+(2)$, $u_-(1)u_-(2)$, can be easily decomposed into $\{\imath_s^2\}_+$ and $\{\imath_s^2\}_-$ as follows: $\{\imath_s^2\}_+$ has the basis

$$u_+(1)u_+(2), \frac{1}{\sqrt{2}}(u_+(1)u_-(2) + u_-(1)u_+(2)), u_-(1)u_-(2).$$

$\{\imath_s^2\}_-$ is one-dimensional with basis vector

$$\frac{1}{\sqrt{2}}(u_+(1)u_-(2) - u_-(1)u_+(2)).$$

In $\imath_s^2$ the product representation of the rotation group is given by $\mathscr{D}_{1/2} \times \mathscr{D}_{1/2} = \mathscr{D}_1 + \mathscr{D}_0$. The two spin angular momenta can therefore be added to obtain a total spin angular momentum of $S = 1$ or $S = 0$. In order that the representation operators for the rotation group commute with permutations in $\imath_s^2$, they must leave the subspaces $\{\imath_s^2\}_+$ and $\{\imath_s^2\}_-$ invariant. Since $\mathscr{D}_{1/2} \times \mathscr{D}_{1/2}$ may be decomposed into two irreducible components $\mathscr{D}_1$ of dimension 3 and $\mathscr{D}_0$ of dimension 1, we obtain the representation $\mathscr{D}_1$ in $\{\imath_s^2\}_+$ and $\mathscr{D}_0$ in $\{\imath_s^2\}_-$.

For an energy level corresponding to an electron configuration consisting of one $1s$ electron together with an electron with principle quantum number n and orbital angular momentum number l we find that the exact eigenvectors of H_0 in $\{\mathscr{H}_b^2\}_+$ must describe the representation $\mathscr{D}_L$ of the rotation group where $L = l$ because the representation will not be affected by "turning on" the perturbation. Therefore there exist $(2L + 1)$ exact eigenvectors

$$\phi_M^+ \ (M = -L, -L + 1, \ldots, L - 1, L)$$

in $\{\mathscr{H}_b^2\}_+$ and ϕ_M^- in $\{\mathscr{H}_b^2\}_-$, respectively, corresponding to the exact eigenvalues; these exact eigenvectors do not take on the approximate form $(1/\sqrt{2})(\phi_1(1)\phi_{nlm}(2) \pm \phi_{nlm}(1)\phi_1(2))$! Let us denote these eigenspaces from $\{\mathscr{H}_b^2\}_\pm$ by $\imath_L^\pm$. Then the complete set of eigenvectors for H_0 will consist of the vectors from the space $\imath_L^+ \times \{\imath_s^2\}_-$ for the symmetric case in $\{\mathscr{H}_b^2\}_+$ and the vectors from the space $\imath_L^- \times \{\imath_s^2\}_+$ for the antisymmetric case in $\{\mathscr{H}_b^2\}_-$. Since $\mathscr{D}_0$ is the representation of the rotation group in $\{\imath_s^2\}_-$, we find that $\mathscr{D}_L \times \mathscr{D}_0 = \mathscr{D}_L$ is the representation in $\imath_L^+ \times \{\imath_s^2\}_-$. The total angular momentum for both electrons, including orbital and spin is usually denoted by J. Therefore we obtain $J = L$. Since the representation of the rotation group in $\imath_L^+ \times \{\imath_s\}_-$ is irreducible, the symmetric energy levels cannot be split by means of the spin dependent (but rotation invariant) term H_1 of the Hamiltonian operator, the symmetric energy levels remain simple. The situation is somewhat different for the case $\imath_L^- \times \{\imath^2\}_+$. Here we find that the representation of the rotation group is $\mathscr{D}_L \times \mathscr{D}_1 = \mathscr{D}_{L+1} + \mathscr{D}_L + \mathscr{D}_{L-1}$ for $L > 1$ and $\mathscr{D}_0 \times \mathscr{D}_1 = \mathscr{D}_1$ for $L = 0$, respectively. The total angular momentum can therefore take on values $J = L + 1, L, L - 1$ and for $L = 0$ only $J = 1$. Therefore the antisymmetric S term (that is $L = 0$) cannot be split by the spin perturbation. In contrast, the other antisymmetric energy levels are split into three fine structure energy levels corresponding to the values $J = L - 1, L, L + 1$.

The number $2S + 1$ is called the multiplicity and is written above and to the left of the term symbol. For $S = 0$, ${}^1P_1$ is therefore a "singlet" P term (that is $L = 1$) with total angular momentum $J = 1$. For example, for $S = 1$, ${}^3P_2$ is a "triplet" P term with total angular momentum $J = 2$. For helium the three terms ${}^3P_2$, ${}^3P_1$, ${}^3P_0$ lie closely spaced. Since the symmetric terms are, according to the Pauli exclusion principle, associated with $S = 0$ and the antisymmetric terms are associated with $S = 1$, it is not necessary to provide a characterization of the terms according to its symmetry under permutation. Although an antisymmetric ${}^3S_1$ term (that is, $L = 0$) is not split, that is, it has only one component ${}^3S_1$, we still call it a triplet term corresponding to $2S + 1$. The number $2S + 1$ above the term symbol on the left therefore specifies both the total spin as well as the symmetry of the vectors in $\mathscr{H}_b^2$.

By introducing the perturbation H_1 we find that the spin angular momentum and the symmetry of the position functions from $\{\mathscr{H}_b^2\}_+$ and $\{\mathscr{H}_b^2\}_-$ are exactly defined only for eigenvectors of the zeroth approximation (that is, for the "correct" linear combinations). The stronger the influence of H_1, the greater the deviation of the exact eigenfunctions from the linear combinations of the zeroth approximation. The value J, that is, the representation $\mathscr{D}_J$ cannot, of course, be changed (see XI, §7).

In order to obtain a quantitative estimate of the magnitude of the splitting it is necessary to know the form of H_1. Since we do not wish to make any explicit quantitative computation of these effects, we shall delay any such discussion about the displacement between the individual fine structure components until we have discussed the case for more than two electrons and the associated multiplicity $2S + 1$.

CHAPTER XIII

Selection Rules and the Intensity of Spectral Lines

1 Intensity of Spectral Lines

According to XI (1.17) the intensity of spectral lines is determined by the effect operator for the emission of a photon of frequency ω_{nm} as follows:

$$\Delta^1 F_{n\to m} = \Delta t \frac{4\omega_{nm}^3}{3} E_n \vec{d}(0) \cdot E_m \vec{d}(0) E_n. \tag{1.1}$$

The probability for the emission of a photon in the time interval Δt for an ensemble described by the operator W is given by

$$\Delta t \frac{4\omega_{nm}^3}{3} \operatorname{tr}(W E_n \vec{d}(0) \cdot E_m \vec{d}(0) E_n). \tag{1.2}$$

Since $E_n^2 = E_n$ we find that

$$\operatorname{tr}(W E_n \vec{d}(0) \cdot E_m \vec{d}(0) E_n) = \operatorname{tr}(E_n W E_n \vec{d}(0) \cdot E_m \vec{d}(0) E_n).$$

We therefore find that we need only be concerned with the components $W_n = E_n W E_n$ of W. We shall often find that W_n is proportional to E_n, that is, $W_n = (1/s_n) E_n \operatorname{tr}(W E_n)$ where s_n is the dimension of E_n; this is indeed the case in XI (2.1) in which

$$W_n = E_n \frac{e^{-\beta\varepsilon_n}}{\operatorname{tr}(e^{-\beta H})}, \tag{1.3}$$

where H is the Hamiltonian operator describing the relative motion with respect to the nucleus. Equation (1.3) describes the case of thermal radiation.

Except for the factor $\operatorname{tr}(WE_n)$ which depends upon W and represents the probability of E_n in the ensemble, the intensity of the spectral line ω_{nm} is essentially determined by the expression (1.2) where we replace W by $(1/s_n)E_n$. We obtain

$$\Delta t \frac{4\omega_{nm}^3}{3s_n} \operatorname{tr}(E_n \vec{d}(0) \cdot E_m \vec{d}(0) E_n). \tag{1.4}$$

Let the three spatial components of $\vec{d}$ be denoted by d_ν; let us introduce a complete orthogonal basis $v_{n\alpha}$ for the subspace determined by E_n. We may then rewrite (1.4) as follows:

$$\Delta t \frac{4\omega_{nm}^3}{3s_n} \sum_{\alpha=1}^{s_n} \sum_{\beta=1}^{s_m} \sum_{\nu=1}^{3} |\langle v_{n\alpha}, d_\nu v_{m\beta} \rangle|^2, \tag{1.5}$$

where we have replaced $d_\nu(0)$ by d_ν.

In order to make assertions concerning the intensity of spectral lines it is therefore necessary to evaluate the transition matrix elements $\langle v_{n\alpha}, d_\nu v_{m\beta} \rangle$.

In (1.5) we have considered the dipole radiation contribution. This is a good approximation provided (see XI, §1) that the wavelength λ of the radiation is large compared with the diameter d of the atom. According to XI (3.6) an estimate of d is given by r_0. It remains to show that, for atoms, λ is in fact substantially greater than the diameter d (λ is approximately a thousand times greater than d).

Higher order effects, such as magnetic dipole radiation (see XI, §1) or quadrupole radiation will only be noticable if all the $\langle v_{n\alpha}, d_\nu v_{m\beta} \rangle$ are identically zero for the transition $n \to m$. Here we note that it is clear that in the case of radiation associated with other moments that we would have to replace $\langle v_{n\alpha}, d_\nu v_{m\beta} \rangle$ by other matrix elements $\langle v_{n\alpha}, A v_{m\beta} \rangle$ where A is the corresponding operator.

2 Representation Theory and Matrix Elements

In a Hilbert space $\mathscr{H}$ let $\{\phi_\mu\}$ be a set of vectors (not necessarily linearly independent) which span a subspace $\mathscr{T}$ of $\mathscr{H}$. Let $\mathscr{H}$ be a completely reducible representation space for a set of operators (see AIV, §14)

$$\mathscr{H} = \imath_1 \oplus \imath_2 \oplus \cdots, \tag{2.1}$$

where the $\imath_\nu$ are irreducible. Let the subspace $\mathscr{T}$ be invariant under the operators of the representation. If φ_ν is a complete normed orthogonal system which is compatible with the decomposition (2.1) then the ϕ_μ may be expressed in terms of the φ_ν as follows:

$$\phi_\mu = \sum_\nu \varphi_\nu a_{\nu\mu}. \tag{2.2}$$

A homomorphic map of $\mathscr{T}$ into $\imath_1$ is defined by $\phi_\mu \to \sum_\nu' \varphi_\nu a_{\nu\mu}$ where the summation may only take place over the φ_ν in $\imath_1$. Thus the irreducible space

$\mathfrak{r}_1$ must be isomorphic to an irreducible component of the space $\mathscr{T}$, or, if $\mathscr{T}$ has no irreducible component isomorphic to $\mathfrak{r}_1$, all the $a_{\nu\mu}$ (where the index ν is one of the φ_ν from $\mathfrak{r}_1$) must be equal to zero (see AIV, §14).

Let $\tilde{\mathscr{H}}$ be another representation space which can be completely reduced as follows:

$$\tilde{\mathscr{H}} = \tilde{\mathfrak{r}}_1 \oplus \tilde{\mathfrak{r}}_2 \oplus \cdots. \tag{2.3}$$

Let ψ_μ be a set of vectors (which are not necessarily linearly independent!) in $\tilde{\mathscr{H}}$ which transforms under the representation with the same matrix as the vectors ϕ_μ, that is, if U is a transformation in $\mathscr{H}$ of the representation operator and V is the operator for the same transformation, we therefore find that

$$U\phi_\mu = \sum_\nu \phi_\nu \rho_{\nu\mu} \quad \text{and} \quad V\psi_\mu = \sum_\nu \psi_\nu \rho_{\nu\mu}. \tag{2.4}$$

Let χ_ν be a complete normalized system of vectors which is compatible with the decomposition (2.3). Then we may express the ψ_μ as follows:

$$\psi_\mu = \sum_\nu \chi_\nu b_{\nu\mu}. \tag{2.5}$$

Let $\mathfrak{r}_\alpha$ and $\tilde{\mathfrak{r}}_\beta$ be a pair of isomorphic representation spaces. Let $\varphi_{\alpha_1}, \varphi_{\alpha_2}, \ldots, \varphi_{\alpha_\nu}$ denote the φ_ν from $\mathfrak{r}_\alpha$ and let $\chi_{\beta_1}, \chi_{\beta_2}, \ldots, \chi_{\beta_\nu}$ denote the χ_ν from $\tilde{\mathfrak{r}}_\beta$ where φ_{α_τ} and χ_{β_τ} are a pair of isomorphic basis for $\mathfrak{r}_\alpha$ and $\tilde{\mathfrak{r}}_\beta$, respectively.

We shall now construct a vector space $\mathscr{V}$ from a linearly independent basis v_μ. The number of the v_μ may be equal to the number of the ϕ_μ and of the ψ_μ. $\mathscr{V}$ may be transformed into a representation space by means of the transformation $v_\mu \to \sum_\nu v_\nu \rho_{\nu\mu}$ where the $\rho_{\nu\mu}$ are defined according to (2.4). A homomorphic mapping of $\mathscr{V}$ onto $\mathscr{T}$ is defined by $v_\mu \to \phi_\mu$. If $\mathscr{V}$ has a decomposition of the form $\mathscr{V} = \mathscr{V}_1 + \mathscr{V}_2 + \cdots$ into irreducible components, then $\mathscr{T}$ must be isomorphic to the sum $\mathscr{V}_{\lambda_1} + \mathscr{V}_{\lambda_2} + \cdots$ of a part of the $\mathscr{V}_1, \mathscr{V}_2, \ldots$. We note that $\mathscr{T}$ need not contain all the $\mathscr{V}_\nu$ because we did not assume that the ϕ_μ are linearly independent.

We will now assume that every irreducible component in $\mathscr{V}$ occurs exactly once. If we then construct a vector space $\mathscr{W}$ which is isomorphic to $\mathfrak{r}_\alpha$ and $\tilde{\mathfrak{r}}_\beta$ using a basis w_τ for $\mathscr{W}$ which is isomorphic to the φ_{α_τ} and χ_{β_τ} then the maps $\mathscr{V}_\mu \to \sum_\tau w_\tau a_{\alpha_\tau\mu}$ and $\mathscr{V}_\mu \to \sum_\tau w_\tau b_{\beta_\tau\mu}$ are homomorphic maps of $\mathscr{V}$ onto $\mathscr{W}$. Since $\mathscr{V}$ contains only different irreducible components $\mathscr{V}_\nu$, all $\mathscr{V}_\nu$ except at most one must be mapped onto the null space (see AIV, §14). If $\mathscr{V}_\eta$ is isomorphic to $\mathscr{W}$ then it can be mapped onto the space $\mathscr{W}$. The map of $\mathscr{V}_\eta$ onto $\mathscr{W}$ is uniquely determined up to a numerical factor (see AIV, §14). Therefore we obtain

$$\lambda_1 a_{\alpha_\tau\mu} = \lambda_2 b_{\beta_\tau\mu}, \tag{2.6}$$

where the ratio $\lambda_1 : \lambda_2$ depends only on $\mathfrak{r}_\alpha$ and $\tilde{\mathfrak{r}}_\beta$ and not upon τ and μ.

The relations derived above hold for the special case in which $\mathscr{H} = \tilde{\mathscr{H}}$ and if $\varphi_\nu = \chi_\nu$ and $\mathfrak{r}_\alpha = \tilde{\mathfrak{r}}_\beta$; here the ϕ_μ and ψ_μ may, of course, be different.

Let A_i be a series of self-adjoint operators such that, for the operator U, the following representation

$$UA_iU^+ = \sum_k A_k\, d_{ki}. \tag{2.7}$$

holds. The representation $\mathscr{D}_A$ is given by the matrix d_{ki}. If $\varphi_1, \varphi_2, \ldots, \varphi_n$ are the basis vectors of $\imath_1$, then the $\phi_\mu = \phi_{ik} = A_i\varphi_k$ $(k = 1, 2, \ldots, n)$ and $\mu = i, k$ span a representation space $\mathscr{T}$ and therefore contains a part of the irreducible components of the product representation $\mathscr{D}_A \times \mathscr{D}_{\imath_1}$. From

$$\phi_{ik} = A_i\varphi_k = \sum_\nu \varphi_\nu\langle\varphi_\nu, A_i\varphi_k\rangle \tag{2.8}$$

we find that all matrix elements $\langle\varphi_\nu, A_i\varphi_k\rangle$ for which ν is the index for the vectors φ_ν in $\imath_\alpha$ where $\imath_\alpha$ is not isomorphic to an irreducible representation in $\mathscr{D}_A \times \mathscr{D}_{\imath_1}$ must vanish. If the representation $\mathscr{D}_A \times \mathscr{D}_{\imath_1}$ contains only different irreducible components then it is possible to determine the matrix elements $\langle\varphi_\nu, A_i\varphi_k\rangle$ up to numerical factors λ_α by means of group theory alone, by investigating the homomorphic maps of a space $\mathscr{V}$ for the product representation $\mathscr{D}_{A_i} \times \mathscr{D}_{\imath_1}$ onto $\imath_\alpha$.

3 Selection Rules for One-Electron Spectra

In addition to the previously discussed representation of the proper rotation group $\mathscr{D}_\mathscr{G}$ (the rotations with determinant $+1$, which are continuously connected to the unit element), the representation of all rotations (including the reflection $r{:}x'_\nu = -x_\nu$ which was introduced in VII, §7) is of great importance. By analogy with the description of $\mathscr{D}_\mathscr{G}$ as the group of rotations "about the nucleus" we will consider r to be a "reflection about the nucleus."

According to VII (7.1) we may represent r by means of a unitary operator R where R is defined in the momentum representation of $\mathscr{H}_b$ by the equation

$$R\langle k_1, k_2, k_3|\varphi\rangle = \langle -k_1, -k_2, -k_3|\varphi\rangle$$

and for the position representation by the equation

$$R\langle x_1, x_2, x_3|\varphi\rangle = \langle -x_1, -x_2, -x_3|\varphi\rangle.$$

In $\mathscr{H}_b$ the $(2l+1)$-dimensional subspaces $\imath_{\nu l}$ with basis vectors $\varphi_\nu(r)Y^l_m(\theta, \varphi)$ $(m = l, l-1, \ldots, -l)$ are also invariant under R; we obtain

$$R\varphi_\nu(r)Y^l_m(\theta, \varphi) = (-1)^l\varphi_\nu(r)Y^l_m(\theta, \varphi).$$

This follows from the fact that the Y^l_m are homogeneous polynomials of degree l in the components $e_1 = \sin\theta\cos\varphi$, $e_2 = \sin\theta\sin\varphi$, $e_3 = \cos\theta$ of the unit vector.

For a fixed nucleus (that is, for electron coordinates relative to the nucleus) the Hamiltonian operator is not only invariant under rotations but also under reflections, that is, $(R \times \mathbf{1})H = H(R \times \mathbf{1})$ also holds for n electrons, where R acts upon the orbit space and $\mathbf{1}$ acts in the spin space of n electrons. Later we will make use of this fact.

If we ignore the spin for a one-electron system, then the eigenspace $\imath_{nl}$ corresponding to the eigenvalue ε_{nl} belongs to the representation $\mathscr{D}_l$ of the rotation group and to the "reflection character" $(-1)^l$. The dipole moment of the electron is $\vec{d} = e\vec{r}$ because $\vec{r}$ is the position vector from the nucleus to the electron. The three components of $\vec{r}$ are the position operators Q_1, Q_2, Q_3. Instead of these components we shall use linear combinations of the Q_i which, under rotations, are transformed in the same way as the spherical harmonics Y_1^1, Y_0^1, Y_{-1}^1.

It is easy to verify that

$$\begin{aligned} R_1 &= \frac{1}{\sqrt{2}}(Q_1 + iQ_2), \\ R_0 &= -Q_3, \\ R_{-1} &= \frac{1}{\sqrt{2}}(-Q_1 + iQ_2) \end{aligned} \tag{3.1}$$

satisfy these conditions, because, according to VII, §3

$$Y_1^1 = \sqrt{\frac{3}{2\pi}\frac{1}{2}}e^{i\varphi}\sin\theta,$$

$$Y_0^1 = -\sqrt{\frac{3}{2\pi}}\frac{1}{\sqrt{2}}\cos\theta,$$

$$Y_{-1}^1 = \sqrt{\frac{3}{2\pi}\frac{1}{2}}\,e^{-i\varphi}\sin\theta.$$

The R_μ therefore transform according to the representation $\mathscr{D}_1$. The reflection character of the components R_1, R_0, R_{-1} is (-1). The only transitions $\varepsilon_{nl} \to \varepsilon_{n'l'}$ which can occur are those for which the reflection character of ε_{nl} multiplied by the reflection character of d_1, d_2, d_3 is equal to reflection character of the energy level $\varepsilon_{n'l'}$ and where $\mathscr{D}_{l'}$ occurs in the product representation of $\mathscr{D}_1 \times \mathscr{D}_l = \mathscr{D}_{l+1} + \mathscr{D}_l + \mathscr{D}_{l-1}$. Therefore we find that $(-1)^{l'} = (-1)^{l+1}$ and $l' = l + 1$ or l or $l - 1$. The value $l' = l$ clearly does not satisfy the first condition; we therefore obtain the selection rule $l \to l \pm 1$. S terms can be combined only with P terms, P terms only with either S or D terms, etc. In the energy level diagram (Figure 11) the slanted lines indicate the possible transitions and their corresponding wavelengths.

It is easy to show that, for quadrupole radiation, the selection rules are $l \to l + 2$, l, and $l - 2$.

If we take spin into consideration we find that the eigenfunctions of a term $\varepsilon_{n'l'j'}$ which arises continuously from ε_{nl} upon "switching on" the spin belong to the representation $\mathscr{D}_j$ of the rotation group and to the reflection character $(-1)^l$. Dipole radiation transitions are possible only for energy levels $\varepsilon_{n'l'j'}$ for which the reflection character $(-1)^{l'}$ is equal to $(-1)^{l+1}$ and the angular momentum quantum number j' is equal to $j + 1$, j or $j - 1$. The above selection rules for l need not apply rigorously because the eigenvectors corresponding to ε_{nlj} under the operator V_D in $\mathscr{H}_b$ need not transform precisely according to $\mathscr{D}_l$. However, since $j = l \pm \frac{1}{2}$, it follows from $j \to j - 1, j, j + 1$ that $l' = l + 2, l + 1, l - 1, l - 2$ are possible. From $(-1)^{l'} = (-1)^{l+1}$ we exclude the values $l' = l - 2, l, l + 2$; we therefore find that $l \to l \pm 1$ is correct.

Using the symmetry group of rotations and reflections it is not difficult to show that the exact eigenfunctions have the same form as the correct linear combinations of the zeroth approximation (see XI, §§11.1 and 11.2), that is,

$$\psi_{nlj}(r)\left(\sqrt{\frac{l+m+1}{2l+1}}\,Y^l_m u_+ + \sqrt{\frac{l-m}{2l+1}}\,Y^l_{m+1} u_-\right) \quad \text{for } j = l + \tfrac{1}{2}$$

and

$$\psi_{nlj}(r)\left(-\sqrt{\frac{l-m}{2l+1}}\,Y^l_m u_+ + \sqrt{\frac{l+m+1}{2l+1}}\,Y^l_{m+1} u_-\right) \quad \text{for } j = l - \tfrac{1}{2},$$

where, in the zeroth approximation $\psi_{nll+1/2} = \varphi_{nl} = \psi_{nll-1/2}$. The eigenvalues ε_{nlj} therefore correspond *exactly* to the representation $\mathscr{D}_l$ with respect to the operator V_D which acts in $\mathscr{H}_b$.

The selection rules determine the qualitative structure of a multiplet spectral line. A transition from an S term (that is, $l = 0$) can only occur to a P term ($l = 1$). The sodium D doublet, a pair of yellow spectral lines, may be explained in this way. The P term consists of two levels with $j = \frac{3}{2}$ and $j = \frac{1}{2}$; for the S term $j = \frac{1}{2}$. Both transitions $j = \frac{3}{2} \to \frac{1}{2}$ and $j = \frac{1}{2} \to j = \frac{1}{2}$ are permitted. If we consider, for another example, a transition between a D term and a P term in which both terms are split, we may then expect four closely

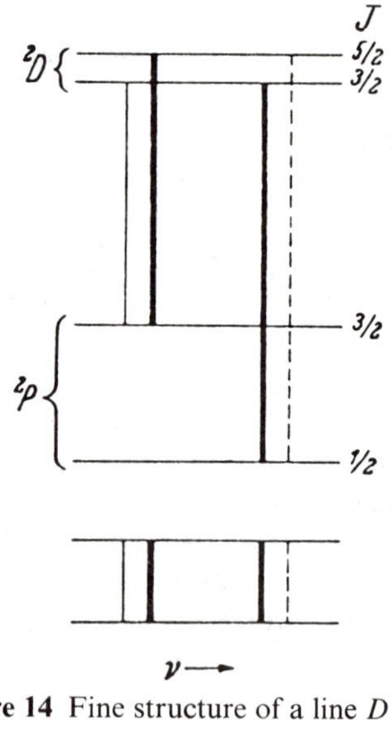

Figure 14 Fine structure of a line $D \to P$.

spaced spectral lines. According to the selection rule for j, the transition $j = \frac{5}{2} \to j = \frac{1}{2}$ does not occur. We therefore only obtain three spectral lines. For the ratios of the intensity of these lines we may make assertions of a pure group theoretical type providing that we assume that the exact eigenfunctions differ only slightly from the correct linear combinations of ϕ_{nljM} in the zeroth approximation (XI, §§11.1 and 11.20). This question will be treated in detail in XIV, §5. In Figure 14 the line thickness provides a crude estimate of the intensity for the transition.

4 Selection Rules for the Helium Spectrum

For the case of dipole radiation a helium atom term with total angular momentum quantum number J can only be combined with a term $J' = J + 1$, J or $J - 1$. The proof is obtained from $\mathscr{D}_1 \times \mathscr{D}_J = \mathscr{D}_{J+1} + \mathscr{D}_J + \mathscr{D}_{J-1}$ directly from the above discussion and from the fact that the eigenspace of a term with quantum number J the representation of the rotation group is $\mathscr{D}_J$.

For the accuracy with which the exact eigenfunctions may be described by the "correct linear combinations" of zeroth approximations with respect to the spin as a perturbation, we may deduce the following additional selection rules: The eigenfunctions of an S, P, D term correspond to the representations $\mathscr{D}_0, \mathscr{D}_1, \mathscr{D}_2, \ldots$ with respect to the operators V_D which act in $\mathscr{H}_b$. Since $\mathscr{D}_L \times \mathscr{D}_1 = \mathscr{D}_{L+1} + \mathscr{D}_L + \mathscr{D}_{L-1}$, it follows from the above considerations that the only possible transitions for dipole radiation are $L \to L - 1, L, L + 1$.

The reflection symmetry of a term consisting of a $1s$ electron and a second electron with orbital angular momentum l is obviously $(-1)^l$. Therefore we must exclude the transition $L \to L$.

Since the operator for the dipole moment $\vec{d} = e(\vec{r}_1 + \vec{r}_2)$ (and similarly for the quadrupole moment) is transformed into himself by permutation of the electrons, it transforms a vector of the antisymmetric or symmetric representation of the permutation group (with respect to the permutation of position coordinates alone) into one of the same type, so that transitions are possible only between the symmetric terms (and correspondingly between the antisymmetric terms). An example of such a term system is given by ortho- and para-helium (see Figure 20). The transitions indicated in Figure 20 illustrate the selection rules for L and for the permutation symmetry. The lowest level of the ^3S terms is therefore said to be metastable. A transition $1s\ 2s\ {}^3S \to (1s)^2\ {}^1S$ (not necessarily by means of dipole radiation) is not absolutely impossible because the exact eigenfunction of $1s\ 2s\ {}^3S$ is not a product of an antisymmetric position function and a symmetric spin function as is the case for the correct linear combinations of zeroth order. Such a transition is, however, seldom seen in the laboratory because an atom in the state $1s\ 2s\ {}^3S$ in a gas will collide with other atoms (because the ideal gas considered in XI, §2 is not really ideal!) and must undergo a transition to the state $(1s)^2\ {}^1S$ before it can emit a photon.

CHAPTER XIV

Spectra of Many-Electron Systems

1 Energy Terms in the Absence of Spin

The Hamiltonian operator for f electrons

$$H_0 = \frac{1}{2m}\sum_{i=1}^{f} \vec{P}_i^2 - \sum_{i=1}^{f}\frac{Ze^2}{r_i} + \sum_{i<k}^{f}\frac{e^2}{r_{ik}}$$

is invariant under the symmetry group of rotations $\mathscr{D}_{\mathscr{G}}$, the group $\mathscr{R}$ of reflections consisting of the elements e and r and the permutation group $\mathbf{S}_f$, that is, under the group $\mathscr{D}_{\mathscr{G}} \times \mathscr{R} \times \mathbf{S}_f$. The eigenspace of H_0 will therefore be the representation spaces of $\mathscr{D}_{\mathscr{G}} \times \mathscr{R} \times \mathbf{S}_f$. Except for "accidental" degeneracies such an eigenspace of H will be irreducible with respect to the group $\mathscr{D}_{\mathscr{G}} \times \mathscr{R} \times \mathbf{S}_f$.

If $\imath$ is such an eigenspace, then it can be reduced with respect to the representation of $\mathscr{D}_{\mathscr{G}}$ as follows:

$$\imath = \imath_1 \oplus \imath_2 \oplus \cdots. \tag{1.1}$$

The representation operators for the group $\mathbf{S}_f$ in $\mathscr{H}_b^f$ commute with the representation operators for the group $\mathscr{D}_{\mathscr{G}}$. For this reason (see AIV, §14) we may choose the basis vectors in $\imath_i$ in such a way that we obtain a matrix of the form

$$\begin{array}{cccccc} & \tilde{\imath}_L & \tilde{\imath}_{L-1} & \cdots & \tilde{\imath}_{-L} & \\ \imath_1 & \psi^L_{L1} & \psi^L_{L-11} & \cdots & \psi^L_{-L1} & \mathscr{D}_L \\ \imath_2 & \psi^L_{L2} & \psi^L_{L-12} & \cdots & \psi^L_{-L2} & \mathscr{D}_L \\ \vdots & \vdots & \vdots & & \vdots & \vdots \\ \imath_r & \psi^L_{Lr} & \psi^L_{L-1r} & \cdots & \psi^L_{-Lr} & \mathscr{D}_L \\ & \Delta & \Delta & \cdots & \Delta & \end{array} \tag{1.2a}$$

where, for fixed μ, the $\psi^L_{M\mu}$ generate the irreducible representation $\mathscr{D}_L$ of $\mathscr{D}_{\mathscr{g}}$, and for fixed M, for each M generates the same irreducible representation Δ of $\mathbf{S}_f$. The above matrix therefore also spans an irreducible subspace $\mathscr{T}$ of $\imath$ with respect to the group $\mathscr{D}_{\mathscr{g}} \times \mathbf{S}_f$. Since r commutes with respect to the elements of $\mathscr{D}_{\mathscr{g}} \times \mathbf{S}_f$, we may choose $\mathscr{T}$ such that the reflection operator results in the multiplication of the vectors by either 1 or -1. Since we assumed that $\imath$ is irreducible with respect to $\mathscr{D}_{\mathscr{g}} \times \mathbf{S}_f \times \mathscr{R}$, we therefore find that $\mathscr{T} = \imath$.

Thus the space $\mathscr{T}$ of the basis vectors (1.2a) can, in a formal sense, be written as a product space (see AIV, §14) as follows:

$$\mathscr{T} = \mathscr{T}_L \times \mathscr{T}_\Delta, \tag{1.2b}$$

where the representation operators of $\mathscr{D}_{\mathscr{g}}$ have the form $U \times \mathbf{1}$ and those of $\mathbf{S}_f$ have the form $\mathbf{1} \times V$ and, in addition, $\mathscr{T}_L$ and $\mathscr{T}_\Delta$ are irreducible with respect to the rotation group and the permutation group, respectively. We will denote the basis vectors of $\mathscr{T}_L$ by $\tilde{\psi}^L_M$ $(M = -L, \ldots, L)$, and the basis vectors of $\mathscr{T}_\Delta$ by $\tilde{\psi}_\nu$ $(\nu = 1, \ldots, r)$; we then obtain

$$\psi^L_{M\nu} = \tilde{\psi}^L_M \tilde{\psi}_\nu. \tag{1.2c}$$

Each energy eigenvalue of H_0 in $\mathscr{H}^f_b$ can therefore be characterized by an orbital angular momentum L corresponding to the representation $\mathscr{D}_L$ and by a representation Δ of $\mathbf{S}_f$.

For the case of helium with two electrons L takes the values $0, 1, 2, \ldots$ where, for each value of L with the exception of $L = 0$, both possible representations Δ of $\mathbf{S}_2$—the symmetric and antisymmetric—occur. Since the Pauli exclusion principle must be satisfied, symmetric eigenvectors of $\mathscr{H}^2_b$ must be combined with the antisymmetric vectors of $\imath^2_s$ and vice versa. The antisymmetric vectors in $\imath^2_s$ form, according to XIII, §4, an irreducible representation space for the rotation group with the representation $\mathscr{D}_0$; the symmetric vectors form a corresponding irreducible representation space with representation $\mathscr{D}_1$. Thus the symmetric eigenvectors in $\mathscr{H}^2_b$ uniquely correspond to the total spin angular momentum quantum number $S = 0$, the antisymmetric with $S = 1$. Similarly, for the case of f electrons we may make the following claim:

In $\imath^f_s$ we may reduce the group $\mathscr{D}_{\mathscr{g}} \times \mathbf{S}_f$, that is, we can choose the basis vectors u^S_{MK} in $\imath^f_s$ such that they can be ordered as a matrix as follows:

$$\begin{matrix}
\mathscr{V}_S & \mathscr{V}_{S-1} & \cdots & \mathscr{V}_{-S} & \\
\downarrow & \downarrow & & \downarrow & \\
u^S_{S1} & u^S_{S-11} & \cdots & u^S_{-S1} & \mathscr{D}_S \\
u^S_{S2} & u^S_{S-1,2} & \cdots & u^S_{-S2} & \mathscr{D}_S \\
\vdots & \vdots & & \vdots & \vdots \\
u^S_{St} & u^S_{S-1t} & \cdots & u^S_{-St} & \mathscr{D}_S \\
\Delta' & \Delta' & \cdots & \Delta' &
\end{matrix} \tag{1.3a}$$

where each row of the matrix generates a representation space of $\mathscr{D}_{\mathscr{g}}$ for the representation $\mathscr{D}_L$ and each column, in the same way, transforms according to a representation Δ' of the symmetric group $\mathbf{S}_f$. The matrix therefore generates a subspace $\mathscr{W}^S_{\Delta'}$ of $\imath^f_s$ which is irreducible with respect to the group $\mathscr{D}_{\mathscr{G}} \times \mathbf{S}_f$. The reflection r need not be described because, according to VII, §7, r behaves like the unit operator in $\imath^f_s$. The representation of $\mathscr{D}_{\mathscr{G}}$ in $\imath^f_s$ is given by $\mathscr{D}_{1/2} \times \mathscr{D}_{1/2} \times \cdots \times \mathscr{D}_{1/2} = (\mathscr{D}_{1/2})^f$. The representations $\mathscr{D}_S$ are therefore obtained by reduction of this product representation. Since $\mathscr{D}_j \times \mathscr{D}_{j'} = \mathscr{D}_{j+j'} + \mathscr{D}_{j+j'-1} + \cdots + \mathscr{D}_{|j-j'|}$, it follows that

$$\mathscr{D}_{1/2} \times \mathscr{D}_{1/2} = \mathscr{D}_1 + \mathscr{D}_0;$$
$$\mathscr{D}_{1/2} \times \mathscr{D}_{1/2} \times \mathscr{D}_{1/2} = \mathscr{D}_{3/2} + \mathscr{D}_{1/2} + \mathscr{D}_{1/2} = \mathscr{D}_{3/2} + 2\mathscr{D}_{1/2};$$
$$\mathscr{D}_{1/2} \times \mathscr{D}_{1/2} \times \mathscr{D}_{1/2} \times \mathscr{D}_{1/2} = (\mathscr{D}_{1/2})^4 = \mathscr{D}_2 + 3\mathscr{D}_1 + 2\mathscr{D}_0;$$

etc. Therefore the possible S values for f electrons are $S = f/2, f/2 - 1, \ldots \frac{1}{2}$ or 0.

In §7 we shall prove that for a particular value of S there is only one subspace $\mathscr{W}^S_{\Delta'}$ of $\imath^f_s$ and that different S correspond to different Δ'. Therefore we obtain:

$$\imath^f_s = \sum_{S=1/2,\,0}^{f/2} \oplus \mathscr{W}^S_{\Delta'}. \tag{1.3b}$$

According to AIV, §14, the subspace $\mathscr{W}^S_{\Delta'}$ with basis vectors (1.3a) can be written as a product space

$$\mathscr{W}^S_{\Delta'} = \mathscr{W}_S \times \mathscr{W}_{\Delta'}, \tag{1.3c}$$

where we find that the rotation group in $\mathscr{W}_S$ and the permutation group in $\mathscr{W}_{\Delta'}$ have irreducible representations. Let the basis vectors of $\mathscr{W}_S$ be denoted by $\tilde{u}^S_M$ $(M = -S, \ldots, S)$, and those of $\mathscr{W}_{\Delta'}$ be denoted by $\tilde{u}_\mu$ $(\mu = 1, \ldots, t)$; then we obtain

$$u^S_{M\nu} = \tilde{u}^S_M \tilde{u}_\nu. \tag{1.3d}$$

We shall now seek the antisymmetric subspace $\{\mathscr{H}^f\}_-$ of $\mathscr{H}^f$ where $\mathscr{H} = \mathscr{H}_b \times \imath_s$, or, what amounts to the same thing, to select the vectors in all the $\mathscr{T} \times \mathscr{W}^S_{\Delta'}$ for the antisymmetric representation of $\mathbf{S}_f$ in $\mathscr{H}^f$. If we use the form (1.2b) and (1.3c) for $\mathscr{T}$ and $\mathscr{W}^S_{\Delta'}$ then we obtain $\mathscr{T} \times \mathscr{W}^S_{\Delta'} = (\mathscr{T}_L \times \mathscr{W}_S) \times (\mathscr{T}_\Delta \times \mathscr{W}_{\Delta'})$. Therefore we need to find those vectors in $\mathscr{T}_\Delta \times \mathscr{W}_{\Delta'}$ which transform according to the antisymmetric representation of $\mathbf{S}_f$. The representation of $\mathbf{S}_f$ in $\mathscr{T}_\Delta \times \mathscr{W}_{\Delta'}$ is $\Delta \times \Delta'$. When and how often does $\Delta \times \Delta'$ contain the antisymmetric representation? To provide an answer to this question we shall consider the following more general question for an arbitrary group: How often does the representation $\Delta_\rho \times \Delta_\sigma$ (the product of two irreducible representations) contain an irreducible representation Δ_ν? The character χ of $\Delta_\rho \times \Delta_\sigma$ clearly satisfies $\chi(a) = \chi_\rho(a)\chi_\sigma(a)$. According to AV, §8 we find that

$$\chi_\rho(a)\chi_\sigma(a) = \sum_\nu c^\nu_{\rho\sigma}\chi_\nu(a), \tag{1.4}$$

where the $c_{\rho\sigma}^{\nu}$ are integers and specify the desired multiplicity of the representation Δ_ν in $\Delta_\rho \times \Delta_\sigma$. For unitary representations, from the orthogonality relations AV, §8 we therefore obtain:

$$c_{\rho\sigma}^{\nu} = \frac{1}{h} \sum_a \overline{\chi_\nu(a)} \chi_\rho(a) \chi_\sigma(a). \tag{1.5}$$

If Δ_ν is the antisymmetric representation Δ_- of $\mathbf{S}_f$, we then have $\chi_-(P) = (-1)^P$, and we therefore obtain

$$c_{\rho\sigma}^{-} = \frac{1}{n!} \sum_P (-1)^P \chi_\rho(P) \chi_\sigma(P). \tag{1.6}$$

To each representation Δ_σ of $\mathbf{S}_f$ there exists an "associated" representation $\Delta_{\sigma'}$, the matrix of which is $(-1)^P$ times the complex conjugate matrix of Δ_σ. For Δ_- as the antisymmetric representation and $\bar{\Delta}_\sigma$ as the conjugate complex representation of Δ_σ we obtain $\Delta_{\sigma'} = \bar{\Delta}_\sigma \times \Delta_-$. Therefore we may rewrite (1.6) as follows:

$$c_{\rho\sigma}^{-} = \frac{1}{n!} \sum_P \chi_\rho(P) \overline{\chi_{\sigma'}(P)}. \tag{1.7}$$

According to the orthogonality relations the right-hand side of the above equation is either 0 or 1 according to whether $\sigma' \neq \rho$ or $\sigma' = \rho$.

Therefore we find that $\mathscr{T}_\Delta \times \mathscr{W}_{\Delta'}$ contains the antisymmetric representation either once or not at all depending whether Δ' is or is not the associated representation for Δ. Since Δ' is uniquely associated with the value of S, this is also the case for the representation Δ which corresponds to an eigenvalue of H_0 permitted by the Pauli exclusion principle. All such $\mathscr{T}_\Delta$ and corresponding terms having a representation Δ which do not have associated representations in $\imath_s^f$ do not occur in nature. For fixed electron number f the permitted representations Δ will be distinguished by the index S. The only vector from $\mathscr{T}_\Delta \times \mathscr{W}_{\Delta'}$ which transforms according to the antisymmetric representation of $\mathbf{S}_f$ is easy to determine. Since Δ' is the associated representation for Δ, we find that $r = t$. For the normalized vector.

$$\tilde{\phi} = \frac{1}{\sqrt{r}} \sum_k \tilde{\psi}_k \tilde{u}_k \tag{1.8}$$

we obtain

$$\begin{aligned} P\tilde{\phi} &= \frac{1}{\sqrt{r}} \sum_k (P\tilde{\psi}_k)(P\tilde{u}_k) \\ &= \frac{1}{\sqrt{r}} \sum_{kln} \tilde{\psi}_l P_{lk} u_n \bar{P}_{nk} (-1)^P, \end{aligned} \tag{1.9}$$

where P_{lk} is the matrix of the representation Δ and $\bar{P}_{nk}(-1)^P$ is the matrix corresponding to the representation Δ'. Since the representation of $\mathbf{S}_f$ is unitary, we obtain $\sum_k P_{lk}\bar{P}_{nk} = \delta_{ln}$, and from (1.9) we therefore obtain

$$P\tilde{\phi} = (-1)^P\tilde{\phi}. \tag{1.10}$$

$\tilde{\phi}$ is the desired vector from $\mathscr{T}_\Delta \times \mathscr{W}_{\Delta'}$. Since $\mathscr{T} \times \mathscr{W}^S_{\Delta'} = (\mathscr{T}_L \times \mathscr{W}_S) \times (\mathscr{T}_\Delta \times \mathscr{W}_{\Delta'})$, the antisymmetric subspace $\{\mathscr{T} \times \mathscr{W}^S_{\Delta'}\}_-$ of $\mathscr{T} \times \mathscr{W}^S_{\Delta'}$ is given by

$$(\mathscr{T}_L \times \mathscr{W}_S) \times \{\tilde{\phi}\}, \tag{1.11}$$

where $\{\tilde{\phi}\}$ is the one-dimensional subspace of $\mathscr{T}_\Delta \times \mathscr{W}_\Delta$ spanned by $\tilde{\phi}$. If we introduce the identification (1.2c), (1.3d), then, for the basis vectors of $\{\mathscr{T} \times \mathscr{W}^S_{\Delta'}\}$ we may use the

$$\phi_{m_L m_S} = \frac{1}{\sqrt{r}} \sum_k \psi^L_{m_L k} u^S_{m_S k}, \tag{1.12}$$

where $m_L = -L, \ldots, L$ and $m_S = -S, \ldots, S$. From (1.11) we may write the basis vectors as follows

$$\tilde{\psi}^L_{m_L} \tilde{u}^S_{m_S} \tilde{\phi} \tag{1.13}$$

from which we may easily recognize their transformation properties under rotations and permutations. $\{\mathscr{T} \times \mathscr{W}^S_{\Delta'}\}$ is the eigenspace corresponding to an eigenvalue of the energy operator H_0 where the influence of the spin upon the energy has been neglected. Such a term is usually denoted by ${}^{2S+1}X$ where $X = S, P, D, \ldots$ depending on whether $L = 0, 1, 2, \ldots, 2S + 1$ and is called the multiplicity of the term; it also characterizes the total spin *and* the representation Δ of the permutation group.

2 Fine Structure Splitting of Spectral Lines

The total energy operator must have the form $H = H_0 + H'$ where H' describes the effect of spin upon the energy. We again consider the operator $H(\lambda) = H_0 + \lambda H'$, where we permit λ to increase from 0 to 1. The eigenvalues of H_0, called *parent* terms, are degenerate of order $(2L + 1)\cdot(2S + 1)$ and correspond to the eigenvectors $\phi_{m_L m_S}$. The representation of the rotation group $\mathscr{D}_{\mathscr{G}}$ in this eigenspace is $\mathscr{D}_L \times \mathscr{D}_S$, a result which directly follows from (1.13). Since $\mathscr{D}_L \times \mathscr{D}_S = \mathscr{D}_{L+S} + \mathscr{D}_{L+S-1} + \cdots + \mathscr{D}_{|L-S|}$, the parent terms will be split into several components corresponding to the total angular momentum quantum numbers $J = L + S, L + S - 1, \ldots, |L - S|$ (that is, for $L > S$, into $2S + 1$ components) because H commutes with the rotation operators $U_D = V_D \times R_D$ (where V_D and R_D act in $\mathscr{H}^f_b$ and r^f_s, respectively). We shall denote the individual components of the energy level ${}^{2S+1}X$ by the symbol ${}^{2S+1}X_J$.

It is easy to determine the correct linear combinations for a perturbation computation. Since the $\phi_{m_L m_S}$ transform like products $\tilde{\psi}_{m_L}\tilde{U}_{m_S}$, from XI (10.14) we obtain:

$$\Gamma^J_M = \sum_{m_L+m_S=M} d^J_{m_L m_S}(L, S)\phi_{m_L m_S}. \tag{2.1}$$

According to VIII (5.8) the operator H has the form: $H = H_0 + H'$ where $H' = \sum_{k=1}^{6} H_k$ providing we exclude external fields. Since H_1, H_2, H_6 commute with V_D, they may only result in a displacement of the energy levels from those of H_0. Only H_3, H_4 and H_5 contribute to splitting. From experience we find that H_4 can, for the most part, be neglected in comparison to H_3 and H_5. Therefore we find that the only component of H' which significantly affects the splitting is an expression of the form $H'' = \sum_{i=1}^{1} H^{ii}$ where the H^{ii} may easily be written as diagonal elements of a tensor H^{ik}. Here H^{ik} is a tensor in the sense that, for fixed k, the H^{ik} behave, under V_D as a vector; similarly, for fixed i, they behave like a vector under R_D. Thus, in the subspace spanned by $\phi_{m_L m_S}$ (or Γ^J_M) we obtain

$$\langle \phi_{m'_L m'_S}, H^{ii}\phi_{m_L m_S}\rangle = \tau\langle \phi_{m'_L m'_S}, L_i S_i \phi_{m_L m_S}\rangle \tag{2.2}$$

because the matrix elements of the operators H^{ik} are, up to a factor, uniquely determined by the transformation properties of the H^{ik}. We shall now prove the above claim.

Under transformations of the form V_D the H^{ik} transform according to $\mathscr{D}_1$, and under transformations of the form R_D according to $\mathscr{D}_1$. In addition to the rotation group $\mathscr{D}_{\mathscr{g}}$ let us consider the group $\mathscr{D}_{\mathscr{g}} \times \mathscr{D}_{\mathscr{g}}$, that is, the outer product of the rotation group with itself. Its elements are clearly all pairs $D_i \times D_k$ of two rotations D_i and D_k (note that $D_i \times D_k$ is not the usual product of two rotations!). We have often investigated the representations of a group $\mathscr{G}_1 \times \mathscr{G}_2$ (see, for example, $\mathscr{D}_{\mathscr{g}} \times \mathbf{S}_f$ in §1). If the irreducible representations of $\mathscr{G}_1$ and $\mathscr{G}_2$ are already known, then we know the irreducible representations of $\mathscr{G}_1 \times \mathscr{G}_2$. For example, if $\imath_1$ is an irreducible representation space of $\mathscr{G}_1$ and $\imath_2$ of $\mathscr{G}_2$ then $\imath_1 \times \imath_2$ is an irreducible representation space for $\mathscr{G}_1 \times \mathscr{G}_2$; conversely, if $\imath$ is an irreducible representation space of $\mathscr{G}_1 \times \mathscr{G}_2$ then, according to theorems in AIV, §14, we find that $\imath$ can be written in the form $\imath_1 \times \imath_2$ (as we have often used earlier) where the representation operators of $\mathscr{G}_1$, $\mathscr{G}_2$ have the form $U \times \mathbf{1}$ and $\mathbf{1} \times V$, respectively. If Δ is the representation of $\mathscr{G}_1$ in $\imath_1$ and Δ' is the representation of $\mathscr{G}_2$ in $\imath_2$ then we denote the representation of $\mathscr{G}_1 \times \mathscr{G}_2$ in $\imath_1 \times \imath_2$ by $\Delta \divideontimes \Delta'$. The irreducible representations of the group $\mathscr{D}_{\mathscr{g}} \times \mathscr{D}_{\mathscr{g}}$ are therefore the $\mathscr{D}_j \divideontimes \mathscr{D}_{j'}$. The H^{ik} are therefore subject to the representation $\mathscr{D}_1 \divideontimes \mathscr{D}_1$; the $\phi_{m_L m_S}$ are subject to the irreducible representation $\mathscr{D}_L \divideontimes \mathscr{D}_S$ (see, for example, (1.13)).

Since $(\mathscr{D}_1 \divideontimes \mathscr{D}_1) \times (\mathscr{D}_L \divideontimes \mathscr{D}_S) = \sum_{S'=S-1}^{S+1} \sum_{L'=L-1}^{L+1} \mathscr{D}_{L'} \divideontimes \mathscr{D}_{S'}$, the expansion coefficients of the $H^{ik}\phi_{m_L m_S}$ with respect to the $\phi_{m_L m_S}$ are identical to those of the $L_i S_k \phi_{m_L m_S}$ up to a factor (see XIII, §2).

If, in the eigenspace we select the basis Γ^J_M (defined in (2.1)) instead of the basis $\phi_{m_L m_S}$, then the matrix of $\sum_{i=1}^{3} H_{ii}$ is diagonal and equal to the

τ multiple of $\sum_{i=1}^{3} L_i S_i = \vec{L} \cdot \vec{S}$. Since $2\vec{L} \cdot \vec{S} = \vec{J} - \vec{L}^2 - \vec{S}^2$ the eigenvalues of H are split, in a first approximation, by a magnitude

$$\Delta E_J = \frac{\tau}{2}[J(J+1) - L(L+1) - S(S+1)]. \tag{2.3}$$

The energy difference between two successive energy levels is given by

$$\Delta E_{J+1} - \Delta E_J = \tau(J+1). \tag{2.4}$$

If the factor τ is positive, we speak of a normal, otherwise of a reversed multiplet.

Here we note that Landé's interval rule, given by (2.3), holds, of course, only for multiplet splitting which is small compared to the separation between the parent terms. For some atoms such is not the case (see §3).

3 Structure Principles

In order to obtain an overview of the different energy levels of a complicated atom it is useful to think of the eigenvalues of the operator

$$H_0 = \frac{1}{2m}\sum_i \vec{P}_i^2 - \sum_i \frac{Ze^2}{r_i} + \sum_{i<k} \frac{e^2}{r_{ik}}$$

as continuously arising from an operator

$$H_{00} = \frac{1}{2m}\sum_i \vec{P}_i^2 + \sum_i V(r_i)$$

where $V(r_i)$ takes into consideration the mutual interaction of the electrons in addition to $-Ze^2/r_i$. If we then think of "switching on" in a continuous manner from H_{00} to the complete mutual interaction of the electrons, then we may easily determine which energy levels of H_0 result from the energy levels of H_{00}. The energy values of H_{00} are equal to $\varepsilon_{n_1 l_1} + \varepsilon_{n_2 l_2} + \cdots + \varepsilon_{n_f l_f}$ where ε_{nl} is an eigenvalue of the operator $\frac{1}{2}\vec{P}^2 + V(r)$ corresponding to the principle quantum number n and the angular momentum quantum number l. This situation is qualitatively analogous to that described in XI, §8. Let $\imath_{nl}$ denote the eigenspace corresponding to the eigenvalue ε_{nl} in $\mathscr{H}_b$. Then the eigenspace $\mathscr{T}$ corresponding to the eigenvalue $\varepsilon_{n_1 l_1} + \cdots + \varepsilon_{n_f l_f}$ of H_{00} in $\mathscr{H}_b^f$ will be spanned by $\imath_{n_1 l_1} \times \imath_{n_2 l_2} \times \cdots \times \imath_{n_f l_f}$ and by all the other spaces obtained from the latter by the application of the permutation operators. Let χ_{nlm} (where $m = -l, -l+1, \ldots, l$) denote the basis vectors of $\imath_{nl}$; then the eigenspace $\mathscr{T}$ for the above eigenvalue is spanned by the set of all $P\chi_{n_1 l_1 m_1}(1), \chi_{n_2 l_2 m_2}(2), \ldots, \chi_{n_f l_f m_f}(f)$ where P denotes all elements of $\mathbf{S}_f$ and m all values of $-l$ to l.

The eigenvalues $\varepsilon_{n_1 l_1} + \cdots + \varepsilon_{n_f l_f}$ are called the "electron configuration," and is often denoted by $(n_1 l_1)(n_2 l_2) \cdots (n_f l_f)$. For the case in which some of the $(n_i l_i)$ are identical, we often abbreviate the above expression as follows: $(n_1 l_1)^{\alpha_1}(n_2 l_2)^{\alpha_2} \cdots (n_\tau l_\tau)^{\alpha_\tau}$ where α_ν is the number of the same $(n_\nu l_\nu)$, that is, the number of "equivalent" electrons.

The reflection character of an electron configuration is clearly $(-1)^{\Sigma_i l_i}$. Therefore all energy levels of H arising continuously from this electron configuration as the mutual electron interaction is "turned on" must have the same reflection character $(-1)^{\Sigma_i l_i}$.

In order to determine the energy levels of H_0 into which a given electron configuration is split by the interaction of the electrons we must reduce the eigenspace $\mathscr{T}$ corresponding to the eigenvalue $\varepsilon_{n_1 l_1} + \cdots + \varepsilon_{n_f l_f}$ of H_{00} according to the group $\mathscr{D}_{\mathscr{G}} \times \mathbf{S}_f$, that is, we must seek basis vectors $\mathring{\psi}^L_{m_L k}$ which may be arranged in a matrix of the form (1.2) where the rows and columns transform according to irreducible representations $\mathscr{D}_L$ of $\mathscr{D}_{\mathscr{G}}$ and Δ of $\mathbf{S}_f$, respectively. Each such matrix will then give an irreducible representation of $\mathscr{D}_{\mathscr{G}} \times \mathbf{S}_f$. By switching on the interaction the basis vectors may be changed, that is, $\mathring{\psi}^L_{M_L k} \to \psi^L_{M_L k}$. Nevertheless, the representations $\mathscr{D}_L$ and Δ are not changed. Therefore the type of terms of H_0 with respect to their representation in the form (1.2a) is exactly known from the electron configuration. A detailed solution of this problem in the manner described above requires detailed knowledge of the representations of the group $\mathbf{S}_f$ and will be carried out in §7. Since each of these matrices (1.2a) (where ψ is replaced by $\mathring{\psi}$) can only be combined with a particular value of the spin quantum number S, we may proceed as follows: Instead of determining the $\mathring{\psi}^L_{M_L k}$ and later establishing the connection with the $U^S_{m_S k}$ by means of (1.12), we immediately seek the

$$\mathring{\phi}_{m_L m_S} = \frac{1}{\sqrt{r}} \sum_k \mathring{\psi}^L_{m_L k} U^S_{m_S k}$$

which lie in $\{\mathscr{H}^f\}_-$. To do this we proceed by making use of the set of eigenvectors of H_{00} in $\{\mathscr{H}^f\}_- = \{(\mathscr{H}_b \times \imath_s)^f\}_-$. In $\mathscr{H}^f$ the eigenspace of the eigenvalues $\varepsilon_{n_1 l_1} + \cdots + \varepsilon_{n_f l_f}$ is spanned by the vectors

$$P\chi_{n_1 l_1 m_1}(1)u_{\mu_1}(1)\chi_{n_2 l_2 m_2}(2)u_{\mu_2}(2) \cdots \chi_{n_f l_f m_f}(f)u_{\mu_f}(f), \tag{3.1}$$

where m_i takes on values $-l_i$ to l_i, μ_i takes on values $-\frac{1}{2}$ and $\frac{1}{2}$ and P runs through all permutations of $\mathbf{S}_f$. The basis vectors for the subspace $\mathscr{T}_-$ corresponding to $\{\mathscr{H}^f\}_-$ are obtained from the Slater determinant XIII (4.8) as follows:

$$\psi_{n_1 l_1 m_1 \mu_1, n_2 l_2 m_2 \mu_2, \ldots, n_f l_f m_f \mu_f}$$
$$= \frac{1}{\sqrt{f!}} \sum_P (-1)^P P\chi_{n_1 l_1 m_1}(1)u_{\mu_1}(1) \cdots \chi_{n_f l_f m_f}(f)u_{\mu_f}(f). \tag{3.2}$$

Here no pairs $(n_i l_i m_i \mu_i)$ may be identical, otherwise the right-hand side of (3.2) would be zero. Thus interchanging the order of the $(n_1 l_1 m_1 \mu_1)$, $(n_2 l_2 m_2 \mu_2), \ldots, (n_f l_f m_f \mu_f)$ as indices for $\psi_{n_1 l_1 m_1 \mu_1, \ldots, n_f l_f m_f \mu_f}$ does not lead to a new vector; we may therefore impose an order, for example, lexiographic order, as follows: $n_i l_i m_i \mu_i$ may be on the left of $n_k l_k m_k \mu_k$ only if $n_i > n_k$ or if $n_i = n_k$ then $l_i > l_k$ or if $n_i = n_k, l_i = l_k, m_i > m_k$ or if $n_i = n_k, l_i = l_k, m_i = m_k$, $\mu_i > \mu_k$.

We now again consider the transformations $V_{D_1} \times R_{D_2}$ (where D_1 is independent of D_2) as a representation of $\mathscr{D}_{\mathscr{G}} \times \mathscr{D}_{\mathscr{G}}$. The desired $\mathring{\phi}_{m_L m_S}$ generates an irreducible representation $\mathscr{D}_L ⋇ \mathscr{D}_S$ with respect to $\mathscr{D}_{\mathscr{G}} \times \mathscr{D}_{\mathscr{G}}$. Since V_D commutes with R_D, the reduction of $\mathscr{T}_-$ can again be expressed in the form of a matrix, where the rows and columns transform according to representations $\mathscr{D}_L$ and $\mathscr{D}_S$ with respect to V_D and U_D, respectively. The matrices obtained in this way yields the desired $\mathring{\phi}_{m_L m_S}$. The fact that we may use the entire group $\mathscr{D}_{\mathscr{G}} \times \mathscr{D}_{\mathscr{G}}$ instead of the group $\mathscr{D}_{\mathscr{G}}$ rests upon the fact that H_0 commutes with all $V_{D_1} \times R_{D_2}$. Including the influence of spin we find that the entire energy operator H commutes only with $U_D = V_D \times R_D$. With respect to U_D the matrix of the $\phi_{m_L m_S}$ satisfies the representation $\mathscr{D}_L \times \mathscr{D}_S = \mathscr{D}_{L+S} + \cdots + \mathscr{D}_{|L-S|}$ where the latter corresponds to a multiplet splitting.

With the help of the characters it is not difficult to determine the irreducible representations $\mathscr{D}_L ⋇ \mathscr{D}_S$ in $\mathscr{T}_-$ and their multiplicity with which they occur. Since all the transformations V_D (as R_D) are equivalent to diagonal transformations we need only consider rotations D about the third axis to find the characters:

$$V_{D_1} \psi_{n_1 l_1 m_1 \mu_1 \ldots n_f l_f m_f \mu_f} = e^{i(m_1 + m_2 + \ldots + m_f)\alpha} \psi_{n_1 l_1 m_1 \mu_1 \ldots n_f l_f m_f \mu_f} \tag{3.3}$$

and

$$R_{D_2} \psi_{n_1 l_1 m_1 \mu_1, \ldots n_f l_f m_f \mu_f} = e^{i(\mu_1 + \mu_2 + \ldots + \mu_f)\beta} \psi_{n_1 l_1 m_1 \mu_1 \ldots n_f l_f m_f \mu_f}. \tag{3.4}$$

The character of the representation of $\mathscr{D}_{\mathscr{G}} \times \mathscr{D}_{\mathscr{G}}$ in $\mathscr{T}_-$ is therefore

$$\sum_{n_i l_i m_i \mu_i} e^{i(\alpha \Sigma_i m_i + \beta \Sigma \mu_i)}, \tag{3.5}$$

where the summation is over permitted values of $n_i l_i m_i \mu_i$. The character of $\mathscr{D}_L ⋇ \mathscr{D}_S$ is given by

$$\sum_{m_L = -L}^{L} \sum_{m_S = -S}^{S} e^{i(\alpha m_L + \beta m_S)}. \tag{3.6}$$

We will now show how we may decompose the character (3.5) according to the characters (3.6). We begin by writing down all possible symbol sequences in lexiographic order (where $n_i l_i$ is fixed, $m_i = -l_i, \ldots, l_i$; $\mu_i = -\frac{1}{2}, \frac{1}{2}$):

$$(n_1 l_1 m_1 \mu_1)(n_2 l_2 m_2 \mu_2) \cdots (n_f l_f m_f \mu_f) \tag{3.7}$$

(here two identical symbol sets may not occur). Each such symbol set characterizes a basis vector for $\mathscr{T}_-$. After each symbol set we may write $\sum_i m_i$ and $\sum_i \mu_i$. This pair of values $\sum_i m_i$, $\sum_i \mu_i$ can be arranged in a table of the form

$\sum_i m_i = m_L$ \ $\sum_i \mu_i = m_S$	σ	σ_{-1}	σ_{-2}	$\cdots$
δ	$\cdots$	$\cdots$	$\cdots$	$\cdots$
$\delta - 1$	$\cdots$	$\times \times \cdots$	$\cdots$	$\cdots$
$\delta - 2$	$\cdots$	$\cdots$	$\times \times \times \cdots$	$\cdots$
$\vdots$				

marked by $\times$'s. For each desired representation $\mathscr{D}_L \circledast \mathscr{D}_S$ at each of the locations m_L, m_S for $m_L = -L, -L+1, \ldots, L$, $m_S = -S, -S+1, \ldots, S$ an $\times$ has to be deleted (that is, in total $(2L+1)(2S+1)$ $\times$'s). We begin with the largest possible value of m_L, and for this value of m_L, the largest possible value of m_S.

EXAMPLE. Three equivalent p electrons (that is, $l = 1$):

$(n$	1	1	$\frac{1}{2})$	$(n$	1	1	$-\frac{1}{2})$	$(n$	1	0	$\frac{1}{2})$	$\sum_i m_i = 2$	$\sum_i \mu_i = \frac{1}{2}$
(		1	$\frac{1}{2})$	(		1	$-\frac{1}{2})$	(		-1	$\frac{1}{2})$	1	$\frac{1}{2}$
(		1	$\frac{1}{2})$	(		0	$\frac{1}{2})$	(		0	$-\frac{1}{2})$	1	$\frac{1}{2}$
(		1	$\frac{1}{2})$	(		0	$\frac{1}{2})$	(		-1	$\frac{1}{2})$	0	$\frac{3}{2}$
(		1	$\frac{1}{2})$	(		0	$\frac{1}{2})$	(		-1	$-\frac{1}{2})$	0	$\frac{1}{2}$
(		1	$\frac{1}{2})$	(		0	$-\frac{1}{2})$	(		-1	$\frac{1}{2})$	0	$\frac{1}{2}$
(		1	$-\frac{1}{2})$	(		0	$\frac{1}{2})$	(		-1	$\frac{1}{2})$	0	$\frac{1}{2}$

It is not necessary to include additional symbol sets because they would result in negative values for the two sums $\sum_i m_i$ or $\sum_i \mu_i$ and do not permit further representations of $\mathscr{D}_L \circledast \mathscr{D}_S$. The table has the form

m_S \ m_L	2	1	0
$\frac{3}{2}$			$\times$
$\frac{1}{2}$	$\times$	$\times\times$	$\times\times\times$

$L = 2$, $S = \frac{1}{2}$ are the largest values of m_L and for $m_L = 2$ of m_S. For each of the $m_L, m_S = 2, \frac{1}{2}; 1, \frac{1}{2}; 0, \frac{1}{2}$ one of the $\times$'s has to be deleted. For the rest the largest values of m_L, m_S are given by $m_L = 1$, $m_S = \frac{1}{2}$, that is, $L = 1$, $S = \frac{1}{2}$ and we finally obtain $L = 0$, $S = \frac{3}{2}$. We therefore obtain three matrices corresponding to representations $\mathscr{D}_2 \circledast \mathscr{D}_{1/2}$; $\mathscr{D}_1 \circledast \mathscr{D}_{1/2}$ and $\mathscr{D}_0 \circledast \mathscr{D}_{3/2}$. From the eigenvalue $3\varepsilon_{n_1}$ of H_{00}, that is, from the electron configuration $(n1)^3$ we obtain three energy levels of H_0—the ${}^2D$, ${}^2P$ and ${}^4S$ terms.

In practice three rules greatly simplify the process of finding the energy levels of H_0 corresponding to a given electron configuration. We note that $2(2l+1)$ electrons with the same (n, l) values are said to form a full shell because it is not possible to have more than $2(2l+1)$ electrons with the same (n, l) value. Thus we obtain the following rule: Full shells in an electron configuration do not increase the term manifold, that is, they may be neglected in the determination of the terms.

The proof of this fact follows from the fact that in each symbol series all possible (n, l, m, μ) occur for a full (n, l) shell and therefore do not contribute to $\sum_i m_i$ and $\sum_i \mu_i$.

The second rule states that in each symbol set we can separately treat groups of equivalent electrons and, afterwards we can combine such groups without taking into account the Pauli exclusion principle.

For example, if we are given two groups of equivalent electrons which correspond to a set of values m'_L, m'_S and m''_L, m''_S (arising from $\sum_i m_i$ and $\sum_i \mu_i$), then it is easy to see that for both groups taken together the pair m_L, m_S can take on all combinations of values $m_L = m'_L + m''_L$ and $m_S = m'_S + m''_S$.

Therefore, if L', S' is a term for the first group and L'', S'' is one for the second group, then, for both groups taken together we obtain all possible pairs L, S for which $L = L' + L'', L' + L'' - 1, \ldots, |L' - L''|$ and $S = S' + S'', S' + S'' - 1, \ldots, |S' - S''|$. We may therefore arbitrarily combine terms of the individual groups.

The third rule states that h and $2(2l + 1) - h$ equivalent (n, l) electrons yield the same term manifold. This results from the fact that to each symbol set $(nlm_1\mu_1)(nlm_2\mu_2)\cdots$ of h electrons we can append a symbol set for the remaining $2(2l + 1) - h$ equivalent electrons. For the series of $2(2l + 1) - h$ for $\sum_i m_i, \sum_i \mu_i$ we obtain the negative value of $\sum_i m_i, \sum_i \mu_i$ for the h electrons, from which it follows that for both cases we obtain the same manifold of pairs $\sum_i m_i$ and $\sum_i \mu_i$.

On the basis of these three rules it is only necessary to specify the terms for equivalent electrons for the most important cases. These are given in Table 1.

Table 1 Terms of equivalent electrons.

Electron configuration	Terms
p^2	^1S ^1D ^3P
p^3	^2P ^2D ^4S
d^2	^1S ^1D ^1G ^3F
d^3	^2P $^2D(2)$ ^2F ^2G ^2H ^4P ^4F
d^4	$^1S(2)$ $^1D(2)$ ^1F $^1G(2)$ ^1I $^3P(2)$ ^3D $^2F(2)$ ^3G ^3H ^5D
d^5	^2S ^2P $^2D(3)$ $^2F(2)$ $^2G(2)$ ^2H ^2I ^4P ^4D ^4F ^4G ^6S

The procedure described above for determining the energy level terms from the electron configuration by first "switching on" the interactions of the electrons and then the effects of spin is called the structure principle, and will be discussed in the next section on the so-called periodic system of the elements. In addition, some examples for term schemes will be discussed. In order to make qualitative statements concerning the location of the terms we only need the fact that the terms of higher multiplicity (that is, large spin values) which arise from an electron configuration have lower energy. We have seen this for the case of two electrons in helium. In some cases the multiplet splitting is not small compared to the separation between energy levels (the so-called parent terms) which arises from an electron configuration taking only the Coulomb interaction into account. Thus the structure principles establish the correct energy manifold but do not necessarily establish the correct location qualitatively. If the effect of spin is small we shall speak of Russell–Saunders coupling. In this case, as described above, the orbital angular momentum of the electrons with the quantum number

l_i are added to the total orbital angular momentum with the quantum number L; similarly, the spins are added to the total spin S. Then the influence of the spin splits the parent terms in different fine structure terms corresponding to different values J of the quantum number for total angular momentum.

In the extreme opposite case the influence of spin is large compared to the Coulomb interaction of electrons. This case occurs for large values of Z (H_5 is, according to VIII (5.8), proportional to Z!) and is called *jj* coupling. In order to estimate the location of terms it is advantageous, as in the case of a single electron, to examine the effect of "turning on" the spin perturbation. The terms ε_{nl} will split into components ε_{nlj} as described for alkali atoms. Then it is necessary to examine the effect of "turning on" the smaller perturbation of the Coulomb interaction between the electrons. This may be carried out by reducing the representation of the rotation group in the eigenspace corresponding to the energy eigenvalue $\varepsilon_{n_1 l_i j_i} + \cdots + \varepsilon_{n_f l_f j_f}$. The term manifold obtained from the electron configuration $(n_1 l_1)(n_2 l_2) \cdots (n_f l_f)$ in the manner described above must be the same as that obtained earlier in this section.

4 The Periodic System of the Elements

In Table 2 the electron configuration of the ground state of atoms and the symbol for the ground state are given. For helium the $1s$ shell is filled; thus the ground state must be an $^1S_0$ term. For lithium one electron occupies the $2s$ shell. The filled $1s$ shell does not contribute to the term manifold; thus, we may treat lithium as a one-electron problem, as we have done in XII, §8. Since the $1s$ shell is occupied, the third electron cannot have the configuration $1s$ (see Figure 10). For beryllium the $2s$ shell is already filled; thus the ground state is again an $^1S_0$ term. We may expect that the term scheme for beryllium will be similar to that of helium, corresponding to the electron configurations $(1s)^2(2s)(ns)$, $(1s)^2(2s)(np)$, $(1s)^2(2s)(nd), \ldots$. With the exception of the $(1s)^2(2s)^2$ term, the terms are singlet and triplet; according to the above rule the triplet terms have lower energy than the corresponding singlet terms. This is shown in the left side of Figure 15.

In addition there are terms (see Figure 15) which arise from the configuration $(1s)^2(2p)(np)$ and similar configurations. In part they lie in the continuum, that is, they are not regular discrete energy eigenvalues (and are therefore called anomolous terms) because a probability for ionization of atoms exists but is very small, and as a result there is at most a radiative transition to a lower energy level.

From boron the $2p$ shell will be filled until we reach neon, a noble gas. Later we shall discuss the term schemes for the elements carbon and oxygen. The further structure of the elements is clearly evident in Table 2. For argon the $3p$ shell will be fully occupied. For potassium instead of a $3d$ electron we find a $4s$ electron which is energetically more favorable. Theoretically this

Table 2 Electron configuration and ground state of the elements.

Element	K	L		M			N				O					Ground state
	1*s*	2*s*	2*p*	3*s*	3*p*	3*d*	4*s*	4*p*	4*d*	4*f*	5*s*	5*p*	5*d*	5*f*	5*g*	
1. H	1															$^2S_{1/2}$
2. He	2															$^1S_0$
3. Li	2	1														$^2S_{1/2}$
4. Be	2	2														$^1S_0$
5. B	2	2	1													$^2P_{1/2}$
6. C	2	2	2													$^3P_0$
7. N	2	2	3													$^4S_{3/2}$
8. O	2	2	4													$^3P_2$
9. F	2	2	5													$^2P_{3/2}$
10. Ne	2	2	6													$^1S_0$
11. Na	2	2	6	1												$^2S_{1/2}$
12. Mg	2	2	6	2												$^1S_0$
13. Al	2	2	6	2	1											$^2P_{1/2}$
14. Si	2	2	6	2	2											$^3P_0$
15. P	2	2	6	2	3											$^4S_{3/2}$
16. S	2	2	6	2	4											$^3P_2$
17 Cl.	2	2	6	2	5											$^2P_{3/2}$
18. A	2	2	6	2	6											$^1S_0$
19. K	2	2	6	2	6		1									$^2S_{1/2}$
20. Ca	2	2	6	2	6		2									$^1S_0$
21. Sc	2	2	6	2	6	1	2									$^2D_{3/2}$
22. Ti	2	2	6	2	6	2	2									$^3F_2$
23. V	2	2	6	2	6	3	2									$^4F_{3/2}$
24. Cr	2	2	6	2	6	5	1									$^7S_3$
25. Mn	2	2	6	2	6	5	2									$^6S_{5/2}$
26. Fe	2	2	6	2	6	6	2									$^5D_4$

Table 2 (*Continued*)

27. Co	2	2	6	2	6	7	2						$^4F_{9/2}$	
28. Ni	2	2	6	2	6	8	2							$^3F_4$
29. Cu	2	2	6	2	6	10	1						$^2S_{1/2}$	
30. Zn	2	2	6	2	6	10	2							$^1S_0$
31. Ga	2	2	6	2	6	10	2	1					$^2P_{1/2}$	
32. Ge	2	2	6	2	6	10	2	2						$^3P_0$
33. As	2	2	6	2	6	10	2	3					$^4S_{3/2}$	
34. Se	2	2	6	2	6	10	2	4						$^3P_2$
35. Br	2	2	6	2	6	10	2	5					$^2P_{3/2}$	
36. Kr	2	2	6	2	6	10	2	6						$^1S_0$
37. Rb	2	2	6	2	6	10	2	6			1		$^2S_{1/2}$	
38. Sr	2	2	6	2	6	10	2	6			2			$^1S_0$
39. Y	2	2	6	2	6	10	2	6	1		2		$^2D_{3/2}$	
40. Zr	2	2	6	2	6	10	2	6	2		2			$^3F_2$
41. Nb(Cb)	2	2	6	2	6	10	2	6	4		1		$^6D_{1/2}$	
42. Mo	2	2	6	2	6	10	2	6	5		1			$^7S_3$
43. Tc	2	2	6	2	6	10	2	6	5		2		$(^6S_{5/2})$	
44. Ru	2	2	6	2	6	10	2	6	7		1			$^5F_5$
45. Rh	2	2	6	2	6	10	2	6	8		1		$^4F_{9/2}$	
46. Pd	2	2	6	2	6	10	2	6	10					$^1S_0$
47. Ag	2	2	6	2	6	10	2	6	10		1		$^2S_{1/2}$	
48. Cd	2	2	6	2	6	10	2	6	10		2			$^1S_0$
49. In	2	2	6	2	6	10	2	6	10		2	1	$^2P_{1/2}$	
50. Sn	2	2	6	2	6	10	2	6	10		2	2		$^3P_0$
51. Sb	2	2	6	2	6	10	2	6	10		2	3	$^4S_{3/2}$	
52. Te	2	2	6	2	6	10	2	6	10		2	4		$^3P_2$
53. I	2	2	6	2	6	10	2	6	10		2	5	$^2P_{3/2}$	
54. Xe	2	2	6	2	6	10	2	6	10		2	6		$^1S_0$
	2	8			18									

Table 2 (*Continued*)

Element	K	L	M	N 4s	N 4p	N 4d	N 4f	O 5s	O 5p	O 5d	O 5f	O 5g	P 6s	P 6p ···	Ground state
55. Cs	2	8	18	2	6	10		2	6				1		$^2S_{1/2}$
56. Ba	2	8	18	2	6	10		2	6				2		$^1S_0$
57. La	2	8	18	2	6	10		2	6	1			2		$^2D_{3/2}$
58. Ce	2	8	18	2	6	10	1	2	6	1			2		$^1G_4$
59. Pr	2	8	18	2	6	10	3	2	6				2		$^4I_{9/2}$
60. Nd	2	8	18	2	6	10	4	2	6				2		$^5I_4$
61. Pm	2	8	18	2	6	10	5	2	6	1			2		$^6H_{5/2}$
62. Sm	2	8	18	2	6	10	6	2	6				2		$^3F_0$
63. Eu	2	8	18	2	6	10	7	2	6				2		$^8S_{7/2}$
64. Gd	2	8	18	2	6	10	7	2	6	1			2		$^9D_2$
65. Tb	2	8	18	2	6	10	(9)	2	6				(2)		$(^6H_{15/2})$
66. Dy	2	8	18	2	6	10	10	2	6				2		$^5I_8$
67. Ho	2	8	18	2	6	10	11	2	6				2		$^4I_{15/2}$
68. Er	2	8	18	2	6	10	12	2	6				2		$^3H_6$
69. Tm	2	8	18	2	6	10	13	2	6				2		$^2F_{7/2}$
70. Yb	2	8	18	2	6	10	14	2	6				2		$^1S_0$
71. Lu	2	8	18	2	6	10	14	2	6	1			2		$^2D_{3/2}$
72. Hf	2	8	18	2	6	10	14	2	6	2			2		$^3F_2$
73. Ta	2	8	18	2	6	10	14	2	6	3			2		$^4F_{3/2}$
74. W	2	8	18	2	6	10	14	2	6	4			2		$^5D_0$
75. Re	2	8	18	2	6	10	14	2	6	5			2		$^6S_{5/2}$
76. Os	2	8	18	2	6	10	14	2	6	6			2		$^5D_4$
77. Ir	2	8	18	2	6	10	14	2	6	7			2		$^4F_{9/2}$
78. Pt	2	8	18	2	6	10	14	2	6	9			1		$^3D_3$
79. Au	2	8	18	2	6	10	14	2	6	10			1		$^3S_{1/2}$

Table 2 (*Continued*)

80.	Hg	2	8	18	2	6	10	14	2	6	10	2		$^1S_0$
81.	Tl	2	8	18	2	6	10	14	2	6	10	2	1	$^2P_{1/2}$
82.	Pb	2	8	18	2	6	10	14	2	6	10	2	2	$^3P_5$
83.	Bi	2	8	18	2	6	10	14	2	6	10	2	3	$^4S_{3/2}$
84.	Po	2	8	18	2	6	10	14	2	6	10	2	4	$^3P_2$
85.	At	2	8	18	2	6	10	14	2	6	10	2	5	$^2P_{3/2}$
86.	Rn	2	8	18	2	6	10	14	2	6	10	2	6	$^1S_0$
		2	8	18			32							

Element		K	L	M	N	O				P		Q	Ground state
						$5s$	$5p$	$5d$	$5f$	$6s$	$6d\cdots$	$7s\cdots$	
87.	Fr	2	8	18	32	2	6	10				1	$^2S^2_{1/2}$
88.	Ra	2	8	18	32	2	6	10				2	$^1S_0$
89.	Ac	2	8	18	32	2	6	10			1	2	$^2D_{3/2}$
90.	Th	2	8	18	32	2	6	10			2	2	$^3F_2$
91.	Pa	2	8	18	32	2	6	10	2		1	2	$^4K_{11/2}$
92.	U	2	8	18	32	2	6	10	3		1	2	$^5L_6$
93.	Np	2	8	18	32	2	6	10	4		1	2	$^6L_{11/2}$
94.	Pu	2	8	18	32	2	6	10	6			2	$^7F_0$
95.	Am	2	8	18	32	2	6	10	7			2	$^8S_{7/2}$
96.	Cm	2	8	18	32	2	6	10	7		1	2	$^9D_2$
97.	Bk	2	8	18	32	2	6	10	(8)		(1)	(2)	$(^8G_{15/2})$
98.	Cf	2	8	18	32	2	6	10	(10)			(2)	$(^5I_8)$
99.	E	2	8	18	32	2	6	10	11			2	$^4I_{15/2}$
100.	Fm	2	8	18	32	2	6	10	12			2	$^3H_6$
101.	Mv	2	8	18	32	2	6	10	13			2	$^2F_{7/2}$
102.	No	2	8	18	32	2	6	10	14			2	$^1S_0$
103.	Lw	2	8	18	32	2	6	10	14		1	2	$^2D_{3/2}$

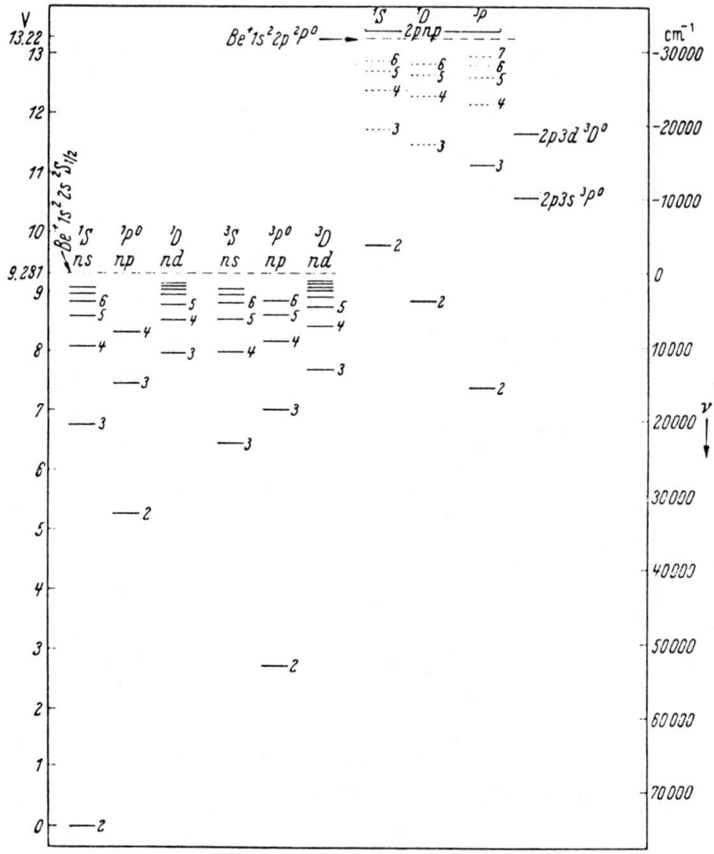

Figure 15 Term scheme of beryllium.

result can be obtained only by numerical calculation from the eigenfunctions of the potassium atom. For calcium the 4*s* shell is fully occupied. For the elements numbered from 21 to 28 the 3*d* shell will become filled. A similar situation exists in the case of the rare earth elements (numbered 58–70) where the 4*f* shell is filled. The chemical behavior is determined by the outer electron shell; this fact explains the chemical similarity of the rare earth elements.

For an example of an energy level diagram we will consider the case of carbon (Figure 16). The electron configuration for the ground state of carbon is $(1s)^2(2s)^2(2p)^2$, which, according to Table 1 for two equivalent *p* electrons results in three terms ^1S, ^1D, ^3P. The latter is the term of largest multiplicity and is therefore the groundstate. For the configuration $(1s)^2(2s)^2(2p)(np)$ (where $n \neq 2$) all possible combinations of orbital and spin angular momentum are permitted, so that for each such configuration there exists three singlet and three triplet states. For $n \geq 3$ the configurations $(1s)^2(2s)^2(2p)(ns)$, $(1s)^2(2s)^2(2p)(nd)$ give rise to terms, which, if observed, are listed in Figure 16. We suppose, however, that the excitation of an electron from the filled (2*s*) shell into the (2*p*) shell will give rise to additional discrete

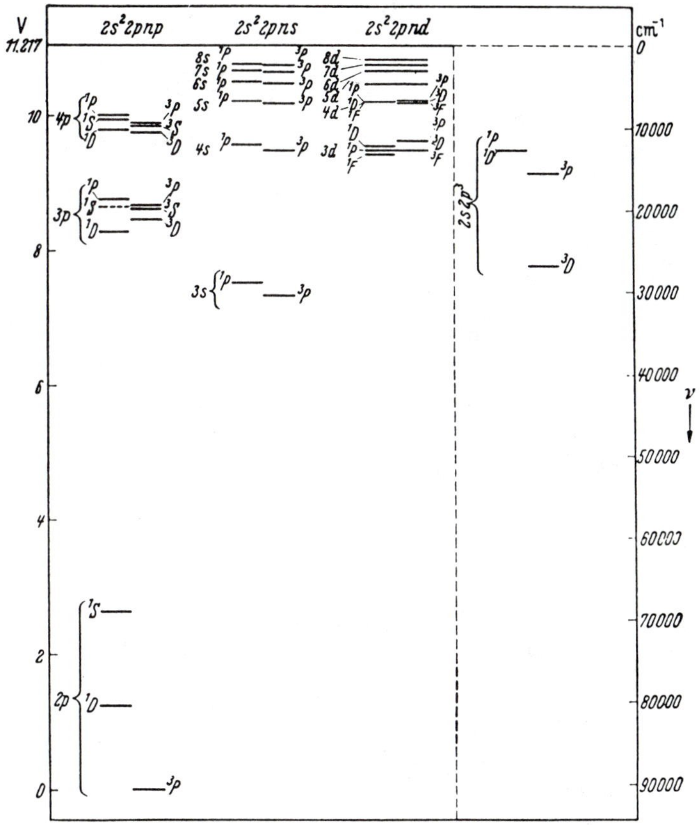

Figure 16 Term scheme of carbon.

terms. In order to determine the terms which arise from the electron configuration $(1s)^2(2s)(2p)^3$ we shall first take note of the possible terms for three equivalent p electrons from Table 1: ${}^2P$, ${}^2D$, ${}^4S$, that is, we have the following L, S pairs: $1, \frac{1}{2}$; $2, \frac{1}{2}$; $0, \frac{3}{2}$. Note that one $2s$ electron gives rise to the term ${}^2S$, that is, to the pair: $0, \frac{1}{2}$. From $0, \frac{1}{2}$ and $1, \frac{1}{2}$ we obtain the possible terms $1, 1$; $1, 0$; from $0, \frac{1}{2}$ and $2, \frac{1}{2}$ we obtain the terms $2, 1$; $2, 0$; from $0, \frac{1}{2}$ and $0, \frac{3}{2}$ we obtain the terms $0, 2$ and $0, 1$. From the configuration $(1s)^2(2s)(2p)^3$ we obtain the six terms ${}^3P$, ${}^1P$, ${}^3D$, ${}^1D$, ${}^5S$, ${}^3S$. The experimentally observed values are given in Figure 16. ${}^5S$ is the lowest energy state for these terms. It plays an important role for the chemical properties of carbon.

If Figure 17 we present the energy level diagram for oxygen atoms. The electron configuration for the lowest energy level is $(1s)^2(2s)^2(2p)^4$ and gives rise to the same term manifold as does p^2, that is, the same as we found for the carbon atom for the ${}^1S$, ${}^1D$, ${}^3P$ terms. Here again the last term is both the term of highest multiplicity and the ground state. These three terms in Figure 24 are not placed in the same column but are integrated within three term groups. These three term groups are easy to understand: Excited states

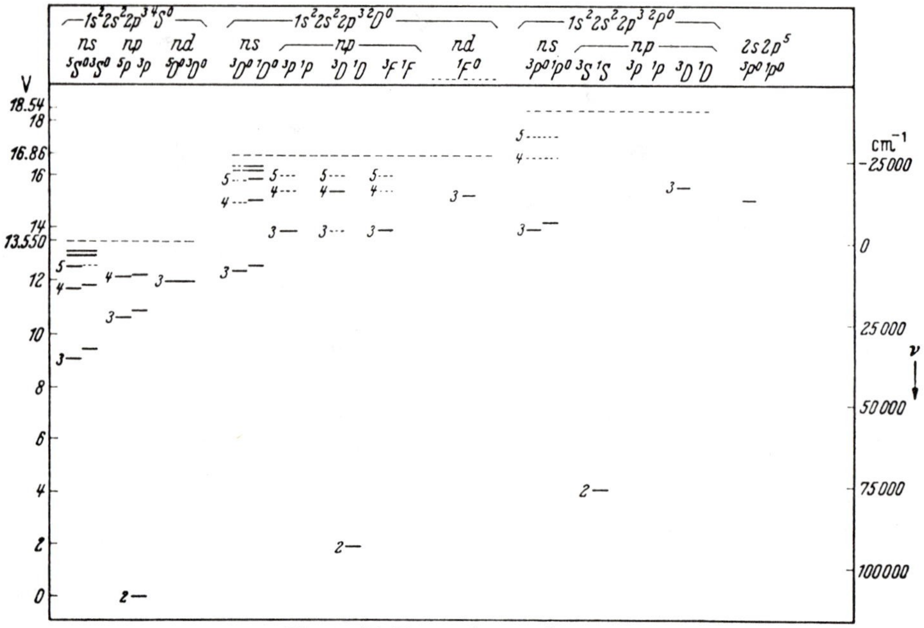

Figure 17 Term scheme of oxygen.

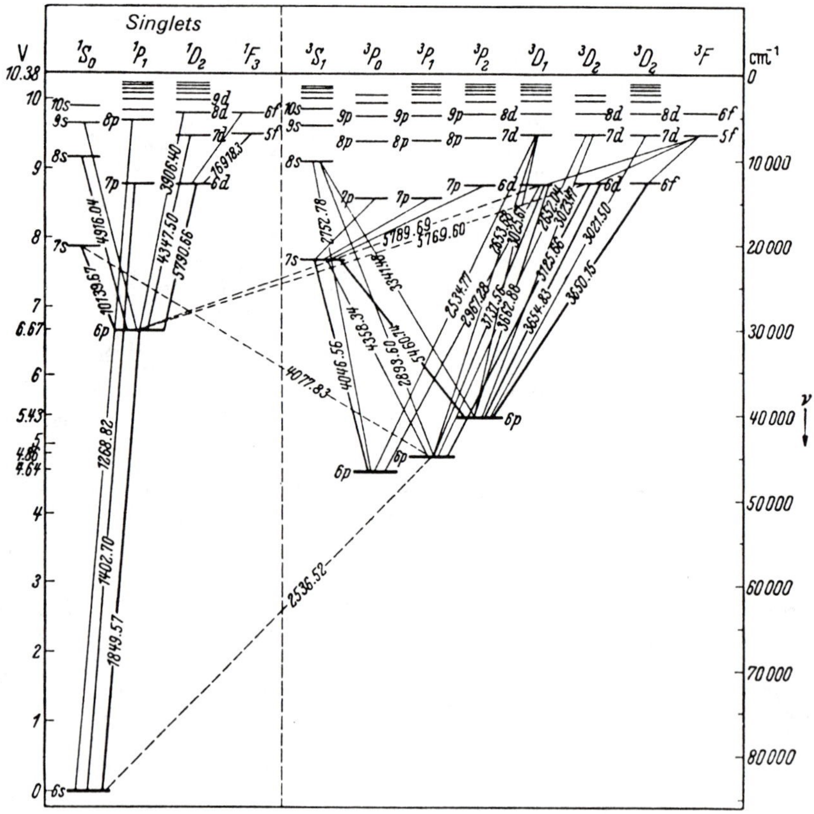

Figure 18 Term scheme of mercury.

of the oxygen atom give rise to an electron configuration where an electron of the $2p$ shell is excited to a higher energy level: $(1s)^2(2s)^2(2p)^3(ns)$, $(1s)^2(2s)^2(2p)^3(np)$, $(1s)^2(2s)^2(2p)^3(nd)$, etc. where $n > 2$. In order to determine terms for such a configuration it is first necessary to obtain the terms for three equivalent p electrons according to the table: ${}^2P$, ${}^2D$, ${}^4S$. These also represent the lowest states for the ogygen ion. ${}^2P$ is the highest and ${}^4S$ is the lowest of these three terms. From the ${}^4S$ term (the left term group in Figure 17) the (ns) electron gives rise to an ${}^5S$ and an ${}^3S$ term; from a (np) electron we obtain a ${}^5P$ and a ${}^3P$ term, etc. From the ${}^2D$ term of the ion the (ns) electron gives rise to a ${}^3D$ and ${}^1D$ term. The (np) electron gives rise to the following six terms: ${}^3P$, ${}^1P$, ${}^3D$, ${}^1D$, ${}^3F$, ${}^1F$. Finally, from the ${}^2P$ term we obtain the group of terms on the right in Figure 17.

In Figure 18 we show the term scheme of mercury, the terms of which are, according to electron configuration given in Table 2, similar to that of helium, which is indeed the case. Here it is evident that the multiplet splitting is so large that it stands out in the energy level diagram, especially for the lowest ${}^3P$ term.

5 Selection and Intensity Rules

From the atom's dipole moment $d = e \sum_{k=1}^{f} \vec{r}_k$ we may obtain the representation $\mathscr{D}_1$ of the rotation group if we introduce the following linear combinations of the components of $\vec{d}$

$$\begin{aligned} R_1 &= \frac{1}{\sqrt{2}}(d_1 + id_2), \\ R_0 &= -d_3, \\ R_{-1} &= \frac{1}{\sqrt{2}}(-d_1 + id_2) \end{aligned} \tag{5.1}$$

as already described in XIII, §3. Using this result, in XIII, §4 we have obtained the selection rule $J \to J - 1, J, J + 1$ for the total angular momentum quantum number and shown that only terms having different reflection symmetry properties, that is, only energy levels for which the corresponding values of $\sum_{i=1}^{f} l_i$ which differ by an odd number can be combined to produce dipole radiation. If the parity of an energy level is (-1) then we append to the term symbol a small "o" (see Figures 15 and 17).

If the multiplet splitting is small, so that for all practical purposes the eigenvectors coincide with linear combinations of zeroth approximation (1.8) then it follows that (in the same manner as described in XIII, §4) the selection rule $L \to L - 1, L, L + 1$ applies, and that only terms of the same multiplicity $2S + 1$ can be combined.

If we may approximate the eigenvector by means of a Slater determinant (3.2) of one-electron functions then we obtain an approximate rule which states that transitions which do not satisfy the selection rule in which only one of the l_i may change by either 1 or -1 will occur with small probability. This result follows from the fact that $\vec{d}$ is obtained additively from the dipole moments $e\vec{r}_i$ of the individual electrons and that, for a transition the matrix elements must have at least one $e\vec{r}_i$ which is nonzero and that from XIII, §3 it follows that l must satisfy $l \to l + 1$ or $l - 1$.

For the case in which the multiplet splitting is small it follows that the selection rules $J \to J - 1, J, J + 1$ and $L \to L - 1, L, L + 1$ and $S \to S$ are satisfied and that the transition intensities for multiplet transitions can be determined on purely group theoretical grounds. We will now consider the transition between two parent terms with quantum numbers L', S' and L, S. Since it is possible that $L = L'$ we will distinguish both terms by the use of indices N and N'. The eigenvectors of both terms will, according to (1.8), have the following form:

$$\phi_{NLSm_Lm_S} = \frac{1}{\sqrt{r}} \sum_k \psi^L_{Nm_Lk}\, u^S_{m_Sk}. \tag{5.2}$$

For the individual fine structure components J the eigenvectors will be given approximately by the correct linear combinations (2.1) as follows:

$$\Gamma^J_{NLSM} = \sum_{m_L + m_S = M} d^J_{m_Lm_S}(L, S)\phi_{NLSm_Lm_S}. \tag{5.3}$$

According to XIII, §2 the matrix elements

$$(R_\mu)^{N'L'S'J'M'}_{NLSJM} = \langle \Gamma^{J'}_{N'L'S'M'}, R_\mu \Gamma^J_{NLSM}\rangle \tag{5.4}$$

must be identical up to a factor $\rho^{N'L'S'J'}_{NLSJ}$ to the expansion coefficient of the product representation of $\mathscr{D}_1 \times \mathscr{D}_J$ with respect to the representation $\mathscr{D}_{J'}$. Therefore, using XI (10.17) we obtain

$$(R_\mu)^{N'L'S'J'M'}_{NLSJM} = \rho^{N'L'S'J'}_{NLSJ}\alpha_{M\mu;M'J'} = \rho^{N'L'S'J'}_{NLSJ} d^{J'}_{M\mu}(J, 1)\delta_{M',M+\mu}. \tag{5.5}$$

The intensity of a multiplet transition $J' \to J, L' \to L$ is, according to XIII (1.5) proportional to $[1/(2J' + 1)] \sum_{\mu,M,M'} |(R_\mu)^{N'L'S'J'M'}_{NLSJM}|^2$. From (5.5) we obtain

$$\sum_{\mu,M,M'} \left|(R_\mu)^{N'L'S'J'M'}_{NLSJM}\right|^2 = \left|\rho^{N'L'S'J'}_{NLSJ}\right|^2 \sum_{\mu,M,M'} |\alpha_{M\mu;M'J'}|^2. \tag{5.6}$$

Since $\alpha_{M\mu;M'J'}$ is a unitary matrix, we obtain $\sum_{M\mu} |\alpha_{M\mu;M'J'}|^2 = 1$ and we therefore obtain

$$\frac{1}{2J' + 1} \sum_{\mu,M,M'} |(R_\mu)^{N'L'S'J'M'}_{NLSJM}|^2 = |\rho^{N'L'S'J'}_{NLSJ}|^2. \tag{5.7}$$

The intensities of the individual lines $J' \to J$ for the same $L' \to L$ are therefore given by the $|\rho^{N'L'SJ'}_{NLSJ}|$ (since $S' = S$).

According to XIII, §2, using factors $\tau^{N'L'S}_{NLS}$ we obtain:

$$\begin{aligned} R_\mu \phi_{NLSm_Lm_S} = \sum_{N'} \{ & d^{L+1}_{\mu m_L}(1L)\tau^{N'L+1S}_{NLS}\phi_{N'L+1S'm_L+\mu M_S} \\ & + d^L_{\mu M_L}(1L)\tau^{N'L-1S}_{NLS}\phi_{N'LSm_L+\mu m_S} \\ & + d^{L-1}_{\mu m_L}(1L)\tau^{N'L-1S}_{NLS}\phi_{N'L-1Sm_L+\mu m_S}\}. \end{aligned} \tag{5.8}$$

We therefore obtain

$$R_\mu \Gamma^J_{NLSM} = \sum_{m_L+m_S=M} \sum_{N'} d^J_{m_L m_S}(L,S) \left\{ \sum_{L'=L-1}^{L+1} d^{L'}_{\mu M_L}(1L) \tau^{N'L'S}_{NLS} \phi_{N'L'Sm_L+\mu m_S} \right\}. \tag{5.9}$$

Since, according to (5.3) and XI (10.19), we have

$$\phi_{N'L'Sm_L+\mu m_S} = \sum_{J'} d^{J'}_{m+\mu m_S}(L'\ S)\Gamma^{J'}_{N'L'SM+\mu} \tag{5.10}$$

from (5.5) it follows that

$$\rho^{N'L'SJ'}_{NLSJ}\ d^{J'}_{M\mu}(J1) = \tau^{N'L'S}_{NLS} \sum_{m_L+m_S=M} d^J_{m_L m_S}(LS)\, d^{L'}_{\mu m_L}(1L)\, d^{J'}_{m_L+\mu m_S}(L', S). \tag{5.11}$$

If, in particular, we choose $M = J$; $M' = J'$ and therefore $\mu = J' - J$ we will find that

$$\rho^{N'L'SJ'}_{NLSJ}\, d^{J'}_{JJ'-J}(J, 1) = \tau^{N'L'S}_{NLS} \sum_{m_L+m_S=J} d^J_{m_L m_S}(LS)\, d^{J'}_{m_L+J'-J, m_S}(L'S)\, d^{L'}_{J'-J, m_L}(1L). \tag{5.12}$$

Thus, from the preceding result and XI (10.15) we obtain

$$\begin{aligned}
&\rho^{N'L-1SJ+1}_{NLSJ} \\
&\quad = \tau^{N'L-1S}_{NLS} \sqrt{\frac{(L+S-J)(L+S-J-1)(J-L+S+2)(J-L+S+1)}{(2J+3)(2J+2)2L(2L+1)}}, \\
&\rho^{N'L-1SJ}_{NLSJ} \\
&\quad = \tau^{N'L-1S}_{NLS} \sqrt{\frac{(L+S-J)(J-L+S+1)(J+L+S+1)(J+L-S)}{L(2L+1)(2J+2)2J}}, \\
&\rho^{N'L-1SJ-1}_{NLSJ} \\
&\quad = \tau^{N'L-1S}_{NLS} \sqrt{\frac{(J+L+S+1)(J+L+S)(J+L-S)(J+L-S-1)}{2L(2L+1)2J(2J-1)}}, \\
&\rho^{N'LSJ+1}_{NLSJ} \\
&\quad = \tau^{N'LS}_{NLS} \sqrt{\frac{(L+S-J)(J+L+S+2)(J+L-S+1)(J-L+S+1)}{L(3L+2)(2J+2)(2J+3)}}, \\
&\rho^{N'LSJ}_{NLSJ} \\
&\quad = \tau^{N'LS}_{NLS} \frac{J(J+1)+L(L+1)-S(S+1)}{2\sqrt{L(L+1)J(J+1)}}, \\
&\rho^{N'LSJ-1}_{NLSJ} \\
&\quad = \tau^{N'LS}_{NLS} \sqrt{\frac{(L+S-J+1)(J+L+S+1)(J+L-S)(J-L+S)}{L(2L+2)(2J-1)2J}}.
\end{aligned} \tag{5.13}$$

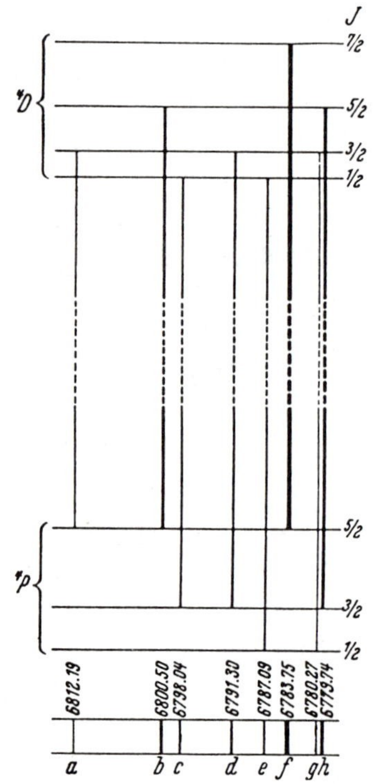

Figure 19 $^4D \to {}^4P$ transition for ionized carbon.

We need not make a special computation of the values $\rho_{NLSJ}^{N'L+1SJ'}$ since from

$$|(R_\mu)_{NLSJM}^{N'L'SJ'M'}| = |(R_\mu)_{N'L'SJ'M'}^{NLSJM}|$$

it follows from (5.5) that

$$|\rho_{NLSJ}^{N'L'SJ'}|\,|d_{M\mu}^{J'}(J1)\delta_{M'M+\mu}| = |\rho_{N'L'SJ'}^{NLSJ}|\,|d_{M'\mu}^{J}(J'1)\delta_{M,M'+\mu}| \tag{5.14}$$

and, for $M = 0$, $M' = 0$ and $\mu = 0$ we obtain

$$|\rho_{NLSJ}^{N'L'SJ'}|\,d_{00}^{J'}(J1) = |\rho_{N'L'SJ}^{NLSJ}|\,d_{00}^{J}(J'1).$$

Thus it follows that

$$|\rho_{NLSJ}^{N'L+1SJ'}| = \sqrt{\frac{2J+1}{2J'+1}}\,|\rho_{N'L+1SJ'}^{NLSJ}|. \tag{5.15}$$

In Figure 19 the $^4D \to {}^4P$ transition of ionized carbon at 6800 Å is described. Here we recognize the separation of the individual fine structure components corresponding to the interval rule (2.5). The relative intensities of the lines are indicated by the corresponding thickness.

6 Zeeman Effect

The rotational symmetry of the Hamiltonian operator for an atom can be broken by an external field. For an atom in an external field there are additional terms obtained from H_0 in VIII (5.8) given by

$$\frac{1}{2m}\sum_{k=1}^{f}\left[\sum_{\nu=1}^{3}(P_\nu^{(k)} - eA_\nu(Q^{(k)}, t))^2 + e\varphi(Q^{(k)}, t)\right]$$

and an additional term from H_5 in VIII (5.8). If we consider a constant magnetic field consisting of only a B_3 component, then the Hamiltonian operator will contain an expression of the form $A_1 = -\frac{1}{2}B_3 x_2$, $A_2 = \frac{1}{2}B_3 x_1$, $A_3 = 0$ for each electron. We therefore obtain

$$\frac{1}{2m}\sum_{k=1}^{f}\left[\left(P_1^{(k)} + \frac{e}{2}B_3 Q_2^{(k)}\right)^2 + \left(P_2^{(k)} - \frac{e}{2}B_3 Q_1^{(k)}\right)^2 + P_3^{(k)2}\right]$$

$$= \sum_{k=1}^{f}\left[\frac{1}{2m}\vec{P}^{(k)2} + \mu_0 B_3 L_3^{(k)} + \frac{e^2}{8m}(Q_1^{(k)^2} + Q_2^{(k)^2})B_3^2\right], \qquad (6.1)$$

where $\mu_0 = -e/2m$. From H_5 in VIII (5.8) we obtain the additional term

$$\mu_0 B_3 = 2S_3. \qquad (6.2)$$

If we neglect the case of large magnetic fields, then it suffices to consider only linear terms in B_3; we then find it necessary to add the term

$$H' = \mu_0 B_3(L_3 + 2S_3) \qquad (6.3)$$

to the Hamiltonian operator for a free atom. If we compare (6.3) to XI (1.10) we find that, for electrons with spin (including the case of magnetic dipoleradiation), it is necessary to replace the magnetic moment $\vec{m} = \mu_0\vec{L}$ by the operator $\mu_0(\vec{L} + 2\vec{S})$.

Let Γ_M^J denote the exact eigenfunctions for an energy level of a free atom. They span an eigenspace $\mathcal{T}_J$ which is invariant under the representation $\mathcal{D}_J$ of the rotation group $\mathcal{D}_{\mathcal{g}}$. The entire Hamiltonian operator including H' is, however, only invariant under the subgroup of rotations about the 3-axis. The eigenspace $\mathcal{T}_J$ is not irreducible under the action of this subgroup. It decomposes into the $2J + 1$ one-dimensional subspaces which are spanned by the individual Γ_M^J, since, under rotations about the 3-axis each Γ_M^J is multiplied by a factor $e^{-iM\alpha}$. Thus the external field breaks the $(2J + 1)$-fold degeneracy of the atomic energy levels.

For the magnitude of the splitting we may only obtain simple assertions in the extreme cases. If the magnetic field is so weak that the splitting is small compared with the separation between individual fine structure components or adjacent energy levels of the free atom, then the splitting can be computed by perturbation theory. Here, for the eigenfunction we choose the "correct" linear combination which must be given by the Γ_M^J. From (6.3) we obtain

$$H' = \mu_0 B_3(J_3 + S_3). \qquad (6.4)$$

We obtain $J_3\Gamma^J_M = M\Gamma^J_M$. The matrix elements $\langle\Gamma^J_M, S_3\Gamma^J_M\rangle$ must be determined by group theoretical methods, because the theorems from XIII, §2 are applicable to the $S_k\Gamma^J_M$ and the $J_k\Gamma^J_M$. Thus we obtain

$$\langle\Gamma^J_{M'}, S_3\Gamma^J_M\rangle = \tau\langle\Gamma^J_{M'}, J_3\Gamma^J_M\rangle = \tau M\delta_{M'M}. \tag{6.5}$$

Therefore H' has the form of a diagonal matrix $(1 + \tau)M\delta_{M'M}$ with respect to Γ^J_M as a basis of the eigenspace $\mathscr{T}_J$, that is, the splitting has the magnitude

$$\varepsilon_M = (1 + \tau)M\mu_0 B_3 = gM\mu_0 B_3. \tag{6.6}$$

The energy level with the total angular momentum quantum number J is therefore split into $2J + 1$ equidistant terms corresponding to the different values M of the third component J_3 of the total angular momentum.

The factor $g = (1 + \tau)$ may be simply determined in the case in which the Γ^J_M are, in good approximation, given by the linear combinations (5.3)

In the following we shall consider the operators $\vec{S}$ and $\vec{L}$ only in $\mathscr{T}_J$ for the computation of the value of g, that is, instead of $\vec{S}$, $\vec{L}$ we shall consider $P_J\vec{S}P_J$, $P_J\vec{L}P_J$ where P_J is the projection operator onto $\mathscr{T}_J$. According to the above result, in $\mathscr{T}_J$ we obtain $P_J\vec{S}P_J = \tau P_J\vec{J}P_J = \tau\vec{J}P_J$. Thus in $\mathscr{T}_J$ we obtain

$$P_J\vec{S}P_J\cdot\vec{J} = \tau\vec{J}\cdot\vec{J}P_J = \vec{J}\cdot P_J\vec{S}P_J = \tau J(J + 1)P_J.$$

On the other hand, we obtain

$$\vec{L}^2 = (\vec{J} - \vec{S})^2 = \vec{J}^2 - \vec{J}\cdot\vec{S} - \vec{S}\cdot\vec{J} + \vec{S}^2.$$

For the Γ^J_M of the form (5.3) we obtain $\vec{L}^2\Gamma^J_M = L(L + 1)\Gamma^J_M$, and $\vec{S}\Gamma^J_M = S(S + 1)\Gamma^J_M$. Since $P_J\vec{J}\cdot\vec{S}P_J = \vec{J}\cdot P_J\vec{S}P_J = \tau J(J + 1)P_J$, we obtain the Lande factor

$$g = 1 + \tau = 1 + \frac{J(J + 1) + S(S + 1) - L(L + 1)}{2J(J + 1)}. \tag{6.7}$$

For singlet terms, that is, for $S = 0$ we obtain $J = L$ and therefore $g = 1$. In Figure 20 the Zeeman effect for the sodium D line is illustrated. Not all the transitions $M' \to M$ are permitted. Since, under rotations about the 3-axis, the R_1, R_0, R_{-1} are multiplied by $e^{-i\alpha}$, 1, $e^{i\alpha}$, only the matrix elements $(R_\mu)_{M'M}$ are different from zero only for $M' = M + \mu$, that is, the selection rule $M \to M - 1, M, M + 1$, a result which we have already obtained in (5.5). For a transition $J' \to J$ we may now easily state the intensity ratios of the individual Zeeman components, because, from (5.5) it follows that

$$(R_\mu)^{N'L'SJ'M'}_{NLSJM} = \rho^{N'L'SJ'}_{NLSJ}\, d^{J'}_{M\mu}(J1)\delta_{M',M+\mu}. \tag{6.8}$$

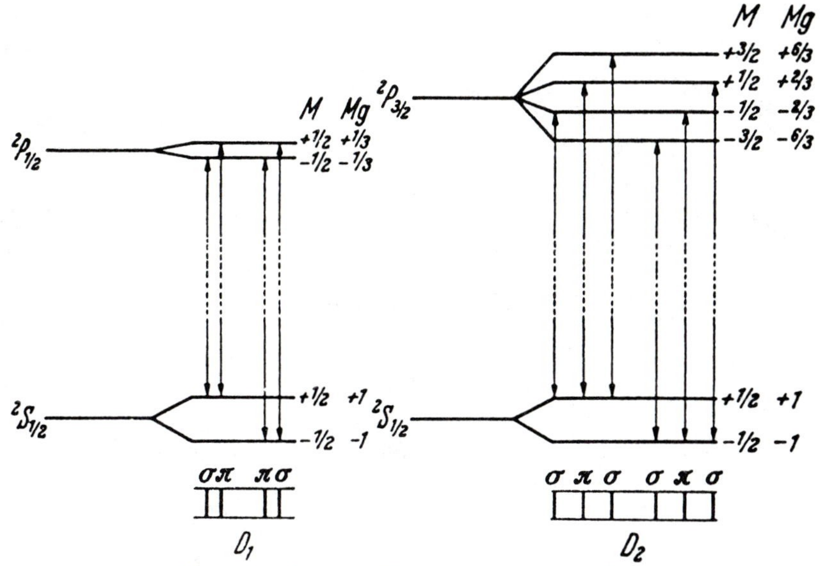

Figure 20 Zeeman effect for the sodium D line.

We therefore obtain

$$
\begin{aligned}
(R_1)_{NLSJM}^{N'L'SJ-1\,M+1} &= \rho_{NLSJ}^{N'L'SJ-1}\sqrt{\frac{(J-M)(J-M+1)}{2J(2J+1)}},\\
(R_0)_{NLSJM}^{N'L'SJ-1\,M} &= -\rho_{NLSJ}^{N'L'SJ-1}\sqrt{\frac{(J+M)(J-M)}{J(2J+1)}},\\
(R_{-1})_{NLSJM}^{N'L'SJ-1\,M-1} &= \rho_{NLSJ}^{N'L'SJ-1}\sqrt{\frac{(J+M)(J+M-1)}{2J(2J+1)}},\\
(R_1)_{NLSJM}^{N'L'SJM+1} &= \rho_{NLSJ}^{N'L'SJ}\sqrt{\frac{(J+M+1)(J-M)}{2J(J+1)}},\\
(R_0)_{NLSJM}^{N'L'SJM} &= \rho_{NLSJ}^{N'L'SJ}\frac{M}{\sqrt{J(J+1)}}, \qquad (6.9)\\
(R_{-1})_{NLSJM}^{N'L'SJM-1} &= \rho_{NLSJ}^{N'L'SJ}\sqrt{\frac{(J+M)(J-M+1)}{2J(J+1)}},\\
(R_1)_{NLSJM}^{N'L'SJ+1\,M+1} &= \rho_{N'L'SJ}^{N'L'SJ+1}\sqrt{\frac{(J+M+2)(J+M+1)}{(2J+1)(2J+2)}},\\
(R_0)_{NLSJM}^{N'L'SJ+1\,M} &= \rho_{NLSJ}^{N'L'SJ+1}\sqrt{\frac{(J+M+1)(J-M+1)}{(J+1)(2J+1)}},\\
(R_{-1})_{NLSJM}^{N'L'SJ+1\,M-1} &= \rho_{NLSJ}^{N'L'SJ+1}\sqrt{\frac{(J-M+2)(J-M+1)}{(2J+1)2(J+2)}}.
\end{aligned}
$$

It is easy to show, on the basis of XI, §1, that the R_1, R_{-1} correspond to circular polarization of the radiation observed in the direction of the 3-axis (σ component) and R_0 corresponds to a linear polarization orhtogonal to the 3-direction observed in the direction of the 3-axis (π components).

If the spin perturbation is so small that the Γ_M^J for the free atom are given by the linear combination (5.3), and the magnetic field B, is so strong that its perturbation is no longer small compared to the spin perturbation, then both perturbations must be handled at once, that is, we must proceed from the parent terms with the eigenspace $\imath_{LS}$ spanned by the $\phi_{m_L m_S}$. Let H'' denote the additional term of the Hamiltonian operator which takes spin into account. For the linear combinations Γ_M^J we therefore obtain $H''\Gamma_M^J = \tilde{\tau}[J(J+1) - L(L+1) - S(S+1)]$ according to (2.3). On the other side, it is clear that

$$H'\phi_{m_L m_S} = \mu_0 B_3(m_L + 2m_S)\phi_{m_L m_S} \tag{6.10}$$

and therefore

$$H'\Gamma_M^J = \mu_0 B_3 \sum_{m_L + m_S = M} (m_L + 2m_S)\, d^J_{m_L, m_S} \phi_{m_L m_S}. \tag{6.11}$$

With

$$\phi_{m_L m_S} = \sum_J d^J_{m_L m_S} \Gamma^J_{m_L + m_S}$$

we therefore obtain

$$H'\Gamma_M^J = \mu_0 B_3 \sum_{m_L + m_S = M} \sum_{J'} (m_L + 2m_S)\, d^J_{m_L m_S}\, d^{J'}_{m_L m_S} \Gamma^{J'}_M. \tag{6.12}$$

For each $M = -J, -J+1, \ldots, J$ it is necessary to solve the secular determinant for the eigenvalues ε_M of the perturbation (with roots ε_{ML+S}, $\varepsilon_{ML+S-1}, \ldots, \varepsilon_{M|L-S|}$ corresponding to the values $J = L+S, \ldots, |L-S|$):

$$\left| \begin{matrix} \{\tilde{\tau}[J(J+1) - L(L+1) - S(S+1) - \varepsilon_M\}\delta_{JJ'} \\ + \mu_0 B_3 \sum\limits_{m_S + m_L = M} (m_L + 2m_S)\, d^J_{m_L m_S}\, d^{J'}_{m_L m_S} \end{matrix} \right| = 0, \tag{6.13}$$

where the matrix has the indices J, J'. If we multiply this matrix on the right and on the left by the unitary matrices $\alpha_{\nu\nu'; MJ'}$ from XI (10.17), we obtain the equivalent secular equation

$$\left| \begin{matrix} [\mu_0 B_3(m_L + 2m_S) - \varepsilon_M]\delta_{m_S \mu_S}\delta_{m_L \mu_L} \\ + \tilde{\tau} \sum\limits_J [J(J+1) - L(L+1) - S(S+1)]\, d^J_{\mu_L \mu_S}\, d^J_{m_L m_S} \end{matrix} \right| = 0, \tag{6.14}$$

where $m_L + m_S = \mu_L + \mu_S = M$ and the pairs $m_L, m_S; \mu_L, \mu_S$ are to be considered as the matrix indices.

For very small B_3 from the first secular equation it follows that

$$\begin{aligned} \varepsilon_{MJ} = {} & \tilde{\tau}[J(J+1) - L(L+1) - S(S+1)] \\ & + \mu_0 B_3 \sum_{m_L + m_S = M} (m_L + 2m_S)\, d^J_{m_L m_S}\, d^J_{m_L m_S}, \end{aligned} \tag{6.15}$$

where the latter again leads us to the result (6.6) with the Landé g factor (6.7). Conversely, if $\tilde{\tau} \ll \mu_0 B_3$, then, from the second form we obtain

$$\begin{aligned}\varepsilon_{Mm_S} = {} & \mu_0 B_3(M + m_S) \\ & + \tilde{\tau}[-L(L+1) - S(S+1) + \sum_J J(J+1)\, d^J_{M-m_S m_S}\, d^J_{M-m_S m_S}]. \end{aligned} \tag{6.16}$$

The essential contribution gives $\mu_0 B_3(M + m)$ for the splitting. Each of the Zeeman components arising from the gross structure energy level exhibits a fine structure splitting $\tilde{\tau}[\cdots]$; without the term proportional to $\tilde{\tau}$, all energy terms with the same sum $M + m_S = m_L + 2m_S$ will coincide.

The transition from the first to the other extreme case is called the Paschen–Back effect.

For the case of the alkali metals ($S = \frac{1}{2}$, and therefore $J = L + \frac{1}{2}, L - \frac{1}{2}$) (6.13) and (6.14) are two-dimensional secular equations. We can solve these in an elementary way, and discuss the splitting as a function of B_3 (for fixed $\tilde{\tau}$).

7 f Electron Problems and the Symmetric Group

Until now we have not investigated the representations of the permutation group in $\mathscr{H}_b^f$ because the coupling of these representations with spin angular momentum has made it possible to solve most problems of interest without requiring an explicit knowledge of these representations. In order to make the behavior of the energy levels more evident for the case of f electrons we will now examine the representations of the permutation group in more detail.

Let $\imath$ be an n-dimensional vector space (not necessarily a Hilbert space). We shall now investigate the representation of $\mathbf{S}_f$ in the space $\mathscr{T} = \imath^f$. If φ_ν is a basis of $\imath$, then the vectors of $\mathscr{T}$ are given by

$$\phi = \sum_{\nu_1 \cdots \nu_f} a_{\nu_1\nu_2\cdots\nu_f} \varphi_{\nu_1}(1)\varphi_{\nu_2}(2)\cdots\varphi_{\nu_f}(f). \tag{7.1}$$

$P\phi$ is defined for a permutation P in XIII, §1. The representation of $\mathbf{S}_f$ in $\mathscr{T}$ may be completely reduced (see AIV, §14 and AV, §10.6). We can adopt the basis vectors of $\mathscr{T}$ for this reduction in the form (see AIV (14.7)):

$$\mathscr{T} = \sum_\nu + \mathscr{p}^{(\nu)} \times \imath^{(\nu)} \tag{7.2}$$

in order that the transformations of $\mathbf{S}_f$ have the following form:

$$P = \sum_\nu Q_\nu(\mathbf{1} \times P^{(\nu)})Q_\nu.$$

The Q_ν are the projections onto the subspaces $\mathscr{p}^{(\nu)} \times \imath^{(\nu)}$ of $\mathscr{T}$. The $P^{(\nu)}$ yield irreducible representations of $\mathbf{S}_f$ in $\imath^{(\nu)}$ which are not isomorphic for different ν. With a basis $u_\rho^{(\nu)}$ of $\mathscr{p}^{(\nu)}$ and $v_\sigma^{(\nu)}$ of $\imath^{(\nu)}$ we have $u_{\rho\sigma}^{(\nu)} = u_\rho^{(\nu)} v_\sigma^{(\nu)}$ as a basis for $\mathscr{T}$. Let $\tilde{\mathscr{p}}_\sigma^{(\nu)} = \mathscr{p}^{(\nu)} v_\sigma^{(\nu)}$ and $\tilde{\imath}_\rho^{(\nu)} = u_\rho^{(\nu)} \imath^{(\nu)}$. $\tilde{\mathscr{p}}_\sigma^{(\nu)}$ is spanned by the $u_{\rho\sigma}^{(\nu)}$ (ρ fixed) and $\tilde{\imath}_\rho^{(\nu)}$ by the $u_{\rho\sigma}^{(\nu)}$ (σ fixed). The transformations A in $\mathscr{T}$ which commute with all the P in $\mathbf{S}_f$ are of the form $\sum_\nu Q_\nu(A^{(\nu)} \times \mathbf{1})Q_\nu$ where the $A^{(\nu)}$ are arbitrary. We shall denote the ring of such transformations A by (LS). Thus $\mathscr{T}$ is completely reduced by (7.2) relative to the ring (LS). According to AV, §7 all operators which commute with all operators from (LS) are given by the elements $\sum_P a(P)P$ of the group ring $\mathscr{R}_{\mathbf{S}_f}$ of $\mathbf{S}_f$.

The transformations from (LS) will be called symmetric transformations. From

$$A\varphi_{\nu_1}(1)\cdots\varphi_{\nu_f}(f) = \sum_\mu \varphi_{\mu_1}(1)\cdots\varphi_{\mu_f}(f)a_{\mu_1\cdots\mu_f,\nu_1\cdots\nu_f} \tag{7.3}$$

we easily see that, for symmetric transformations,

$$a_{\mu_1\cdots\mu_f,\nu_1\cdots\nu_f} = a_{\mu_{\alpha 1}\cdots\mu_{\alpha f};\nu_{\alpha 1}\cdots\nu_{\alpha f}} \tag{7.4}$$

holds, that is, the matrix elements are not changed if we subject the left and right sets of indices to the same permutation.

Let $\mathscr{L}_n$ denote the ring of all linear transformations in $\imath$. Then we get a representation of $\mathscr{L}_n$ in $\imath^f$ which we denote by $\mathscr{L}_n^f$. $\mathscr{L}_n^f$ is then a part of (LS) because if a is an element of $\mathscr{L}_n$ which satisfies $\varphi_\nu = \sum_\mu \varphi_\mu a_{\mu\nu}$ then a is an operator in $\imath^f$ defined by

$$a\varphi_{\nu_1}(1)\cdots\varphi_{\nu_f}(f) = \sum_\mu \varphi_{\mu_1}(1)\cdots\varphi_{\mu_f}(f)a_{\mu_1\nu_1}a_{\mu_2\nu_2}\cdots a_{\mu f\,\nu f} \tag{7.5}$$

and (7.4) is satisfied. We now assert that the operators in (LS) can be expressed as linear combinations of operators in $\mathscr{L}_n^f$. Instead of the two indices μ_i, ν_i we shall write l_i as an abbreviation. We shall abbreviate the sequence $l_1, l_2, \ldots, l_f$ by the index l. The matrix coefficients $a_{\mu_1\cdots\mu_f,\nu_1\cdots\nu_f} = a_{l_1\cdots l_f} = a_l$ of all symmetric transformations (as a sequence of numbers) form a finite-dimensional vector space, a subspace of the vector space of all transformations $a_{\mu_1\cdots\mu_f,\nu_1\cdots\nu_f}$. The assertion is therefore that the special vectors $\alpha_l = a_{l_1}a_{l_2}\cdots a_{ln}$ of the transformations from $\mathscr{L}_n^f$ span the subspace of all symmetric transformations. The subspace spanned by all the α_l is given by all the vectors b_l which satisfy $\sum_l x_l b_l = 0$ if $\sum_l x_l\alpha_l = 0$ is satisfied for all $\alpha_l \in \mathscr{L}_n^f$. If we show that $\sum_l x_l\alpha_l = 0$ implies that $\sum_l x_l a_l = 0$, the theorem is proven.

From

$$\sum_l x_l\alpha_l = \sum_{l_1\cdots l_f} x_{l_1l_2\cdots l_f}a_{l_1}a_{l_2}\cdots a_{l_f} = 0 \tag{7.6}$$

for arbitrary a_l it follows that all coefficients of the same power in a_{l_i} in (7.6) must vanish. A particular member $a_{l_1}a_{l_2}\cdots a_{l_i}$ repeatedly occurs in (7.6), and its coefficient is equal to

$$\sum_P Px_{l_1\cdots l_f}. \tag{7.7}$$

This coefficient must be equal to zero. The elements $a_{l_1l_2\cdots l_f}$ of (LS) satisfy $Pa_{l_1\cdots l_f} = a_{l_1\cdots l_f}$ and we obtain

$$\sum_{l_1\cdots l_f} x_{l_1\cdots l_f}a_{l_1\cdots l_f} = 0. \tag{7.8}$$

Since all elements of (LS) are linear combinations of elements from $\mathscr{L}_n^f$ the $\not{h}^{(\nu)}$ are also irreducible as representation spaces of $\mathscr{L}_n^f$.

The elements of (LS) are not only linear combinations of elements of $\mathscr{L}_n^f$ but also of U_n^f where U_n is the group of all unitary transformations in $\imath$. We can also impose the condition that the determinants of the elements of U_n are equal to 1; these groups are then called SU_n (the special unitary groups).

PROOF. Let $a_{\nu\mu}$ be the matrix of an element from U_n. The condition $|a_{\nu\mu}| = 1$ does not play an essential role; if $|a_{\nu\mu}| = \alpha$ then $|\alpha^{-1/n} a_{\nu\mu}| = 1$. Since the form $\sum_l x_l a_l$ is homogeneous, we may neglect the factor $\alpha^{1/n}$. From

$$\begin{aligned}\varphi &= \sum_l x_{l_1 l_2 \cdots l_f} a_{l_1} a_{l_2} \cdots a_{l_f} = \sum_{\mu\nu} x_{\mu_1\nu_1\mu_2\nu_2\cdots\mu_f\nu_f} a_{\mu_1\nu_1} a_{\mu_2\nu_2} \cdots a_{\mu_f\nu_f} \\ &= 0 \end{aligned} \tag{7.9}$$

it follows that

$$\sum_{\mu\nu} \frac{\partial\varphi}{\partial a_{\mu\nu}} da_{\mu\nu} = 0. \tag{7.10}$$

Since the matrix $a = (a_{\mu\nu})$ is unitary, from $(a + da) = a(1 + a^+ \, da)$ and $(a + da)^+(a + da) = 1$ where $da = ia\delta a$, it follows that $(1 - i\delta a^+)(1 + i\delta a) = 1$ and that δa is a Hermitian matrix. From (7.10) it follows that

$$\sum_{\nu\mu\rho} \frac{\partial\varphi}{\partial a_{\nu\mu}} a_{\nu\rho} \delta a_{\rho\mu} = 0, \tag{7.11}$$

where $\delta a = (\delta a_{\mu\nu})$. If $B_{\rho\mu} = \sum_\nu (\partial\varphi/\partial a_{\nu\mu}) a_{\nu\rho}$, then for arbitrary Hermitian $\delta\beta_{\rho\mu}$ it follows that

$$\sum_{\rho\mu} B_{\rho\mu} \delta\alpha_{\rho\mu} = 0.$$

If the $\delta\beta_{\rho\mu}$ are completely arbitrary we find that

$$\delta^1\alpha_{\rho\mu} = \delta\beta_{\rho\mu} + \overline{\delta\beta}_{\mu\rho} \quad \text{and} \quad \delta^2\alpha_{\rho\mu} = i(\delta\beta_{\rho\mu} - \delta\overline{\beta}_{\mu\rho})$$

are Hermitian matrices. Thus, for arbitrary $\delta\beta_{\rho\mu}$ it follows that

$$\sum_{\rho\mu} B_{\rho\mu} \delta\beta_{\rho\mu} = 0,$$

that is, the $B_{\rho\mu} = 0$. We therefore obtain

$$\sum_\nu \frac{\partial\varphi}{\partial a_{\nu\mu}} a_{\nu\rho} = 0. \tag{7.12}$$

Since the matrix $a_{\nu\rho}$ (as a unitary matrix) is not singular, it follows that

$$\frac{\partial\varphi}{\partial a_{\nu\mu}} = 0. \tag{7.13}$$

Here we note that $\partial\varphi/\partial a_{\nu\mu}$ is homogeneous polynomial (like φ) but of degree $f - 1$. By induction it finally follows that $\varphi = 0$, that is, the coefficients $x_{l_1 \cdots l_f}$ must satisfy the same relationships (7.7).

The $\mathscr{k}^{(\nu)}$ are therefore also irreducible subspaces for the representation $(SU_n)^f$ of SU_n. If the space $\imath$ is a spin space, then $(SU_2)^f$ is the representation $(\mathscr{D}_{1/2})^f$ of the rotation group. Thus we have proven the theorem used in §§1 and 3 that for the spaces $\mathscr{W}_{\Delta'}^S$ in (1.3c) for f electrons the representation Δ' of the permutation group uniquely corresponds to the spin angular momentum quantum number S.

The reduction of $\mathscr{T}$ in the form (7.2) is closely connected to the group ring $\mathscr{R}_{\mathbf{S}_f}$. Instead of $\mathscr{R}_{\mathbf{S}_f}$ we shall consider the ring $\mathscr{R}^*$ of all transformations which commute with (LS). $\mathscr{R}_{\mathbf{S}_f}$ is homomorphically or isomorphically mapped onto $\mathscr{R}^*$. $\mathscr{R}^*$ is the ring of all transformations of the form $\sum_\nu Q_\nu(\mathbf{1} \times B^{(\nu)})Q_\nu$. The ring $\mathscr{R}^*$ may be easily decomposed (for the general case see AV, §7) into left and right ideals as follows: All elements of $\mathscr{R}^*$ may be written as follows:

$$\sum_{\nu\rho\sigma} \tau_{\rho\sigma}^{(\nu)} e_{\rho\sigma}^{*(\nu)}, \tag{7.14}$$

where $e_{\rho\sigma}^{*(\nu)} u_{\rho'\sigma'}^{(\mu)} = \delta_{\nu\mu}\delta_{\sigma\sigma'} u_{\rho'\rho}^{(\nu)}$. For fixed ν and σ the $e_{\rho\sigma}^{*(\nu)}$ form a basis for an irreducible left ideal $\ell_\sigma^{*(\nu)}$ of $\mathscr{R}^*$; for fixed ν and ρ the $e_{\rho\sigma}^{*(\nu)}$ form a basis for an irreducible right ideal $\imath_\rho^{*(\nu)}$. The $e_{\rho\rho}^{*(\nu)}$ are idempotent elements and $\ell_\sigma^{*(\nu)} = \mathscr{R}^* e_\sigma^{*(\nu)}$, $\imath_\rho^{*(\nu)} = e_{\rho\rho}^{*(\nu)}\mathscr{R}^*$. The $\imath_\rho^{*(\nu)}$ are uniquely related to the $\not{p}_\rho^{(\nu)} = \not{p}^{(\nu)} v_\rho^{(\nu)} = \imath_\rho^{*(\nu)}\mathscr{T} = e_{\rho\rho}^{*(\nu)}\mathscr{T}$, a result which follows directly from the operators $e_{\rho\sigma}^{*(\nu)}$ described above. The decomposition of the unit element in $\mathscr{R}^*$ into irreducible idempotent elements $e = \sum_{\rho,\nu} e_{\rho\rho}^{*(\nu)}$ yields such a decomposition of $\mathscr{T}$ into invariant irreducible $\tilde{\not{p}}_\rho^{(\nu)}$ with respect to (LS), $\mathscr{L}_n^f$, U_n^f and $(SU_n)^f$, respectively.

If the dimension $n \geq f$ then the vectors $P\varphi_{\nu_1}(1)\varphi_{\nu_2}(2)\cdots\varphi_{\nu_f}(f)$ $(\nu_i \neq \nu_k)$ (where the P are all elements of $\mathbf{S}_f$) span an $(f\,!)$-dimensional subspace of $\mathscr{T}$ in which the representation of $\mathbf{S}_f$ is isomorphic to the regular representations of $\mathbf{S}_f$ in $\mathscr{R}_{\mathbf{S}_f}$. $\mathscr{R}_{\mathbf{S}_f}$ is therefore isomorphic to the ring $\mathscr{R}^*$ because not more than one element of $\mathbf{S}_f$ can be mapped onto the unit element of $\mathscr{R}^*$. Thus it follows that $\tilde{\not{p}}_\rho^{(\nu)} = \imath_\rho^{(\nu)}\mathscr{T} = e_{\rho\rho}^{(\nu)}\mathscr{T} \neq (0)$ where $e_{\rho\rho}^{(\nu)}$ are irreducible idempotent elements and $\imath_\rho^{(\nu)}$ are irreducible right ideals of $\mathscr{R}_{\mathbf{S}_f}$.

If $n < f$ then $\mathscr{R}_{\mathbf{S}_f}$ is only homomorphically mapped onto $\mathscr{R}^*$. The elements of $\mathscr{R}_{\mathbf{S}_f}$ which are mapped onto the null element of $\mathscr{R}^*$ clearly form a two-sided ideal $\not{q}$. From $\mathscr{R}_{\mathbf{S}_f}$ as an additive group with operators we may construct the factor group (factor space) $\mathscr{R}_{\mathbf{S}_f}/\not{q}$. It easily follows that $\mathscr{R}_{\mathbf{S}_f}/\not{q} \cong \mathscr{R}^*$. $\not{q}$ is characterized by $\not{q}\mathscr{T} = (0)$. Since $\mathscr{R}_{\mathbf{S}_f}$ is completely reducible it follows that

$$\mathscr{R}_{\mathbf{S}_f} = \not{q} + \ell,$$

where ℓ is clearly a two-sided ideal and $\ell \cong \mathscr{R}_{\mathbf{S}_f}/\not{q}$. Only the irreducible right ideals lying in ℓ and its idempotent elements $e_{\rho\rho}^{(\nu)}$ gives rise to invariant subspaces $\tilde{\not{p}}_\rho^{(\nu)} = e_{\rho\rho}^{(\nu)}\mathscr{T}$. If the irreducible right ideal $\imath_\rho^{(\nu)}$ lies in ℓ then all the $\imath_\rho^{(\nu)} e_{\rho\sigma}^{(\nu)} = \imath_\sigma^{(\nu)}$ lie in ℓ because ℓ is a two-sided ideal. Thus for a given ν either all equivalent irreducible right ideals (and therefore all $e_{\rho\rho}^{(\nu)}$ (ν fixed)) lie in ℓ or none do.

The inequivalent idempotent elements of $\mathscr{R}_{\mathbf{S}_f}$ are, according to AV, §9, given by a tableau T_k as $\gamma_k I_k K_k$. We may set

$$e_{11}^{(k)} = \gamma_k I_k K_k. \tag{7.15}$$

(The two-sided ideal spanned by the $\imath_\rho^{(k)}$ (k fixed) consists of all elements of the form $\mathscr{R}_{\mathbf{S}_f} e_{11}^{(k)} \mathscr{R}_{\mathbf{S}_f}$.) From the meaning of I_n, K_n it follows that $e_{11}^{(k)}\mathscr{T} = 0$ if, in the tableau T_k there are more rows than the dimension of $\imath$ since we would then obtain $K_k \mathscr{T} = 0$. We therefore obtain a sufficient system of idempotent elements if we restrict ourselves to tableaux with not more than n rows. Other representations of $\mathbf{S}_f$ which do not belong to these tableaux cannot occur in $\mathscr{T}$.

If $\imath$ is the two-dimensional spin space, then in $\imath^f$ we find only the representation of $\mathbf{S}_f$ which belong to tableaux with at most two rows. We shall now describe how we may explicitly obtain the spin function matrix of the $u_{\rho\sigma}^{(\nu)}$ for f electrons. For this purpose we shall consider such a tableau with at most two rows. The first row contains g positions more than the second, that is, the first row has $l + g$, the second has l where $2l + g = f$. We seek to obtain the subspace $\tilde{\not{\kappa}}_\rho^{(\nu)}$ of $\imath^f$ which corresponds to this tableau. Such a $\tilde{\not{\kappa}}_\rho^{(\nu)}$ can be obtained from the idempotent element which corresponds to the tableau. On the basis of the structure of this element, as described in AV, §9, we are led to the conjecture that a space $\tilde{\not{\kappa}}_\rho^{(\nu)}$ is obtained from the "spin invariants"

$$A = \prod_{v=1}^{l} [u_+(v)u_-(l+g+v) - u_-(v)u_+(l+g+v)] \cdot \prod_{\mu=1}^{g} [u_+(l+\mu)x + u_-(l+\mu)y] \tag{7.16}$$

by means of the coefficients v_M^S of

$$X_M^S = \frac{x^{S+M}y^{S-M}}{\sqrt{(S+M)!\,(S-M)!}}, \tag{7.17}$$

where $2S = g$, $M = -S, -S+1, \ldots, S$; for if we apply the permutation s (where s is given by AV, §9) we then obtain $sA = (-1)^s A$ and we find that KA is a multiple of A. If we then apply a permutation r then we find that A is not mapped into itself, and that we may apply a symmetrization $I = \sum_r r$ to A. This is, however, not necessary because, by analogy with XI, §9, it is easy to see that under rotations the coefficients of X_M^S transform according to $\mathscr{D}_S$. Thus we have indeed found a space $\tilde{\not{\kappa}}_\rho^{(\nu)}$. From the v_M^S (M fixed) we obtain an entire space $\tilde{\imath}_M^{(\nu)}$ which, transforms according to the corresponding representation of $\mathbf{S}_f$ when we subject v_M^S to all permutations. This follows from the fact that the vector $Iv_M^S = (\sum_r r)v_M^S$ must also be in such a space $\tilde{\imath}_M^{(\nu)}$ and that the $\tilde{\imath}_M^{(\nu)}$ must be irreducible, because they correspond to a spin quantum number S. If we wish to find the representation $\mathscr{D}_S$ and the corresponding representation Δ of $\mathbf{S}_f$ only in a formal way, then in the expression (7.16) for A we may formally set $u_\alpha(1) = u_\alpha(2) = \cdots u_\alpha(l+g) = u_\alpha$ and $u_\alpha(l+g+1) = u_\alpha(l+g+2) = \cdots = u_\alpha(2l+g) = v_\alpha$. If we then apply $\sum_r r$ to A we will obtain a multiple of A.

We will now use the above results to discuss the coarse structure of energy levels associated with an electron configuration, without using the spin functions, as we have done in §3. We will directly investigate which spaces

$\mathscr{p}^{(\nu)} \times \imath^{(\nu)}$ in (7.2) can arise from an electron configuration. Let the initial electron configuration energy level be given by

$$\varepsilon = \alpha_1 \varepsilon_{n_1 l_1} + \alpha_2 \varepsilon_{n_2 l_2} + \cdots + \alpha_s \varepsilon_{n_s l_s}, \tag{7.18}$$

and the corresponding electron configuration is given by

$$(n_1 l_1)^{\alpha_1} (n_2 l_2)^{\alpha_2} \cdots (n_s l_s)^{\alpha_s}, \tag{7.19}$$

where $\alpha_1 + \alpha_2 + \cdots + \alpha_s = f$. In the Hilbert space of an electron (without spin) let

$$\imath = \imath_{n_1 l_1} \oplus \imath_{n_2 l_2} \oplus \cdots \oplus \imath_{n_s l_s}, \tag{7.20}$$

where $\imath_{n_i l_i}$ is the eigenspace corresponding to the eigenvalue $\varepsilon_{n_i l_i}$. We now choose $\mathscr{T} = \imath^f$. $\mathscr{T}$ does not only include the above term ε in (7.18) but also other combinations of the $\varepsilon_{n_i l_i}$. We will now see that it is easy to choose the subspace in $\mathscr{T}$ which is the eigenspace corresponding to the energy level (7.18). In $\mathscr{T}$ we may think of $\mathbf{S}_f$ as being reduced so that we obtain the form (7.2). The spaces $\mathscr{p}^{(\nu)} \times \imath^{(\nu)}$ are also uniquely characterized by the representations of U_n. All vectors of the same $\tilde{\imath}_\rho^{(\nu)}$ correspond to the same eigenvalue of H_0 (neglecting the Coulomb interaction of the electrons), that is, a $\tilde{\imath}_\rho^{(\nu)}$ corresponds either to the entire energy level (7.18) or does not correspond at all. What is the multiplicity of the energy level ε (7.18), that is, how many of the $\tilde{\imath}_\rho^{(\nu)}$ correspond to ε? To answer this question we consider the group U_n in $\imath$. Let u_ν be a basis in $\imath$ which corresponds to the decomposition (7.20). All elements of U_n are conjugate to the diagonal transformation

$$A u_\nu = \varepsilon_\nu u_\nu. \tag{7.21}$$

According to (7.18) all of the eigenvectors $u_{\nu_1}(1) u_{\nu_2}(2) \ldots u_{\nu_f}(f)$ with eigenvalue ε will be multiplied by $\prod_\nu \varepsilon_\nu$ in $\imath^f$ where the ε_ν corresponding to the vectors u_ν in $\imath_{n_i l_i}$ occur with multiplicity α_i. If $X^{(\nu)}(\varepsilon_1, \varepsilon_2, \ldots)$ is the character of the representation of U_n in the $\mathscr{p}^{(\nu)}$, that is, in the $\mathscr{p}^{(\nu)}$ of $\mathscr{p}^{(\nu)} \times \imath^{(\nu)}$ then $X^{(\nu)}$ is a homogeneous polynomial of fth order in the ε_ν. Thus we may uniquely decompose $X^{(\nu)}$ into $X_1^{(\nu)} + X_2^{(\nu)}$ where $X_1^{(\nu)}$ contains all members of $X^{(\nu)}$ which are of order α_i in the ε_ν corresponding to the vectors u_ν in $\imath_{n_i l_i}$. Thus $X_1^{(\nu)}(1, 1, \ldots)$ is the multiplicity with which the term ε occurs in $\mathscr{p}^{(\nu)}$.

We will now proceed from U_n to the subgroup of rotations, the representation of which decomposes in the irreducible representation

$$\mathscr{D}_{l_1} + \mathscr{D}_{l_2} + \cdots + \mathscr{D}_{l_s},$$

then the $\mathscr{p}^{(\nu)}$ are no longer irreducible. The subspace $\mathscr{s}^{(\nu)}$ of $\mathscr{p}^{(\nu)}$ corresponding to the energy level ε is an invariant subspace with respect to the representation of the rotation group because, under rotations, the vectors of the $\imath_{n_i l_i}$ transform only among themselves. The character for the representation of the rotation group in $\mathscr{s}^{(\nu)}$ is then equal to $X_1^{(\nu)}(\varepsilon_1, \varepsilon_2, \ldots, \varepsilon_s)$ if we replace the ε_{ν_k} corresponding to the vectors $u_{\nu_1}, u_{\nu_2}, \ldots, u_{\nu_{2l+1}}$ of $\imath_{n_i l_i}$ by $e^{il_i\alpha}, e^{i(l_i - 1)\alpha}, \ldots, e^{-il_i\alpha}$. Setting $\varepsilon = e^{i\alpha}$ we then obtain $X_1^{(\nu)}(\varepsilon_1, \varepsilon_2, \ldots, \varepsilon_s) = Y^{(\nu)}(\varepsilon)$. Now, in order to reduce the rotation group in $\not{h}^{(\nu)}$ we need only decompose $Y^{(\nu)}(\varepsilon)$ into irreducible characters

$$\sum_{m=-L}^{L} \varepsilon^m = \frac{\varepsilon^{L+1} - \varepsilon^{-L}}{\varepsilon - 1}. \tag{7.22}$$

We set $(\varepsilon - 1)\sum_{m=-L}^{L} \varepsilon^m = \varepsilon^{L+1} - \varepsilon^{-L} = \langle L \rangle$. It must therefore be possible to write

$$(\varepsilon - 1)Y^{(\nu)}(\varepsilon) = \sum_L a_L \langle L \rangle, \tag{7.23}$$

where the a_L are integers and $a_L > 0$. Then the a_L are the multiplicities with which the energy level with orbital angular momentum L occurs in the $\not{h}^{(\nu)}$ corresponding to the (ν)th representation of $\mathbf{S}_f$. From (7.23) it follows that

$$\sum_L a_L \langle L \rangle = \sum_L a_L(\varepsilon^{L+1} - \varepsilon^{-L})$$
$$= a_0 \varepsilon + a_1 \varepsilon^2 + a_2 \varepsilon^3 + \cdots - a_0 - a_1 \varepsilon^{-1} - a_2 \varepsilon^{-2} - \cdots.$$

Therefore, for $Y^{(\nu)}(\varepsilon) = \sum_\nu b_\nu \varepsilon^\nu$ we obtain

$$a_L = b_L - b_{L+1} = b_{-L} - b_{-(L+1)}. \tag{7.24}$$

The determination of the gross structure terms (the so-called parent terms) arising from an electron configuration is, in principle, solved in this way if we know the characters of the irreducible representations of U_n in $\not{h}^{(\nu)}$ which uniquely correspond to an irreducible representation of $\mathbf{S}_f$, that is, uniquely correspond to a tableau in AV, §9. We will seek these characters in the next section. We will now show that the representations in $\imath^{(\nu)}$ associated with the representation in spin space are given by a tableau (AV, §9) having two columns—more generally: Two associated representations belong to two tableaux which are obtained from each other by exchanging rows and columns.

Proof. If $e = \sum_p e(p)p$ is idempotent and generates the irreducible left ideal l in $\mathscr{R}_{\mathbf{S}_f}$ by means of the equation $\mathscr{R}_{\mathbf{S}_f} e = l$, then the trace of an element q in the representation given by l is equal to the trace of the transformation $x \to qxe$ in the entire $\mathscr{R}_{\mathbf{S}_f}$ because it is identical to the null transformation outside of l and within l from $xe = x$ is identical to the transformation $x \to qx$. We therefore obtain:

$$\chi(q) = \sum_t e(t^{-1}q^{-1}t). \tag{7.25}$$

According to AV, §9 the idempotent element corresponding to a tableau has the form $e = \gamma \sum_r r \sum_s (-1)^s s = \gamma \sum_{r,s} (-1)^s rs$. The idempotent element corresponding to the tableau obtained from the previous one by interchanging rows and columns can be chosen as follows:

$$e^* = \gamma \sum_{rs} (-1)^r sr = \gamma \sum_{r's'} (-1)^{r'} s'^{-1} r'^{-1} = \gamma \sum_{rs} (-1)^r (rs)^{-1}. \tag{7.26}$$

Thus we find that $e^*(p) = (-1)^p e(p^{-1})$; from (7.25) we obtain:

$$\chi^*(q^{-1}) = \chi^*(q) = (-1)^q \chi(q).$$

We will now simplify the search for the gross structure of the energy levels by considering the composition of two groups of electrons. Let us consider two groups consisting of f_1 and f_2 electrons where $f = f_1 + f_2$. We shall decompose $\mathscr{v}^f$ as follows: $\mathscr{v}^f = \mathscr{v}^{f_1} \times \mathscr{v}^{f_2}$. In the group $\mathbf{S}_f$ we shall consider the two subgroups σ_1, the permutation group for the f_1 electrons and σ_2 the permutation group for the f_2 electrons, respectively, and the subgroup $\sigma_1 \times \sigma_2$ containing all mutually independent permutations of the f_1 and f_2 electrons.

In (7.2) the spaces $\mathscr{v}^{(\nu)}$, under the restriction of the group $\mathbf{S}_f$ to the subgroup $\sigma_1 \times \sigma_2$ are no longer irreducible; instead, the representation of $\sigma_1 \times \sigma_2$ in $\mathscr{v}^{(\nu)}$ is split into irreducible components, where each component is of the form $\Delta_1 \circledast \Delta_2$ where Δ_1 and Δ_2 are irreducible representations of σ_1 and σ_2, respectively. Not all transformations which commute with the operators in $\sigma_1 \times \sigma_2$ are of the form $\sum_\nu Q_\nu(A^{(\nu)} \times \mathbf{1})Q_\nu$, because equivalent representations of $\sigma_1 \times \sigma_2$ can occur in different $\mathscr{v}^{(\nu)}$. For this reason we find it necessary to consider the groups σ_1 and $U_n^{f_1}$ in $\mathscr{T}_1 = \mathscr{v}^{f_1}$ which permit a decomposition $\mathscr{T}_1 = \sum_\rho + \mathscr{p}_1^{(\rho)} \times \mathscr{v}_1^{(\rho)}$ similar to that in (7.2). The $\mathscr{p}_1^{(\rho)}$ are irreducible relative to U_n and the $\mathscr{v}_1^{(\rho)}$ are irreducible relative to σ_1. We shall also do the same for $\mathscr{v}^{f_2} = \mathscr{T}_2 = \sum_\sigma + \mathscr{p}_2^{(\sigma)} \times \mathscr{v}_2^{(\sigma)}$. It follows that the decomposition of $\mathscr{T}$ is given by

$$\mathscr{T} = \sum_{\rho_1\sigma} + (\mathscr{p}_1^{(\rho)} \times \mathscr{p}_2^{(\sigma)}) \times (\mathscr{v}_1^{(\sigma)} \times \mathscr{v}_2^{(\sigma)}). \tag{7.27}$$

The representation of $\sigma_1 \times \sigma_2$ in $\mathscr{v}_1^{(\rho)} \times \mathscr{v}_2^{(\sigma)}$ is irreducible and has the form $\Delta_1 \circledast \Delta_2$. All A which commute with $\sigma_1 \times \sigma_2$ are of the form

$$A = \sum_{\rho\sigma} Q_{\rho\sigma}(A^{(\rho_1\sigma)} \times \mathbf{1})Q_{\rho\sigma}$$

where $A^{(\rho\sigma)}$ are arbitrary transformations in $\mathscr{p}^{(\rho)} \times \mathscr{p}^{(\sigma)}$. U_n is represented in $\mathscr{p}^{(\rho)} \times \mathscr{p}^{(\sigma)}$ by the representation $\tilde{\mathscr{H}}^{(\rho)} \times \tilde{\mathscr{H}}^{(\sigma)}$ where $\tilde{\mathscr{H}}^{(\rho)}$, $\tilde{\mathscr{H}}^{(\sigma)}$ are the irreducible representations of U_n in $\mathscr{p}_1^{(\rho)}$, $\mathscr{p}_2^{(\sigma)}$, respectively. $\tilde{\mathscr{H}}^{(\rho)} \times \tilde{\mathscr{H}}^{(\sigma)}$ is not necessarily irreducible; if reducible then it can be decomposed into several irreducible components, each of which is uniquely related to a representation of $\mathbf{S}_f$ in $\mathscr{T}$ as follows: $\mathscr{p}_1^{(\rho)} \times \mathscr{p}_2^{(\sigma)} = \sum_\nu' \mathscr{p}^{(\nu)}$ with some of the $\mathscr{p}^{(\nu)}$ in (7.2).

If we compare the two decompositions of $\mathscr{T}$ (7.2) and (7.27) we can make the following conclusion: If the irreducible representation $\Delta^{(\nu)}$ of $\mathbf{S}_f$ in $\mathscr{v}^{(\nu)}$ contains the irreducible representation $\Delta_1^{(\rho)} \circledast \Delta_2^{(\sigma)}$ of $\sigma_1 \times \sigma_2$ b times, then the reducible representation $\tilde{\mathscr{H}}_1^{(\rho)} \times \tilde{\mathscr{H}}_2^{(\sigma)}$ of U_n corresponding to $\Delta_1^{(\rho)} \times \Delta_2^{(\sigma)}$ contains the irreducible representation of U_n in $\mathscr{p}^{(\nu)}$ corresponding to $\Delta^{(\nu)}$ exactly b times.

We will now show that this fact permits us to obtain the energy levels arising from a given electron configuration in several steps (in §3 a different method was outlined). In the first step we treat a collection of equivalent electrons as a whole. In the second step we combine groups of equivalent electrons.

If, for a given electron configuration, we have two groups of electrons for which no electron from the first group is equivalent to an electron of the other group, we may then make use of the subgroup $\sigma_1 \times \sigma_2$ instead of the group $\mathbf{S}_f$. An irreducible representation $\Delta_1 \divideontimes \Delta_2$ of $\sigma_1 \times \sigma_3$ in $\imath_s^f$ occurs in the representation Δ corresponding to the spin quantum number S either once or not at all, depending whether S occurs in the values

$$S_1 + S_2, S_1 + S_2 - 1, \ldots, |S_1 - S_2|,$$

where S_1, S_2 are the spin quantum numbers corresponding to Δ_1 and Δ_2, respectively. Since the associated representations are essentially the same, except for a factor of (-1), it follows that the representation Δ' of $\mathbf{S}_f$ in the orbit function space $\imath^f$ contains $\Delta_1' \times \Delta_2'$ with the same frequency as Δ contains the representation $\Delta_1 \times \Delta_2$.

We may now use this result to find spaces of the form $\mathscr{p}_1^{(v)} \times \imath_1^{(v)}$ and $\mathscr{p}_2^{(\mu)} \times \imath_2^{(\mu)}$ for the two groups of electrons separately which correspond to the representations Δ_1', $\mathscr{D}_{L_1}$ and Δ_2', $\mathscr{D}_{L_2}$. Since none of the electrons from the first group is equivalent to any of the electrons in the second group, we may construct the product space $\mathscr{p}_1^{(v)} \times \mathscr{p}_2^{(\mu)} \times \imath_1^{(v)} \times \imath_2^{(\mu)}$ corresponding to the representations $\Delta_1' \divideontimes \Delta_2'$ and $\mathscr{D}_{L_1} \times \mathscr{D}_{L_2}$. To Δ_1' and Δ_2' there are corresponding representations Δ_1 and Δ_2 in spin space with spin quantum numbers S_1 and S_2. As we have seen, $\Delta_1' \divideontimes \Delta_2'$ occurs exactly once in the representation Δ' of $\mathbf{S}_f$ providing Δ corresponds to one of the values

$$S = S_1 + S_2, S_1 + S_2 - 1, \ldots, |S_1 - S_2|.$$

For a given Δ the energy levels are split by the Coulomb interaction into different energy levels with different orbital angular momentum

$$L = L_1 + L_2, L_1 + L_2 - 1, \ldots, |L_1 - L_2|.$$

Thus we have shown that the study of atomic energy levels may be reduced to the study of groups of equivalent electrons, as described in §3.

The above considerations permit us to obtain a rough picture of the nature of the chemical bond, that is, of the processes by which atoms bind to form a molecule. Let us consider the atoms at large distances being slowly brought together. The energy levels of the electrons in the field of the nuclei depend on the relative positions of the nuclei. If, for a certain position of the nuclei there is a relative minimum for the energy, then the atoms may then combine to form a molecule. A more precise determination concerning the possibility and the type of such a bonding requires a detailed quantitative evaluation of the energy levels (see XV, §5). We shall now provide a rough description:

Suppose that the ground states of the atoms belong to the representations $\Delta_1', \Delta_2', \ldots$ of the permutation groups of its electrons. Let their corresponding spin functions be characterized by their spin values S_1, $S_2, \ldots$, and their representations Δ_1, Δ_2 of the permutation groups. Then, as the atoms are brought together these ground states give rise to energy levels of differing symmetry Δ', namely those Δ' found in the representations $\Delta_1' \divideontimes \Delta_2' \divideontimes \cdots$. According to the results obtained above, only such Δ' can be found which are associated in the spin space with the representation Δ corresponding to the spin numbers S which are obtained from the reduction of $\mathscr{D}_{S_1} \times \mathscr{D}_{S_2} \times \cdots$. If we call $v_1 = 2S$, $v_2 = 2S_2, \ldots$ the valence of the atomic states, then we obtain the possible valences of the molecule directly from the combination possibilities for the valence lines, a problem which is equivalent to the problem of the reduction of $\mathscr{D}_{S_1} \times \mathscr{D}_{S_2} \times \cdots$. If we then apply a perturbation computation by analogy with the computation presented in §9, we will find that, for the most part, the lowest energy level is the state in which there are no free valence lines $v = 0$. We will present a detailed discussion of this topic in *XV*, §7.

8 The Characters for the Representations of $\mathbf{S}_f$ and U_n

In the space $\mathscr{T} = \imath^f$ we shall now compute the trace of a transformation PA where A, P are elements of U_n and $\mathbf{S}_f$, respectively. For A let us choose a diagonal representation

$$A\varphi_\nu = \varepsilon_\nu \varphi_\nu, \tag{8.1}$$

where φ_ν is a basis for $\imath$. For $\mathscr{T}$ we may choose two different systems of basis vectors: First the $\phi_{\nu_1\nu_2\cdots\nu_f} = \varphi_{\nu_1}(1)\varphi_{\nu_2}(2)\cdots\varphi_{\nu_f}(f)$ and second, the $u_\rho^{(\nu)}v_\sigma^{(\nu)}$ obtained from the decomposition (7.2). Here we obtain

$$PA\phi_{\nu_1\cdots\nu_f} = \varepsilon_{\nu_1}\varepsilon_{\nu_2}\cdots\varepsilon_{\nu_f}\phi_{\nu_{\alpha_1}\nu_{\alpha_2}\cdots\nu_{\alpha_f}}. \tag{8.2}$$

We will obtain diagonal elements only if the permutation P of the indices $\nu_1 \cdots \nu_f$ does not change the sequence of the indices. If $P \neq e$ this can be only if some of the indices are identical. Which of the index sets $\nu_1 \cdots \nu_f$ will be transformed into itself under P? Let P be described in terms of a cyclic representation (see AV, §9) where $\gamma_1, \gamma_2, \ldots$ are the lengths of the cycles. Here we find that P transforms an index set into itself if the individual cycles permute identical indices. The vectors $\phi_{\nu_1\cdots\nu_f}$ which are transformed by P into themselves form a basis for the vector space

$$\imath_{\gamma_1} \times \imath_{\gamma_2} \times \cdots, \tag{8.3}$$

where, for example, $\imath_{\gamma_1}$ is spanned by the vectors $\varphi_\nu(x_1)\varphi_\nu(x_2)\cdots\varphi_\nu(x_{\gamma_1})$ where $x_1, x_2, \ldots, x_{\gamma_1}$ are the numbers in the cycle γ_1 of P. Therefore we obtain

$$\operatorname{tr}(PA) = \operatorname{tr}_{\gamma_1}(A)\operatorname{tr}_{\gamma_2}(A)\cdots, \tag{8.4}$$

where $\operatorname{tr}_{\gamma_1}(A)$ is the trace of A in $\imath_{\gamma_1}$. It is easy to see that this is equal to $s_{\gamma_1} = \sum_{\nu=1}^{n} \varepsilon_\nu^{\gamma_1}$. Thus we find that

$$\operatorname{tr}(PA) = s_{\gamma_1}s_{\gamma_2}\cdots. \tag{8.5}$$

If we use the basis $u_\mu^{(\nu)}v_\rho^{(\nu)}$, we obtain

$$Au_\mu^{(\nu)}v_\sigma^{(\nu)} = \sum_\lambda u_\lambda^{(\nu)}a_{\lambda\mu}^{(\nu)}v_\sigma^{(\nu)} \tag{8.6}$$

and we find that

$$PAu_\mu^{(\nu)}v_\sigma^{(\nu)} = \sum_{\lambda\rho} u_\lambda^{(\nu)}a_{\lambda\mu}^{(\nu)}v_\rho^{(\nu)}P_{\rho\sigma}^{(\nu)}. \tag{8.7}$$

From these results, we find that

$$\operatorname{tr}(PA) = \sum_{\nu}\sum_{\sigma\mu} a_{\mu\mu}^{(\nu)} P_{\sigma\sigma}^{(\nu)} = \sum_{\nu} x^{(\nu)}(P) X^{(\nu)}(\varepsilon_1 \cdots \varepsilon_n), \tag{8.8}$$

where $\chi^{(\nu)}$ is the character of the representation (ν) of $\mathbf{S}_f$ and $X^{(\nu)}$ is the character of the corresponding representation of U_n.

From (8.4) and (8.8) we therefore obtain the important relationship:

$$\sum_{\nu} \chi^{(\nu)}(P) X^{(\nu)}(\varepsilon_1 \cdots \varepsilon_n) = s_{\gamma_1} s_{\gamma_2} \cdots. \tag{8.9}$$

If we are successful in computing the $X^{(\nu)}(\varepsilon_1 \cdots \varepsilon_n)$ then we may also compute the $\chi^{(\nu)}(P)$ from (8.9) by using the orthogonality relations (AV, §10.5). Since A is unitary, $\varepsilon_\nu = e^{i\alpha_\nu}$ where α_ν is real. Since all A of the form (8.1) are diagonal matrices, they commute; thus we may choose the basis vectors $u_k^{(\nu)}$ in $\mathscr{p}^{(\nu)}$ (7.2) such that the operators A are also diagonal in the representation in $\mathscr{p}^{(\nu)}$. Instead of (8.6) we may express the operators A as follows:

$$Au_k^{(\nu)} = f_k^{(\nu)}(\alpha_1 \cdots \alpha_n) u_k^{(\nu)}. \tag{8.10}$$

with α_ν from (8.1).

Since this provides a representation of A, it follows that if $B\varphi_\nu = e^{i\beta_\nu}\varphi_\nu$, then from $AB\varphi_\nu = e^{i(\alpha_\nu + \beta_\nu)}\varphi_\nu$ it follows that

$$f_k^{(\nu)}(\alpha_1 \cdots \alpha_n) f_k^{(\nu)}(\beta_1 \cdots \beta_n) = f_k^{(\nu)}(\alpha_1 + \beta_1, \ldots, \alpha_n + \beta_n). \tag{8.11}$$

Thus it follows that if

$$g_{kl}^{(\nu)}(\alpha_l) = f_k^{(\nu)}(0, 0, \ldots, \alpha_l, 0 \cdots 0),$$

$$f_k^{(\nu)}(\alpha_1, \ldots, \alpha_n) = \prod_{l=1}^{n} g_{kl}^{(\nu)}(\alpha_l),$$

$$g_{kl}^{(\nu)}(\alpha_l + \beta_l) = g_{kl}^{(\nu)}(\alpha_l) g_{kl}^{(\nu)}(\beta_l), \tag{8.12}$$

and

$$g_{kl}^{(\nu)}(0) = 1.$$

Since the $g_{kl}^{(\nu)}(\alpha_l)$ are periodic in α_l and have magnitude 1 it follows that $g_{kl}^{(\nu)}(\alpha_l) = e^{ih_l\alpha_l}$ where h_l is an integer; we therefore obtain

$$f_k^{(\nu)}(\alpha_1, \ldots, \alpha_n) = e^{i\Sigma_{l=1}^n h_l \alpha_l}. \tag{8.13}$$

$X^{(\nu)}(\varepsilon_1 \cdots \varepsilon_n)$ is therefore a sum of positive and negative powers in the $\varepsilon_\nu = e^{i\alpha_\nu}$ with positive integer coefficients. Since two elements for which $\{\varepsilon_\nu\}$ differ only in the order belong to the same class of conjugate elements, $X^{(\nu)}(\varepsilon_1 \cdots \varepsilon_n)$ must be a symmetric functions in the arguments ε_ν.

As we have shown in AV, §10.5, the "volume" element for a class of conjugate elements for α_k between α_k and $\alpha_k + d\alpha_k$ is given by

$$c|\Delta|^2 \, d\alpha_1 \, d\alpha_2 \cdots d\alpha_n, \tag{8.14}$$

where $\Delta = \prod_{i<k} (\varepsilon_i - \varepsilon_k)$ and c is the normalization factor, satisfying

$$c \int |\Delta|^2 \, d\alpha_1 \cdots d\alpha_n = 1. \tag{8.15}$$

We obtain

$$\Delta = \begin{vmatrix} \varepsilon_1^{n-1} & \varepsilon_1^{n-2} & \cdots & \varepsilon_1 & 1 \\ \varepsilon_2^{n-1} & \varepsilon_2^{n-2} & \cdots & \varepsilon_2 & 1 \\ \vdots & & & & \vdots \\ \varepsilon_n^{n-1} & \varepsilon_n^{n-2} & \cdots & \varepsilon_n & 1 \end{vmatrix} = \sum_P (-1)^P P e^{i[(n-1)\alpha_1 + (n-2)\alpha_2 + \cdots + \alpha_{n-1} + 0\alpha_n]}, \tag{8.16}$$

where P permutes the $\alpha_1 \cdots \alpha_n$. If we have an arbitrary sum of powers of the form

$$p_{h_1 \cdots h_n}(\alpha_1 \cdots \alpha_n) = \sum_P (-1)^P P e^{i\Sigma_{k=1}^n h_k \alpha_k}, \tag{8.17}$$

where $h_1 > h_2 > \cdots > h_n$, then, it is easy to see that

$$\int \overline{p_{h_1 \cdots h_n}(\alpha_1 \cdots \alpha_n)} p_{h'_1 \cdots h'_n}(\alpha_1 \cdots \alpha_n)\, d\alpha_1 \cdots d\alpha_n = \begin{cases} n!\,(2\pi)^n & \text{if } h_k = h_{k'}, \\ 0 & \text{otherwise.} \end{cases} \tag{8.18}$$

Thus it follows that $c = (n!)^{-1}(2\pi)^{-n}$ in (8.15). According to AV, §10.5 for the characters $X^{(\nu)}(\varepsilon_1 \cdots \varepsilon_n)$ we obtain the orthogonality relations:

$$\frac{1}{n!\,(2\pi)^n} \int \overline{\Delta X^{(\nu)}} \Delta X^{(\mu)}\, d\alpha_1 \cdots d\alpha_k = \delta_{\nu\mu}. \tag{8.19}$$

Instead of using $X^{(\nu)}$ we shall use

$$\Gamma^{(\nu)} = \Delta X^{(\nu)}. \tag{8.20}$$

The $\Gamma^{(\nu)}(\varepsilon_1 \cdots \varepsilon_n)$ are therefore antisymmetric power sums in the ε_ν with integer coefficients $a_{h_1 \cdots h_n}$. We obtain

$$\Gamma^{(\nu)}(\varepsilon_1 \cdots \varepsilon_n) = \sum_{h_1 \cdots h_n} a_{h_1 \cdots h_n} p_{h_1 \cdots h_n}(\alpha_1 \cdots \alpha_n). \tag{8.21}$$

From

$$\int |\Gamma^{(\nu)}|^2\, d\alpha_1 \cdots d\alpha_n = n!\,(2\pi)^n$$

it follows that

$$\sum_{h_1 \cdots h_n} |a_{h_1 \cdots h_n}|^2 = 1. \tag{8.22}$$

Since all the $a_{h_1 \cdots h_n}$ are integers, only one of these $a_{h_1 \cdots h_n} \neq 0$ and is therefore equal to 1 or -1, so that for a suitable series $h_1 \cdots h_n$:

$$\Gamma^{(\nu)}(\varepsilon_1 \cdots \varepsilon_n) = \pm\, p_{h_1 \cdots h_n}(\alpha_1 \cdots \alpha_n). \tag{8.23}$$

The coefficient of the term $e^{i\Sigma_{k=1}^n h_k \alpha_k}$ in $\Gamma^{(\nu)}$ must be positive, because the term arises as the product of the term $e^{i\Sigma_{k=1}^n (n-k)\alpha_k}$ of Δ and a term of the power sum

$X^{(\nu)}(\varepsilon_1 \cdots \varepsilon_n)$ having only positive coefficients. Thus we find that we must use the + sign in (8.23). Thus we obtain

$$X^{(\nu)}(\varepsilon_1 \cdots \varepsilon_n) = \Delta^{-1} p_{h_1 \cdots h_n}(\alpha_1 \cdots \alpha_n). \tag{8.24}$$

Since the above results (from (8.10) on) hold for any irreducible representation of U_n (because in (8.10) we have not made any assumption about the irreducible representation space $\mathscr{h}^{(\nu)}$) all irreducible characters have the form (8.24).

Since, according to (8.17), (8.18) the $\Delta^{-1}p_{h_1 \cdots h_n}$ are orthogonal for different $h_1 > h_2 > \cdots > h_n$, the $\Delta^{-1}p_{h_1 \cdots h_n}$ comprise all possible irreducible characters for the group U_n. The $\Delta^{-1}p_{h_1 \cdots h_n}$ must therefore form a complete system of functions with respect to the classes (that is, as functions of the $\alpha_1 \cdots \alpha_n$) (see AV, §10.5). This general group theoretical fact is another proof of the Fourier theorem: the set of all $e^{-i\Sigma_l n_l \alpha_l}$ with integer n_l form a complete system.

Have we obtained all possible irreducible representations of U_n by the process of reducing the representation U_n^f (for all f)? To answer this question we must return to (8.10) and ask how the sequence $h_1 > h_2 > \cdots > h_n$ depends upon the index (ν), that is, upon the corresponding representation of the permutation group and therefore upon the tableau corresponding to this representation (see AV, §9).

We shall now seek a basis vector $u_k^{(\nu)}$ in $\mathscr{h}^{(\nu)}$ ($\tilde{\mathscr{h}}^{(\nu)}$ may, according to §7, correspond to an indempotent element of $\mathscr{R}_{\mathbf{S}_f}$ belonging to a tableau with row length $n_1, n_2, \ldots, n_k$) for which the exponents k_v in

$$Au_k^{(\nu)} = \varepsilon_1^{k_1}\varepsilon_2^{k_2} \cdots \varepsilon_n^{k_n} u_k^{(\nu)}$$

form a "largest" possible sequence $k_1, k_2, \ldots, k_n$ [$k_1, k_2, \ldots, k_n$ is said to be "larger" than $k_1', k_2', \ldots, k_n'$ if $k_1 > k_1'$ or if $k_2 > k_2'$ for $k_1 = k_1'$, etc.]. Since the vectors in $\tilde{\mathscr{h}}^{(\nu)}$ may be obtained by application of the idempotent element γJK to the vectors of $\mathscr{T}$, $\tilde{\mathscr{h}}_\rho^{(\nu)}$ is spanned only by vectors of the form

$$\gamma IK\varphi_{\nu_1}(1) \cdots \varphi_{\nu_f}(f),$$

where the groups of the same ν_i do not have lengths longer than $n_1, n_2, \ldots, n_k$, otherwise the application of K would result in 0. Thus it follows that the "largest" possible sequence $k_1, k_2, \ldots, k_n$ is given by $n_1, n_2, \ldots, n_k$, that is, $k_i = n_i$.

Since $\Gamma^{(\nu)} = \Delta X^{(\nu)}$ then from (8.16) and (8.24) it follows that the following relations hold for the h_i:

$$h_1 = n_1 + (n-1), h_2 = n_2 + (n-2), \cdots, h_n = n_n, \tag{8.25}$$

where we set $n_\nu = 0$ for $\nu > k$.

According to §7, for sufficiently large n, we obtain subspaces $\tilde{\mathscr{h}}_\rho^{(\nu)}$ which are nonzero in $\mathscr{r}^f$ for all tableaux and we therefore obtain all possible decompositions $f = n_1 + n_2 + \cdots + n_k$ where $n_1 \geq n_2 \geq \cdots \geq n_k$. The values $h_l = n_l + (n - l)$ do not, however, run through all possible sequences $h_1 > h_2 > \cdots > h_n$.

If we replace the representation $A \to B \in U_n^f$ by $A \to |A|^l B$ where $|A|$ is the determinant of A and l is an integer ($l = 0, \pm 1, \pm 2, \ldots$), then in the space $\not{h}^{(\nu)}$ we obtain (for fixed l) a representation with the character $|A|^l X^{(\nu)}(\varepsilon_1 \cdots \varepsilon_n)$. If the highest powers of $X^{(\nu)}$ are given by the $k_1, k_2 \ldots, k_n$ then the $|A|^l X^{(\nu)}$ may be obtained from the series $k_1 + l$, $k_2 + l, \ldots, k_n + l$. By a suitable choice for l we can therefore always obtain positive values of the k_i and, for sufficiently large f we may set the $k_i = n_i$.

We therefore obtain all irreducible representations of U_n, reducing U_n^f (for all $f \geq 0$) and multiplying the irreducible parts of U_n^f by powers of the determinant. In particular, we find that the reduction of $(SU_n)^f$ contains all irreducible representations of SU_n, because the determinants of the transformations from SU_n are equal to 1. The characters for the irreducible representations of SU_n are obtained from (8.24), where we note that $\prod_{\nu=1}^n \varepsilon_\nu = e^{i\Sigma\nu\alpha_\nu} = 1$. According to (8.24) we find that two characters with index sets $h_1, h_2, \ldots, h_n$ and $h_1', \ldots, h_n'$ are equal if there exists a number l for which $h_i' = h_i + l$. For SU_n we can, for example, set $h_n = 0$ and we need only use all index sets $h_1 > h_2 > \cdots > h_n = 0$ in order to obtain all irreducible characters.

We will now compute the degree with which the representation of U_n occurs in $\not{h}^{(\nu)}$. Since this degree is the same for SU_n, all representations for which $h_i' = h_i + l$ have the same degree as that with index set h_i. The degree $N^{(\nu)}$ of the representation is

$$N^{(\nu)} = X^{(\nu)}(1, 1, \ldots, 1).$$

From the formula (8.24) it follows that (in a formal sense) we would obtain 0/0 for $N^{(\nu)}$. Therefore we shall now consider the special case $\alpha_\nu = (n - \nu)\omega$. Here we obtain

$$p_{h_1 \cdots h_n} = \begin{vmatrix} e^{ih_1(n-1)\omega} & e^{ih_1(n-2)\omega} & \cdots & 1 \\ e^{ih_2(n-1)\omega} & e^{ih_2(n-2)\omega} & \cdots & 1 \\ \vdots & & & \vdots \\ e^{ih_n(n-1)\omega} & e^{ih_n(n-2)\omega} & \cdots & 1 \end{vmatrix} = \prod_{i<k} (e^{ih_i\omega} - e^{ih_k\omega}) \sim \prod_{i<k} [(h_i - h_k)i\omega],$$

where the last expression gives the smallest power in ω. In the same way we obtain

$$\Delta = \prod_{j<k} (e^{i(n-j)\omega} - e^{i(n-k)\omega}) \sim \prod_{j<k} ([(n-j) - (n-k)]i\omega).$$

Thus in the limit $\omega \to 0$ we obtain:

$$N^{(\nu)} = \frac{\prod_{i<k} (h_i - h_k)}{\prod_{i<k} (k - i)} = \frac{\Delta(h_1, h_2, \ldots, h_n)}{\Delta(n-1, n-2, \ldots, 0)}$$

with

$$\Delta(a_1, a_2, \ldots, a_n) = \begin{vmatrix} a_1^{n-1} & a_1^{n-2} & \cdots & a_1 & 1 \\ a_2^{n-1} & a_2^{n-1} & \cdots & a_2 & 1 \\ \vdots & & & & \\ a_n^{n-1} & a_n^{n-2} & \cdots & a_n & 1 \end{vmatrix}.$$

Since the $X^{(\nu)}$ are known, from (8.9) we may compute the $\chi^{(\mu)}(P) = \chi^{(\mu)}(\gamma_1, \gamma_2, \ldots)$. From (8.24) we obtain:

$$\sum_{(n)} \chi_{n_1 n_2 \cdots n_k}(\gamma_1, \gamma_2, \ldots) p_{h_1 \cdots h_n} = \Delta s_{\gamma_1} s_{\gamma_2} \cdots, \tag{8.26}$$

where, instead of (ν) we use the row lengths $n_1 n_2 \cdots n_k$ of the corresponding tableau. From (8.26) it follows that $\chi_{n_1 n_2 \cdots n_k}(\gamma_1, \gamma_2, \ldots)$ is the coefficient of $\varepsilon_1^{h_1} \varepsilon_2^{h_2} \cdots \varepsilon_n^{h_n}$ in the expansion of $\Delta s_{\gamma_1} s_{\gamma_2} \cdots$ in powers of $\varepsilon_1 \cdots \varepsilon_n$. The degree g of the representation corresponding to the tableau $n_1 \cdots n_k$ is obtained by substituting $\gamma_1 = \gamma_2 = \cdots = \gamma_f = 1$ in $\chi_{n_1 n_2 \cdots n_k}(\gamma_1, \gamma_2 \cdots)$, and for this case we find that $s_{\gamma_1} s_{\gamma_2} \cdots = s_1^f = (\sum_{\nu=1}^n \varepsilon_\nu)^f$. Therefore we find that g is the coefficient of $\varepsilon_1^{h_1} \varepsilon_2^{h_2} \cdots \varepsilon_n^{h_n}$ in the expansion of

$$\left(\sum_{\nu=1} \varepsilon_\nu\right)^f \sum_P (-1)^P P \varepsilon_1^{n-1} \varepsilon_2^{n-2} \cdots \varepsilon_n^0 \tag{8.27}$$

in powers of the $\varepsilon_1 \cdots \varepsilon_n$. (8.27) is, however, equal to

$$\sum_P (-1)^P \sum_{\nu_1 \cdots \nu_n} \frac{f!}{\nu_1! \, \nu_2! \cdots \nu_n!} \varepsilon_1^{\nu_1} \varepsilon_2^{\nu_2} \cdots \varepsilon_n^{\nu_n} P \varepsilon_1^{n-1} \cdots \varepsilon_n^0.$$

If $P \varepsilon_1^{n-1} \cdots \varepsilon_n^0 = \varepsilon_1^{k_1} \varepsilon_2^{k_2} \cdots \varepsilon_n^{k_n}$, we therefore obtain

$$g = f! \sum_P (-1)^P [(h_1 - k_1)! \, (h_2 - k_2)! \cdots (h_n - k_n)!]^{-1}$$

$$= f! \begin{vmatrix} \dfrac{1}{[h_1 - (n-1)]!} & \dfrac{1}{[h_1 - (n-2)]!} & \cdots & \dfrac{1}{h_1!} \\ \dfrac{1}{[h_2 - (n-1)]!} & \dfrac{1}{[h_2 - (n-2)]!} & \cdots & \dfrac{1}{h_2!} \\ \vdots & \vdots & & \vdots \\ \dfrac{1}{[h_n - (n-1)]!} & \dfrac{1}{[h_n - (n-2)]!} & \cdots & \dfrac{1}{h_n!} \end{vmatrix}$$

$$= \frac{f}{h_1! \, h_2! \cdots h_n!}$$

$$\begin{vmatrix} h_1(h_1 - 1) \cdots (h_1 - n + 2) & h_1(h_1 - 1) \cdots (h_1 - n + 3) & \cdots & h_1 & 1 \\ \vdots & \vdots & & \vdots & \vdots \\ h_n(h_n - 1) \cdots (h_n - n + 2) & h_n(h_n - 1) \cdots (h_n - n + 3) & \cdots & h_n & 1 \end{vmatrix}.$$

By selective subtraction of the columns, for the determinant we obtain

$$\begin{vmatrix} h_1^{n-1} & h_1^{n-2} & \cdots & h_1 & 1 \\ \vdots & & & & \vdots \\ h_n^{n-1} & h_n^{n-2} & \cdots & h_n & 1 \end{vmatrix}.$$

Therefore we obtain:

$$g = \frac{f! \prod_{i<k}(h_i - h_k)}{h_1! \, h_2! \cdots h_n!}. \tag{8.28}$$

For the group SU_2 which is the representation of the rotation group in spin space we may set $h_1 > h_2 = 0$ in $p_{h_1 h_2}(\varepsilon_1, \varepsilon_2)$, as shown above for the general case of SU_n. If we introduce the quantity S by $h_1 = 2S + 1$ then S can take on the values $0, \frac{1}{2}, 1, \frac{3}{2}, \ldots$. Thus, setting $\varepsilon_1 = x$ and $\varepsilon_2 = 1/x$ we obtain

$$p_{2S+1,0}(\varepsilon_1, \varepsilon_2) = x^{2S+1} - x^{-(2S+1)}$$

and

$$\Delta = x - x^{-1}.$$

Thus, from (8.24) it follows that (using S instead of v)

$$X^{(S)}(\varepsilon_1, \varepsilon_2) = \frac{x^{2S+1} - x^{-(2S+1)}}{x - x^{-1}} = x^{2S} + x^{2S-2} + \cdots x^{-2S}.$$

If we use SU_2 as a representation of the rotation group, then we obtain $x = e^{i\varphi/2}$ where φ is the rotation angle and find that

$$X^{(S)}(\varepsilon_1, \varepsilon_2) = e^{iS\varphi} + e^{i(S-1)\varphi} + \cdots + e^{-iS\varphi}.$$

Thus we have again obtained the well-known character of the representation $\mathscr{D}_S$ of the rotation group. The degree of the representation is therefore given by

$$N^{(S)} = (X^{(S)})_{\varphi=0} = 2S + 1.$$

For the corresponding character $\chi_S(P)$ of the representation of $\mathbf{S}_f$ in $\imath^f$ ($\imath$ is the spin space), it follows from (8.9) that

$$\sum_S \chi_S(P)(x^{2S} + x^{2S-2} + \cdots + x^{-2S}) = (x^{\gamma_1} + x^{-\gamma_1})(x^{\gamma_2} + x^{-\gamma_2})\cdots. \tag{8.29}$$

Since in $\imath^f$ $S \leq 2f$ it follows that $S = 2f - k$ where k is an integer ≥ 0. By multiplication with $x^f(1 - x^2)$ it follows that for $z = x^2$:

$$\sum_S \chi_S(P)(z^k - z^{f+1-k}) = (1 - z)(1 + z^{\gamma_1})(1 + z^{\gamma_2})\cdots.$$

Since $k \leq f/2$ it follows that $\chi_S(P)$ is the coefficient of z^k on the right side of this equation. In particular, the degree g of the representation is the coefficient of z^k in $(1 - z)(1 + z)^f$, that is,

$$g = \binom{f}{k} - \binom{f}{k-1} = \frac{f!\,(f - 2k + 1)}{k!\,(f - k + 1)!} = \binom{f}{k}\frac{f - 2k + 1}{f - k + 1}. \tag{8.30}$$

The characters for the associated representations are given by $\chi'_S = (-1)^P \chi_S$.

The group SU_3 plays an important role in elementary particle physics; we shall therefore discuss the group SU_3 as an additional example. We will not, however, discuss its physical meaning for elementary particle physics.

At this point we could evaluate the formula (8.24) subject to the additional condition $h_n = 0$ for the case of SU_3 (see the above discussion for the general case of SU_n). Actually, it is preferable to make use of the so-called branching theorems, which we shall now state and prove.

Branching Theorem for the Permutation Group

The irreducible representation of $\mathbf{S}_f$ corresponding to a tableau with row lengths $n_1, n_2, \ldots, n_k$ is decomposed into irreducible representations with respect to the subgroup $\mathbf{S}_{f-1}$ corresponding to tables with row lengths

$$\begin{matrix} n_1 - 1 & n_2 & \cdots & n_k, \\ n_1 & n_2 - 1 & \cdots & n_k, \\ \vdots & \vdots & & \vdots \\ n_1 & n_2 & \cdots & n_k - 1, \end{matrix}$$

where we omit those tables for which the line lengths are not decreasing. Each such irreducible representation of $\mathbf{S}_{f-1}$ occurs exactly once.

Proof. Let P be a permutation of $\mathbf{S}_{f-1}$, that is, a permutation of $\mathbf{S}_f$ which leaves the last number f fixed. As a permutation of $\mathbf{S}_f$, P contains a cycle of length 1 more than it does as an element of $\mathbf{S}_{f-1}$. From (8.26) we have concluded that in the expansion

$$\Delta s_{\gamma_1} s_{\gamma_2} \cdots = \sum a_{h_1 \cdots h_n} \varepsilon_1^{h_1} \cdots \varepsilon_n^{h_n}$$

the coefficient $a_{h_1 \cdots h_n}$ is equal to $\chi_{n_1 \cdots n_k}(\gamma_1, \gamma_2, \ldots)$. If we consider P to be an element of $\mathbf{S}_{f-1}$, then with the cycles $\gamma_1, \gamma_2, \ldots$ of P (as an element of $\mathbf{S}_{f-1}$) we obtain:

$$\Delta s_{\gamma_1} s_{\gamma_2} \cdots = \sum \chi_{n_1 \cdots n_k}(P) \varepsilon_1^{h_1} \cdots \varepsilon_n^{h_n}. \tag{8.31}$$

Here $\chi_{n_1 \cdots n_k}(P)$ is the character of P in the irreducible representation of $\mathbf{S}_{f-1}$ corresponding to $n_1 \cdots n_k$.

On the other hand, if we consider P as an element of $\mathbf{S}_f$, since $s_1 = \sum_{\nu=1}^n \varepsilon_\nu$ (that is, s_γ for a cycle length $\gamma = 1$), we obtain:

$$\Delta\left(\sum_{\nu=1}^{n} \varepsilon_\nu\right) s_{\gamma_1} s_{\gamma_2} \cdots = \sum \chi_{n'_1 \cdots n'_k}(P) \varepsilon_1^{h'_1} \cdots \varepsilon_n^{h'_n}. \tag{8.32}$$

Here $\chi_{n'_1 \cdots n'_k}(P)$ is the character of P in the irreducible representation of $\mathbf{S}_f$ corresponding to $n'_1 \cdots n'_k$.

If we multiply (8.31) by $\sum_{\nu=1}^n \varepsilon_\nu$ and then compare the result to (8.32) using (8.25) it follows that $\chi_{n'_1 \cdots n'_k}(P)$ is the sum of the $\chi_{n_1 \cdots n_k}$, as given by the branching rule.

Branching Theorem for U_n

If we restrict U_n to a subgroup U_{n-1} of a $(n-1)$-dimensional subspace of $\imath$ we find that an irreducible representation of U_n corresponding to a tableau with row lengths $n_1, n_2, \ldots, n_k$ (or to the values $h_1, h_2, \ldots, h_k$ in (8.25)) is decomposed with respect to U_{n-1} into irreducible representations corresponding to tableaux with row lengths $n'_1, n'_2, \ldots, n'_{k-1}$ (or to the values $h'_1, h'_2, \ldots, h'_k$) where

$$n_1 \geq n'_1 \geq n_2 \geq n'_2 \geq \cdots \geq n'_{k-1} \geq n_k \tag{8.33}$$

or

$$h_1 > h'_1 \geq h_2 > h'_2 \geq h_3 > \cdots > h'_{n-1} \geq h_n, \tag{8.34}$$

respectively. Each of the components corresponding to (8.33) (or to (8.34)) occurs exactly once.

Proof. We obtain the mappings of U_{n-1} if we map, for example, φ_n into φ_n, that is, set $\varepsilon_n = 1$. According to (8.24) and (8.17) we need to compute $p_{h_1 \cdots h_n}(\alpha_1, \ldots, 0)$ (that is, set $\alpha_n = 0$):

$$p_{h_1 \cdots h_n}(\alpha_1, \ldots, 0) = \begin{vmatrix} e^{ih_1\alpha_1} & e^{ih_1\alpha_2} & \cdots & 1 \\ e^{ih_2\alpha_1} & e^{ih_2\alpha_2} & \cdots & 1 \\ \vdots & \vdots & & \\ e^{ih_n\alpha_1} & e^{ih_n\alpha_2} & \cdots & 1 \end{vmatrix} = \begin{vmatrix} \varepsilon_1^{h_1} & \varepsilon_2^{h_1} & \cdots & 1 \\ \varepsilon_1^{h_2} & \varepsilon_2^{h_2} & \cdots & 1 \\ \vdots & \vdots & & \vdots \\ \varepsilon_1^{h_n} & \varepsilon_2^{h_n} & \cdots & 1 \end{vmatrix}$$

$$= \begin{vmatrix} \varepsilon_1^{h_1} - \varepsilon_1^{h_2} & \varepsilon_2^{h_1} - \varepsilon_2^{h_1} & \cdots & 0 \\ \varepsilon_1^{h_2} - \varepsilon_1^{h_3} & \varepsilon_2^{h_2} - \varepsilon_2^{h_3} & \cdots & 0 \\ \vdots & \vdots & & \vdots \\ \varepsilon_1^{h_n} & \varepsilon_2^{h_n} & \cdots & 1 \end{vmatrix}$$

$$= \begin{vmatrix} \varepsilon_1^{h_1} - \varepsilon_1^{h_2} & \varepsilon_2^{h_1} - \varepsilon_2^{h_2} & \cdots & \varepsilon_{n-1}^{h_1} - \varepsilon_{n-1}^{h_2} \\ \varepsilon_1^{h_2} - \varepsilon_1^{h_3} & \varepsilon_2^{h_2} - \varepsilon_2^{h_3} & \cdots & \varepsilon_{n-1}^{h_2} - \varepsilon_{n-1}^{h_3} \\ \vdots & \vdots & & \vdots \\ \varepsilon_1^{h_{n-1}} - \varepsilon_1^{h_n} & \varepsilon_2^{h_{n-1}} - \varepsilon_2^{h_n} & \cdots & \varepsilon_{n-1}^{h_{n-1}} - \varepsilon_{n-1}^{h_n} \end{vmatrix}$$

and similarly

$$\Delta(\varepsilon_1, \ldots, 1) = \begin{vmatrix} \varepsilon_1^{n-2} & \varepsilon_1^{n-2} & \cdots & \varepsilon_1 & 1 \\ \varepsilon_2^{n-1} & \varepsilon_2^{n-2} & \cdots & \varepsilon_2 & 1 \\ \vdots & \vdots & & \vdots & \vdots \\ 1 & 1 & & 1 & 1 \end{vmatrix}$$

$$= \begin{vmatrix} \varepsilon_1^{n-2} & \varepsilon_1^{n-3} & \cdots & 1 \\ \varepsilon_2^{n-2} & \varepsilon_2^{n-3} & \cdots & 1 \\ \vdots & \vdots & & \vdots \\ \varepsilon_{n-1}^{n-2} & \varepsilon_{n-1}^{n-3} & \cdots & 1 \end{vmatrix} (\varepsilon_1 - 1)(\varepsilon_2 - 1) \cdots (\varepsilon_{n-1} - 1)$$

$$= \Delta(\varepsilon_n, \ldots, \varepsilon_{n-1})(\varepsilon_1 - 1)(\varepsilon_2 - 1) \cdots (\varepsilon_{n-1} - 1)$$

and we then may compute $\Delta^{-1} p_{h_1 \cdots h_n}$ by using (8.24).

Using the formula;

$$\frac{\varepsilon^h - \varepsilon^{h'}}{\varepsilon - 1} = \varepsilon^{h-1} + \cdots + \varepsilon^{h'}$$

from (8.24) we obtain

$$X^{(\nu)}(\varepsilon_1, \ldots, 1) = \Delta^{-1}(\varepsilon_1, \ldots, \varepsilon_{n-1}) \begin{vmatrix} \varepsilon_1^{h_1-1} + \cdots + \varepsilon_1^{h_2} & \varepsilon_2^{h_1-1} + \cdots + \varepsilon_2^{h_2} & \cdots \\ \varepsilon_1^{h_2-1} + \cdots + \varepsilon_1^{h_3} & \varepsilon_2^{h_2-1} + \cdots + \varepsilon_2^{h_3} & \cdots \\ \vdots & \vdots & \vdots \\ \varepsilon_1^{h_{n-1}-1} + \cdots + \varepsilon_1^{h_n} & \varepsilon_2^{h_{n-1}-1} + \cdots + \varepsilon_2^{h_n} & \cdots \end{vmatrix}.$$

The last determinant is, however, the sum of all determinants of the form

$$\begin{vmatrix} \varepsilon_1^{h_1'} & \varepsilon_2^{h_1'} & \cdots \\ \varepsilon_1^{h_2'} & \varepsilon_2^{h_1'} & \cdots \\ \vdots & \vdots & \vdots \\ \varepsilon_1^{h_{n-1}'} & \varepsilon_2^{h_{n-1}'} & \cdots \end{vmatrix},$$

where $h_1', h_2', \ldots$ are all the values for which (8.34) is satisfied. From (8.25) we find that (8.34) is equivalent to (8.33).

Branching Theorem for SU_n

From the branching theorem for U_n we immediately obtain a corresponding branching theorem for SU_n, except that only those representations described by $h_1', h_2', \ldots$ for which the h_i' differ by a fixed integer are equivalent, as we have proven earlier.

We will now apply this theorem to SU_3! An irreducible representation of SU_3 is characterized by two numbers l_1, l_2, because we can set $h_3 = 0$, as we have shown earlier. According to (8.25)

$$h_1 = n_1 + 2, \qquad h_2 = n_2 + 1, \qquad h_3 = n_3$$

here we can set $h_3 = n_3 = 0$.

A representation corresponding to the numbers

$$h_1 > h_2 > h_3 = 0$$

is decomposed, as a representation of SU_2 into those representations h_1', h_2' where

$$h_1 > h_1' \geq h_2 > h_2' \geq 0.$$

All representations with the same difference $h_1' - h_2' = 2S + 1$ are equivalent to the representation $h_1' = 2S + 1$, $h_2' = 0$, that is, to the representation of SU_2 with "spin quantum number" S. Thus it follows that the representation of SU_3 characterized by h_1, h_2, as a representation of SU_2, decomposes into representations corresponding to spin quantum numbers S where $1 \geq 2S + 1 \leq h_1 - 1$, where the representation for the spin quantum number S has the following multiplicities:

(1) $2S + 1 \geq h_2$ and $2S + 1 \geq h_1 - h_2$: $h_1 - (2S + 1)$.
(2) $2S + 1 \geq h_2$ and $2S + 1 \leq h_1 - h_2$: h_2.
(3) $2S + 1 \leq h_2$ and $2S + 1 \geq h_1 - h_2$: $(h_1 - h_2)$.
(4) $2S + 1 \leq h_2$ and $2S + 1 \leq h_1 - h_2$: $(2S + 1)$.

Since $n_1 = h_1 - 2$, $n_2 = h_2 - 1$ and $n_3 = 0$ the irreducible representation of SU_3 in r^f characterized by h_1, h_2 occurs in r^f with a multiplicity of $f = h_1 + h_2 - 3$.

For the case in which $h_1 = 3$, $h_2 = 1$ we find that $f = 1$, that is, SU_3 is represented by itself in $\mathfrak{r}$. As a representation of the subgroup SU_2 it is decomposed in the representations for $S = \frac{1}{2}$ and $S = 0$ which occurs once.

For the case in which $h_1 = 3, h_2 = 2$ we find that $f = 2$. The representation occurs as an irreducible part in $\mathfrak{r} \times \mathfrak{r}$. It decomposes as a representation of SU_2 in both representations for $S = 0, \frac{1}{2}$, exactly once each. This representation is therefore $1 + 2 = 3$ dimensional.

For $h_1 = 4$, $h_2 = 1$ it follows that $f = 2$, and the decomposition is into $S = 0, \frac{1}{2}, 1$ (each simply). The representation is therefore $1 + 2 + 3 = 6$ dimensional.

For $h_1 = 4, h_2 = 2$ it follows that $f = 3$. It decomposes into $S = 0$ (simple), $S = \frac{1}{2}$ (double) and $S = 1$ (simple). The order of the representation is therefore $1 + 2 \cdot 2 + 3 = 8$.

We will now briefly state a few additional cases. Here d will represent the order of the representation.

$h_1 = 4$, $h_2 = 3$, $f = 4$; $S = 0$ (simple), $S = \frac{1}{2}$ (simple), $S = 1$ (simple); $d = 1 + 2 + 3 = 6$.

$h_1 = 5, h_2 = 1, f = 3$; $S = 0, \frac{1}{2}, 1, \frac{3}{2}$ (all simple); $d = 1 + 2 + 3 + 4 = 10$.

$h_1 = 5$, $h_2 = 2$, $f = 4$; $S = 0$ (simple), $S = \frac{1}{2}$ (double), $S = 1$ (double), $S = \frac{3}{2}$ (simple); $d = 1 + 2 \cdot 2 + 2 \cdot 3 + 4 = 15$.

$h_1 = 5$, $h_2 = 3$, $f = 5$; $S = 0$ (simple), $S = \frac{1}{2}$ (double), $S = 1$ (double), $S = \frac{3}{2}$ (simple); $d = 15$.

$h_1 = 4, h_2 = 4, f = 6$; $S = 0, \frac{1}{2}, 1, \frac{3}{2}$ (all simple); $d = 10$.

9 Perturbation Computations

In order to make qualitative statements concerning the position of the parent terms (the multiplet energy levels) it is necessary, in a first approximation, to use perturbation theory to compute the energy changes for the general case in the same way as we have used in the case of the helium atom. We will assume that the Hamiltonian operator is given in the following form.

$$H = \sum_{k=1}^{f} \mathring{H}_k + \sum_{i<k} V_{ik}, \tag{9.1}$$

where the $\mathring{H}_k$ act only on the coordinates of the kth electron. Here V_{ik} represents the interaction between the ith and kth electron. V_{ik} need not take into account the entire Coulomb interaction; a part of this interaction, the shielding of the Coulomb field of the nucleus may be taken into account by the $\mathring{H}_k$. We note that f need not be the total electron number of the atom because in the vicinity of the nucleus the closed electron shells contribute primarily to the shielding.

The initial configuration $\mathring{\varepsilon} = \varepsilon_1 + \varepsilon_2 + \cdots + \varepsilon_n$ may consist of inequivalent nondegenerate terms ε_ν. The eigenvectors of $\mathring{H}_k$ corresponding to the eigenvalue ε_k may be φ_{ν_k}. The eigenvalue problem in zeroth approximation is expressed in terms of the space spanned by the $P\varphi_{\nu_1}(1)\cdots\varphi_{\nu_f}(f)$ as follows:

$$\begin{aligned}\langle Q\varphi_{\nu_1}(1)\cdots\varphi_{\nu_f}(f), H\sum_P x(P)P\varphi_{\nu_1}(1)\cdots\varphi_{\nu_f}(f)\rangle \\ = \varepsilon\langle Q\varphi_{\nu_1}(1)\cdots\varphi_{\nu_f}(f), \sum_P x(P)P\varphi_\nu(1)\cdots\varphi_{\nu_f}(f)\rangle,\end{aligned} \tag{9.2}$$

where Q and P are permutations. Since H commutes with Q, from the fact that the φ_{ν_i} are mutually orthogonal and normalized, it follows that

$$\sum_P x(P)\langle\varphi_{\nu_1}(1)\cdots\varphi_{\nu_f}(f), \sum_{i<k} V_{ik}Q^{-1}P\varphi_{\nu_1}(1)\cdots\varphi_{\nu_f}(f)\rangle = \eta x(Q), \tag{9.3}$$

where $\eta = \varepsilon - \varepsilon^0 = \varepsilon - \sum_k \varepsilon_k$.

If we set

$$a(P) = \langle\varphi_{\nu_1}(1)\cdots\varphi_{\nu_f}(f), \sum_{i<k} V_{ik}P^{-1}\varphi_{\nu_1}(1)\cdots\varphi_{\nu_f}(f)\rangle \tag{9.4}$$

then the equation can be written in the form

$$xa = \eta x \tag{9.5}$$

using the elements

$$a = \sum_P a(P)P \quad \text{and} \quad x = \sum_P x(P)P \tag{9.6}$$

from the ring of groups $\mathscr{R}_{\mathbf{S}_f}$. The solution of (9.5) can be greatly simplified by the reduction of the group ring. We need only solve (9.5) in one of the repective equivalent right ideals. If x_1 is a solution corresponding to the eigenvalue η_1 which corresponds to the idempotent element $e^{(\nu)}$ then the vector

$$\phi = x_1\varphi_{\nu_1}(1)\cdots\varphi_{\nu_f}(f)$$

is the corresponding solution of (9.2). Since $x_1 = e^{(\nu)}x_1$, we obtain $e^{(\nu)}\phi = \phi$. Therefore ϕ is the desired solution for the subspace $\not{p}^{(\nu)}$ corresponding to $e^{(\nu)}$ in (7.2). We need only solve (9.5) for those idempotent elements $e^{(\nu)}$ (where $e^{(\nu)}x = x$) for which the corresponding tableau has at most two columns, otherwise the Pauli exclusion principle cannot be satisfied using spin functions.

In our case (e denotes the unit element, (ik) denotes the exchange of the ith and kth electrons) we obtain

$$\begin{aligned}a &= \sum_P \langle P\varphi_{\nu_1}(1)\cdots\varphi_{\nu_f}(f), \sum_{i<k} V_{ik}\varphi_{\nu_1}(1)\cdots\varphi_{\nu_f}(f)\rangle P \\ &= e\sum_{i<k} C_{ik} + \sum_{i<k}(ik)A_{ik},\end{aligned} \tag{9.7}$$

where

$$C_{ik} = \langle \varphi_{v_i}(i)\varphi_{v_k}(k), V_{ik}\varphi_{v_i}(i)\varphi_{v_k}(k)\rangle,$$
$$A_{ik} = \langle \varphi_{v_k}(i)\varphi_{v_i}(k), V_{ik}\varphi_{v_i}(i)\varphi_{v_k}(k)\rangle. \tag{9.8}$$

We may formulate the problem (9.2) in such a way as to include the spin vectors and use the Slater determinant. Then the desired eigenvector of zeroth approximation will have the following form:

$$\sum_{\alpha_1\cdots\alpha_f} x(\alpha_1,\cdots,\alpha_f)\sum_P(-1)^P P[\varphi_{v_1}(1)\cdots\varphi_{v_f}(f)u_{\alpha_1}(1)\cdots u_{\alpha_f}(f)] \tag{9.9}$$

or, what is the same,

$$\sum_P(-1)^P[P\varphi_{v_1}(1)\cdots\varphi_{v_f}(f)]PU, \tag{9.10}$$

where

$$U = \sum_{\alpha_1\cdots\alpha_f} x(\alpha_1\cdots\alpha_f)u_{\alpha_1}(1)\cdots u_{\alpha_f}(f).$$

Here U is a vector from the spin space of the f electrons. Here it is clear that, on the basis of the Pauli exclusion principle, we may transfer the eigenvalue problem to the spin space. According to (9.2) the eigenvalue problem in the zeroth approximation is described by the equation

$$\sum_P(-1)^P(Q\varphi_{v_1}(1)\cdots\varphi_{v_f}(f), \sum_{i<k} V_{ik}P\varphi_{v_1}(1)\cdots\varphi_{v_f}(f))PU$$
$$= \eta\sum_P(-1)^P\langle Q\varphi_{v_1}(1)\cdots\varphi_{v_f}(f)\rangle, P\varphi_{v_1}(1)\cdots\varphi_{v_f}(f)\rangle PU. \tag{9.11}$$

We may set $Q = e$; the other permutations do not lead to any new problems. Thus, for

$$a'(P) = (-1)^P\langle\varphi_{v_1}(1)\cdots\varphi_{v_f}(f), \sum_{i<k} V_{ik}P\varphi_{v_1}(1)\cdots\varphi_{v_f}(f)\rangle \tag{9.12}$$

and for $a' = \sum_P a'(P)P$ in spin space we need to solve the eigenvalue equation

$$a'U = \eta U. \tag{9.13}$$

The solution of the two equivalent problems (9.5) and (9.13) can be solved with the aid of the characters of the permutation group as follows:

According to the general derivations in AV, §7 the trace of the right transformation (9.5) is equal to that of the left transformation ax. If we restrict our considerations in (9.5) to the subspace given by a single $e^{(v)}$, then we obtain the following trace relationships for the sum of the eigenvalues η_i:

$$\sum_P a(P)\chi^{(v)}(P) = \sum_{i=1}^{r}\eta_i, \tag{9.14}$$

where r is the dimension of the (v) representation. If we then use the equation $xa^m = \eta^m x$, then from

$$a^m = \left(\sum_P a(P)P\right)^m = \sum_P a_m(P)P \tag{9.15}$$

it follows that

$$\sum_P a_m(P)\chi^{(\nu)}(P) = \sum_{i=1}^{r} \eta_i^m. \tag{9.16}$$

From these equations it is, in principle, possible to determine the η_i.

The problem (9.13) can be treated in a similar manner. The spin space can be decomposed into products $\mathscr{W}_S \times \mathscr{W}_{\Delta^{(\nu)}}$ (see (1.3c)) for the different values of S which belong to different representations $\Delta^{(\nu)}$ of $\mathbf{S}_f$, the characters of which we shall denote by $\chi_S(P)$ (see §8). For a given value of S from (9.13) it follows that:

$$\sum_P a'(P)\chi_S(P) = \sum_{i=1}^{r} \eta_i \tag{9.17}$$

and, more generally:

$$\sum_P a'_m(P)\chi_S(P) = \sum_{i=1}^{r} \eta_i^m \quad \text{where} \quad a'^m = \sum_P a'_m(P)P. \tag{9.18}$$

Since the representations of $\mathbf{S}_f$ in orbital space are associated with those in spin space, it is easy to see that both problems lead to the same result.

The problem (9.13) can be treated in an elementary way with the aid of the spin invariants (7.16). The spin invariants characterize all vectors which transform in the same way under permutation, that is, an entire column of the matrix (1.3a). If the representation of $\mathbf{S}_f$ corresponding to S has dimension g then there are g linearly independent spin invariants $A_1, \ldots, A_g$. We may then set $U = \sum_{\nu=1}^{g} x_\nu A_\nu$; we then obtain the problem

$$\sum_{P,\nu} x_\nu a'(P)PA_\nu = \eta \sum_\nu x_\nu A_\nu. \tag{9.19}$$

We may then compute, in an elementary way (see, for example, XV, §7)

$$PA_\nu = \sum_\mu A_\mu P_{\mu\nu}.$$

We then have the following problem to solve:

$$\sum_P a'(P)P_{\mu\nu}x_\nu = \eta x_\mu.$$

If, for a', we use the special form (obtained from (9.12) using the same procedures used in the derivation of (9.7))

$$a' = e \sum_{i<k} C_{ik} - \sum_{i<k} (ik)A_{ik} \tag{9.20}$$

then in spin space we may transform the problem (9.13) into the following remarkable form:

Let $\vec{S}_i$ and $\vec{S}_k$ be the spin angular momentum operators for the ith and kth electrons, then if we choose the basis such that $(\vec{S}_i + \vec{S}_k)^2$ and $(\vec{S}_i + \vec{S}_k)_3$ are in diagonal form, then $\vec{S}_i \cdot \vec{S}_k$ must be in diagonal form. According to XII, §4 there are two cases: The vector in spin space is antisymmetric in i and k and

$(\vec{S}_i + \vec{S}_k)^2$ has the eigenvalue 0 or the vectors in spinspace are symmetric and $(\vec{S}_i + \vec{S}_k)^2$ has the eigenvalue 2. From $(\vec{S}_i + \vec{S}_k)^2 = \vec{S}_i^2 + \vec{S}_i \cdot \vec{S}_k + \vec{S}_k^2$ it follows that, in the first case, $2\vec{S}_i \cdot \vec{S}_k$ has the eigenvalue $-\frac{3}{2}$ and, in the second case, $2 - \frac{3}{2} = \frac{1}{2}$. Thus $\frac{1}{2} + 2\vec{S}_i \cdot \vec{S}_k$ has the possible eigenvalues -1 and 1. Thus we have shown that

$$(ik) = \tfrac{1}{2} + 2\vec{S}_i \cdot \vec{S}_k.$$

Thus, from (9.13) we obtain

$$\left[e \sum_{i<k} (C_{ik} - \tfrac{1}{2}A_{ik}) - 2 \sum_{i<k} A_{ik}\vec{S}_i \cdot S_k\right] U = \eta U. \tag{9.21}$$

The Coulomb interaction of the electrons is such that they behave as if their spin is elastically coupled with the potential energy $-2 \sum_{i<k} A_{ik}\vec{S}_i \cdot \vec{S}_k$. If the "exchange integral" A_{ik} is positive, as we would expect for the correlation energy (XII, §3), then the parallel orientation of the spin would lead to lower energy levels. Classically this fact has the following intuitive explanation: a change in the state from the parallel position requires work against the Coulomb(!) interaction, because, on the basis of the Pauli exclusion principle, a change of the spin is automatically affected by a change in the correlation in the position probability. If the spin is parallel, then both electrons "avoid each other's company"; if antiparallel they "prefer each other's company."

At the beginning we assumed that the initial configuration consists of nonequivalent nondegenerate energy levels. We shall now present an outline of how computations are to be made in the event that the above condition is not satisfied.

Let the electron configuration be given by $(n_1 l_1)(n_2 l_2) \cdots (n_f l_f)$. Suppose that the groups of equivalent electrons contain $k_1, k_2, \ldots$ electrons, so that $k_1 + k_2 + \cdots = f$. Let σ denote the subgroups of $\mathbf{S}_f$ which permutes only equivalent electrons. Two index sets $m_1, m_2, \ldots, m_f$ and $m'_1, m'_2, \ldots, m'_f$ are said to be congruent with respect to σ if there exists a permutation in σ which permutes one into the other. We can then classify index sets $m_1, m_2, \ldots, m_f$ into congruence classes, and choose a representative for each of these. For the perturbation computation we shall now consider the following eigenfunctions:

$$\begin{aligned}&\sum_m{}' \sum_\alpha x(m_1 m_2 \cdots m_f, \alpha_1 \cdots \alpha_f)\\ &\cdot \sum_P (-1)^P P[\varphi_{n_1 l_1 m_1}(1) \cdots \varphi_{n_f l_f m_f}(f) u_{\alpha_1}(1) \cdots u_{\alpha_f}(f)]\\ &\qquad = \sum_m{}' \sum_P (-1)^P P[\varphi_{n_1 l_1 m_1}(1) \cdots \varphi_{n_f l_f m_f}(f)] P U_{m_1 m_2 \cdots m_f},\end{aligned} \tag{9.22}$$

where

$$U_{m_1 m_2 \cdots m_f} = \sum_\alpha x(m_1, m_2, \ldots m_f, \alpha_1 \cdots \alpha_f) u_{\alpha_1}(1) \cdots u_{\alpha_f}(f).$$

Here $\sum'$ means that the sum is to take place only over the representative index sets $m_1 \cdots m_f$. Congruent index sets will lead to linearly dependent functions (see §3). Let $V = \sum_{i<k} V_{ik}$; the eigenvalue problem then can be written as follows:

$$\begin{aligned}\sum_{m'}{}' \sum_{P} (-1)^P \langle \varphi_{n_1 l_1 m_1}(1) \cdots \varphi_{n_f l_f m_f}(f), VP\varphi_{n_1 l_1 m_1}(1) \cdots \varphi_{n_f l_f m_f}(f)\rangle PU_{m'_1 \cdots m'_f} \\ = \eta \sum_{m'}{}' \sum_{P} (-1)^P \langle \varphi_{n_1 l_1 m_1}(1) \cdots \varphi_{n_f l_f m_f}(f), P\varphi_{n_1 l_1 m_1}(1) \cdots \varphi_{n_f l_f m_f}(f)\rangle \\ \cdot PU_{m'_1 \cdots m'_f}. \end{aligned} \tag{9.23}$$

Since, for $m_1 \cdots m_f$ and $m'_1 \cdots m'_f$, only one representative from each congruence class is present, we find that

$$\langle \varphi_{n_1 l_1 m_1}(1) \cdots \varphi_{n_f l_f m_f}(f), P\varphi_{n_1 l_1 m'_1}(1) \cdots \varphi_{n_f l_f m'_f}(f)\rangle = 1 \tag{9.24}$$

if and only if $m'_1 \cdots m'_f = m_1 \cdots m_f$ and P is a permutation from σ which leaves the index set $m_1 \cdots m_f$ invariant. We shall denote such permutations by P_m. Thus the right side of (9.23) is equal to

$$\eta \sum_{P_m} (-1)^{P_m} P_m U_{m_1 \cdots m_f} = \eta W_{m_1 \cdots m_f}. \tag{9.25}$$

We shall denote the subgroup formed by the elements P_m by σ_m.

Since

$$\begin{aligned}\langle \varphi_{n_1 l_1 m_1}(1) \cdots \varphi_{n_f l_f m_f}(f), VQP_{m'}\varphi_{n_1 l_1 m'_1}(1) \cdots \varphi_{n_f l_f m'_f}(f)\rangle \\ = \langle \varphi_{n_1 l_1 m_1}(1) \cdots \varphi_{n_f l_f m_f}(f), VQ\varphi_{n_1 l_1 m'_1}(1) \cdots \varphi_{n_f l_f m'_f}(f)\rangle \end{aligned} \tag{9.26}$$

we can write the left side of (9.23) as follows:

$$\begin{aligned}\sum_{m}{}' \sum_{Q}{}^{(m')} \sum_{P_{m'}} (-1)^Q (-1)^{P_{m'}} \\ \cdot \langle \varphi_{n_1 l_1 m_1}(1) \cdots \varphi_{n_f l_f m_f}(f), VQ\varphi_{n_1 l_1 m'_1}(1) \cdots \varphi_{n_f l_f m'_f}(f)\rangle \\ \cdot QP_{m'} U_{m'_1 \cdots m'_f} \\ = \sum_{m'}{}' \sum_{Q}{}^{(m')} a(Q; m_1 \cdots m_f, m'_1 \cdots m'_f) Q W_{m'_1 \cdots m'_f}, \end{aligned} \tag{9.27}$$

where the sum $\sum^{(m')}$ takes place only over the permutations Q of each left coset of $\sigma_{m'}$ and where

$$\begin{aligned} a(Q; m_1 \cdots m_f, m'_1 \cdots m'_f) \\ = (-1)^Q \langle \varphi_{n_1 l_1 m_1}(1) \cdots \varphi_{n_f l_f m_f}(f), VQ\varphi_{n_1 l_1 m'_1}(1) \cdots \varphi_{n_f l_f m'_f}(f)\rangle. \end{aligned} \tag{9.28}$$

Using the following notation

$$a_{m_1 \cdots m_f, m'_1 \cdots m'_f} = \sum_{Q}{}^{(m')} a(Q; m_1 \cdots m_f, m'_1 \cdots m'_f) Q \tag{9.29}$$

we may rewrite the eigenvalue equation (9.23) as follows:

$$\sum_{m'}{}' a_{m_1 \cdots m_f, m'_1 \cdots m'_f} W_{m'_1 \cdots m'_f} = \eta W_{m_1 \cdots m_f}. \tag{9.30}$$

If we obtain a η and $W_{m'_1 \cdots m'_f}$ from the above equation, then from (9.22) we obtain the entire eigenfunction in zeroth approximation as follows:

$$\sum_m{}' \sum_P (-1)^P P[\varphi_{n_1 l_1 m_1}(1) \cdots \varphi_{n_f l_f m_f}(f)] P U_{m_1 \cdots m_f}$$
$$= \sum_m{}' \sum_Q{}^{(m)} (-1)^Q Q[\varphi_{n_1 l_1 m_1}(1) \cdots \varphi_{n_f l_f m_f}(f) W_{m_1 \cdots m_f}].$$

We may reduce equation (9.30) by considering its application to a subspace of spin space corresponding to the spin quantum number S. We may make the transition from the $m_1 \cdots m_f$ to the different quantum numbers L, m_L of the total orbital angular momentum, where the latter are obtained from the $m_1 \cdots m_f$ in the manner described in §3 used to obtain the various energy levels. For the chosen value of S we may, on the basis of the discussions in §§3 and 7, obtain energy levels with L values $L_1, L_2, \ldots$ where the terms $L_1 S$ and $L_2 S \cdots$ may occur with multiplicities λ_1 and $\lambda_2 \ldots$, respectively. Then, from (9.30) it follows that, with the eigenvalues $\eta_{L_i}^{(v)}$ $(v = 1, 2, \ldots, \lambda_i)$, by taking the trace, we obtain

$$\sum_{\substack{m \\ \left(\text{with} \left|\sum_{i=1}^f m_i\right| \le M\right)}}{}' \sum_Q{}^{(m)} a(Q; m_1 \cdots m_f, m_1 \cdots m_f) \chi_S(Q)$$
$$= \sum_{\substack{L_i, v \\ (\text{with } L_i \le M)}} (2L_i + 1)\eta_{L_i}^{(v)} + (2M + 1) \sum_{\substack{L_i, v \\ (\text{with } L_i > M)}} \eta_{L_i}^{(v)}. \tag{9.31}$$

If the $\lambda_i = 1$, then we have already obtained the eigenvalue η_{L_i}; otherwise we only obtain the sum $\sum_{v=1}^{\lambda_i} \eta_{L_i}^{(v)}$. Then we may use the iterated equation of (9.30)

$$\sum_{m'}{}' a^{(k)}_{m_1 \cdots m_f, m'_1 \cdots m'_f} W_{m'_1 \cdots m'_f} = \eta^k W_{m_1 \cdots m_f}, \tag{9.32}$$

where

$$a^{(k)}_{m_1 \cdots m_f, m'_1 \cdots m'_f} = \sum_{m''}{}' a_{m_1 \cdots m_f, m''_1 \cdots m''_f} a^{(k-1)}_{m''_1 \cdots m''_f, m'_1 \cdots m'_f}$$

from which we can obtain relationships similar to (9.31) for the kth power of the eigenvalues $\eta_{L_i}^{(v)}$.

CHAPTER XV

Molecular Spectra and the Chemical Bond

1 The Hamiltonian Operator for a Molecule

Let $\vec{r}_k$ and $\vec{p}_k$ (using Latin indices) denote the positions and momenta of the nuclei and let $\vec{r}_\kappa$ and $\vec{p}_\kappa$ (using Greek indices) denote the positions and momenta of the electrons, respectively. The Hamiltonian operator for the nuclei and electrons is then given by

$$H = \frac{1}{2} \sum_{k=1}^{h} \frac{1}{m_k} \vec{p}_k^{\,2} + \frac{1}{2m} \sum_{\kappa=1}^{f} \vec{p}_\kappa^{\,2} + U(\vec{r}_k, \vec{r}_\kappa) + H_s, \tag{1.1}$$

where m_k are the masses of the nuclei, m is the mass of an electron and H_s are the terms involving the spin of the electrons. U is the potential energy

$$U(\vec{r}_k, \vec{r}_\kappa) = \frac{1}{2} \sum_{\substack{kl \\ k \neq l}} \frac{Z_k Z_l e^2}{r_{kl}} + \frac{1}{2} \sum_{\substack{\kappa\lambda \\ \kappa \neq \lambda}} \frac{e^2}{r_{\kappa\lambda}} - \sum_{k\lambda} \frac{Z_k e^2}{r_{k\lambda}},$$

where $Z_k e$ is the charge of the kth nucleus. Let $\vec{p}_k = (1/i)\text{grad}_k$ and $\vec{p}_\kappa = (1/i)\text{grad}_\kappa$. Then we may rewrite (1.1) as follows:

$$H = -\frac{1}{2} \sum_k \frac{1}{m_k} \Delta_k - \frac{1}{2m} \sum_\kappa \Delta_\kappa + U + H_s. \tag{1.2}$$

In order to separate the center of mass motion we shall now introduce new coordinates (see also VIII, §§1 and 2) as follows:

$$\begin{aligned} M\vec{R} &= \sum_{k=1}^{h} m_k \vec{r}_k + m \sum_{\kappa=1}^{f} \vec{r}_\kappa; \qquad M = \sum_{k=1}^{h} m_k + fm; \\ \vec{r}_k' &= \vec{r}_k - \vec{r}_1, \\ \vec{r}_\kappa' &= \vec{r}_\kappa - \vec{R}, \end{aligned} \tag{1.3}$$

where $k = 2, 3, \ldots, h$ and $\vec{r}_1$ is the position of the first nucleus.

We therefore find that (1.2) is transformed into

$$\begin{aligned} H = &-\frac{1}{2M}\Delta - \frac{1}{2m_1}\left(\sum_{k=2}^{h} \operatorname{grad}_k'\right)^2 + \frac{1}{2M}\left(\sum_{\kappa=1}^{f} \operatorname{grad}_\kappa'\right)^2 \\ &- \sum_{k=2}^{h} \frac{1}{2m_k}\Delta_k' - \frac{1}{2m}\sum_{\kappa=1}^{f} \Delta_\kappa' + U + H_s, \end{aligned} \tag{1.4}$$

where Δ acts upon the coordinates of $\vec{R}$, grad_k' and Δ_k' upon $\vec{r}_k'$, and Δ_κ', $\operatorname{grad}_\kappa'$ upon $\vec{r}_\kappa'$, respectively. Since U and H_s do not depend upon the absolute position of the particles, that is, they are invariant under translations of all particles, in the new coordinates (1.3) they must be independent of $\vec{R}$. The only term in (1.1) which depends upon $\vec{R}$ represents the kinetic energy of the center of mass. We may neglect this term since we are only interested in the internal energy of the molecule (because we are interested in the spectra and the binding energy). Therefore we shall set

$$\begin{aligned} H = &-\sum_{k=2}^{h} \frac{1}{2m_k}\Delta_k' - \frac{1}{2m_1}\left(\sum_{k=2}^{h} \operatorname{grad}_k'\right)^2 \\ &- \frac{1}{2m}\sum_{\kappa=1}^{f} \Delta_\kappa' + U + H_s. \end{aligned} \tag{1.5}$$

The first two terms represent the kinetic energy of the nuclei; the second term is the kinetic energy of the first atomic nucleus, the momentum of which (in center of mass coordinates) is equal to the negative sum of the momenta of the other nuclei. The following term in (1.4)

$$\frac{1}{2M}\left(\sum_{\kappa=1}^{f} \operatorname{grad}_\kappa'\right)^2 \tag{1.6}$$

which represents the effect of the electron motion upon the nucleus should be taken into account. However, since the mass of the electron is small compared to that of a nucleus, we find that this term is small compared to the third term in (1.5), we shall neglect this term as in the case of atoms.

2 The Form of the Eigenfunctions

Here, as in the case of the atomic spectra, we shall for the present neglect the term H_s in (1.5). We shall investigate the effect of this term later in §9. The first two terms in (1.5) contain the masses of the nuclei in the denominator.

Since the latter are more than a thousand times the mass of the electron, the corresponding kinetic energy of the nuclei will be small compared to that of the electrons (see §8). This fact follows directly from the Heisenberg uncertainty relations. The localization of electrons and nuclei in the molecule to a volume of radius d will result in momenta of the order of $p \sim 1/d$. For the nuclei we will obtain a smaller kinetic energy than we will obtain for the electrons. If we neglect these two terms, then we obtain the following Hamiltonian operator:

$$H_e = -\frac{1}{2m}\sum_{\kappa} \Delta'_\kappa + U. \tag{2.1}$$

Here H_e is to be applied to functions of $\vec{r}'_k$ and $\vec{r}'_\kappa$. The eigenfunctions may, however, be written in the form $\delta(\vec{r}'_k - \mathring{\vec{r}}'_k)\varphi(\mathring{\vec{r}}'_k, \vec{r}'_\kappa)$ where

$$H_e\varphi(\mathring{\vec{r}}'_k, \vec{r}'_\kappa) = E(\mathring{\vec{r}}'_k)\varphi(\mathring{\vec{r}}'_k, \vec{r}'_\kappa) \tag{2.2}$$

which, intuitively, means that the nuclei are fixed at locations $\mathring{\vec{r}}'_k$ and are therefore fixed centers of force for the electrons. $E(\mathring{\vec{r}}'_k)$ is the energy of the electrons and is dependent on the positions of the nuclei (including the mutual potential energy of the nuclei). We may suppose that we would obtain a very good approximation for the energy eigenvalues for (1.5), that is, for

$$H = -\sum_{k=2}^{h}\frac{1}{2m_k}\Delta'_k - \frac{1}{2m_1}\left(\sum_{k=2}^{k}\mathrm{grad}'_k\right)^2 + H_e = H_K + H_e \tag{2.3}$$

if we substitute $E(\vec{r}'_k)$ (obtained from (2.2)) (here, for simplicity, we replace $\mathring{\vec{r}}'_k$ by $\vec{r}'_k$) and use the eigenvalues of the equation

$$\{H_K + E(\vec{r}'_k)\}\psi(\vec{r}'_k) = \varepsilon\psi(\vec{r}'_k). \tag{2.4}$$

Thus, we assume that, because of the large masses of the nuclei, the electrons move as if the nuclei were at rest, and that the nuclei move under the influence of the position-dependent energy $E(\vec{r}'_k)$. Mathematically this means we approximate the eigenfunctions of (2.3) by the form:

$$\phi = \psi(\vec{r}'_k)\varphi(\vec{r}'_k, \vec{r}'_\kappa), \tag{2.5}$$

where φ satisfies (2.2), and assume that the dependence of φ upon $\vec{r}'_k$ is much weaker than that of ψ, so that, in the differentiation of the function ϕ with respect to $\vec{r}'_k$ we may assume, in first approximation, that φ is constant. We will later justify this assertion.

The approach (2.5) is reasonable only if the eigenvalue $E(\vec{r}'_k)$ is nondegenerate [as is in general the case of molecules having more than two atoms because degeneracy shall occur only as a result of spatial symmetry of H_e]. For the case of two nuclei there is a rotational symmetry. Here we find it necessary to replace (2.5) by linear combinations of the form (2.5) with different φ corresponding to the same value of $E(\vec{r}'_k)$—see §8.

For more than two atoms we obtain a better approximation for ε then (2.4) if we use the form (2.5) where we obtain φ from (2.2), and we determine the function ψ by means of the Ritz variational principle (VIII, §3).

Then, instead of (2.4) we obtain

$$\{H_K + \bar{E}(\vec{r}_K')\}\psi = \varepsilon\psi, \tag{2.6}$$

where

$$\bar{E}(\vec{r}_k') = E(\vec{r}_k') + \int \varphi(r_k', r_\kappa')H_K\varphi(\vec{r}_k', \vec{r}_\kappa')(d\vec{r}_\kappa')^f, \tag{2.7}$$

where $(d\vec{r}_\kappa')^f$ is integrated over all of the electron coordinates. Note that (2.7) is a better approximation than (2.4) with respect to the dependence of φ upon the coordinates of the nuclei.

Equation (2.6) has, in addition to a continuous spectrum, discrete eigenvalues for the lowest energy eigenvalues only if $\bar{E}(\vec{r}_k')$—or, to a good approximation $E(\vec{r}_k')$—has an absolute minimum; only then will we have bound states. If the lowest energy eigenvalue of (2.2) has this property, then the ground state is bound, that is, we have a chemically bound molecule. In the following section we shall consider the eigenvalues of (2.2) and their dependence upon the positions of the nuclei. The minimum for the ground state of (2.2) yields the equilibrium position of the nuclei, that is, the chemical structure of the molecule.

3 The Ionized Hydrogen Molecule

The simplest example is the ionized hydrogen molecule in which an electron moves in the field of two nuclei. Equation (2.2) takes on the following form

$$\left(-\frac{1}{2m}\Delta - \frac{e^2}{r_1} - \frac{e^2}{r_2} + \frac{e^2}{a}\right)\varphi(\vec{r}) = E\varphi(\vec{r}), \tag{3.1}$$

where $\vec{r}$ is the position vector of the electron, r_1 and r_2 are the distances from the electron to the nuclei 1 and 2, respectively, and a is the separation between the two nuclei. Note that φ and E depend upon a. $E' = E - e^2/a$ is the energy of the electron.

Equation (3.1) can be separated and solved by introducing elliptical coordinates (see [14]). We will now seek to obtain, using symmetry considerations, a qualitative description of the dependence of the energy E' upon the distance a. All problems with two nuclei are invariant under rotations about the line joining the nuclei and under reflections in planes containing the above line. These symmetries form a group which we shall denote by $\mathscr{A}$. If we choose the line joining the nuclei as the 3-axis we can describe a special reflection as follows:

$$s_2\colon\quad \begin{aligned} x_1' &= x_1, \\ x_2' &= -x_2, \\ x_3' &= x_3. \end{aligned} \tag{3.2}$$

If D_α is a rotation about the 3-axis by the angle α then it easily follows that (e = unit element of the group):

$$s_2 D_\alpha = D_{-\alpha} s_2, \qquad (s_2)^2 = e. \tag{3.3}$$

Each element of $\mathscr{d}$ can be described either in the form D_α or $s_2 D_\alpha$.

The representations of $\mathscr{d}$ are easy to find. The rotations D_α form an abelian subgroup of $\mathscr{d}$, the irreducible representations of which are one-dimensional and are already familiar to us as representations of a subgroup of the entire spatial rotation group (VII, §3):

$$D_\alpha u_m = e^{-im\alpha} u_m \qquad (m = 0, \pm\tfrac{1}{2}, \pm 1, \pm\tfrac{3}{2}, \ldots). \tag{3.4}$$

According to (VII, §3), in the space $\mathscr{H}_b$ we need only consider the values $m = 0, 1, 2, \ldots$.

Therefore we may choose basis vectors of the form (3.4) for any (not only for the irreducible case) representation space $\mathscr{R}$ of $\mathscr{d}$. Since, according to (3.3) we obtain

$$D_\alpha(s_2 u_m) = s_2 D_{-\alpha} u_m = e^{im\alpha}(s_2 u_m)$$

the two vectors u_m (with $m > 0$) and $s_2 u_m$ span an invariant two-dimensional subspace (with respect to $\mathscr{d}$) of $\mathscr{R}$. This subspace is, by construction, clearly irreducible. The set of all vectors v which are invariant under rotations D_α (that is, $D_\alpha v = v$) forms an invariant subspace $\imath_0$ with respect to $\mathscr{d}$. The basis vectors in $\imath_0$ can be chosen such that they are eigenvectors of s_2.

According to the second relation in (3.3) we therefore obtain either $s_2 v = v$ or $s_2 v = -v$. In this manner we obtain all irreducible representations of $\mathscr{d}$. They can be characterized by a scalar $\Lambda \geq 0$. For $\Lambda > 0$ they are two-dimensional:

$$\begin{aligned} D_\alpha v_\Lambda &= e^{-i\Lambda\alpha} v_\Lambda, \\ D_\alpha v_{-\Lambda} &= e^{i\Lambda\alpha} v_{-\Lambda}, \end{aligned} \qquad s_2 v_\Lambda = v_{-\Lambda} \tag{3.5}$$

and, for $\Lambda = 0$ they are one-dimensional. Here it is necessary to distinguish the two cases O^+:

$$D_\alpha v_0 = v_0, \qquad s_2 v_0 = v_0 \tag{3.6a}$$

and O^-:

$$D_\alpha v_0 = v_0, \qquad s_2 v_0 = -v_0. \tag{3.6b}$$

We shall denote these representations by $\mathscr{A}_1, \mathscr{A}_{0^+}, \mathscr{A}_{0^-}$.

Since, in the case of the ionized hydrogen molecule (H_2^+), both nuclei have the same charge, there is an additional symmetry—that of reflections about the midpoint of the line joining the two nuclei. Let us choose this point as the origin of coordinates. We then obtain:

$$s\colon x_1' = -x_1, x_2' = -x_2, x_3' = -x_3. \tag{3.7}$$

Since s commutes with $\mathscr{d}$, it is possible to choose the irreducible representation of the group generated by s and $\mathscr{d}$ such that, for all vectors either

$sv = v$ or $sv = -v$, that is, in (3.5) we have two cases; the first, we denote by Λ_e satisfies

$$sv_\Lambda = v_\Lambda, \qquad sv_{-\Lambda} = v_{-\Lambda},$$

the other, which we denote by Λ_o satisfies

$$sv_\Lambda = -v_\Lambda, \qquad sv_{-\Lambda} = -v_{-\Lambda}.$$

Similar expressions can be found corresponding to (3.6a), (3.6b) for the cases

$$O_e^+, O_o^+, O_e^-, O_o^-.$$

For the case of H_2^+, φ depends only upon the coordinates of a single electron; thus for $\Lambda = 0$, φ does not depend upon the polar angle about the 3-axis and is therefore only a function of x_3 and $\sqrt{x_1^2 + x_2^2}$. From this it follows that $s_2\varphi = \varphi$, and we may therefore rule out the case O^-.

In order to obtain information about the qualitative behavior of $E'(a)$ we shall consider the two limiting cases $a \sim 0$ and $a \sim \infty$. For $a = 0$ we obtain the same energy levels for $E'(0)$ we found in the case of He^+ and we therefore obtain the same structure as we found in the case of the hydrogen atom (Figure 21). $E'(0)$ is therefore characterized by a principal quantum number n. For each n the orbital angular momentum can take on values $l = 0, 1, 2, \ldots, n - 1$. To each l there are $2l + 1$ linearly independent eigenvectors. If we

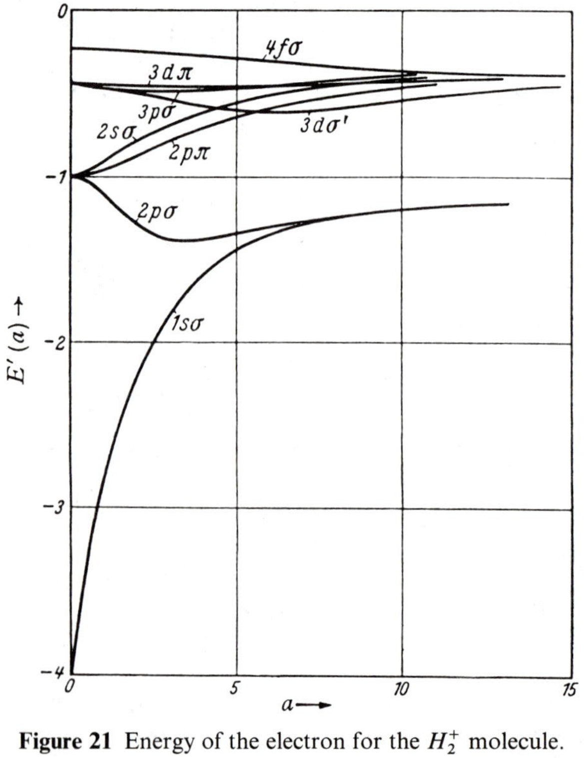

Figure 21 Energy of the electron for the H_2^+ molecule.

permit a to increase, then the Hamiltonian operator is no longer invariant with respect to the entire rotation group, but only with respect to s and d. The energy levels will therefore be split. From XIII, §3 we know that under reflection s the eigenfunctions of orbital angular momentum l are multiplied by $(-1)^l$; they are even (e) or odd (o) depending on whether l is even or odd. For continuous changes in a this property cannot change. For a single electron we shall write λ instead of Λ.

The $2l + 1$ eigenfunctions for the quantum number l fall with respect to s and d in the irreducible representation $\lambda = l, l - 1, \ldots, O^+$ with (e) or (o) depending on whether l is even or odd. Instead of $\lambda = 0, 1, 2, \ldots$ we often write $\sigma, \pi, \delta, \ldots$.

The ground state $n = 1$ is nondegenerate and therefore cannot be split. It corresponds to the energy level σ_e^+ (Figure 21). The energy level $n = 2$ is split, since $l = 0$ or 1, into the energy levels σ_e^+, σ_o^+ and π_o. Similarly, the energy level $n = 3$ is split, since $l = 0, 1, 2$, into energy levels $\sigma_e^+, \sigma_o^+, \pi_o; \sigma_e^+$, π_e, δ_e. The quantitative behavior of the splitting for increasing a can be obtained by a perturbation calculation. The latter, however, will not provide a measure of the positions of the energy levels for intermediate distances.

For this purpose it is necessary to consider the opposite case—$a \sim \infty$. Here we find that the electron is in the vicinity of one of the nuclei, and has the energy levels associated with the hydrogen atom, as characterized by the principle quantum number n. To each n there correspond the l values $l = 0, 1, \ldots, n - 1$. Here we note that to each value of n and l there exist $2(2l + 1)$ linearly independent eigenfunctions instead of $(2l + 1)$ because the electron can be found either around the first nucleus $\varphi_{(1)m}^l$ or the second nucleus $\varphi_{(2)m}^l$. Since under reflection about the nucleus, φ_m^l becomes multiplied by $(-1)^l$, it is easy to show that for reflections about the midpoint between the nuclei we obtain

$$s\varphi_{(1)m}^l = (-1)^l \varphi_{(2)m}^l, \qquad s\varphi_{(2)m}^l = (-1)^l \varphi_{(1)m}^l.$$

We therefore obtain the functions

$$\begin{aligned} \varphi_{em}^l &= \varphi_{(1)m}^l + (-1)^l \varphi_{(2)m}^l, \\ \varphi_{om}^l &= \varphi_{(1)m}^l - (-1)^l \varphi_{(2)m}^l, \end{aligned} \tag{3.8}$$

which are either even (e) or odd (o) with respect to s. Since the $\varphi_{e\Lambda}^l$ and $\varphi_{e-\Lambda}^l$ form a representation of $\mathscr{d}$, it directly follows that the eigenfunctions corresponding to an energy level n, l are associated with the irreducible representations Λ_e, Λ_o where $\Lambda = l, l - 1, \ldots, 0$; for the case $\Lambda = 0$ the only possible cases are O_e^+ and O_o^+. As the nuclei approach each other the corresponding energy levels are split into irreducible representations, as is shown in the right side of Figure 21. Here we may obtain a numerical estimate of the initial splitting by means of perturbation theory.

How do the energy levels for $a \sim 0$ transform into those for $a \sim \infty$ as the distance a changes continuously? It is clear that the representation properties of the eigenfunctions cannot be changed in a continuous change in a. Thus, only energy levels having the same representation class (pedigree) can be transformed into each other. We shall show below that, in general, energy levels of the same pedigree cannot overlap.

On the basis of the above two rules it is possible to uniquely determine which energy levels are transformed into each other as follows: For each pedigree we need only begin with the lowest energy level and sequentially identify corresponding pairs of energy levels. In Figure 21 the behavior of $E'(a)$ is illustrated using precise calculation. The most striking case is that of the strong decrease in the lowest σ_e^+ energy level with decreasing a. If we consider $E = E' + e^2/a$ (Figure 22) we will find that there is a strong binding of the two nuclei because there is a pronounced minimum of $E(a)$.

We may now treat the many electron molecule in the same way as we have in the case of many electron atoms in terms of one electron moving in the average field generated by the other electrons (Hartree method, XII, §3). In this way we obtain a one-electron problem with a non-Coulomb central force which is analogous to the case of the alkali metals described earlier. Here we find that for different values of l the energy levels no longer coincide (See Figure 23, left- and right-hand sides). The nature of the splitting and the meaning of the connection lines in Figure 23 is clear from the above explanation for the case of two identical force centers. If the two force centers are not

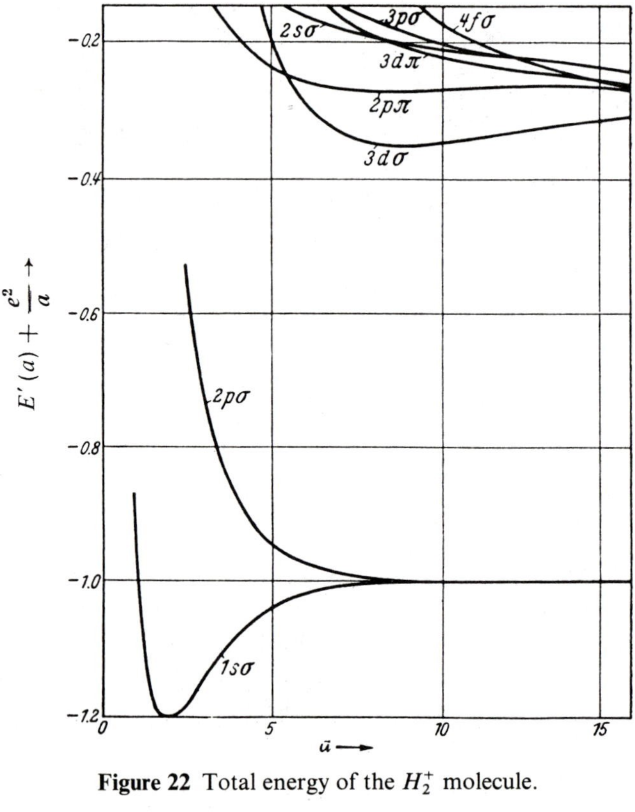

Figure 22 Total energy of the H_2^+ molecule.

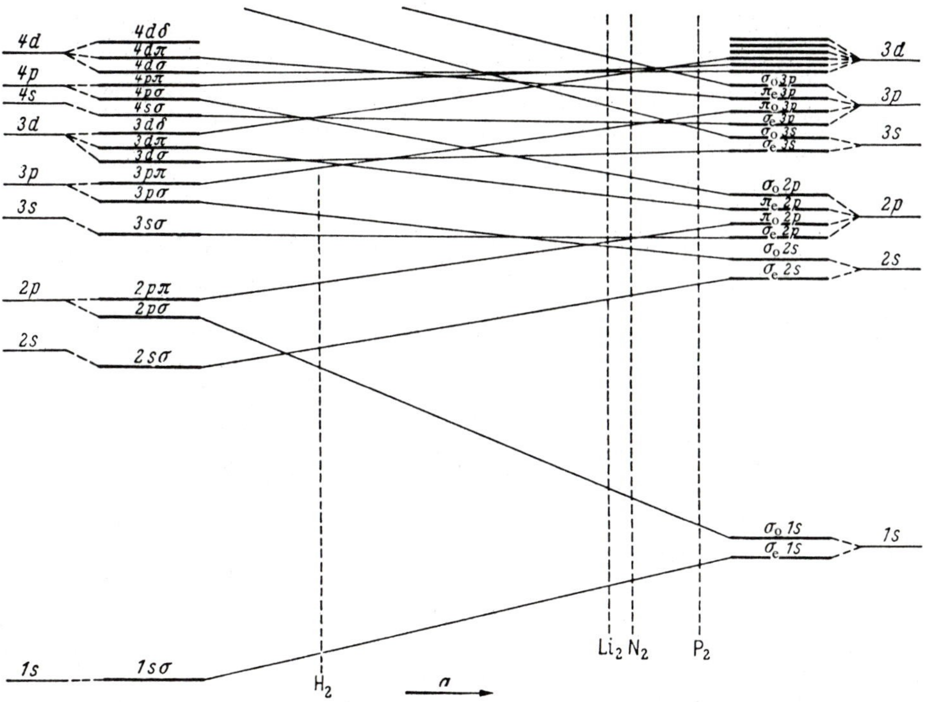

Figure 23 Energy levels of the electron for the case of two identical force centers.

identical, then we obtain a different situation as illustrated in Figure 24. Here the reflection operator s is not a symmetry operator, and the designations odd and even are no longer meaningful. For the case $a \sim \infty$ we have twice as many energy levels as we did in the case of identical centers because the values E' for the cases that the electrons are found near center (1) or (2) are different. Thus Figure 24 is understandable without additional explanation.

In closing we will now verify the result used often above, that two energy levels having the same pedigree do not overlap. Let $\varphi_1(a), \varphi_2, \ldots, \varphi_n$ and $\varphi_{n+1}, \varphi_{n+2}, \ldots, \varphi_m$ are the eigenvectors of $H(a)$ corresponding to the eigenvalues $\varepsilon_1(a)$ and $\varepsilon_2(a)$ which may intersect at $a = a_0$ (that is, $\varepsilon_1(a_0) = \varepsilon_2(a_0)$). Here we assume that $\varphi_1, \varphi_2, \ldots, \varphi_n$ and $\varphi_{n+1}, \varphi_{n+2}, \ldots, \varphi_m$ (where $m = 2n$) are two equivalent irreducible representations of the symmetry group of $H(a)$. If we change $H(a)$ by adding a small term $\lambda V(a)$ where $V(a)$ belongs to the same symmetry group as does $H(a)$, then, from XI, §§6 and 7 we obtain approximate values for the new eigenvalues E for small λ from the secular equation:

$$\begin{vmatrix} (\varepsilon_1(a) - E)\delta_{ik} + \lambda V_{ik}(a) & \lambda V_{il}(a) \\ \lambda V_{rk}(a) & (\varepsilon_2(a) - E)\delta_{rl} + \lambda V_{rl}(a) \end{vmatrix} = 0, \tag{3.9}$$

where $i, k = 1, 2, \ldots, n$; $r, l = n + 1, \ldots, 2n$ and $V_{sl} = \langle \varphi_s, V\varphi_l \rangle$. Since $V(a)$ admits the same symmetry operation, the "correct linear combinations"

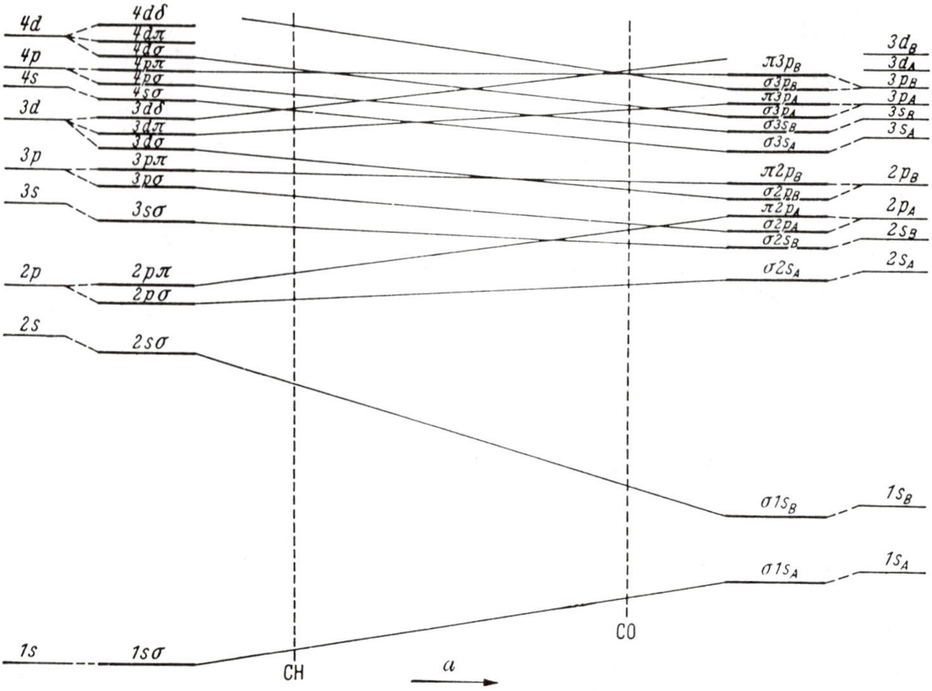

Figure 24 Energy levels of the electron for the case of two nonidentical force centers.

can, according to XI, §7, be completely specified as follows: $x\varphi_1 + y\varphi_{n+1}$, $x\varphi_2 + y\varphi_{n+2}, \ldots$, where $x\varphi_1 + y\varphi_{n+1}$ and $x\varphi_2 + y\varphi_{n+2}$ correspond to the same eigenvalue, etc. Thus we find that (3.9) can be replaced by an equivalent set of two-dimensional secular equations:

$$\begin{vmatrix} \varepsilon_1(a) - E + \lambda V_{1,1}(a) & \lambda V_{1,n+1}(a) \\ \lambda V_{n+1,1}(a) & \varepsilon_2(a) - E + \lambda V_{n+1,n+1}(a) \end{vmatrix} = 0. \qquad (3.10)$$

The difference in the roots E_1, E_2 of (3.10) are given by

$$E_1 - E_2 = \sqrt{(\varepsilon_1 - \varepsilon_2 + \lambda V_{1,1} - \lambda V_{n+1,n+1})^2 + 4\lambda^2 |V_{1,n+1}|^2}.$$

Thus we find that $E_1 - E_2 \neq 0$ whenever $V_{1,n+1} \neq 0$. The terms E_1 and E_2 cannot, therefore, coincide, as we have assumed (see Figure 25a). If the $\varphi_1, \ldots, \varphi_n$ and $\varphi_{n+1}, \ldots, \varphi_m$ belong to different irreducible representations, then all nondiagonal elements satisfy $V_{ik} = 0$, then $E_1 - E_2 = \varepsilon_1 - \varepsilon_2 + \lambda V_{1,1} - \lambda V_{n+1,n+1}$. Both terms will therefore be somewhat displaced (by $\lambda V_{1,1}$ or $\lambda V_{n+1,n+1}$) and may intersect (see Figure 25b). Since a small perturbation which does not affect the symmetry of the system gives rise to a separation of two terms of the same pedigree but keeps two terms of different pedigrees intersecting, the intersection of terms of the same pedigree in complicated systems is highly unlikely.

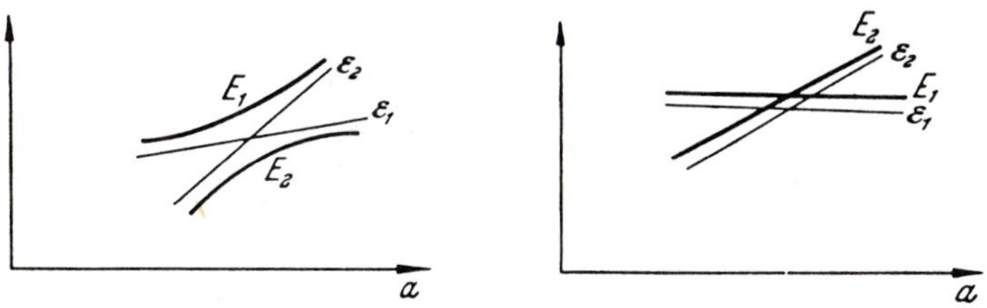

a b

Figure 25 The behavior of two terms (a) of the same pedigree and (b) of different pedigree.

4 Structure Principles for Molecular Energy Levels

As in the case of atoms, we shall now seek to obtain an overview of the electron energy levels for molecules by considering the one-electron eigenfunctions and "turning on" the mutual interaction of the electrons. For two atom molecules the invariance groups are the permutation (symmetric) group $\mathbf{S}_f$ for the f electrons and the rotation and reflection groups $\mathscr{d}$. If both nuclei have the same charge, then we need also include the reflection operator s given in (3.7). The only difference between this case and the case of the atom described in XIV, §3 is that the rotation group $\mathscr{D}_{\mathscr{g}}$ is replaced by $\mathscr{d}$ (or by $\mathscr{d}$ and s). The following results obtained earlier remain applicable: Full shells do not affect the term manifold; n equivalent electrons are described by the same manifold as is described by a full shell missing n electrons; equivalent electrons can be separately treated and then joined together without taking the Pauli exclusion principle into account.

For the atomic case we found it necessary to reduce the representation $\mathscr{D}_j \times \mathscr{D}_{j'}$. For two atom molecules it will be necessary to do the same for $\mathscr{A}_\lambda \times \mathscr{A}_{\lambda'}$. If $\lambda > 0$ and $\lambda' > 0$ then $\mathscr{A}_\lambda \times \mathscr{A}_{\lambda'}$ will be generated by four vectors (using (3.5)) $v_\lambda u_{\lambda'}, v_\lambda u_{-\lambda'}, v_{-\lambda} u_{\lambda'}, v_{-\lambda} v_{-\lambda'}$. $v_\lambda u_{\lambda'}$ and $v_{-\lambda} u_{-\lambda'}$ span a subspace of the representation $\mathscr{A}_{\lambda+\lambda'}$ and $v_\lambda u_{-\lambda'}$, $v_{-\lambda} u_{\lambda'}$ for the representation $\mathscr{A}_{|\lambda'-\lambda|}$ (providing $\lambda \neq \lambda'$). If $\lambda = \lambda'$ then $v_\lambda u_{-\lambda} + v_{-\lambda} u_{\lambda'}$ belong to the representation $\mathscr{A}_{O^+}$ and $v_\lambda u_{-\lambda} - v_{-\lambda} u_\lambda$ belong to the representation $\mathscr{A}_{O^-}$. Thus, for $\lambda > 0$, $\lambda' > 0$ we obtain

$$\mathscr{A}_\lambda \times \mathscr{A}_{\lambda'} = \begin{cases} \mathscr{A}_{\lambda+\lambda'} + \mathscr{A}_{|\lambda-\lambda'|} & \text{for } \lambda \neq \lambda', \\ \mathscr{A}_{2\lambda} + \mathscr{A}_{O^+} + \mathscr{A}_{O^-} & \text{for } \lambda = \lambda'. \end{cases} \tag{4.1}$$

In the same way, for $\lambda > 0$, it is easy to show that

$$\mathscr{A}_\lambda \times \mathscr{A}_{O^\pm} = \mathscr{A}_\lambda \tag{4.2}$$

and, trivially,

$$\mathscr{A}_{O^+} \times \mathscr{A}_{O^+} = \mathscr{A}_{O^-} \times \mathscr{A}_{O^-} = \mathscr{A}_{O^+}; \qquad \mathscr{A}_{O^+} \times \mathscr{A}_{O^-} = \mathscr{A}_{O^-}. \tag{4.3}$$

The formulas (4.1)–(4.3) can be used directly only in cases in which several inequivalent groups are joined together, each of which consists of equivalent electrons. The case of a set of equivalent electrons must be treated separately.

We may now develop a rule for eigenfunctions which is similar to that developed in XIV, §3, where we list the values $\pm\lambda$, $\pm\frac{1}{2}$ for each electron. For example, for two equivalent π electrons, we obtain the following table:

		Λ	S	
$(1, \frac{1}{2})$	$(1, \frac{1}{2})$	2	0	
$(1, \frac{1}{2})$	$(-1, \frac{1}{2})$	0	1	-1
$(1, \frac{1}{2})$	$(-1, -\frac{1}{2})$	0	0	
$(1, -\frac{1}{2})$	$(-1, \frac{1}{2})$	0	0	
$(1, -\frac{1}{2})$	$(-1, -\frac{1}{2})$	0	-1	-1
$(-1, \frac{1}{2})$	$(-1, -\frac{1}{2})$	-2	0	

In the last column there is, at two places, a (-1). Here we have written $(+1)$ or (-1) depending on whether the vector on the left is multiplied by the factor $+1$ or -1 under the reflection s_2.

In cases in which the vector is not reproduced up to a factor upon multiplication by s_2, the last column was left blank. The factors $+1$ and -1 are obtained, according to the above definition, from $(-1)^{p+\nu}$ (from the Pauli exclusion principle—Slater Determinant) where ν is the number of times O^- occurs and $p = 0$ or 1 depending on whether an odd or even permutation is required in order to return the symbol sequence $(+\lambda, \pm\frac{1}{2})$ altered by s_2 to its original order. On the basis of the important trace relations (see AV, §§8 and 10.5) for this example it is easy to see that there will be three energy levels $\Lambda = 2, S = 0$; $\Lambda = 0, S = 1$ and $\Lambda = 0, S = 1$. Are these O^+ or O^- energy levels? In the space for the equivalent electrons under consideration we find that the trace of s_2 is, on the basis of the last column, equal to -2. The character of s_2 in $\mathscr{A}_\lambda$ $(\lambda \neq 0)$ is equal to 0, in $\mathscr{A}_{O^+}$ it is equal to 1, and in $\mathscr{A}_{O^-}$ is equal to -1. Therefore, if x, y denote the characters of s_2 for the energy level $\Lambda = 0$, $S = 1$ and $\Lambda = 0$, $S = 0$, respectively, then we find that the following equation holds: $3x + y = -2$.

Since $x = \pm 1$, $y = \pm 1$, the only solution is $x = -1$, $y = 1$, from which we find that the terms are O^-, $S = 1$ and O^+, $S = 0$. For $\Lambda = 0, 1, 2, \ldots$ we shall write $\Sigma, \Pi, \Delta, \ldots$,; thus, in conventional notation we have obtained the following result: Two equivalent π electrons (written π^2) correspond to the energy levels ${}^3\Sigma^-$, ${}^1\Sigma^+$, ${}^1\Delta$. In §6 we shall investigate the energy levels for the case of the hydrogen molecule.

In order to determine whether a state results in the binding of the two atoms, that is, whether $E(a)$ has an (absolute) minimum, it is necessary to examine the behavior of $E(a)$ as a function of a.

5 Formation of a Molecule from Two Atoms

While the structure principle described above together with the Hartree method leads to excellent quantitative results for intermediate distances a, for large values of a it is preferable to use perturbation theory beginning with the "zeroth" approximation of separated atoms in order to determine the behavior of the energy levels. Using group theory we may precisely determine the possible terms for the molecule which arise from the energy levels of the two separated atoms by studying their behavior as the quantity a varies continuously from ∞. Of course, we will obtain the same energy manifold as we have obtained from §4; here, however, we will determine which molecular energy levels continuously transform into which atomic energy levels of the separated atoms.

If we begin with a pair of separated atoms, then we would be concerned with combining the electrons of one atom with those of another in order to obtain a new system. Since for $a \sim \infty$ we are concerned with two inequivalent electron groups, we may, according to XIV, §3 obtain the corresponding energy levels by adding the spins corresponding to the representations of the permutation group; here we do not have to take the Pauli exclusion principle into account. For example, if we have two atoms with quantum numbers L_1, S_1 and L_2, S_2, respectively (where L_1, L_2 are the orbital angular momentum and S_1, S_2 are the spin angular momentum quantum numbers corresponding to the representations of the permutation group—note that we do not consider the effect of spin on the energy), then we obtain energy levels with $S = S_1 + S_2, S_1 + S_2 - 1, \ldots, |S_1 - S_2|$. To each of these S values we can associate each possible value of Λ.

Under restriction to the subgroup $\mathscr{d}$ the representation $\mathscr{D}_L$ may be decomposed as follows:

$$\mathscr{D}_L = \mathscr{A}_L + \mathscr{A}_{L-1} + \cdots + \mathscr{A}_0. \tag{5.1}$$

We may, therefore, obtain all possible Λ values providing we reduce $\mathscr{A}_{\Lambda_1} \times \mathscr{A}_{\Lambda_2}$ by using (4.1)–(4.3) where $\Lambda = L_1, L_1 - 1, \ldots, 0$; $\Lambda_2 = L_2, L_2 - 1, \ldots, 0$. For $\Lambda_1 = 0$ and $\Lambda_2 = 0$ do we have the representations $\mathscr{A}_{O^+}$ or $\mathscr{A}_{O^-}$? To answer this question we make use of the reflection operator s_2. Let s be the reflection about the origin of the coordinate system. Then we obtain

$$sD_{2(\pi)} = D_{2(\pi)}s = s_2, \tag{5.2}$$

where $D_{2(\pi)}$ is the rotation operator about the 2-axis by the angle π.

From (VII, §3) we find that the vector corresponding to $M = 0$ is multiplied by $(-1)^L$ when it is transformed by $D_{2(\pi)}$. Therefore we obtain $\mathscr{A}_{O^+}$ or $\mathscr{A}_{O^-}$ depending on whether $(-1)^L w$ is equal to $+1$ or -1 where w depends on whether the atom energy level L is even or odd (that is, $w = -1$ for energy levels of the form ${}^{2S+1}X^{\mathrm{o}}$; o = odd, see XIV, §5).

If both nuclei are identical, then we need to introduce the character of the eigenfunction under reflection s about the midpoint as part of the characterization of the energy level as an index e and o. Here it is evident that for two different energy levels L_1, S_1; L_2, S_2 of the separated atoms the resulting energy levels are doubled, once as even (e) and once as odd (o), since we only need to construct linear combinations as we have in (3.8):

$$(1 + s)\varphi_{(1)}\varphi_{(2)};\ (1 - s)\varphi_{(1)}\varphi_{(2)}, \tag{5.3}$$

where $\varphi_{(1)}$ is an eigenfunction for atom (1) for the energy level L_1, S_1 and $\varphi_{(2)}$ is such for atom (2). For the case of identical atomic energy levels

$$(L_1 = L_2, S_1 = S_2)$$

each term occurs only once because $s\varphi_{(1)}\varphi_{(2)}$ is not linearly independent with respect to $\varphi_{(1)}\varphi_{(2)}$ and the eigenfunctions obtained from the latter by permutation of the electrons. If we reduce $\mathscr{A}_{\Lambda_1} \times \mathscr{A}_{\Lambda_2}$ as above for the case $\Lambda_1 \neq \Lambda_2$ we would find that the above procedure (even in the case of identical atomic energy levels) is equivalent to the energy level manifold for inequivalent electrons; thus, in making linear combinations (5.3) we would find that each energy level contains both even and odd terms. However, for $\Lambda_1' = \Lambda_2$ and $\Lambda_2' = \Lambda_1$ we find that no new terms are introduced by $\mathscr{A}_{\Lambda_1'} \times \mathscr{A}_{\Lambda_2'}(!)$ because they are already included in the linear combinations (5.3). For $\Lambda_1 \neq \Lambda_2$ we therefore obtain all terms, both the even and odd terms, but, independent of the order of the Λ_1, Λ_2, each only once.

For the case in which $\Lambda_1 = \Lambda_2$ the corresponding molecular energy levels $\Lambda = 2\Lambda_1, O^+$ or O^- have only the following reflection characters ε which we shall now derive: For each L, S term of an atom the only position dependent eigenfunctions (according to XIV (1.2*a*); here we shall use the symbol φ instead of ψ) $\varphi^L_{M,v}$ correspond to a definite spin value; here, for fixed M, the index v corresponds to representations Δ of $\mathbf{S}_f$. Thus, for the energy levels $\Lambda = 2\Lambda_1$ the eigenfunctions are given by

$$P\varphi^{L_1}_{\Lambda_1 v(1)}(1, 2, \ldots, f)\varphi^{L_1}_{\Lambda_1 v'(2)}(f + 1, \ldots, 2f);$$

the same holds also if we replace Λ_1 by $-\Lambda_1$. Here (1) refers to the eigenfunction for the first nucleus (a notation which we have often used); $1, 2, \ldots, f, f + 1, \ldots, 2f$ refer to the coordinates of the $2f$ electrons. Here P refers to the permutation operators for the electrons. Here we find that the reflection s is identical to reflecting the eigenfunctions φ^L_{Mv} about the nuclei and exchanging the electrons of the first atom with those of the second. Since we are considering the same atomic energy levels, for the product $\varphi_{(1)}\varphi_{(2)}$ the reflection s is identical to exchanging the electrons of one atom with the other because the reflection of the electron coordinates with respect to the nuclei results in the same factor $+1$ or -1 for both $\varphi_{(1)}$ and $\varphi_{(2)}$. It is therefore necessary to determine the factor which multiplies the eigenfunctions for the energy level $\Lambda = 2\Lambda_1, S$ under interchange of the f electrons of one nucleus with those of the other nucleus. The representation Δ of $\mathbf{S}_{2f}$ is characterized by S. According to XIV, §1 the corresponding representation Δ' in spin space differs only by a factor $(-1)^P$; under the exchange of the f electrons of one nucleus with those of the other we obtain the factor $(-1)^f$. The factor in the representation Δ' is obtained from the formula for the addition of the angular momentum S_1 and $S_2 = S_1$ to obtain the total spin S is, according to XI, §10, equal to $(-1)^f (-1)^S$. Therefore we obtain $\varepsilon = (-1)^S$.

For the case in which $\Lambda = O^{\pm}$ the eigenfunctions are

$$P[\varphi^{L_1}_{\Lambda_1 k(1)}(1, 2, \ldots, f)\varphi^{L_1}_{-\Lambda_1 k'(2)}(f + 1, \ldots, 2f) \pm \varphi^{L_1}_{-\Lambda_1 k(1)}(1, 2, \ldots, f)\varphi^{L_1}_{\Lambda_1 k'(2)}(f + 1, \ldots, 2f)].$$

For the case $\Lambda = O^-$ in addition to the above considerations for the case in which $\Lambda = 2\Lambda_1$ we get an additional factor (-1) in order that for $\Lambda = O^-$ the reflection character is $\varepsilon = (-1)^{S+1}$ and, only for the case $\Lambda = O^+$ is simply $(-1)^S$.

After we determine the molecular energy levels corresponding to the energy levels of the separated atoms, the question remains concerning the quantitative dependence as a function of the distance a between the nuclei. To begin, for large values of a, it can be obtained by means of perturbation computations, as we shall see in the following sections. We may only determine whether the two atoms are chemically bound from the behavior of $E(a)$. Since $E(a) = z_1 z_2 e^2/a + E'(a)$, $E'(a)$ must decrease with decreasing a. This condition may often be tested by considering the limiting case $a = 0$. For $a = 0$ we obtain an atom with nuclear charge $Z_1 + Z_2$, the energy levels of which may be determined from XIV, §3. Here we may easily determine the molecular energy levels into which a particular energy level for an atom L, S (with reflection character w) (where $Z = Z_1 + Z_2$) will be split as a increases as follows: We have to take the subgroup $\mathscr{d}$ of the entire rotation group $\mathscr{D}_{\mathscr{g}}$; thus we find it necessary to reduce the representation $\mathscr{D}_L$ according to (5.1). Since the permutation group is not affected by changes in a, S is conserved. For different atoms the L, S energy levels split into energy levels $\Lambda = L$, $L - 1, \ldots, O; S$. In order to determine whether we obtain a O^+ or an O^- energy level we must consider the reflection operator s_2. As we have already found in the discussion following (5.2), $(-1)^l w$ describes the reflection character of $\Lambda = 0$ under s_2. For identical nuclei we have to include the reflection s with respect to the midpoint of the two nuclei as a symmetry operation. For $a = 0$ this is identical to a reflection about the nucleus. For this reason the reflection character must be given by $\varepsilon = w$.

From the rule that energy levels of the same pedigree cannot intersect we find that, as we have found for the case of a single electron, it is possible to determine which energy level for $a \sim 0$ is transformed into a given energy level for $a \sim \infty$ as a changes continuously from 0 to ∞. Atoms having closed shells have particularly low-energy (singlet) ground states. If both atoms have the same spin S as $a \to \infty$ then the singlet molecular energy level will have a particularly strong bound state since, for decreasing a, as $a \to 0$, the ground state must transform into a low-energy singlet ground state (see the example in §6). We shall call $v = 2S$ the valence of the atomic state. We then can state the following rule: The saturation of the valences (for singlet states) leads to chemical binding. The extent to which this rule is applicable to more than two atoms will be discussed in §7.

6 The Hydrogen Molecule

We will now consider the hydrogen molecule in more detail as an example of the energy level scheme for the electrons of a molecule. In Figure 26 the behavior of the energy levels are qualitatively illustrated as the distance a is

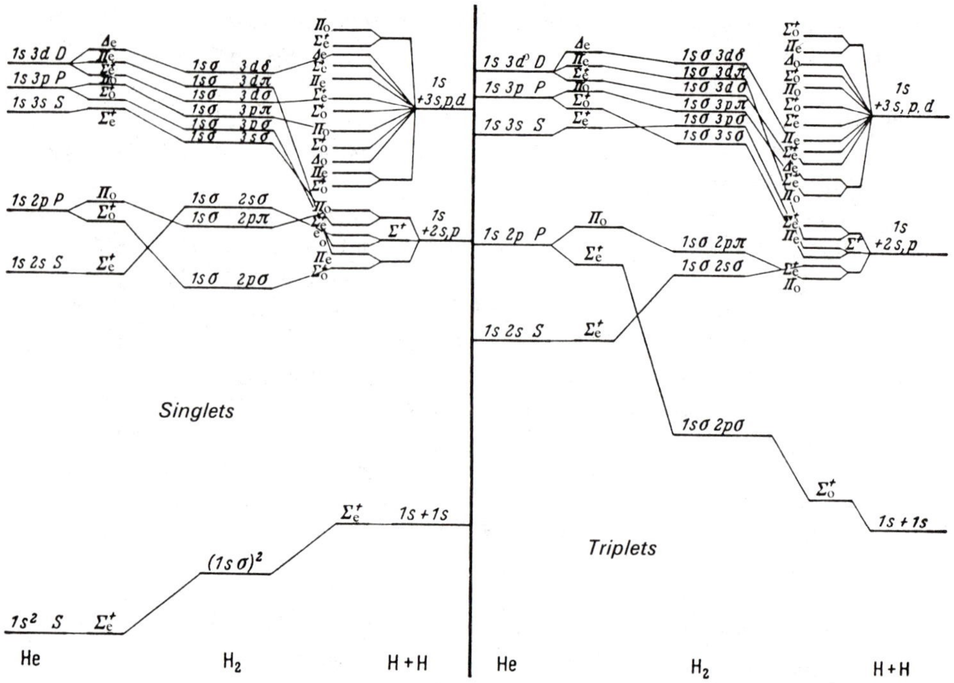

Figure 26 The behavior of the energy levels of the H_2 molecule as the distance a is varied.

varied. For clarity we have sketched the singlet and triplet terms separately. On the left-hand side ($a = 0$) we have the terms of the He atom and their splitting as the nuclei are separated. The S levels cannot be split; they have the symmetry character ${}^{1,3}\Sigma_e^+$, as given by the rules of §5. The P levels must be split into the levels ${}^{1,3}\Sigma_e^+$, ${}^{1,3}\Pi_\sigma$; the D levels into ${}^{1,3}\Sigma_e^+$, ${}^{1,3}\Pi_e$, ${}^{1,3}\Delta_e$. The energy levels of the H_2 molecule may, on the basis of the structure principles (§5), be considered to be derived from the one-electron terms. $(1s\sigma)^2$ leads to a single ${}^1\Sigma_e^+$ level; $1s\sigma\ 2p\sigma$ leads to two levels ${}^{1,3}\Sigma_o^+$. Since, according to Figure 22, the level $2p\sigma$ of an electron strongly decreases with increasing a, for H_2 it follows that the two levels arising from $1s\sigma\ 2p\sigma$ and the ground state are the lowest energy states; the ${}^3\Sigma_o^+$ level is unstable because it transforms into the ground state of the separated atoms. $1s\sigma\ 2p\pi$ leads to ${}^{1,3}\Pi_0$ levels, $1s\sigma\ 2s\sigma$ to ${}^{1,3}\Sigma_e^+$ levels, etc. On the right side ($a \sim \infty$) we obtain the levels of the separated atoms H + H, where one of the atoms is in the ground state and the other is in one of the states $1s$; $2s$, p; $3s$, p, d, etc. With the help of §5 we may easily show into which levels the terms $1s + ns, p, \ldots$ are split as the atoms approach each other. The results are illustrated on the right-hand side of Figure 26. In conclusion we may then connect energy levels having the same pedigree.

From these qualitative considerations it is already clear that the ground state ${}^1\Sigma_e^+$ leads to a strong chemical bond. We will examine this fact in greater detail using perturbation theory. There are basically two methods: The first begins with separate atoms and considers the mutual interaction of the electrons and the interaction of the electron with the other nucleus as the perturbation. The other uses the Hartree method, that is, one electron functions in the field of both nuclei, and the interaction of the two electrons is taken into account only globally.

In the first procedure we seek to approximate the eigenfunctions of the molecule by taking linear combinations of the two functions $\varphi_a(1)\varphi_b(2)$ and $\varphi_b(1)\varphi_a(2)$ where φ_α and φ_b are the normed eigenfunctions of the ground states of the hydrogen atoms a and b, respectively. Since the Hamiltonian operator is invariant under exchange of the electrons, $\phi^\pm = \varphi_a(1)\varphi_b(2) \pm \varphi_b(1)\varphi_a(2)$ must be the correct linear combinations for the perturbation calculation. The energy is therefore given by

$$E = \frac{\langle \phi^\pm, H\phi^\pm \rangle}{\|\phi^\pm\|^2}, \tag{6.1}$$

where

$$H = \frac{1}{2m}(\vec{P}_1^2 + \vec{P}_2^2) + \frac{e^2}{r_{12}} - \frac{e^2}{r_{a1}} - \frac{e^2}{r_{a2}} - \frac{e^2}{r_{b1}} - \frac{e^2}{r_{b2}}.$$

Here r_{12} is the distance between the two electrons, r_{a1} is the distance of electron 1 from nucleus a, etc. Since, for example

$$\left[\frac{1}{2m}\vec{P}_1^2 - \frac{e^2}{r_{a1}}\right]\varphi_a(1) = E_0\,\varphi_a(1)$$

it therefore follows from (6.1) that

$$E^+ = 2E_0 + \frac{C \pm A}{1 \pm S^2},$$

where

$$C = \int |\varphi_a(1)|^2\,|\varphi_b(2)|^2 \left(\frac{e^2}{r_{12}} - \frac{e^2}{r_{b1}} - \frac{e^2}{r_{a2}}\right) d^3\vec{r}_1\, d^3\vec{r}_2,$$

$$A = \int \varphi_a(1)\overline{\varphi_a(2)}\overline{\varphi_b(1)}\varphi_b(2) \left(\frac{e^2}{r_{12}} - \frac{e^2}{r_{b1}} - \frac{e^2}{r_{a2}}\right) d^3\vec{r}_1\, d^3\vec{r}_2,$$

$$S = \int \overline{\varphi_a(1)}\varphi_b(1)\, d^3\vec{r}_1.$$

E^+ is a singlet, E^- is a triplet. If $A < 0$ then the singlet energy level will lie below the triplet level. The expression e^2/r_{12} in A yields a positive contribution (for atoms this term is decisive; for example, for helium the triplet levels lie below the singlet level!). Here, however, the sign of A is determined by the two terms $-e^2/r_{b1} - e^2/r_{a2}$. (For a precise determination of the integrals A, C, S see [15].)

For the case of an atom (XII, §2) we were able to intuitively justify the energy differences between the singlet and triplet states by the fact that, for

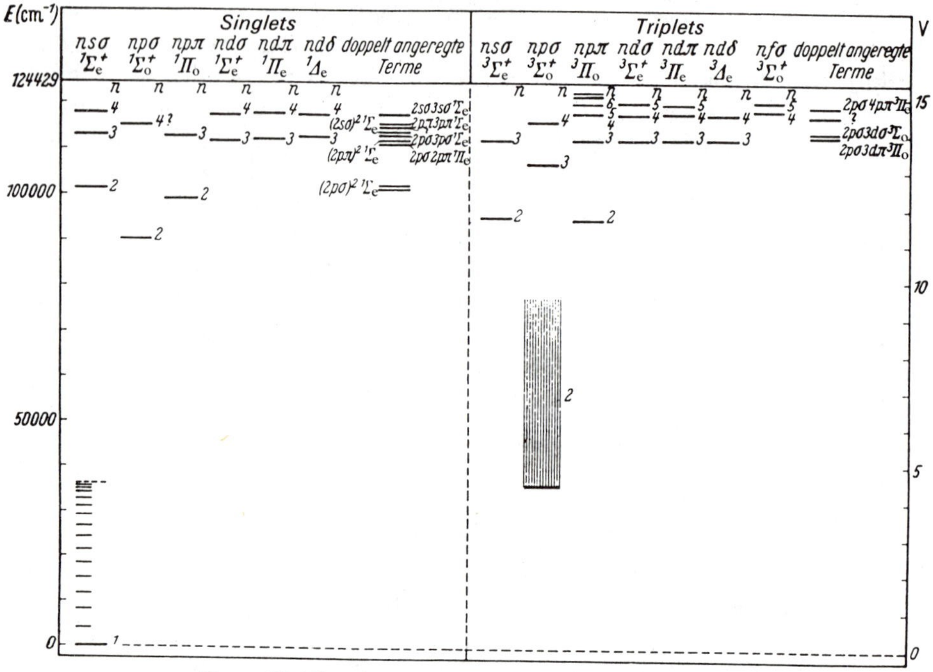

Figure 27 The observed energy level scheme for H_2.

symmetric electron position eigenfunctions, the electrons are closer than they would be for antisymmetric functions, where the electrons avoid each other. For the case of a molecule the effect of a symmetric eigenfunction does not only mean that the electrons would be closer, but that they would also be found more frequently in the vicinity *between* the nuclei than would be the case of an antisymmetric function. However, in the vicinity between the two nuclei the negative potential energy of the electron in the field of the *two* nuclei is very strong, and therefore the presence of the electrons between the nuclei will be more favorable energetically.

The second procedure simply uses the one-electron function of the H_2^+ molecule. The filling of the $1s\sigma$ state with two electrons leads to a binding, because, according to Figure 23, both electrons contribute to the binding. Here we speak of a binding electron pair. Here it is possible to carry out approximations for the ground state by using the Ritz variation principle (see XII, §3).

In Figure 27 the observed energy level scheme for H_2 is displayed.

7 The Chemical Bond

In order to obtain a theoretical basis for chemistry it is necessary to be able to determine whether the lowest energy state of the electrons (including the Coulomb repulsion energy of the nuclei), that is, the energy $E(\vec{r}_k)$ defined in (2.2) is a minimum for a particular spatial configuration of the nuclei. Since an exact solution of the many electron problem for equation (2.2) is not possible it is necessary to make use of qualitative considerations and approximation methods. Here we shall not attempt to present a summary of the methods which can be used in particular cases in order to obtain usable results [16]. Instead, we shall only examine the general characteristics of the chemical bond.

For more than two atoms the spatial symmetries are generally lost, and the only remaining symmetry is that of the permutation of the electrons. In §5 we have already found that the multiplet characteristics of a molecular energy level which arise continuously from the energy levels of the separate atoms are simply obtained by addition of the spins of the individual atoms. The energy levels which arise from the ground state of the separated atoms are, in general, not uniquely determined by the multiplet character. Nevertheless, the following rule is often valid: The addition of spin resulting in a total spin of zero leads to a state of stronger binding. This is the case because, as the atoms approach each other, they begin to resemble an atom with a closed shell, a situation which we have already discussed for the case of two atoms in §§5 and 6. It is clear that this can only be a rough rule. If, from each atom we draw the number of lines corresponding to its valence $v = 2S$, then there is a parallelism between the reduction of the representation $\mathscr{D}_{S_1} \times \mathscr{D}_{S_2} \times \cdots \times \mathscr{D}_{S_n}$ for n atoms with spin quantum numbers S_1, $S_2, \ldots, S_n$ and the combinatorics of the valence lines provided that we draw all "valence diagrams" in such a way that some pairs of the valence lines are bound together to single lines which run from one atom to another. The "valence diagrams" arising in this way correspond uniquely to the spin invariants in XIV, §7. If $v = 2S$ is the valence of an atom, then we choose a basis for the representation space for $\mathscr{D}_S$ as follows

$$U_M^S = \frac{u_+^{S+M} u_-^{S-M}}{\sqrt{(S+M)!\,(S-M)!}},$$

where u_+, u_- are the basis vectors for the representation $\mathscr{D}_{1/2}$. We construct the spin invariants

$$A = \prod_{\nu<\mu} [u_+(\nu)v_-(\mu) - u_-(\nu)v_+(\mu)]^{g_{\nu\mu}} \prod_{\nu} [u_+(\nu)x + u_-(\nu)y]^{f_\nu}, \quad (7.1)$$

where $g_{\nu\mu}$ is equal to the number of valence lines between the νth and μth atom (the bound valences) and f_ν is equal to the number of free valence lines of the νth atom. (For the valence scheme given in Figure 28 we obtain $g_{13} = 2$, $g_{23} = 1$, $g_{12} = 0$, $f_1 = 1$, $f_2 = f_3 = 0$.) The coefficients of

$$X_M^S = \frac{x^{S+M} y^{S-M}}{\sqrt{(S+M)!\,(S-M)!}} \quad (7.2)$$

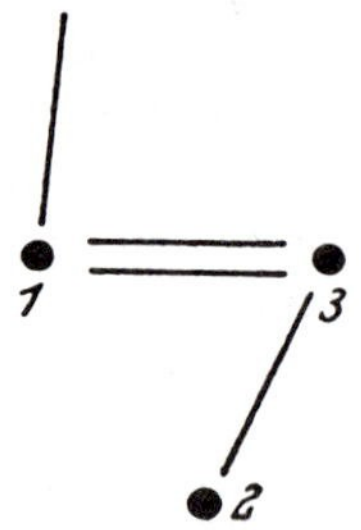

Figure 28 Valence scheme.

in A transform then according to $\mathscr{D}_S$. We therefore obtain $2S = \sum_\nu f_\nu$ and

$$f_\nu + \sum_\mu g_{\nu\mu} = v_\nu \tag{7.3}$$

which is equal to the valence of the νth atom.

If we then construct all possible spin invariants subject to the constraint (7.3) then we obtain basis vectors for all representations $\mathscr{D}_S$ which occur in the reduction of $\mathscr{D}_{S_1} \times \mathscr{D}_{S_2} \times \cdots \times \mathscr{D}_{S_n}$. For a given S there exist as many spin invariants as there are different valence schemes with free valences given by $\sum_\nu f_\nu = 2S$. It is easy to give an example which illustrates the fact that it is possible to have more valence schemes for the same value of S than is permitted by the $\mathscr{D}_S$ in the reduction of the $\mathscr{D}_{S_1} \times \mathscr{D}_{S_2} \times \cdots$. This is based on the fact that not all the spin invariants corresponding to S are linearly independent. The number of complete linearly independent spin invariants for S must, of course, be equal to the multiplicity of $\mathscr{D}_S$.

As an example, let us consider three atoms, each with valence $v = 1$. We then obtain $\mathscr{D}_{1/2} \times \mathscr{D}_{1/2} \times \mathscr{D}_{1/2} = 2\mathscr{D}_{1/2} + \mathscr{D}_{3/2}$. For $S = \frac{3}{2}$ there exists a single spin invariant.

$$A = \prod_{\nu=1}^{3} [u_+(\nu)x + u_-(\nu)y]. \tag{7.4}$$

For $S = \frac{1}{2}$ there are three spin invariants

$$\begin{aligned} A_1 &= [u_+(1)x + u_-(1)y][u_+(2)u_-(3) - u_-(2)u_+(3)], \\ A_2 &= [u_+(2)x + u_-(2)y][u_+(1)u_-(3) - u_-(1)u_+(2)], \\ A_3 &= [u_+(3)x + u_-(3)y][u_+(1)u_-(2) - u_-(1)u_+(2)]. \end{aligned} \tag{7.5}$$

Note that

$$A_3 = A_2 - A_1. \tag{7.6}$$

Of the three invariants in (7.5) only two are linearly independent.

A much more difficult problem is that of finding the actual bound states and estimating their binding energies. The valence scheme of chemistry is only a first qualitative approximation of the real state of affairs. Many atoms exhibit multiple valences in their chemical behavior. It is not difficult to explain this situation. If the interaction energy of an atom with other atoms in the molecule is not small compared to the difference between one of its excited states of a different pedigree and the groundstate, then we cannot exclude the possibility that the energy of a state arising from one of the excited states of the separate atoms, under continuous change of the separation, will lie below the molecular energy level arising from the ground states of the atoms. One of the most important examples is the carbon atom, the binding possibilities of which form the basis of all of organic chemistry. The ground state is (see XIV, §4) a triplet state and is therefore a state of valence 2. There we have found that there is also a lower lying quintet state, that is, $v = 4$ arising from the configuration $2s\ 2p^3$. How can such a strong bound state arise from this quintet state under continuous change of the separation of the atoms? To answer this question we need to present a more precise quantitative discussion.

One method, which is based on the Heitler–London theory of the H_2 molecule, seeks to estimate the binding energy by using perturbation theory, beginning with separate atoms. The calculations are completely analogous to those presented in XIV, §9, so that they need not be repeated here. We could begin by using the ground states of the free atoms as the zeroth approximation. It is advantageous, however, to begin with the one-electron eigenfunctions of the ground state electron configuration as the starting point of the perturbation calculation where the mutual interaction of the electrons in an atom are taken into account by means of average fields (see XIV, §9). Closed shells can, for the most part, be ignored because they do not play a significant role in binding because the closed shells of the individual atoms interact only slightly at an equilibrium distance.

In contrast to XIV, §9 the initial one-electron eigenfunctions belong to different atoms, and therefore, they need not be mutually orthogonal. Furthermore, the rotation group is not a symmetry group of the molecule.

For our first example let us choose f atoms, each with an electron outside its closed shell. Then, we may use the computations in XIV (9.2) to (9.21) directly. For $x(Q)$ we obtain the equation

$$xa = \varepsilon x, \tag{7.7}$$

where

$$a = \sum_P a(P)P = e \sum_{i<k} C_{ik} + \sum_{i<k} (ik)A_{ik}, \tag{7.8}$$

where in (7.8) we have neglected terms which are small compared to C_{ik} and A_{ik} and the nonorthogonality of the φ_{ν_i}. By analogy with the H_2 atom the C_{ik} and A_{ik} are given by

$$\begin{aligned} C_{ik} &= \left\langle \varphi_{\nu_i}(i)\varphi_{\nu_k}(k), \left\{\frac{e^2}{r_{ik}} - \sum_{j\neq k}\frac{e^2}{r_{a_j k}} - \sum_{j\neq i}\frac{e^2}{r_{a_j i}}\right\}\varphi_{\nu_i}(i)\varphi_{\nu_k}(k)\right\rangle, \\ A_{ik} &= \left\langle \varphi_{\nu_i}(k)\varphi_{\nu_k}(i), \left\{\frac{e^2}{r_{ik}} - \sum_{j\neq k}\frac{e^2}{r_{a_j k}} - \sum_{j\neq i}\frac{e^2}{r_{a_j i}}\right\}\varphi_{\nu_i}(i)\varphi_{\nu_k}(k)\right\rangle, \end{aligned} \tag{7.9}$$

where $r_{a_j i}$ is the distance between the ith electron and the jth nucleus.

The solution can be carried out using the method of spin invariants described in XIV, §9. The computation is similar to the valence scheme, as we shall later show for a more complicated case. The lowest state corresponds to a certain linear combination $\sum_\nu x_\nu A_\nu$ of the spin invariants, but not necessarily to a specific valence scheme! (Consider, for example, three H atoms, and show that the formation of a H_3 molecule is impossible!)

If the atoms have several electrons in the outer shells, then we must use equations XIV (9.22) to (9.30), except that the simplification in which a transition to a definite total orbital angular momentum is made is not possible because the molecule is not rotationally invariant.

Let φ_{nk_n} denote the one-electron functions, where $n = 1, 2, \ldots, N$ and $k_n = 1, \ldots, r_n$, where closed shells are not taken into account. For fixed n the φ_{nk_n} may correspond to the same energy value for an electron on the nth atom. Here we have used k instead of m since we can start with other linear combinations as we have done in XIV (9.22). The secular equation XIV (9.30) can be written in terms of spin invariants for the case of f electrons as follows:

$$\sum_{n'_1k'_{n'1}\cdots n'_fk'_{n'f}} \Bigg[C_{n_1k_{n1}\cdots n_fk_{nf};\, n'_1k'_{n'1}\cdots n'_fk'_{n'f}} - \sum_{\nu<\mu}{}^{(k')}(\nu\mu) A_{(\nu\mu)n_1k_{n1}\cdots n_fk_{nf};\, n'_1k'_{n'_1}\cdots n'_fk'_{n'_f}} \Bigg] \phi_{n'_1k'_{n'_1}\cdots n'_fk'_{n'_f}}$$
$$= (E - E_{n_1k_{n_1}\cdots n_fk_{n_f}})\phi_{n_1k_{n_1}\cdots n_fk_{n_f}}. \tag{7.10}$$

In (7.10) we have neglected all terms for which $P \neq e$ and $P \neq (\nu\mu)$ and the nonorthogonality of the φ_{nk_n}; the $E_{n_1k_{n_1}\cdots n_fk_{n_f}}$ differ only slightly (relatively to the perturbation). Equation (7.10) is already too complicated in order to be able to solve for the various positions of the nuclei. It is often possible, however, to choose linear combinations of the φ_{nk_n} for which the C and the $A_{(\nu\mu)}$ are diagonal (at least approximately) with respect to the $n_1k_{n_1}\cdots n_fk_{n_f}$; $n'_1k'_{n'_1}\cdots n'_fk'_{n'_f}$ in order that (7.10) can be decomposed into a number of subproblems of the form (7.7) and (7.8). Each of these subproblems is characterized by a well-defined distribution of the f electrons (independent of their labeling) in the states $\varphi_{n_1k_{n_1}}\cdots\varphi_{n_Nk_{n_N}}$. Often we may guess which of these problems leads to the lowest binding energy (see, however, the remarks about polar bonds below). We will provide a more detailed explanation following the example of the benzene ring.

For a subproblem we shall replace the indices $n_1k_{n_1}, \ldots, n_fk_{n_f}$ by $1, 2, \ldots, f$. According to the Pauli exclusion principle we may have at most two equal φ_i, thus we find that the φ_i occur only either in identical pairs $\varphi_i = \varphi_j$ or in simply occupied eigenfunctions.

We may organize the φ_i such that the φ_i which occur in pairs can be numbered such that $\varphi_1 = \varphi_2, \varphi_3 = \varphi_4, \ldots, \varphi_{2\nu-1} = \varphi_{2\nu}$ and the $\varphi_{2\nu+1}, \ldots, \varphi_f$ occur only once. Then a subproblem obtained from (7.10) can be expressed as follows:

$$\Bigg[C - \sum_{(i,k)} A_{ik}(i,k)\Bigg]\phi_S = (E - \bar{E})\phi_S \tag{7.11}$$
$$(i,k) \neq (1,2); (3,4)\cdots; (2\nu-1, 2\nu).$$

Consider the exchange (1, 2). Since $\varphi_1 = \varphi_2$ we find that $A_{1k} = A_{2k}$, $A_{k1} = A_{k2}$. Since $(1, 2)^2 = 1$ and $(1, 2)(2, k)(1, 2) = (i, k)$ for $i, k \neq 1, 2$ and $(1, 2)(1, k)(1, 2) = (2, k)$, $(1, 2)(2, k)(1, 2) = (1, k)$ we therefore obtain

$$(1, 2)\left[C - \sum_{(i,k)} A_{ik}(i, k)\right](1, 2) = \left[C - \sum_{(i,k)} A_{ik}(i, k)\right],$$

that is, the permutation (1, 2) commutes with the operator

$$\left[C - \sum_{(i,k)} A_{ik}(i, k)\right].$$

Therefore the ϕ_S can be chosen such that they are eigenvectors for (1, 2). The eigenvalue $+1$ can be excluded because of the Pauli exclusion principle (because $\varphi_1 = \varphi_2$!); therefore the ϕ_S must be antisymmetric in 1, 2; 3, 4; 5, 6, etc. The quantity $\frac{1}{2}[1 - (1, 2)]$ and the product $e = \frac{1}{2}[1 - (1, 2)]$ $\frac{1}{2}[1 - (3, 4)] \cdots \frac{1}{2}[1 - (2v - 1, 2v)]$ reproduce ϕ_S: $e\phi_S = \phi_S$. Since e commutes with $[C - \sum_{(i,k)} A_{ik}(i, k)]$ we therefore obtain

$$\begin{aligned}\left[C - \sum_{(i,k)} A_{ik}(i, k)\right]\phi_S &= \left[C - \sum_{(i,k)} A_{ik}(i, k)\right]e\phi_S \\ &= e\left[C - \sum_{(i,k)} A_{ik}(i, k)\right]\phi_S = (E - \bar{E})\phi_S.\end{aligned}$$

For the spin invariants we shall use the following simplified notation:

$$\begin{aligned}u_+(1)u_-(2) - u_-(1)u_+(2) &= [1, 2], \\ u_+(1)x + u_-(1)y &= [1, z].\end{aligned} \tag{7.12}$$

Therefore ϕ_S must have the form

$$\phi_S = [1, 2][3, 4] \cdots [2v - 1, 2v] \cdots [k, l] \cdots [m, z], \tag{7.13}$$

where $k, l, m \cdots > 2v$.

Thus it is easy to show that

$$\begin{aligned}e(1, 2)\phi_S &= -\phi_S, \\ e(1, k)\phi_S &= \tfrac{1}{2}\phi_S, \\ e(1, m)\phi_S &= \tfrac{1}{2}\phi_S, \\ e(2, 3)\phi_S &= \tfrac{1}{2}\phi_S.\end{aligned}$$

Therefore (7.11) can be written as follows:

$$\left[\bar{C} - \sum_{\substack{(i,k) \\ i,k > 2v}} A_{ik}(i, k)\right]\phi_S = (E - \bar{E})\phi_S, \tag{7.14}$$

where

$$\bar{C} = C - \frac{1}{2}\sum_{(a,b)} A_{ab} - \frac{1}{2}\sum_{\substack{i \leq 2v \\ k > 2v}} A_{ik}, \tag{7.15}$$

where $\Sigma_{(a,b)}$ is taken over all pairs $a \leq 2v$, $b \leq 2v$ for which $\varphi_a \neq \varphi_b$. Since $\bar{C}$ is independent of the different possible choices of Φ_S, its magnitude is unimportant for the determination of the lowest bound state, that is, the lowest eigenvalues of (7.14). Equation (7.14) has a form in which the pairwise φ_i are no longer present, and therefore we omit the first factors $[1, 2][3, 4]\ldots$ in ϕ_S. Therefore we may also omit the constraint $i, k > 2v$ under the summation sign in (7.14).

As we have suggested earlier, here we may also characterize the spin invariants by a valence scheme. To this purpose we place a point on a circle for each of the remaining functions φ_i. Here it is advantageous to line up the points for φ_i of the same atom (see Figure 29; for the benzene ring see Figure 32). Then we add a particular point O. Then we draw a line connecting pairs of points such that, with the exception of O, only one line connects each point; from O there shall be $2S$ lines. For each such valence scheme there corresponds a spin invariant ϕ_S, because a line between the ith and kth point corresponds to a factor $[i, k]$ in ϕ_S and a line between the mth point and O corresponds to a factor $[m, z]$. All the ϕ_S formed in this way are not, however, linearly independent. We obtain a complete system of linearly independent spin invariants ϕ_S if we use only those valence schemes in which no lines cross. Here, on the basis of

$$[i, k][l, m] = [i, l][k, m] + [i, m][l, k] \tag{7.16}$$

we observe that valence schemes which cross can be replaced, step by step, by noncrossing valence schemes. Thus spin invariants with crossing can be linearly combined by noncrossing valence schemes. On the other hand, it is easy to see that the number of noncrossing valence schemes is equal to the dimension of the representation of the permutation group corresponding to S. We can also use (7.16) to obtain the expressions $(i, k)\phi_S$ in an elementary way.

It is possible that the problem (7.14) can be reduced further—if the subspace of all the $\phi_S = [a, b]\ldots$ (which contains the same factor $[a, b]$) is transformed into itself by $\bar{C} - \sum A_{ik}(i, k)$ and the lowest eigenvalue for an ϕ_S lies in this subspace. We shall then refer to $[a, b]$ as a localized valence.

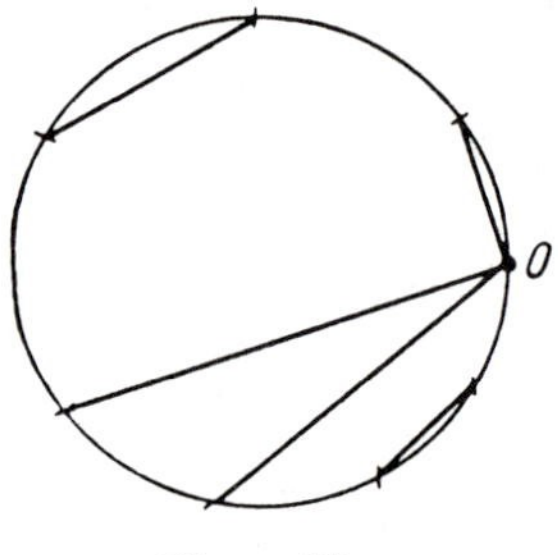

Figure 29

This is therefore certainly the case if all $A_{ak} = 0$, $A_{bk} = 0$ $(k \neq a, b)$. Then (7.14) is transformed into

$$\left[\bar{\bar{C}} - \sum_{\substack{(i,k) \\ i,k \neq a,b}} A_{ik}(ik)\right]\phi_S = (E - \bar{E})\phi_S, \qquad (7.17)$$

where

$$\bar{\bar{C}} = \bar{C} + A_{ab}.$$

In (7.17) we can neglect the factor $[a, b]$ in ϕ_S. In this way we may eliminate all localized valences step by step. We then obtain a problem of the same form (7.14) with $\bar{\bar{C}}$ instead of $\bar{C}$.

We shall now replace the operator (ik) by $\frac{1}{2}(\mathbf{1} + 4\vec{S}^{(i)} \cdot \vec{S}^{(k)})$ as we have in XIV (9.21). Then we obtain the equivalent problem

$$\left[\bar{\bar{C}} - \frac{1}{2}\sum_{(i,k)} A_{ik} - 2\sum_{(i,k)} A_{ik}\vec{S}^{(i)} \cdot \vec{S}^{(k)}\right]\phi_S = (E - \bar{E})\phi_S. \qquad (7.18)$$

If the ith and kth spin in ϕ_S is coupled by a valence as follows

$$[i, k] = u_+(i)u_-(k) - u_-(i)u_+(k)$$

then we find that $\vec{S}^{(i)} \cdot \vec{S}^{(k)}\phi_S = -\frac{3}{4}\phi_S$. If, on the contrary, i is not connected by the valence scheme to k then the expectation value

$$M(\vec{S}^{(i)} \cdot \vec{S}^{(k)}) = \langle \phi_S, \vec{S}^{(i)} \cdot \vec{S}^{(k)}\phi_S \rangle = 0.$$

For ϕ_S as the solution of (7.18) with smallest eigenvalue (ϕ_S need not belong to a particular valence scheme, but may be a linear combination!) we may consider the expectation value of $\Omega_{ik} = -\frac{1}{3}\vec{S}^{(i)} \cdot \vec{S}^{(k)}$ to be a measure for the occurrence of the valence $i - k$ because, under the localization of valence it takes on a value of 1, and for noncoupled valence takes on the value 0.

We will now consider the example of the benzene ring, the usual valence scheme of which is illustrated in Figure 30. We shall shortly find out whether

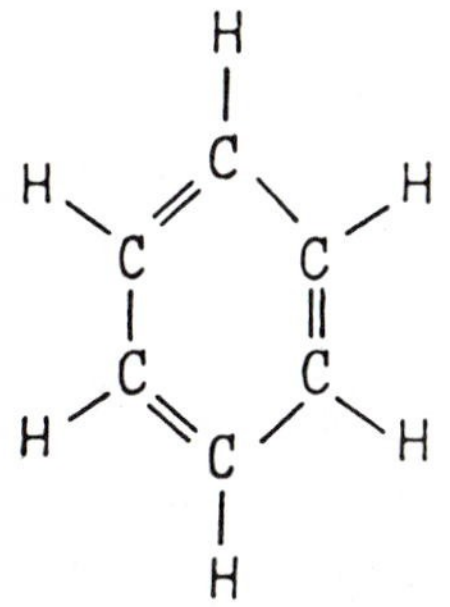

Figure 30 The usual valence scheme of the benzene ring.

this valence scheme reflects reality. The valence of carbon in this scheme is four. As we have already mentioned above, the electron configuration $1s^2\, 2s\, 2p^3$ plays a decisive role for the binding, although the ground state of the C atom arises from $1s^2\, 2s^2\, 2p^2$. We shall replace the four eigenfunctions which are available for the s and p electrons $\psi_0(r)$, $Y_m^1\psi_1(r)$ with $m = 1, 0, -1$ by other linear combinations. We now select one of the C atoms and choose a coordinate system for which the third axis is perpendicular to the plane of the benzene ring. Let e_1, e_2, e_3 be the three components of a unit vector, let $\chi_0 = \psi_0(r)$, $\chi_1 = e_1\psi_1(r)\alpha_1$, $\chi_2 = e_2\psi_1(r)\alpha_2$, $\chi_3 = e_3\psi_1(r)\alpha_3$ where the α_i are normalization constants; we shall choose the following linear combinations:

$$
\begin{aligned}
\varphi_0 &= \sqrt{\tfrac{1}{3}}\,\chi_0 + \phantom{\sqrt{\tfrac{1}{2}}\,\chi_1 -{}} \sqrt{\tfrac{2}{3}}\,\chi_2, \\
\varphi_1 &= \sqrt{\tfrac{1}{3}}\,\chi_0 + \sqrt{\tfrac{1}{2}}\,\chi_1 - \sqrt{\tfrac{1}{6}}\,\chi_2, \\
\varphi_2 &= \sqrt{\tfrac{1}{3}}\,\chi_0 - \sqrt{\tfrac{1}{2}}\,\chi_1 - \sqrt{\tfrac{1}{6}}\,\chi_2, \\
\varphi_3 &= \chi_3.
\end{aligned}
\tag{7.19}
$$

The $\varphi_0, \varphi_1, \varphi_2$ have significant charge densities in three directions in the (12) plane, separated by angles of 120°. The $\varphi_0, \varphi_1, \varphi_2$ are invariant with respect to rotation about these axes. If we choose the initial functions for each C atom in this way, then for each C atom we obtain appreciable exchange integrals for the φ_i $(i = 0, 1, 2)$ only for neighboring C or H atoms which lie in the three directions, and we find that the chemical bonds illustrated in Figure 31 are easily localized. Thus we can limit our consideration to the remaining six electrons which correspond to the eigenfunction φ_3. If we number the C atoms cyclically, then only the exchange integrals $A_{12} = A_{23} = A_{34} = A_{45} = A_{56} = A_{61} = A$ are significant, reducing the problem to the solution of the following equation

$$
\{\bar{\bar{C}} - A[(1,2) + (2,3) + (3,4) + (4,5) + (5,6) + (6,1)]\}\phi_S = (E - \bar{E})\phi_S. \tag{7.20}
$$

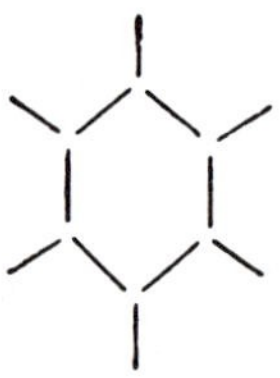

Figure 31 The localized bonds of the benzene ring.

The lowest energy level occurs for the case in which $S = 0$. The five possible linearly independent spin invariants are given by the following diagrams:

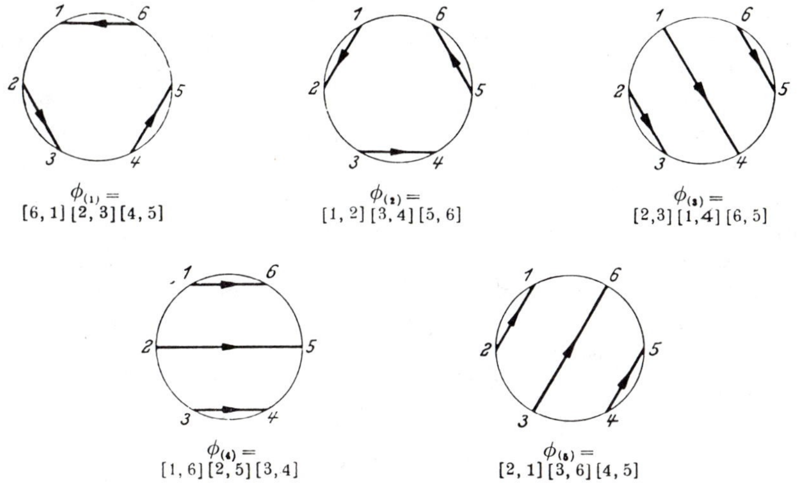

Figure 32

Under the cyclic perturbation $P = (1, 2, 3, 4, 5, 6)$ we obtain

$$P\phi_{(1)} = \phi_{(2)}, \qquad P\phi_{(2)} = \phi_{(1)}, \qquad P\phi_{(3)} = \phi_{(4)}, \qquad P\phi_{(4)} = \phi_{(5)}$$
$$\text{and} \quad P\phi_{(5)} = -\phi_{(3)},$$

$$[(1, 2) + (2, 3) + (3, 4) + (4, 5) + (5, 6) + (6, 1)] = \sum_{n=0}^{5} P^n(1, 2)P^{-n}.$$

Using (7.16) we may readily calculate $(1, 2)\phi_{(k)}$ and therefore also $\sum_{n=0}^{5} P^n(1, 2)P^{-n}\phi_{(k)}$. The smallest eigenvalue of (7.20) is given by

$$E - \bar{E} = \bar{\bar{C}} + A(-1 + \sqrt{13}) = \bar{\bar{C}} + 2.6\ \mathrm{A}. \tag{7.21}$$

For the spin invariant (a linear combination of the $\phi_{(k)}$) belonging to this eigenvalue the expectation values $M((1, 2)) = M((2, 3))$ are given by $\frac{1}{6}(1 - \sqrt{13})$, that is,

$$\begin{aligned} M(\Omega_{12}) = M(\Omega_{23}) &= \cdots = -\tfrac{4}{3}M(\vec{S}^{(1)} \cdot \vec{S}^{(2)}) \\ &= \tfrac{1}{3}M(-2(1, 2) + 1) = \tfrac{1}{9}(2 + \sqrt{13}) \\ &= 0.62. \end{aligned}$$

Without taking the valence coupling of the atoms, that is, for

$$M(\Omega_{12}) = \cdots = 0$$

we obtain the energy from (7.18) as follows:

$$E - \bar{E} = \bar{\bar{C}} - 3\ \mathrm{A}.$$

For localized valences, that is, for example, $M(\Omega_{12}) = 1$, $M(\Omega_{23}) = 0$, $M(\Omega_{34}) = 1$, $M(\Omega_{45}) = 0$, $M(\Omega_{56}) = 1$, $M(\Omega_{61}) = 0$, the energy is equal to

$$E - \bar{E} = \bar{\bar{\bar{C}}} + \tfrac{3}{2}\,\mathrm{A} = \bar{\bar{\bar{C}}} + 1.5\ \mathrm{A}.$$

The actual binding energy is therefore approximately $A(6 \cdot 0.62 - 3) = 0.7$ A greater than that for local binding. In this case we speak of a resonance of the valence diagrams, and 0.7 A is called the resonance energy. Using the computation methods described above we find that additional computation is required in order to show that the case of the six-sided benzene ring is substantially preferred than similar cases of five- or seven-sided rings. We shall use another method to show that this is the case. See Figure 33 for a comparison of the above results with experiments. The digram illustrates the distance between two neighboring C atoms for simple σ binding and increasing π binding (for the concept of σ and π binding, see below (page 205)). Therefore, for simple σ binding the distance is 1.54 Å; for σ + localized π binding the distance is 1.33 Å. The measured distance in benzene is 1.39 Å, which, according to Figure 33 corresponds to approximately

$$M(\Omega_{12}) = \cdots = 0.66,$$

from which the fact of resonance of valence diagram is proven.

Whether it is possible to obtain a satisfactory approximation from beginning with neutral atoms depends upon whether the eigenfunctions for a bound molecule actually have, to some degree, the form of those of the separate atoms. This is certainly not the case for the so-called polar bond such as HCl. As the atoms approach from $a \sim \infty$ the eigenfunctions change so drastically that, in the case of a bound molecule, the electron from the hydrogen atom is, for all practical purposes, completely attached to the chlorine atom, Here it is also correct that the bound singlet state of the molecule is obtained continuously from the state of the separate atoms

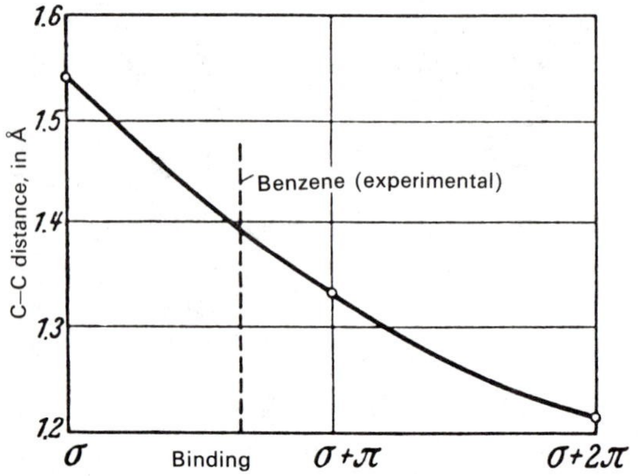

Figure 33 The distance between two neighboring C atoms for simple σ binding and increasing π binding.

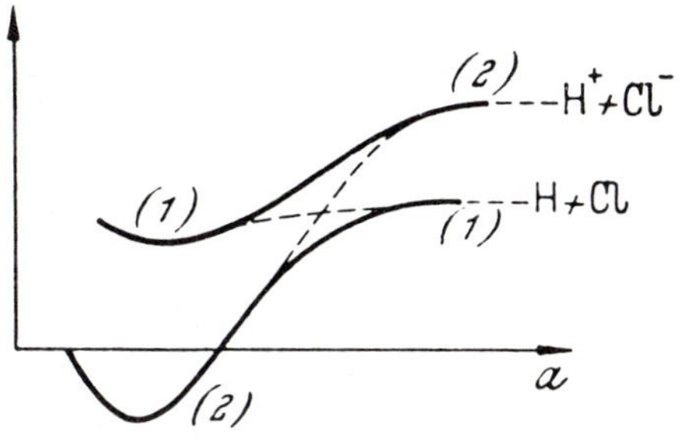

Figure 34

(Figure 34); however, a better approximation is obtained from perturbation theory by beginning with the eigenfunctions of the separate ions Cl^- and H^+, which separately have a higher energy than that of the neutral atoms (Figure 34). The perturbation calculation beginning with neutral atoms yields a curve (1) as a function of a (with the transition indicated by a dashed curve), while that for the ions yields the curve (2). By simultaneous consideration of both initial states we obtain the solid curves, because terms of the same pedigree cannot cross over.

A second method which may be used to attack the problem of the chemical bond is equivalent to the structure principle described in §4 for two molecules. The mutual interaction of the electrons is taken into account only in the sense of the Hartree procedure. Here we begin with the one-electron functions, the corresponding energies of which are qualitatively given in Figures 23 and 24. These one-electron functions are approximated in terms of linear combinations of the eigenfunctions for the separated atoms. If we can construct such linear combinations such that we obtain an essential increase in the electron charge density between the atoms, then we obtain a decrease in the associated energy (as we have seen in the case of the H_2 molecule in §6), that is, we obtain a contribution to the binding energy of the molecule. According to the Pauli exclusion principle, such a state can, on account of the spin, be doubly occupied. If, in a multi-atom molecule, the linear combination which corresponds to the lowest energy state arises from two functions from each of the atoms, then we may represent these two electrons by a valence line. Here we speak of a localized valence and of a bounding electron pair. In the H_2 molecule we have seen such a bounding electron pair. If the eigenfunctions are rotationally symmetric about the line joining the two atoms, we then speak of a σ bond. If the eigenfunctions correspond to the representation $\Lambda = 1$ of the rotation-reflection group $\mathscr{d}$ then we speak of a π bond.

For an example we again consider the case of the benzene ring. We need only consider the electrons which lie outside the ($n = 1$) shell, because the electrons in the ($n = 1$) shell are less perturbed as the atoms approach. For the four electrons with $n = 2$ we again use the atom functions given in (7.19). The functions φ_0, φ_1, φ_2 between neighboring atoms (C—C and C—H) result in localized σ valences. It now remains to specify one electron and one atomic function φ for each C atom, which, for the kth atom, we shall denote by ψ_k. Let us consider a ring of N C atoms. Let V_k be the potential of the kth atom, which, for large distances from the center of the atom must behave like $\sim -e^2/r_k$ because the shielding is such that only a single charge remains. The ψ_k will satisfy the equation

$$H_k\psi_k = E_0\psi_k \quad \text{with} \quad H_k = K + V_k, \tag{7.22}$$

where K is the kinetic energy $(1/2m)\vec{p}^2$. We must determine the eigenfunctions of the equation

$$H\phi = E\phi, \quad \text{where} \quad H = K + \sum_{k=1}^{N} V_k. \tag{7.23}$$

For an approximation to ϕ we set

$$\phi = \sum_{k=1}^{N} a_k\psi_k \tag{7.24}$$

and we seek extreme values of $\langle\phi, H\phi\rangle/\langle\phi, \phi\rangle$. We obtain the system of equations

$$\sum_{k=1}^{N} H_{ik}a_k = Ea_i, \tag{7.25}$$

where $\langle\psi_i, \psi_k\rangle \sim \delta_{ik}$ and

$$H_{ik} = \langle\psi_i, H\psi_k\rangle = \langle\psi_i, H_k\psi_k\rangle + \langle\psi_i, \sum_{j\neq k} V_j\psi_k\rangle.$$

Neglecting the nonorthogonality of the ψ_h for $i \neq k$ we obtain

$$H_{ik} = \left\langle\psi_i, \sum_{j\neq k} V_j\psi_k\right\rangle. \tag{7.26}$$

H_{ik} is, for all practical purposes, nonzero only for neighboring atoms i, $i + 1$. As a result we need only consider the terms in the sum $\sum_{j\neq k} V_j$ for which $j = i$, and we obtain

$$H_{i,i+1} = \langle\psi_i, V_i\psi_{i+1}\rangle.$$

Since all the H_{ik} are the same, we set $H_{i,i+1} = -A\ (A > 0)$. For $i = k$, in the same approximation we obtain

$$H_{ii} = \langle\psi_i, H\psi_i\rangle = E_0 + \left\langle\psi_i, \sum_{j\neq i} V_j\psi_i\right\rangle \approx E_0.$$

We define the operator T as follows:

$$T\psi_k = \psi_{k+1} \quad \text{and} \quad T\psi_N = \psi_1. \tag{7.27}$$

We may therefore write the system of equations (7.25) in the form

$$[E_0 - A(T + T^{-1})]\phi = E\phi. \tag{7.28}$$

From $T^N = 1$ it follows that the eigenvalues of T are the Nth roots of unity $e^{(2\pi i/N)\nu}$. The eigenvalues of (7.28) are therefore given by

$$E_\nu = E_0 - 2A \cos \frac{2\pi}{N} \nu \quad \text{with } \nu = 0, 1, 2, \ldots, N-1.$$

In order to obtain the eigenvectors ϕ_ν we consider the condition

$$T\phi_\nu = \sum_{k=1}^{N} a_k^{(\nu)} \psi_{k+1} = e^{(2\pi i\nu/N)} \sum_{k=1}^{N} a_k^{(\nu)} \psi_k$$

from which it follows that

$$a_k^{(\nu)} = e^{-(2\pi i/N)k\nu}$$

The N electrons are to be assigned to the eigenfunctions ϕ_ν in such a way that pairs in the same state are found in the lowest possible energy levels E_ν. The energy level $E_0 - 2$ A is doubly occupied, $E_0 - 2$ A $\cos 2\pi/N$ has occupancy 4, the others also have occupancy 4. Rings with $N = 6, 10, 14$ therefore have closed shells and are energetically more favorable. In the case of localization of valences each of the electrons have an energy of $E_0 - A$, that is, the total energy would be $E = N(E_0 - A)$. Let $E = N(E_0 - \kappa A)$. In Figure 35 we exhibit the dependence of κ as a function of N. For the benzene ring we obtain $\kappa = \frac{4}{3}$. A single localized bond corresponds to a contribution of -2 A to the energy. Therefore $\kappa/2 = 2/3 = 0.67$ is the strength of the π bond in the benzene ring, compared with the value 0.5 neglecting the resonance of the valence diagrams and 0.62 according to the Heitler–London procedure (Figure 33).

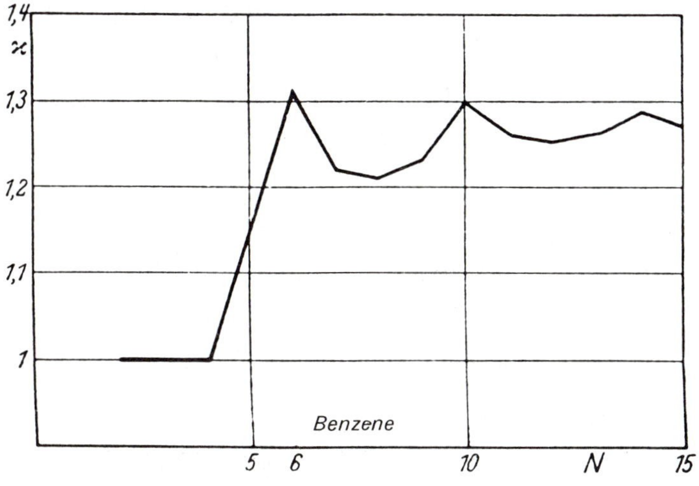

Figure 35 The binding energy in a ring of N C atoms.

8 Spectra of Diatomic Molecules

We will only discuss the actual energy levels of the entire molecule for the case of diatomic molecules. The derivations of §2, in particular, the expression (2.5) cannot be used without modification, because the energy levels for $\Lambda \neq 0$ are doubly degenerate. For $\Lambda = 0$, using the expression (2.5), and the Ritz variation method, we obtain (2.6) and (2.7). For diatomic molecules (2.6) takes on the special form

$$-\frac{1}{2M}\Delta\psi + \bar{E}(r) = \varepsilon\psi, \tag{8.1}$$

where $1/M = 1/m_1 + 1/m_2$ and $\vec{r}$ replaces $\vec{r}'_k$.

According to the contribution of XI, §3, the solutions must take on the form

$$\psi(\vec{r}) = Y^l_m((\theta, \varphi)g(r), \tag{8.2}$$

where g satisfies the equation

$$-\frac{1}{2M}\left(\frac{1}{r}\frac{d^2}{dr^2}(rg) - \frac{l(l+1)}{r^2}g\right) + \bar{E}(r)g = \varepsilon g. \tag{8.3}$$

We will postpone the solution of this equation until later, where we shall also consider the case $\Lambda \neq 0$.

The exact eigenvectors of the total Hamiltonian operator must correspond to a definite total angular momentum quantum number K:

$$\phi^K_M(M = K, K-1, \ldots, -K).$$

Since the ϕ^K_M are basis vectors corresponding to the representation $\mathscr{D}_K$ of the rotation group, the entire set of values for ϕ^K_M are already determined providing they are known for a particular position of the vector joining the nuclei. Let D be a rotation which rotates the vector $\vec{r}$ (in polar coordinates θ, φ, r) in the direction of the 3-axis, that is, in $\theta = 0, \varphi = \cdots, r$). Then we obtain

$$\begin{aligned}\phi^K_M(\vec{r}; \vec{r}_\kappa) &= D\phi^K_M(D\vec{r}; D\vec{r}_\kappa)\\ &= \sum_{M'} \phi^K_{M'}(D\vec{r}; D\vec{r}_\kappa)D_{M'M} = \sum_{M'} \phi^K_{M'}(0, \ldots, \vec{r}; D\vec{r}_\kappa)D_{M'M},\end{aligned} \tag{8.4}$$

where $D_{M'M}$ is the representation matrix in the $\mathscr{D}_K$ representation. The rotation D is a function of θ and φ; it is, however, not uniquely determined, because if $D_{3\alpha}$ is a rotation about the 3-axis by an angle α, then $D_{3\alpha}D$ also satisfies the condition imposed on D. Thus we obtain

$$\begin{aligned}\phi^K_M(\vec{r}; \vec{r}_\kappa) &= \sum_{M'} \phi^K_{M'}(0, \ldots, r; D_{3\alpha}D\vec{r}_\kappa)e^{-iM'\alpha}D_{M'M}\\ &= \sum_{M'} D^{-1}_{3\alpha}\phi^K_{M'}(0, \ldots, r; D\vec{r}_\kappa)e^{-iM'\alpha}D_{M'M}\\ &= \sum_{M'} \phi^K_{M'}(0, \ldots, r; D\vec{r}_\kappa)D_{M'M},\end{aligned} \tag{8.5}$$

where the latter is obtained from $D_{3\alpha}^{-1}\phi_{M'}^{K} = e^{iM'\alpha}\phi_{M'}^{K}$, with the result that the ambiguity of the reaction D plays no role.

Under the assumption that the eigenfunctions can be written, at least approximately, as a product (2.5) or as a linear combination of such products, we may therefore write, as an approximation,

$$\phi_{M'}^{K}(0, \ldots, r; \vec{r}_\kappa') = g_{M'}(r)\varphi_r(\vec{r}_\kappa'),$$

where $D\vec{r}_\kappa = \vec{r}_\kappa'$, where the $\varphi_r(\vec{r}_\kappa')$ belongs only to a single value of the electron energy $E(r)$. Since $\phi_{M'}^{K}$, upon multiplication by $D_{3\alpha}$, is multiplied by the factor $e^{-iM'\alpha}$, and, since, for a eigenvalue $E(r)$ of the electron energy it must correspond to a definite quantum number Λ we find that $\phi_{M'}^{K}(0, \ldots, r, \vec{r}_\kappa')$ must (approximately) vanish for $M' \neq \pm\Lambda$. For $\Lambda = 0$ we therefore obtain

$$\phi_0^K(0, \ldots, r, \vec{r}_\kappa') = g(r)\varphi(\vec{r}_\kappa'),$$

$$\phi_{M,}^{K} = 0 \quad \text{for } M' \neq 0.$$

Thus we obtain

$$\phi_M^K(\vec{r}, \vec{r}_\kappa) = g(r)\varphi_r(D\vec{r}_\kappa)D_{0M}. \tag{8.6}$$

This is, however, identical to (8.2) and (2.5), because D_{0M} is, up to an unimportant factor, identical to the spherical harmonics $Y_M^K(M, \varphi)$. The D_{0M} are functions of $\vec{r}$: $D_{0M}(\vec{r})$. Since D must be such that $D\vec{r}$ lies along the 3-axis, the rotation $DD_{(1)}$ corresponds to the vector $D_{(1)}^{-1}\vec{r}$, that is, rotates it to the 3-axis. It follows that

$$D_{0M}(D_{(1)}^{-1}\vec{r}) = (DD_{(1)})_{0M} = \sum_{M'} D_{0M'}D_{(1)M'M}.$$

As functions of $\vec{r}$ the D_{0M} belong to the representation $\mathscr{D}_K$. Therefore they must be equal to (up to a factor) Y_M^K. Therefore, in the case $\Lambda = 0$ we may identify K (according to (8.2)) with l.

The Hamiltonian operator is also invariant under reflection s of all coordinates (nuclei and electrons). Thus (8.6) must belong to a definite reflection character. In order to determine the latter, we observe that $D_{2(\pi)}s = s_2$ where $D_{2(\pi)}$ is the rotation about the 2-axis of angle π. Thus it follows that

$$s\phi_M^K(\vec{r}; \vec{r}_\kappa) = \phi_M^K(s\vec{r}; s\vec{r}_\kappa) = g(r)\varphi(D's\vec{r}_\kappa)D'_{0M},$$

where D' is the rotation which rotates $s\vec{r}$ in the direction of the 3-axis; therefore $D' = D_{2(\pi)}D$. Therefore, since $D_{2(\pi)}s = s_2$, we obtain

$$s\phi_M^K(\vec{r}; \vec{r}_\kappa) = g(r)\varphi(s_2 D\vec{r}_\kappa) \sum_{M'} D_{2(\pi)0M'}D_{M'M}.$$

According to (VII, §3) the $D_{2(\pi)0M}$ have the form $(-1)^K\delta_{0M}$; thus, from $s_2\varphi(D\vec{r}_\kappa) = \varepsilon\varphi(D\vec{r}_\kappa)$ it follows that

$$s\phi_M^K(\vec{r}, \vec{r}_\kappa) = \varepsilon(-1)^K g(r)\varphi(D\vec{r}_\kappa)D_{0M} = \varepsilon(-1)^K\phi_M^K.$$

The reflection character with respect to s for (8.6) is therefore $(-1)^K$ for O^+ and $(-1)^{K+1}$ for O^- terms.

For $\Lambda \neq 0$, instead of (8.6), it follows that the approximation for the eigenvectors is given by

$$\phi_M^K(\vec{r}; \vec{r}_\kappa) = g_+(r)\varphi_\Lambda(D\vec{r}_\kappa)D_{\Lambda M} + g_-(r)\varphi_{-\Lambda}(D\vec{r}_\kappa)D_{-\Lambda M}. \tag{8.7}$$

Since the eigenvectors must belong to a specific reflection character with respect to s, we may determine the ratio of g_+ to g_-: As above, we obtain

$$\begin{aligned} s\phi_M^K(\vec{r}; \vec{r}_\kappa) &= g_+(r)\varphi_\Lambda(D's\vec{r}_\kappa)D'_{\Lambda M} + g_-(r)\varphi_{-\Lambda}(D's\vec{r}_\kappa)D'_{-\Lambda M} \\ &= g_+(r)\varphi_\Lambda(s_2 D\vec{r}_\kappa) \sum_{M'} D_{2(\pi)\Lambda M'} D_{M'M} \\ &\quad + g_-(r)\varphi_{-\Lambda}(s_2 D\vec{r}_\kappa) \sum_{M'} D_{2(\pi)-\Lambda M'} D_{M'M} \\ &= g_+(r)\varphi_{-\Lambda}(D\vec{r}_\kappa)(-1)^{K-\Lambda}D_{-\Lambda M} + g_-(r)\varphi_\Lambda(D\vec{r}_\kappa)(-1)^{K+\Lambda}D_{\Lambda M'}. \end{aligned}$$

Since $s\,\phi_M^K = w\phi_M^K$, we must therefore have

$$g_-(r) = (-1)^{K+\Lambda} w g_+(r).$$

Both values $w = \pm 1$ are therefore also possible for $\Lambda \neq 0$ for the approximate eigenvectors (here we set $g_+ = g$):

$$\phi_M^K(\vec{r}, \vec{r}_\kappa) = g(r)\varphi_\Lambda(D\vec{r}_\kappa)D_{\Lambda M} + (-1)^{K+\Lambda} w g(r)\varphi_{-\Lambda}(D\vec{r}_\kappa)D_{-\Lambda M}. \tag{8.8}$$

Using the eigenvectors (8.6) or (8.8) we may apply the Ritz variation principle in order to determine $g(r)$.

For $g(r)$ we obtain the following differential equation

$$-\frac{1}{2M}\frac{1}{r}\frac{\partial^2}{\partial r^2}(rg) - \frac{1}{Mr}\frac{\partial}{\partial r}(rg)W(r) + U(r)g = \varepsilon g, \tag{8.9}$$

where

$$U(r) = E(r) + \frac{1}{2Mr^2}\langle \vec{M}^2 \rangle - \frac{1}{2M}\int \overline{\varphi_\Lambda(\vec{r}_\kappa)} \frac{\partial^2}{\partial r^2}\varphi_\Lambda(\vec{r}_\kappa)(d^3\vec{r}_\kappa)^f \tag{8.10}$$

and

$$W(r) = \int \overline{\varphi_\Lambda(\vec{r}_\kappa)} \frac{\partial}{\partial r}\varphi_\Lambda(\vec{r}_\kappa)(d^3\vec{r}_\kappa)^f, \tag{8.11}$$

where $\vec{M}$ is the orbital angular momentum of the nuclei and

$$\begin{aligned} \langle \vec{M}^2 \rangle = \tfrac{1}{2}\langle &\varphi_\Lambda(D\vec{r}_\kappa)D_{\Lambda M} + (-1)^{K+\Lambda} w\varphi_\Lambda(D\vec{r}_\kappa)D_{-\Lambda M}, \\ &\vec{M}^2[\varphi_\Lambda(D\vec{r}_\kappa)D_{\Lambda M} + (-1)^{K+\Lambda} w\varphi_{-\Lambda}(D\vec{r}_\kappa)D_{-\Lambda M}]\rangle. \end{aligned} \tag{8.12}$$

The third summand in (8.10) represents only a slight change in $E(r)$ and will therefore be ignored in the following. The same is also true for $W(r)$. It only remains to compute the term (8.12). With $\vec{L}$ as the orbital angular momentum of the electrons and $\vec{K}$ as the total orbital angular momentum, we obtain $\vec{K} = \vec{M} + \vec{L}$. The L_i component commutes with the ith component of $\vec{K}$, from which we find that

$$\vec{M}^2 = (\vec{K} - \vec{L})^2 = \vec{K}^2 + \vec{L}^2 - 2\vec{K}\cdot\vec{L} = \vec{K}^2 + \vec{L}^2 - 2\vec{L}\cdot\vec{K}.$$

Thus for $\vec{K}^2\phi_M^K = K(K+1)\phi_M^K$, we obtain

$$\langle \vec{M}^2 \rangle = K(K+1) + \langle \vec{L}^2 \rangle - 2\langle \vec{L}\cdot\vec{K}\rangle. \tag{8.13}$$

In the last term we shall, according to VII (3.17) and (3.18), apply the following substitutions:

$$\begin{aligned}
K_1\phi_M^K &= \tfrac{1}{2}\sqrt{(K+M)(K-M+1)}\,\phi_{M-1}^K \\
&\quad + \tfrac{1}{2}\sqrt{(K-M)(K+M+1)}\,\phi_{M+1}^K, \\
K_2\phi_M^K &= -\frac{1}{2i}\sqrt{(K+M)(K-M+1)}\,\phi_{M-1}^K, \\
&\quad + \frac{1}{2i}\sqrt{(K-M)(K+M+1)}\,\phi_{M+1}^K, \\
K_3\phi_M^K &= M\phi_M^K.
\end{aligned}$$

Thus, we find that all the operators in (8.13) which act on the nuclei coordinates have disappeared. Since (8.13) is invariant under rotations, we may choose the position of the line joining the nuclei along the 3-axis (here the rotation D in (8.8) is equal to 1) such that, according to (8.8) we obtain $D_{\Lambda M} = \delta_{\Lambda M}$ and $D_{-\Lambda M} = \delta_{-\Lambda M}$, and, accordingly, only the ϕ_M^K for which $M = \pm\Lambda$ are nonzero. Then, since $\phi_{\Lambda-1}^K = \phi_{\Lambda+1}^K = 0$, we obtain

$$\begin{aligned}
\langle \vec{L}\cdot\vec{K}\rangle &= \Lambda\int \overline{\varphi_\Lambda(\vec{r}_\kappa)}L_3\varphi_\Lambda(\vec{r}_\kappa)(d^3\vec{r}_\kappa)^f = \Lambda^2, \\
\langle \vec{L}^2\rangle &= \Lambda^2 + \int \overline{\varphi_\Lambda(\vec{r}_\kappa)}(L_1^2 + L_2^2)\varphi_\Lambda(\vec{r}_\kappa)(d^3\vec{r}_\kappa)^f.
\end{aligned} \tag{8.14}$$

The last term in (8.14) cannot, in general, be computed. If we neglect it, as we have above for the last term in (8.10) and for $W(r)$, then for (8.9) we finally obtain

$$-\frac{1}{2M}\frac{1}{r}\frac{\partial^2}{\partial r^2}(rg) + \frac{1}{2Mr^2}[K(K+1) - \Lambda^2]g + E(r)g = \varepsilon g. \tag{8.15}$$

Here $[K(K+1) - \Lambda^2]$ has a very intuitive meaning as the value of $\vec{M}^2$. If the electron angular momentum lies (for all practical purposes) along the line joining the nuclei [this is equivalent to neglecting the last term in (8.14)], it then follows that $\vec{M}$ is perpendicular to the line joining the nuclei, that is, $\vec{K}^2 = \vec{M}^2 + \vec{L}^2$ or $\vec{M}^2 = \vec{K}^2 - \vec{L}^2$. With $\vec{K}^2 \to K(K+1)$ and $\vec{L}^2 \to \Lambda^2$ we obtain the above result.

For $\Lambda \neq 0$ the equation (8.15) is independent of the reflection character w. The approximate value ε obtained in the above manner is therefore the same for both solutions (8.8). In higher approximations there must be a small splitting between the energy levels $w = +1$ and -1, which has been observed for large angular momenta.

Equation (8.15) is the eigenvalue equation for an anharmonic oscillator which oscillates about the position ρ of the minimum of the function $U(r) = (1/2M)(K(K+1) - \Lambda^2)/r^2 + E(r)$. From the large value of the mass M, for moderate values of $\vec{M}^2$ we may approximate $U(r)$ (where $r = \rho_0$ is the position of the minimum of $E(r)$) as follows:

$$U(r) = \frac{1}{2M}\frac{(K(K+1) - \Lambda^2)}{r^2} + \frac{(r-\rho_0)^2}{2}E''(\rho_0) + E(\rho_0).$$

Under the above assumptions the position ρ of the minimum of $U(r)$ lies near ρ_0; making a series expansion of $U(r)$ about ρ_0 we obtain

$$U(r) \approx E(\rho_0) + \frac{1}{2M}\frac{K(K+1) - \Lambda^2}{\rho_0^2} + A[K(K+1) - \Lambda^2]^2 + \tfrac{1}{2}(r-\rho_0)^2 E''(\rho_0)$$

and we therefore obtain the eigenvalues

$$\varepsilon_{Kn} = E(\rho_0) + \frac{1}{2M}\frac{K(K+1) - \Lambda^2}{\rho_0^2} + A[K(K+1) - \Lambda^2]^2 + \sqrt{\frac{E''(P_0)}{M}}(n + \tfrac{1}{2}), \tag{8.16}$$

where

$$A = -\frac{1}{M^2}\frac{1}{E''(\rho_0)\rho_0^6}.$$

Here ε_{Kn} is obtained by adding the electron energy $E(\rho_0)$, the oscillator energy $\omega(n + 1/2)$ where $\omega = \sqrt{E''/M}$ and the rotation energy

$$\left(\frac{1}{2M}\right)\left(\frac{(K(K+1) - \Lambda^2}{\rho_0^2}\right).$$

The energy level scheme has the structure illustrated in Figure 36. The transitions between the energy levels produce the spectrum. If the electron energy and the oscillator energy are held fixed, then by changing the rotation energy alone we obtain the rotation energy lines in long-wave infrared, the so-called rotation spectrum. By changing the oscillation and rotation energies we obtain an infrared spectrum which has fine structure due to rotation. If we change the electron energy we obtain the band-structure in the visible part of the spectrum.

The reader may wish to show that the following selection rules are applicable for dipole radiation:

$$K \to K \pm 1, \qquad K \text{ but not } O \to O,$$

$$w \to -w,$$

$$\Lambda \to \Lambda \pm 1, \qquad \text{but not } O^+ \to O^- \quad \text{and } O^- \to O^+,$$

where, for $O^+ \to O^+$ and $O^- \to O^-$ and $K \to K$ the transition (since $w \to -w$) is forbidden.

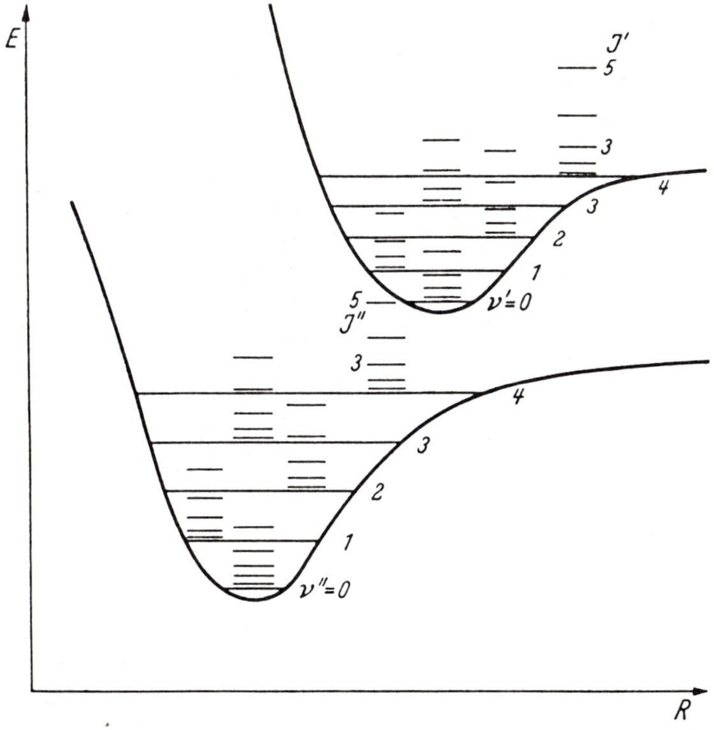

Figure 36 Energy level scheme of a diatomic molecule.

The reader will be able to examine the behavior of the eigenfunctions for the case of molecules having two identical nuclei under exchange of the nuclei, also after we have taken into account the electron spin in the next section. How will the eigenfunctions behave in this case of identical nuclei when both nuclei have the same spin ($\frac{1}{2}$) and the nuclei must therefore satisfy the Pauli exclusion principle (para and ortho hydrogen)?

9 The Effect of Electron Spin on Molecular Energy Levels

The term H_s in (1.1) does not alter the symmetry group. We may therefore obtain the possible molecular energy levels by "switching on" H_s in the form λH_s (where λ changes from 0 to 1). Each molecular energy level belongs to a particular representation with respect to the permutation of the electrons which is uniquely related to the spin quantum number S. We therefore need only add the total orbital angular momentum (quantum number K) and the spin angular momentum (quantum number S) for the corresponding energy level in order to obtain the total angular momentum J. This addition therefore leads to a splitting of energy levels with the increasing influence of λH_s into components with total angular momentum quantum numbers $J = K + S, K + S - 1, \ldots, |K - S|$.

If, under complete inclusion of H_s (that is, for $\lambda = 1$) the splitting of a molecular energy level remains small compared to the separation of two neighboring rotation energy levels, then we have obtained a qualitative description of the position of the energy levels and we may then apply the selection rules.

If, on the other hand, the spin splitting is larger than the separation of the rotation energy levels, we do not obtain a qualitative overview of the energy levels in this way. Here it would be preferable to take the spin into account in computing the electron energy levels. The electron energy levels under consideration with the quantum number Λ belong to a representation of the permutation group of the electrons which correspond to a spin quantum number S. Let $U_S, U_{S-1}, \ldots, U_{-S}$ be the $2S + 1$ spin functions (the eigenfunctions for the spin component about the line joining the nuclei), then, according to the scheme (5.1) we need only reduce the representations $\mathscr{A}_\Lambda \times \mathscr{A}_S, \mathscr{A}_\Lambda \times \mathscr{A}_{S-1}, \ldots, \mathscr{A}_\Lambda \times \mathscr{A}_{1/2}$ or $\mathscr{A}_\Lambda \times \mathscr{A}_0$. In order to determine whether the function U_0 belongs to the representation $\mathscr{A}_{0^+}$ or $\mathscr{A}_{0^-}$ we need to compute $s_2 U_0$. Since $s_2 = D_{2(\pi)} s$ and $sU_0 = U_0$ (from VII, §7) from the remark at the end of (5.2) it follows that $s_2 U_0 = (-1)^S U_0$, from which it is possible to decide whether we have the case $\mathscr{A}_{0^+}$ or $\mathscr{A}_{0^-}$.

Let $\mathscr{A}_\Omega$ denote the representation which arises from this reduction, then all considerations from §8 are easily applicable if we replace Λ by Ω. The electron energy levels are characterized by Ω instead of by Λ.

In closing this chapter, we note that more comprehensive methods have been developed which make the structure of the chemical bond more understandable (see [16]) and which makes it possible to investigate the structure of molecules by investigating their spectra [17].

CHAPTER XVI

Scattering Theory

In XI we have given several examples of operators W which correspond to ensembles, and F which correspond to effects that occur in the investigation of the spectrum of atoms and molecules. Since we do not yet have a more comprehensive theory which permits us to obtain the maps $\mathscr{Q}' \xrightarrow{\varphi} K$ and $\mathscr{F} \xrightarrow{\psi} L$ from information about the construction of the preparation and registration apparatuses, in XI we have "guessed" the form of the operators $W \in K$ and $F \in L$ on the basis of partly theoretical and partly heuristic considerations, in order to formally introduce them as axioms in $\mathscr{M}\mathscr{T}$. This procedure shall now be extended to the more comprehensive domain of experimental results: We shall now consider the large domain of scattering experiments.

In §1 we shall formulate the general properties of the ensembles W used in these experiments. In §2 we will do the same for the effects F. Later, in §6 we shall exhibit special operators W and F in order to define the so-called effective cross section.

1 General Properties of Ensembles Used in Scattering Experiments

In a scattering experiment different microsystems are shot at each other. Here we need to recall the considerations in VII concerning composite systems and interactions. We shall not, however, resort to the route taken there, but, proceed in a different fashion which assumes the description of composite systems set forth in VIII and draw conclusions for the case of scattering experiments. In order not to complicate matters we shall not begin with the most general case, but restrict our initial consideration to composite systems consisting of two elementary systems. According to VIII (2.12) the Hamiltonian operator of the system can be written in the form

$$H = H_0 + H_J, \tag{1.1}$$

where

$$H_0 = \frac{1}{2m_1}\vec{P}_1^2 + \frac{1}{2m_2}\vec{P}_2^2 \tag{1.2}$$

is the kinetic energies of the two elementary systems and H_J is the "interaction". We shall at first make no assumptions concerning the form of H_J; later we shall consider the special case in which H_J describes the potential energy.

Furthermore we shall assume that the two systems (1) and (2) are different in order that we may use the structure described in VIII, §2. In §4.8 we shall briefly consider scattering processes for "identical" systems.

In the following we shall prove theorems which are so general that, in some cases, they can be applied to cases in which the assumptions made above are not satisfied. For an understanding of these theorems it is desirable to assume that the assumptions made above are satisfied.

According to VIII and IX a composite system consisting of two elementary systems (1) and (2) may be most simply described by means of either the Schrödinger picture or the interaction picture, since in both the Hilbert space $\mathcal{H}$ can be described as a product space $\mathcal{H} = \mathcal{H}_1 \times \mathcal{H}_2$ independent of the time. Here the effects associated with systems (1) and (2) have the form $F_1 \times \mathbf{1}$ and $\mathbf{1} \times F_2$ and are time independent in the Schrödinger picture. In the Schrödinger picture the operators for the ensembles are time dependent and have the form (X (2.2))

$$W_t = U_t^+ W_0 U_t, \tag{1.3a}$$

where

$$U_t = e^{iHt}. \tag{1.3b}$$

In the interaction picture (X (3.4), (3.5)) we find that

$$\mathbf{W}_t = V_t^+ W_0 V_t, \tag{1.4a}$$

where

$$V_t = U_t e^{-iH_0 t} = e^{iHt} e^{-iH_0 t}. \tag{1.4b}$$

The Schrödinger picture is particularly suited for the determination of the form of the operator W for an ensemble from the experimental configuration of the preparation apparatus. This "intuitiveness" arises from the fact that the mode of operation of the position operators of both particles is time independent and is easy to specify in a position representation, and from the

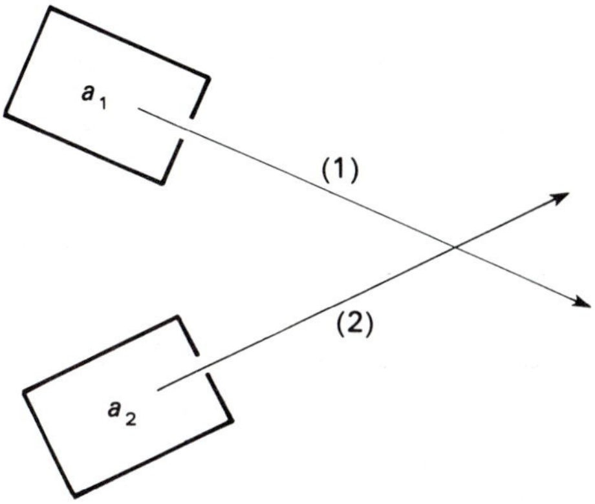

Figure 37

fact that, experimentally, we know that $H_J \sim 0$ when the distance r_{12} of the two particles is large compared to the dimensions of an atom.

A scattering experiment is carried out using a preparation procedure a which consists of two parts (see Figure 37) in which the first part produces system (1) and the other produces system (2) and in which both systems are "shot" at each other. We shall now consider the problem of finding a mathematical structure to describe the physical situation described above and illustrated in Figure 37.

We must therefore introduce a new structure into our mathematical theory. To this purpose we shall introduce a map γ of a subset $\pi \subset \mathcal{Q}' \times \mathcal{Q}'$ into $\mathcal{Q}'$ together with the following physical meaning (that is, with the following mapping principle): Let a_1 and a_2 be two preparation procedures for which the apparatuses can be jointly arranged (that is, $(a_1, a_2) \in \pi$) in such a way as to construct a new preparation procedure $a = \gamma(a_1, a_2)$.

We wish to warn the reader (without thinking carefully about the physical situation) to introduce the axiom that γ maps equivalent preparation procedures (in the sense of III, D (1.1) into equivalent preparation procedures. We may easily have reservations about such an axiom. Here we shall not, however, consider a general setting for such a combination of two preparation procedures. Instead, we shall only consider a subset π_c of π where π_c contains (intuitively speaking) only those pairs (a_1, a_2) of preparation procedures which can be combined in scattering experiments, that is, where the prepared microsystems will interact (see Figure 37) and therefore will not include pairs (see Figure 38) for which the microsystems are unable to interact.

Our next problem is that of discovering axioms from physical intuition about the connection between $\varphi\gamma(a_1, a_2)$ and $\varphi(a_1)$, $\varphi(a_2)$ for the $(a_1, a_2) \in \pi_c$.

We will now present some of these considerations: To this purpose we shall consider the special case in which the preparation procedure a_1 prepares systems of type (1), a_2 prepares systems of type (2), that is, $\varphi(a_1)$ and $\varphi(a_2)$ are nonzero only for components of type (1) and (2), respectively. For this situation we write (not quite rigorously) $\varphi(a_1) = W_1$ and $\varphi(a_2) = W_2$

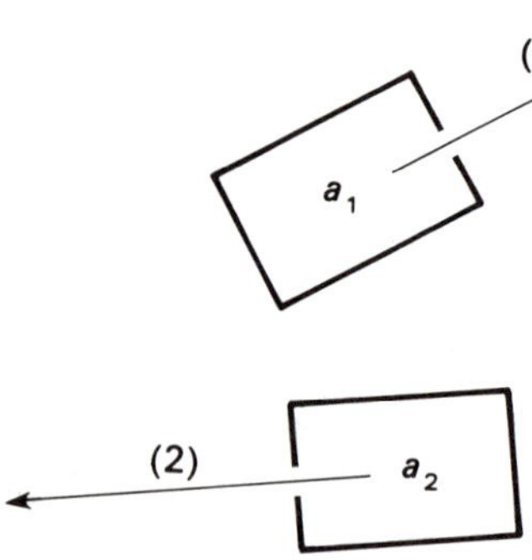

Figure 38

where W_1 and W_2 are operators in the Hilbert space $\mathscr{H}_1$ for (1) and $\mathscr{H}_2$ for (2), respectively.

According to VIII, §2 the effects of systems (1) in the Hilbert space $\mathscr{H} = \mathscr{H}_1 \times \mathscr{H}_2$ of the composite systems will be described by $F_1 \times \mathbf{1}$; similarly, for systems (2) will be described by $\mathbf{1} \times F_2$ (and, in the Schrödinger or interaction picture this form is time independent (see X, §§2 and 3)).

If F_1 and F_2 are decision effects E_1 and E_2, respectively (it suffices when at least one is a decision effect), then the effect that both E_1 and E_2 respond together is uniquely determined as the product $(E_1 \times \mathbf{1})(\mathbf{1} \times E_2) = E_1 \times E_2$ (see IV, §1.3).

If the registration apparatus which registers the effects $E_1 \times \mathbf{1}, \mathbf{1} \times E_2$ and $E_1 \times E_2$ is placed with respect to the preparation apparatus $\gamma(a_1, a_2)$ that the interaction of the systems (1) and (2) with the registration apparatus is such that, for all practical purposes, the interaction between (1) and (2) has not yet occurred, then from experience we may conclude that, in physical approximation, for $W = \varphi(a)$ where $a = \gamma(a_1, a_2)$

$$\operatorname{tr}(W(E_1 \times E_2)) = \operatorname{tr}(W(E_1 \times \mathbf{1})) \operatorname{tr}(W(\mathbf{1} \times E_2). \tag{1.5}$$

Here we say that the response of the two effects $(E_1 \times \mathbf{1})$ and $(\mathbf{1} \times E_2)$ are "statistically independent" of each other, that is, the conditional probability for the response of $(\mathbf{1} \times E_2)$ provided that $(E_1 \times \mathbf{1})$ has responded is the same as the probability for the response of $(\mathbf{1} \times E_2)$ without any condition:

$$\lambda(a \cap b_1, a \cap b_1 \cap b_2) = \lambda(a \cap b_0, a \cap b_2), \tag{1.6}$$

where $\psi(b_0, b_1) = E_1 \times \mathbf{1}$, $\psi(b_0, b_2) = \mathbf{1} \times E_2$. From the coexistence of $\psi(b_0, b_1)$, $\psi(b_0, b_2)$ and $\psi(b_0, b_1 \cap b_2)$ (see IV, Th. 1.3.4) it follows that the relation

$$\psi(b_0, b_1 \cap b_2) = \psi(b_0, b_1)\psi(b_0, b_2) = E_1 \times E_2$$

holds. Thus, from (AS 2.3 in II, §3) it is easy to see that

$$\lambda(a \cap b_0, a \cap b_1 \cap b_2) = \lambda(a \cap b_0, a \cap b_1)\lambda(a \cap b_1, a \cap b_1 \cap b_2)$$

from which we conclude that (1.5) and (1.6) are equivalent.

We now define the following reduction operators R_1 and R_2 which will be of considerable importance later. $F_1 \rightarrow F_1 \times \mathbf{1}$ defines a $\mathscr{B}$-continuous effect morphism $L(\mathscr{H}_1) \overset{T_1}{\rightarrow} L(\mathscr{H}_1 \times \mathscr{H}_2)$ (see V, D4.2.1 and VIII (2.2)). Therefore there exists a mi-morphism $K(\mathscr{H}_1 \times \mathscr{H}_2) \overset{R_1}{\rightarrow} K(\mathscr{H}_1)$ where $T_1 = R_1'$ (for the notion of a mi-morphism see V, D 4.1.1). R_1 is called the reduction operator of the composite system onto subsystem (1). By analogy with R_1 R_2 is defined as the reduction operator onto subsystem (2).

If we require that (1.5) holds for all E_1, E_2, then for $W_1 = R_1 W$, $W_2 = R_2 W$, $W' = W_1 \times W_2$ it follows that

$$\operatorname{tr}((W - W')(E_1 \times E_2)) = 0.$$

For the special case in which $E_1 = P_\varphi$, $E_2 = P_\psi$, $\phi = \varphi\psi$ it follows that $P_\varphi \times P_\psi = P_\phi$ and $\langle \phi, (W - W')\phi \rangle = 0$. If we replace φ by $\varphi_1 + \varphi_2$ and $\varphi_1 + i\varphi_2$, and ψ by $\psi_1 + \psi_2$ and $\psi_1 + i\psi_2$ it follows that $\langle \phi', (W - W')\phi \rangle = 0$ for all $\phi' = \varphi'\psi'$ and $\phi = \varphi\psi$. Since vectors of the form $\varphi\psi$ span the entire Hilbert space, we obtain $W - W' = 0$, that is:

$$W = W_1 \times W_2. \tag{1.7}$$

In the Schrödinger picture, near the time of preparation the ensemble is given by (1.7) where $W = \varphi(a) = \varphi\gamma(a_1, a_2)$. Thus we are now able to formulate our first postulate:

Postulate 1. There exists a time T_i for which (at least in physical approximation)

$$W_{T_i} = W_{1T_i} \times W_{2T_i}$$

holds (in the Schrödinger picture).

We shall later return to the question: Which is the topology for which W_{T_i} is supposed to be "approximately equal" to $W_{1T_i} \times W_{2T_i}$ in the case in which Postulate 1 is not applicable at a fixed time T_i but (as is often assumed for reasons of producing a simpler mathematical formulation) in the limit $T_i \rightarrow -\infty$?

Postulate 1 expresses the experimental fact that systems (1) and (2) are generated by means of two "separate" preparation procedures a_1, a_2; that is, the preparation procedure $a = \gamma(a_1, a_2)$ for a scattering experiment is composed of two separate preparation procedures a_1, a_2. It is now necessary to formulate, in mathematical terms, the idea that both systems are "shot at each other", as intuitively illustrated by Figure 37. First we wish to exclude the case which is intuitively illustrated by Figure 38. First we require (both for the case of Figure 37 and Figure 38) that the preparation procedures are so chosen that at the designated time T_i the prepared systems (1) and (2) are so widely separated that the interaction H_J is, for all practical purposes, zero, that is,

$$\operatorname{tr}(W_{T_i} H_J^2) \sim 0. \tag{1.8}$$

We could require that equality holds in (1.8), but, for mathematical reasons, such an assignment is awkward. It is preferable to require that the interaction

H_J is such that for large separations r_{12} of the two systems H_J is, for all practical purposes, approximately (but not equal to) zero.

We could at this point introduce a second postulate using (1.8), but we shall first take an additional step because equation (1.8) does not express the condition that the systems are "colliding" (as in Figure 37) rather than "separating from each other" (as in Figure 38).

In the situation described in Figure 37 the systems may (they need not necessarily) interact at times $t > T_i$, that is, (1.8) need not hold for $t > T_i$ or

$$\operatorname{tr}(W_t H_J^2) \sim 0 \quad \text{for } t > T_i \tag{1.9}$$

need not be satisfied. In the situation of Figure 38 (1.9) remains valid for times $t > T_i$. This formulation of the distinction between these two cases is unsatisfactory because in both cases (1.9) is satisfied for t sufficiently large (that is, for $t \to \infty$) (see §2). Therefore it is preferable to consider times $t < T_i$. Then the situations described by Figures 37 and 38 are reversed if we consider the time evolution in the direction $t \to -\infty$. For $t < T_i$ (this is possible only in the case of gedanken experiments—see the considerations of time translation in VII, §§1 and 2 and the transformation into the Schrödinger picture given in X, §2) the situation of the "colliding systems" of Figure 37 results in the condition that

$$\operatorname{tr}(W_t H_J^2) \sim 0 \quad \text{for } t < T_i \tag{1.10}$$

since (intuitively) the systems before time T_i are more separated than at time T_i. Now we come to the formulation of the second postulate:

Postulate 2. There exists a time T_i such that (at least in physical approximation) the following condition is satisfied:

$$\operatorname{tr}(W H_J^2) \sim 0 \quad \text{for } t < T_i. \tag{1.11}$$

For the notion of "physical approximation" see the remarks at the end of the formulation of Postulate 1.

Since the mathematical requirement of equality in Postulates 1 and 2 is impossible (that is, leads to contradictions) we shall now reformulate both postulates, imposing postulates only in the limit as $t \to -\infty$. Then, for experiments, practical questions remain to establish finite values of time t in individual cases (in the laboratory scale!) for which the postulates are satisfied in "physical approximation".

For the formulation of the physical problem presented here it is meaningless to investigate the limit of W_t for $t \to -\infty$. We can show that in the case in which the spectrum of H is continuous that $\operatorname{tr}(W_t A) \to 0$ for $t \to \pm\infty$ (such limits are considered in ergodic theory). Physically we are less interested in the limit of W_t as $t \to -\infty$ than we are in the question of whether it is possible to satisfy (1.11). Certainly we can investigate the limit for $\operatorname{tr}(W_t H_J^2)$ for $t \to -\infty$, but it is somewhat impractical. Instead of considering the operator (not necessarily bounded) H_J we shall consider the time evolution of W_t in the Schrödinger picture:

$$W_t = e^{-iHt} W_0 e^{iHt}. \tag{1.12}$$

Since (1.11) holds for $t < T_i$ we should expect that in the following equation (which follows from (1.12)) for $t < T_i$

$$W_t = e^{-iH(t-T_i)} W_{T_i} e^{iH(t-T_i)}$$

we may approximately replace H by H_0 in the experiment, that is, we obtain

$$\begin{aligned} e^{iH_0t} W_t e^{-iH_0t} &= e^{iH_0t} e^{-iH(t-T_i)} W_{T_i} e^{iH(t-T_i)} e^{-iH_0t} \\ &\sim e^{iH_0T_i} W_{T_i} e^{-iH_0T_i} \end{aligned}$$

and therefore find that for $t < T_i$ the operator

$$e^{iH_0t} W_t e^{-iH_0t} \tag{1.13}$$

is, for all practical purposes, time independent. Equation (1.13) is nothing other than the operator $\mathbf{W}_t$ from (1.4a) in the interaction picture.

It is therefore reasonable to mathematically express Postulate 2 by using (1.4b) by means of the following limit:

Coll. 2. For $t \to -\infty$ for $W = \varphi\gamma(a_1, a_2)$ the following limit exists

$$\mathbf{W}_t = V_t^+ W_0 V_t = e^{iH_0t} e^{-iHt} W_0 e^{iHt} e^{-iH_0t} \to W_0^i \tag{1.14}$$

in the $\sigma(\mathscr{B}, \mathscr{D})$ topology where $W_0^i \in K$.

With the help of Coll. 2, Postulate 1 may be reformulated in a more satisfying way as follows: Since for $t = T_i$ we obtain

$$\mathbf{W}_{T_i} \sim W_0^i, \quad \text{that is, } \mathbf{W}_{T_i} \sim e^{iH_0T_i} W_{T_i} e^{-iH_0T_i},$$

where W_{T_i} satisfies Postulate 1 and since e^{iH_0t} leaves the product representation of an operator invariant, we may replace Postulate 1 as follows:

Coll. 1 The operator W_0^i defined according (1.14) has the form

$$W_0^i = W_{10}^i \times W_{20}^i \tag{1.15}$$

W_{10}^i and W_{20}^i are identical to the operators $\varphi(a_1) \in K(\mathscr{H}_1)$ and $\varphi(a_2) \in K(\mathscr{H}_2)$ of the Heisenberg picture of systems (1) and (2), as $Q' \xrightarrow{\varphi} K$ was defined initially! (See III and X.)

Coll. 1 and 2 do not, of course, guarantee that there exist any $\mathbf{W}_t$ which satisfy (1.14) and (1.15). Certainly this will depend strongly on H_J. A portion of the discussion in this chapter is dedicated to this problem.

We therefore consider Coll. 1 and 2 to be valid axioms for preparation procedures describing certain "scattering experiments", as we have described above.

Coll. 1 is nothing other than the introduction of an assertion about the ensembles of prepared pairs with respect to effects of systems (1) and (2), respectively, that is, with respect to the product representation $\mathscr{H} = \mathscr{H}_1 \times \mathscr{H}_2$ of the Hilbert space which, according to VIII, §2, refers to the maps $T_1^{(i)}$, $T_2^{(i)}$. In the formulation of Coll. 1 we have assumed that the maps $T_1^{(i)}$, $T_2^{(i)}$ which have the meaning given in VIII, §2 exist "before" the scattering.

Often we find that the limit (1.14) exists not only in the $\sigma(\mathscr{B}, \mathscr{D})$ topology but also in the norm topology of $\mathscr{B}$. If $\mathbf{W}_t$ converges in the norm topology then it follows that $W_0^i \in K$. Whether (1.14) converges in the norm topology depends upon whether H_J decreases sufficiently rapidly with the separation r_{12} of the two systems. For the case of Coulomb interaction which behaves like $1/r_{12}$ norm convergence does not occur. In this book we shall not consider this case further; the reader is referred to [18].

The norm in $\mathscr{B}$ is the trace norm, which we denote by $\| \cdots \|_s$. We therefore obtain

$$\|\mathbf{W}_t - W_0^i\|_s \to 0 \quad \text{for } t \to -\infty \tag{1.16}$$

from which it also follows that

$$\|W_t - W_t^i\|_s \to 0 \quad \text{for } t \to -\infty, \tag{1.17}$$

where

$$W_t^i = e^{-iH_0t} W_0^i e^{iH_0t} \tag{1.18}$$

because for unitary operators U the trace norm is invariant

$$\|A\|_s = \|UAU^+\|_s.$$

Using Coll. 1 and using the form (1.2) of H_0, that is, for

$$\begin{aligned} H_0 &= H_{01} \times \mathbf{1} + \mathbf{1} \times H_{02} \\ H_{01} &= \frac{1}{2m_1} \vec{P}_1^2, \qquad H_{02} = \frac{1}{2m_2} \vec{P}_2^2 \end{aligned} \tag{1.19}$$

it follows that

$$W_t^i = W_{1t}^i \times W_{2t}^i, \tag{1.20}$$

where

$$W_{1t}^i = e^{-iH_{01}t} W_{10}^i e^{iH_{01}t}, \qquad W_{2t}^i = e^{-iH_{02}t} W_{20}^i e^{iH_{02}t}.$$

2 General Properties of Effects Used in Scattering Experiments

Behind the scattering zone, in a region in which after the scattering systems (1) and (2) are no longer in interaction the systems encounter detectors (counters, etc.) which register the outgoing systems (1) and/or (2). Therefore effects (such as those defined in III, D 1.2) will be registered. What is the special structure of these effects which are registered *after* the scattering?

Here we do not need to introduce any new structure in the theory because in VIII, §2 we have already introduced the appropriate structure which permits us to characterize the effects produced separately by systems (1) and (2). Here we must only answer the question of what we mean by the expression "after the scattering" and provide a mathematical description using previously introduced structures. That is, we wish to develop a mathematical formulation for the experimental situation (using the theory presented earlier in this book) in which the registration apparatus is so constructed and positioned that systems (1) and (2) will interact with the registration apparatus only after no further interaction between (1) and (2) will occur. This means that we may compute the probabilities for the activation of the effect responses for the ensemble W_t (in the Schrödinger picture) for times t in which the interaction between (1) and (2) is no longer active.

The conditions required on W_t in order to compute the effect responses "after scattering" are therefore completely similar to that imposed on W_t before the scattering. While Coll. 2 requires a specific behavior for $t \to -\infty$, we must therefore assume (!) that for $t \to \infty$ there exists a limit as in (1.14) for which

$$\mathbf{W}_t = V_t^+ W_0 V_t \to W_0^f \tag{2.1}$$

provided that W_t also satisfies the conditions Coll. 1 and Coll. 2. If such an assumption is not made, it is not possible to easily use a scattering theory such as presented here. If (2.1) is not satisfied then the simplifications of scattering theory presented here are somewhat unrealistic. In fact it is necessary to make modifications in the case in which we use composite systems (such as atoms) instead of elementary systems.

If scattering theory is applicable, then for a preparation according to a given preparation procedure $a = \gamma(a_1, a_2)$ the corresponding W_t^i defined in (1.8) will be approximately equal to the correct W_t (of (1.12)) for $-\infty < t < T_i$. The time T_i will, of course, depend on the particular preparation procedure a, that is, upon W_t^i. From time T_i to time T_f the quantity W_t may be significantly different from W_t^i and from

$$W_t^f = e^{-iH_0t} W_0^f e^{iH_0t}. \tag{2.2}$$

For $t > T_f$, W_t will be approximately equal to W_t^f. Clearly T_f will of course also depend upon the prepared W_t^i.

Using the Schrödinger picture we have sought to derive, in an intuitive way, the postulates Coll. 1 and Coll. 2 and the assumption (2.1). Here the time parameter t plays an important role. In fact the use of the limits $t \to \pm\infty$ is only an intuitive tool. The actual times for the preparations and registrations clearly are not represented by $t \to -\infty$ and $t \to \infty$, respectively. Since the difference $T_f - T_i$ are, in practice, very small (a fraction of a second) it is understandable that the experimental physicists are not concerned whether the preparation occurred before T_i and the registration occurred after T_f. They "believe" that the spatial arrangement of the preparation and registration apparatus guarantees the condition that W_0^i is determined by $a = \gamma(a_1, a_2)$ and that the ensemble W_0^f may be used to compute the frequency distribution for the registered effects. Mathematically this problem is a problem about the relationship between the asymptotic behavior as $t \to \pm\infty$ and the asymptotic behavior for large spatial separations. This question will be discussed in §4.7.

In summary we may therefore say that the simplification of scattering theory consists first in the fact that we may easily obtain, on the basis of the construction of the preparation apparatus in the form $a = \gamma(a_1, a_2)$ and the arrangement of the two parts a_1, a_2, a mathematical expression for W_0^i if $\varphi(a_1)$ and $\varphi(a_2)$ are known and second, that for the computation of the probability for the registered effects we need only use W_0^f. Thus we may use either an "undisturbed Schrödinger picture" or an "undisturbed Heisenberg picture" for the investigation of the registered effects in the following way: For the composite system for an effect procedure (b_0, b) there corresponds an operator $F = \psi(b_0, b)$, and the probability is to be computed for an exact $W = W_0$ (see X, §2) using $\mathrm{tr}(W_0 F)$. If F was registered after the scattering (which is experimentally guaranteed by the spatial arrangement of the registration apparatus corresponding to b_0 relative to the preparation apparatus corresponding to a) then $\mathrm{tr}(W_0 F) \approx \mathrm{tr}(W_0^f F)$ will be a very good approximation. If we translate the effect procedure by a time τ then we will obtain the approximations

$$\mathrm{tr}(W_0 e^{iH\tau} F e^{-iH\tau}) \approx \mathrm{tr}(W_0^f e^{iH_0\tau} F e^{-iH_0\tau}) \tag{2.3}$$

and

$$\mathrm{tr}(W_0 e^{iH\tau} F e^{-iH\tau}) = \mathrm{tr}(W_\tau F) \approx \mathrm{tr}(W^f F), \tag{2.4}$$

where

$$W_\tau^f = e^{-iH_0\tau} W_0^f e^{iH_0\tau}.$$

Equations (2.3) and (2.4) hold for all $\tau > 0$; for $\tau < 0$ they will, of course, be valid only if the effects $e^{iH_0\tau} F e^{-iH_0\tau}$ displaced by τ satisfy the assumption that the registration occurs only after the scattering, or, in experimental situations, what amounts to the same thing, that $T_\tau b_0$ *may be combined* with a in the sense of III, §1 because the spatial arrangement (far from the interaction area of the two systems) of b_0 to a guarantees that either $T_\tau b_0$ is always registering after the scattering or cannot be combined with a (see III, §1 and [6], XVI and [7], III, §5). For an experimental physicist there is no such problem because intuitively, no registration procedure will be used until the preparation apparatus is "turned on".

3 Separation of Center of Mass Motion

Since the interaction H_J is translation invariant, the motion of the center of mass can be separated (see VIII, §2). Here we will provide a list of the formulas which are necessary for scattering theory.

In the Hamiltonian operator (VIII (2.12))

$$H = \frac{1}{2m_1} \vec{P}_1^2 + \frac{1}{2m_2} \vec{P}_2^2 + H_J \tag{3.1}$$

we shall, according to VIII (2.9) and VIII (2.18), replace the operators $\vec{Q}_1$ and $\vec{Q}_2$ by new operators as follows

$$\vec{Q} = \frac{m_1\vec{Q}_1 + m_2\vec{Q}_2}{m_1 + m_2} \tag{3.2}$$

and

$$\vec{Q}_r = \vec{Q}_1 - \vec{Q}_2; \tag{3.3}$$

where $\vec{Q}$ is the position of the center of mass and $\vec{Q}_r$ is the "relative position". We shall now obtain the canonically conjugate momentum corresponding to $\vec{Q}$ and $\vec{Q}_r$. According to VIII (2.8) and VIII (2.13) the latter are given by

$$\begin{aligned} \vec{P} &= \vec{P}_1 + \vec{P}_2, \\ \vec{P}_r &= \frac{m_2}{M}\vec{P}_1 - \frac{m_1}{M}\vec{P}_2, \end{aligned} \tag{3.4}$$

where $M = m_1 + m_2$. Defining the reduced mass m by the equation

$$\frac{1}{m} = \frac{1}{m_1} + \frac{1}{m_2} = \frac{M}{m_1 m_2} \tag{3.5}$$

it follows that

$$\vec{P}_r = m\left(\frac{1}{m_1}\vec{P}_1 - \frac{1}{m_2}\vec{P}_2\right). \tag{3.6}$$

Defining the "velocity" operators by $(1/m_1)\vec{P}_1 = \vec{V}_1$ and $(1/m_2)\vec{P}_2 = \vec{V}_2$ we find that $(1/m)\vec{P}_r = \vec{V}_1 - \vec{V}_2$ has the form of the velocity vector $\vec{V}$ of an "elementary system" with reduced mass m corresponding to $\vec{Q}_r$. The reduced mass may also be obtained if we rewrite $(1/2m_1)\vec{P}_1^2 + (1/2m_2)\vec{P}_2^2$ using $\vec{P}$, $\vec{P}_r$ (see VIII (2.14)); we find that

$$H = \frac{1}{2M}\vec{P}^2 + \frac{1}{2m}\vec{P}_r^2 + H_J. \tag{3.7}$$

According to VIII (1.2) and VIII (2.17) we may write with $V = H_J$

$$H = H_b + H_i = H_0 + H_J, \tag{3.8}$$

where

$$H_i = \mathring{H}_i + V, \qquad \mathring{H}_i = \frac{1}{2m}\vec{P}_r^2 \tag{3.9}$$

and

$$H_0 = H_b + \mathring{H}_i, \qquad H_b = \frac{1}{2M}\vec{P}^2. \tag{3.10}$$

In the sense of the product representation VIII (1.1) we therefore obtain

$$U_t = e^{iHt} = e^{iH_bt} \times e^{iH_it} = U_t^{(b)} \times U_t^{(i)} \tag{3.11}$$

and

$$\mathring{U}_t = e^{iH_0t} = e^{iH_bt} \times e^{i\mathring{H}_it} = U_t^{(b)} \times \mathring{U}_t^{(i)}, \tag{3.12}$$

where

$$U_t^{(b)} = e^{iH_bt}, \tag{3.13}$$

$$U_t^{(i)} = e^{iH_it}, \tag{3.14}$$

and

$$\mathring{U}_t^{(i)} = e^{i\mathring{H}_it}. \tag{3.15}$$

4 Wave Operators and the Scattering Operator

In this section we shall be concerned with the mathematical question of whether the limits considered in §§1 and 2 exist, and how the "jump" from W_0^i to W_0^f is to be calculated. We can only present a brief insight into a mathematically well-developed theory. We refer readers who are interested in the mathematical structure of the theory to [10] and [18].

4.1 Definition of the Wave Operators

Our first problem is to investigate the limits considered in (1.14) and (2.1) as $t \to \pm\infty$. We shall consider both cases $t \to -\infty$ and $t \to \infty$ at the same time, and we shall distinguish between them by using the indices $-$, $+$ (and eventually i, f). Multiple indices appearing in the same equation mean that the formula is valid for both indices $(-, i)$ and $(+, f)$. We shall here investigate the stronger condition of convergence in the trace norm topology:

$$\|U_t^+ W_0 U_t - \mathring{U}_t^+ W_0^{f,i} \mathring{U}_t\|_s \to 0 \quad \text{for } t \to \pm\infty. \tag{4.1.1}$$

Since the trace norm is invariant under unitary transformations, we may rewrite (4.1.1) as follows:

$$\|W_0 - V_t W_0^{f,i} V_t^+\|_s \to 0 \tag{4.1.2}$$

or

$$\|V_t^+ W_0 V_t - W_0^{f,i}\|_s \to 0$$

for $t \to \pm\infty$ where $V_t = U\mathring{U}_t^+$.

The behavior of the operator

$$V_t = U_t \mathring{U}_t^+ = e^{iHt}e^{-iH_0t} \tag{4.1.3}$$

is of crucial importance for the validity of relationships of the form (4.1.2). This operator V_t is, as we have already mentioned in §1, the operator which describes the time dependence of the ensemble in the interaction picture.

From (4.1.3) and (3.11), (3.12) it follows that:

$$
\begin{aligned}
V_t &= (e^{iH_b t} \times e^{iH_i t})(e^{-iH_b t} \times e^{-i\mathring{H}_i T}),\\
V_t &= \mathbf{1} \times e^{iH_i t} e^{-i\mathring{H}_i t}.
\end{aligned}
\tag{4.1.4}
$$

Equation (4.1.4) means that in the computation of V_t we need only take into account relative motion in the center of mass system for the Hamiltonian operator $H_i = \mathring{H}_i + H_J$. In $\mathscr{H}_b$ (from the representation of $\mathscr{H} = \mathscr{H}_b \times \mathscr{H}_i$)$V_t$ acts like the unit operator. With respect to the product representation $\mathscr{H} = \mathscr{H}_b \times \mathscr{H}_i$ according to (4.1.4) we obtain

$$
V_t = \mathbf{1} \times \Omega(t) \quad \text{with} \quad \Omega(t) = e^{iH_i t} e^{-i\mathring{H}_i t}. \tag{4.1.5}
$$

This condition permits us to restrict our discussion and proofs to consideration of $\Omega(t)$.

Conditions (4.1.2) are satisfied if, in the sense of the strong topology (that is, $A_n \to A$ for $A_n\varphi \to A\varphi$ for all $\varphi \in \mathscr{H}$) the following equation holds

$$
\Omega(t) \to \Omega_\pm \quad \text{for } t \to \pm\infty \tag{4.1.6}
$$

a fact which will be proven in Th. 4.2.2,

In order not to excessively restrict our investigation we shall only require that $\Omega(t)$ converges only in a subspace of $\mathscr{H}_i$; then, since $\Omega(t)$ is unitary it easily follows that the set of $\varphi \in \mathscr{H}_i$ for which $\Omega(t)\varphi$ converges is a subspace of $\mathscr{H}_i$. Thus, we may write (somewhat more generally than as in (4.1.6):

$$
\Omega(t)P_\pm \to \Omega_\pm \quad \text{for } t \to \pm\infty, \tag{4.1.7}
$$

where $P_\pm$ is the projection operator onto the subspace where the limit of $\Omega(t)$ exists for $t \to \pm\infty$.

In order to make the formulas in the next sections more tractible we shall omit the indices i for $\mathscr{H}_i$, H_i and $\mathring{H}_i$. Then we may, according to (4.1.5), write

$$
\Omega(t) = e^{iHt} e^{-i\mathring{H}t}. \tag{4.1.8}
$$

In order to make the dependence on $\mathring{H}$, H more noticeable it is desirable to rewrite (4.1.7) using (4.1.8) as follows:

$$
\lim_{t \to \pm\infty} \Omega(t)P_\pm = \Omega_\pm(H, \mathring{H}). \tag{4.1.9}
$$

$\Omega_\pm$ are called the "wave operators" or "Møller operators".

The projection onto the subspace upon which $\Omega^+(t)$ is convergent will be denoted by $Q_\pm$. In the notation of (4.1.9) we therefore obtain

$$
\begin{aligned}
\Omega(t)P_\pm &\to \Omega_\pm(H, \mathring{H}) \quad \text{for } t \to \pm\infty,\\
\Omega^+(t)Q_\pm &\to \Omega_\pm(\mathring{H}, H) \quad \text{for } t \to \pm\infty.
\end{aligned}
\tag{4.1.10}
$$

In physics instead of $\mathbf{1} \times \Omega$ (as an operator in $\mathscr{H}_b \times \mathscr{H}_i$) it is common to write Ω; similarly, for $\mathbf{1} \times P_\pm$ and $\mathbf{1} \times Q_\pm$ it is common to write $P_\pm$ and $Q_\pm$. In this sense then (4.1.10) holds also if $\Omega(t)$ is replaced by V_t.

4.2 Some General Properties of Wave Operators

In order that we need not deal with unnecessarily many indices, we shall abbreviate (4.1.9) as follows:

$$\Omega(t)P \to \Omega \tag{4.2.1}$$

from which it follows that $\Omega P = \Omega$. Since $\Omega(t)$ is unitary, for $\varphi \in P\mathscr{H}$ it follows that

$$\|\varphi\| = \|\Omega(t)\varphi\| \to \|\Omega\varphi\|, \quad \text{that is,} \quad \|\Omega\varphi\| = \|\varphi\|.$$

Ω is therefore an isometric mapping of $P\mathscr{H}$ onto a subspace of $\mathscr{H}$; we shall let E denote the projection onto this subspace. Thus from (AIV, §7) it follows that

$$\Omega^+\Omega = P \quad \text{and} \quad \Omega\Omega^+ = E. \tag{4.2.2}$$

From

$$\|\Omega(t)\varphi - \tilde{\varphi}\| = \|\varphi - \Omega^+(t)\tilde{\varphi}\|$$

it follows that $\Omega^+(t)$ converges in $E\mathscr{H}$. In addition, for $\varphi \in P\mathscr{H}$ (since $\tilde{\varphi} = \Omega\varphi$, that is, $\varphi = \Omega^+\tilde{\varphi}$) we find that $\Omega^+(t)E$ converges to Ω^+E.

For Q defined by Equation (4.1.10) we find that $\Omega^+(t)Q$ converges, that is, $\Omega^+(t)$ converges in all of $Q\mathscr{H}$. Therefore $Q \geq E$. If $Q \neq E$ then $\Omega(t)$ would converge in a larger space than $P\mathscr{H} = \Omega^+E\mathscr{H}$. Therefore $Q = E$, and we obtain

$$\Omega^+(t)Q \to \Omega^+ \quad \text{and} \quad \Omega\Omega^+ = Q. \tag{4.2.3}$$

Thus we have proven the following theorem:

Th. 4.2.1. *Let $\Omega(t)$ be a unitary operator, and let P be the projection operator onto the convergence space of $\Omega(t)$. Then the limit $\Omega = \lim[\Omega(t)P]$ is an isometric mapping of $P\mathscr{H}$ onto a subspace $Q\mathscr{H}$. In addition $\Omega P = \Omega$, $\Omega^+\Omega = P, \Omega\Omega^+ = Q, \Omega^+Q = \Omega^+$. $Q\mathscr{H}$ is the convergence space of $\Omega(t)$ and $\Omega^+(t)Q \to \Omega^+$.*

Th. 4.2.2. *Let $W \geq 0$, $\operatorname{tr}(W) = 1$ and $PWP = W$. Then in the trace-norm topology $\Omega(t)W\Omega^+(t) \to \tilde{W}$ where $\tilde{W} > 0$ and $\operatorname{tr}(\tilde{W}) = 1$, and $\tilde{W}$ is given by $\tilde{W} = Q\tilde{W}Q = \Omega W\Omega^+$. For each $\tilde{W}$ satisfying $Q\tilde{W}Q$ the following condition is satisfied: $\Omega^+(t)\tilde{W}\Omega(t) \to \Omega^+\tilde{W}\Omega$.*

PROOF. We first consider the special case in which $W = P_\varphi$. Since $PP_\varphi P = P_\varphi$ we obtain $P\varphi = \varphi$ and we therefore obtain $\Omega(t)\varphi \to \tilde{\varphi} = \Omega\varphi$. Since Ω is isometric, we obtain $\|\tilde{\varphi}\| = 1$. For $\varphi(t) = \Omega(t)\varphi$ we obtain $\Omega(t)P_\varphi\Omega^+(t) = P_{\varphi(t)}$. We shall now compute $\|P_{\varphi(t)} - P_\varphi\|_s$.

The eigenvalues for the operator $P_{\varphi(t)} - P_\varphi$ may be obtained from the following equation (here we write φ as an abbreviation for $\varphi(t)$):

$$P_\varphi f - P_{\tilde{\varphi}} f = \varphi\langle\varphi, f\rangle - \tilde{\varphi}\langle\tilde{\varphi}, f\rangle = \lambda f.$$

If we set $\varphi = \tilde{\varphi}\langle\tilde{\varphi}, \varphi\rangle + r$ (therefore $1 = |\langle\tilde{\varphi}, \varphi\rangle|^2 + \|r\|^2$) then, for the pair of nonzero eigenvalues $\lambda_{1,2} = \pm\|r\|$ and we therefore obtain $\|P_{\varphi(t)} - P_{\tilde{\varphi}}\|_s = 2\|r\|$.

Since $\|r\|$ is the shortest distance from $\varphi(t)$ to the line $\lambda\tilde{\varphi}$, we obtain

$$\|\varphi(t) - \tilde{\varphi}\| \geq \|r\|$$

and we therefore obtain

$$\|P_{\varphi(t)} - P_\varphi\|_s \leq 2\|\varphi(t) - \tilde{\varphi}\|.$$

Since $\|\varphi(t) - \tilde{\varphi}\| \to 0$ we therefore obtain

$$\|\Omega(t)P_\varphi\Omega^+(t) - P_{\tilde{\varphi}}\|_s \to 0.$$

From $\tilde{\varphi} = \Omega\varphi$ it follows that $P_{\tilde{\varphi}} = \Omega P_\varphi \Omega^+$.

For arbitrary W satisfying $W = PWP$ it follows that

$$W = \sum_\nu w_\nu P_{\varphi_\nu},$$

where $w_\nu \geq 0$, $\sum_\nu w_\nu = 1$ and $P\varphi_\nu = \varphi_\nu$. For $\Omega(t)\varphi_\nu = \varphi_\nu(t)$: $\varphi_\nu(t) \to \tilde{\varphi}_\nu = \Omega\varphi_\nu$. For $\overline{W} = \sum_\nu w_\nu P_{\tilde{\varphi}_\nu}$ we obtain $\tilde{W} \geq 0$, $\operatorname{tr}(\tilde{W}) = 1$, $\tilde{W} = \Omega W \Omega^+$ and

$$\begin{aligned}\|\Omega(t)W\Omega^+(t) - \tilde{W}\|_s &= \left\|\sum_\nu w_\nu(P_{\varphi_\nu(t)} - P_{\tilde{\varphi}_\nu})\right\|_s \\ &\leq \sum_\nu w_\nu\|P_{\varphi_\nu(t)} - P_{\varphi_\nu}\|_s \\ &\leq \sum_{\nu=1}^{N} w_\nu\|P_{\varphi_\nu(t)} - P_{\tilde{\varphi}_\nu}\|_s + 2\sum_{\nu=N+1}^{\infty} w_\nu.\end{aligned}$$

Since $\sum_\nu w_\nu = 1$, we may first choose N and then choose t sufficiently large in order to obtain

$$\|\Omega(t)W\Omega^+(t) - W\|_s < \varepsilon.$$

Th. 4.2.2 shows that in (4.1.2) $W_0^{f,i} = \Omega_\pm^+ W_0 \Omega_\pm$ and $W_0 = \Omega_\pm W_0^{f,i}\Omega_\pm^+$. Thus (4.1.2) is satisfied provided that $P_\pm W_0^{f,i} P_\pm = W_0^{f,i}$, which is equivalent to $Q_\pm W_0 Q_\pm = W_0$. According to III, §6, using the notation in III, D 3.1, the condition $PWP = W$ is equivalent to the condition $W \in K_0(\mathbf{1} - P) = K_1(P)$. Thus $W \to \Omega W \Omega^+$ is a bijective affine map of $K_1(P)$ onto $K_1(Q)$ and $W \to \Omega^+ W \Omega$ is the corresponding inverse map.

From $W_0 = \Omega_- W_0^i \Omega_-^+$ it follows that the following equation holds in the Schrödinger picture for all t:

$$W_t = e^{-iHt}\Omega_- W_0^i \Omega_-^+ e^{iHt}.$$

With the help of (1.15) we obtain

$$W_t = e^{-iHt}\Omega_-(W_{10}^i \times W_{20}^i)\Omega_-^+ e^{iHt}. \tag{4.2.4}$$

Therefore (4.2.4) is applicable providing that $P_-(W_{10}^i \times W_{20}^i)P_- = W_{10}^i \times W_{20}^i$. Later we will find that "normally" $P_- = \mathbf{1}$.

Therefore $\Omega_-(W_{10}^i \times W_{20}^i)\Omega_-^+$ is the correct operator in the Heisenberg picture, that is, for $\varphi(a_1) = W_{10}^i$, $\varphi(a_2) = W_{20}^i$ we obtain $\varphi\gamma(a_1, a_2) = \Omega_-(W_{10}^i \times W_{20}^i)\Omega_-^+$. The correct operator in the Schrödinger picture is given by (4.2.4). Both may therefore be used for the computation for the probabilities for arbitrary effects, regardless of the time at which the registration method is applied.

Th. 4.2.3. *If $\Omega(t) = e^{iHt}e^{-i\mathring{H}t}$ then $P\mathscr{H}$ is an invariant subspace of the group $e^{i\mathring{H}t}$ and $Q\mathscr{H}$ is an invariant subspace of the group e^{iHt}. Then the following relationships are satisfied: $Pe^{i\mathring{H}t} = e^{i\mathring{H}t}P$, $Qe^{iHt} = e^{iHt}Q$, $\Omega e^{i\mathring{H}t} = e^{iHt}\Omega$ and $\Omega^+ e^{iHt} = e^{i\mathring{H}t}\Omega^+$.*

PROOF. From $\Omega(t + \tau) = e^{iH\tau}e^{iHt}e^{-i\mathring{H}t}e^{-i\mathring{H}\tau}$ it follows that $\Omega(t + \tau) = e^{iH\tau}\Omega(t)e^{-i\mathring{H}\tau}$ and we therefore obtain

$$\Omega(t + \tau)e^{i\mathring{H}\tau}P = e^{iH\tau}\Omega(t)P. \tag{4.2.5}$$

The right side converges with t (for fixed τ) towards $e^{iH\tau}\Omega$. The left side must therefore also converge; thus $e^{i\mathring{H}\tau}P$ must lie in $P\mathscr{H}$, that is,

$$Pe^{i\mathring{H}\tau}P = e^{i\mathring{H}\tau}P \tag{4.2.6}$$

must be satisfied for all τ. Thus it follows that $P\mathscr{H}$ is an invariant subspace for the group $e^{i\mathring{H}\tau}$. If we take the adjoint operator corresponding to (4.2.6) and replace τ by $-\tau$, we then obtain

$$e^{i\mathring{H}\tau}P = Pe^{i\mathring{H}\tau}. \tag{4.2.7}$$

With the aid of (4.2.7) equation (4.2.5) is transformed into

$$\Omega(t + \tau)Pe^{i\mathring{H}\tau} = e^{iH\tau}\Omega(t)P$$

from which we conclude that, for fixed τ, in the limits $t \to \infty$ and $t \to -\infty$

$$\Omega e^{i\mathring{H}\tau} = e^{iH\tau}\Omega. \tag{4.2.8}$$

We may obtain equivalent relationships if we replace H by H_0, P by Q and Ω by Ω^+.

If $P_1 \leq P$ and P_1 (as P) commutes with $e^{i\mathring{H}t}$ and if we require that Q_1 be the projector onto $\Omega P_1\mathscr{H}$, it then follows that $Q_1 \leq Q$. We therefore obtain $\Omega^+\Omega P_1 = P_1$ and $\Omega\Omega^+Q_1 = Q_1$.

From (4.2.8) it follows that

$$\Omega e^{i\mathring{H}\tau}P_1 = e^{iH\tau}\Omega P_1 = e^{iH\tau}Q_1\Omega P_1,$$

$$\Omega e^{i\mathring{H}\tau}P_1 = \Omega P_1 e^{i\mathring{H}\tau} = Q_1\Omega e^{i\mathring{H}\tau}P_1 = Q_1 e^{iH\tau}\Omega P_1.$$

Since $\Omega P_1 \mathscr{H} = Q_1 \mathscr{H}$, $e^{iH\tau}$ leaves the subspace $Q_1 \mathscr{H}$ invariant, and we therefore obtain

$$e^{iH\tau} Q_1 = Q_1 e^{iH\tau}.$$

Thus we have proven the following theorem:

Th. 4.2.4. *If $P_1 \leq P$ and P_1 commutes with $e^{i\mathring{H}t}$ (for all t) (which, for the case in which $P_1 = P$ is satisfied according to Th. 4.2.3) and if Q_1 is the projection on $\Omega P_1 \mathscr{H}$, then it follows that $Q_1 \leq Q$ and $Q_1 \mathscr{H}$ is invariant with respect to e^{iHt} and it follows that $e^{iHt} Q_1 = Q_1 e^{iHt}$.*

Th. 4.2.5. *For $P_1 \leq P$ and $Q_1 \leq Q$ defined as in Th. 4.2.4, the following converge in the strong topology:*

(a) $e^{iHt} e^{-i\mathring{H}t} P \to \Omega$, $\quad e^{i\mathring{H}t} e^{-iHt} Q \to \Omega^+$;

(b) $e^{-iHt} \Omega - e^{-i\mathring{H}t} P \to 0$, $\quad e^{i\mathring{H}t} e^{-iHt} \Omega \to P$;

(c) $(\Omega - 1) e^{-i\mathring{H}t} P \to 0$, $\quad (\Omega^+ - 1) e^{-i\mathring{H}t} P \to 0$;

(d) $(\Omega P_1 - 1) e^{-i\mathring{H}t} P_1 \to 0$, $\quad (\Omega^+ Q_1 - 1) e^{-iHt} P_1 \to 0$;

(e) $e^{i\mathring{H}t} \Omega e^{-i\mathring{H}t} \to P$, $\quad e^{i\mathring{H}t} \Omega^+ e^{-i\mathring{H}t} P \to P$;

(f) $e^{i\mathring{H}t} \Omega P_1 e^{-i\mathring{H}t} \to P_1$, $\quad e^{i\mathring{H}t} \Omega^+ Q_1 e^{-i\mathring{H}t} P_1 \to P_1$.

For a projector $Q_2 \geq Q_1$, that is, for a Q_2 satisfying $Q_2 \Omega P_1 = \Omega P_1$ the following convergence properties are satisfied:

(g) $(1 - Q_2) e^{-i\mathring{H}t} P_1 \to 0$, $\quad e^{i\mathring{H}t} Q_2 e^{-i\mathring{H}t} P_1 \to P_1$.

PROOF. (a) follows directly from (4.2.1) and (4.2.3). The relations (a) can be multiplied on the left (not necessarily on the right!) by a uniformly bounded operator $A(t)$. (b) then follows from the first relation of (a) by multiplication from the left by e^{-iHt} and then by $e^{i\mathring{H}t}$.

If we then use the equation (4.2.8) in the first relation of (b), then we obtain the first relation of (c) since $\Omega = \Omega P$. If we multiply this first relation on the left by Ω^+, then, using (4.2.2) we obtain:

$$\begin{aligned}(\Omega^+ \Omega - \Omega^+) e^{-i\mathring{H}t} P &= (P - Q^+) e^{-i\mathring{H}t} P \\ &= (P - \Omega^+) P e^{-i\mathring{H}t} = (1 - \Omega^+) P e^{-i\mathring{H}t} \\ &= (1 - \Omega^+) e^{-i\mathring{H}t} P \to 0.\end{aligned}$$

All relations (b) and (c) can be multiplied on the right by P_1; the same is true for the first relations from (a). Thus we may multiply the second relation from (a) on the right by Q_1. The relations obtained in the above manner from the relations (a) and (c), are so simple that we shall not take note of them.

If we multiply the first relations in (c) on the right by P_1, we obtain

$$\begin{aligned}(\Omega - 1) e^{-i\mathring{H}t} P_1 &= (\Omega - 1) P_1 e^{-i\mathring{H}t} \\ &= (\Omega P_1 - 1) P e^{-i\mathring{H}t} = (\Omega P_1 - 1) e^{-i\mathring{H}t} P_1 \to 0.\end{aligned}$$

If we then multiply on the left by $\Omega^+ Q_1$, then from $\Omega^+ Q_1 \Omega P_1 = \Omega^+ \Omega P_1 = P_1$ we obtain

$$(P_1 - \Omega^+ Q_1) e^{-i\mathring{H}t} P_1 = (P_1 - \Omega^+ Q_1) P_1 e^{-i\mathring{H}t}$$
$$= (\mathbf{1} - \Omega^+ Q_1) P_1 e^{-i\mathring{H}t} = (\mathbf{1} - \Omega^+ Q_1) e^{-i\mathring{H}t} P_1 \to 0.$$

Thus we have proven the two relations (d).

If we multiply the relations (c) and (d) on the left by $e^{i\mathring{H}t}$ we then obtain the relations (e) and (f).

If we then multiply the first relation of (d) on the left by a $Q_2 \geq Q_1$ (that is, a Q_2 satisfying $Q_2 \Omega P_1 = \Omega P_1$), we then obtain:

$$(\Omega P_1 - Q_2) e^{-i\mathring{H}t} P_1 \to 0.$$

If we subtract this relation from the first relation of (d), we obtain the first relation of (g). If we multiply this first relation of (g) on the left by $e^{i\mathring{H}t}$, we obtain the second relation of (g).

Th. 4.2.6. *Let $\Omega(H, \mathring{H})$, P and Q be defined as before; similarly, let $\Omega(H_1, H)$, $\tilde{P}$ and $\tilde{Q}$. Suppose that P_1 (where $P_1 \leq P$) commute with $e^{i\mathring{H}t}$ (then, according to Th. 4.2.4, $Q_1 = \Omega P_1 \Omega^+$ commutes with e^{iHt}) and let Q_1 commute with $\tilde{P}$. Then $\tilde{P} Q_1 = Q_2$ will be a projector which commutes with e^{iHt}. For $P_2 = \Omega^+ Q_2 \Omega$ we obtain $P_2 \leq P_1$, and, according to Th. 4.2.4 P_2 commutes with $e^{i\mathring{H}t}$.*

Then

$$\Omega(H_1, H)\Omega(H, \mathring{H}) P_1 = \Omega(H_1, \mathring{H}) P_2.$$

In particular, $e^{iH_1 t} e^{-i\mathring{H}t}$ converges in $P_2 \mathscr{H}$.

Proof. Since $e^{iH_1 t} e^{-iHt} \tilde{P} \to \Omega(H_1, H)$, since $Q_1 e^{iHt} e^{-i\mathring{H}t} P_1 \to Q_1 \Omega(H, \mathring{H}) = \Omega(H, \mathring{H}) P_1$ and since the left-hand side of these two relations are uniformly bounded in norm, we obtain

$$e^{iH_1 t} e^{-iHt} \tilde{P} Q_1 e^{iHt} e^{-i\mathring{H}t} P_1 \to \Omega(H_1, H)\Omega(H, \mathring{H}) P_1.$$

Using P_2 defined above, from $\tilde{P} Q_1 = Q_2$ it follows that

$$Q_2 e^{iHt} e^{-i\mathring{H}t} P_2 - Q_2 e^{iHt} e^{-i\mathring{H}t} P_1$$
$$\to Q_2 \Omega(H, \mathring{H}) P_2 - Q_2 \Omega(H, \mathring{H}) P_1 = 0$$

and we obtain

$$e^{iH_1 t} e^{-iHt} Q_2 e^{iHt} e^{-i\mathring{H}t} P_2 \to \Omega(H_1, H)\Omega(H, \mathring{H}) P_1.$$

Since Q_2 commutes with e^{iHt} it follows that

$$e^{iH_1 t} Q_2 e^{-i\mathring{H}t} P_2 \to \Omega(H_1, H)\Omega(H, \mathring{H}) P_1.$$

Since $Q_2 \Omega P_2 = \Omega P_2$, from the first equation of (g) Th. 4.2.5 it follows that

$$(\mathbf{1} - Q_2) e^{-i\mathring{H}t} P_2 \to 0$$

and we therefore obtain

$$e^{iH_1 t}(\mathbf{1} - Q_2) e^{-i\mathring{H}t} P_2 \to 0.$$

Finally, we obtain

$$e^{iH_1t}e^{-i\mathring{H}t}P_2 \to \Omega(H_1, H)\Omega(H, \mathring{H})P_1.$$

Therefore we find that $e^{iH_1t}e^{-i\mathring{H}t}P_2$ is convergent. Therefore, in the limit, we obtain

$$\Omega(H_1, \mathring{H})P_2 = \Omega(H_1, H)\Omega(H, \mathring{H})P_1.$$

Th. 4.2.6 may be applied to the following special case in which P_1 is defined as follows: First, we define $Q_1 = \tilde{P} \wedge Q$; then Q_1 commutes with e^{iHt} since $\tilde{P}\mathscr{H}$ and $\tilde{Q}\mathscr{H}$ are invariant subspaces with respect to e^{iHt}. Let $P_1 = \Omega^+ Q_1 \Omega$; then P_1 commutes with $e^{i\mathring{H}t}$. Since $\tilde{P}Q_1 = Q_1$, we obtain $P_2 = P_1$, and with this P_1 we therefore obtain $\Omega(H_1, \mathring{H})P_1 = \Omega(H_1, H)\Omega(H, \mathring{H})P_1$.

In the special case in which $\tilde{P}$ commutes with Q, then we may set $P_1 = P$ and we obtain $P_2 = \Omega^+ \tilde{P} Q \Omega$, and we obtain

$$\Omega(H_1, \mathring{H})P_2 = \Omega(H_1, H)\Omega(H, \mathring{H}),$$

where we have already used the relation:

$$\Omega(H, \mathring{H})P = \Omega(H, \mathring{H}).$$

4.3 Wave Operators and the Spectral Representation of the Hamiltonian Operators

In §4.2 we have discussed the general properties of the wave operator which follow directly from its definition. We will now consider additional properties of the wave operator which are related to the spectral representation of the Hamiltonian operator. For this purpose we shall use the theorem (proven in AIV, §10; see also VII, §2) that H (and similarly $\mathring{H}$) may be represented as follows

$$H = \int_{-\infty}^{\infty} \varepsilon \, dE(\varepsilon), \qquad e^{iHt} = \int_{-\infty}^{\infty} e^{i\varepsilon t} \, dE(\varepsilon), \tag{4.3.1}$$

whereby $E(\varepsilon)$ is uniquely determined by the operator H.

From (4.3.1) it follows that, for $\varphi \in \mathscr{D}(H)$ (where $\mathscr{D}(H)$ is the domain of definition for H; see, AIV, §10), the following limit exists:

$$\frac{d}{dt} e^{iHt}\varphi = e^{iHt}H\varphi = He^{iHt}\varphi.$$

Thus it follows that the spectral family $E(\varepsilon)$ is also uniquely determined by the operators e^{iHt} (for all t!). If a bounded operator B commutes with all the e^{iHt}, then for $\varphi \in \mathscr{D}(H)$, it follows from

$$Be^{iHt}\varphi = e^{iHt}B\varphi$$

that the left (and therefore also the right) side is differentiable, from which we conclude that $B\varphi \in \mathscr{D}(H)$ and that $BH = HB$. Thus, together with the results from AIV, §§8 and 10, it follows that:

$$e^{iHt}B = Be^{iHt} \text{ (for all } t) \quad \text{is equivalent to} \quad E(\varepsilon)B = BE(\varepsilon) \text{ (for all } \varepsilon). \tag{4.3.2}$$

Thus it follows that B commutes with each E from the complete Boolean ring generated by the spectral family (IV, §1.3).

Examples of such operators B are given by the operators $P_\pm$, $Q_\pm$ defined in §4.1. $P_\pm$, $Q_\pm$ commute with the spectral family of $\mathring{H}$ and H, respectively.

From (4.2.8), using (4.2.3) we obtain

$$\Omega e^{i\mathring{H}t}\Omega^+ = e^{iHt}\Omega\Omega^+ = e^{iHt}Q. \tag{4.3.3}$$

Let $E(\varepsilon)$, $\mathring{E}(\varepsilon)$ denote the spectral families of H and $\mathring{H}$, respectively. Let

$$\tilde{E}(\varepsilon) = \Omega\mathring{E}(\varepsilon)\Omega^+ = \Omega\mathring{E}(\varepsilon)P\Omega^+$$

then $\tilde{E}(\varepsilon)$ defines a spectral family in the space $Q\mathscr{H}$ because Ω is an isometric mapping of $P\mathscr{H}$ onto $Q\mathscr{H}$ and $\mathring{E}(\varepsilon)P$ is a spectral family in $P\mathscr{H}$ (P commutes with all $\mathring{E}(\varepsilon)$!).

According to equation (4.3.3) we obtain

$$\int e^{i\varepsilon t}\, d\tilde{E}(\varepsilon) = \int e^{i\varepsilon t}\, d(E(\varepsilon)Q), \qquad \int \varepsilon\, d\tilde{E}(\varepsilon) = \int \varepsilon\, d(E(\varepsilon)Q).$$

Since the spectral family is uniquely determined (See AIV, §10), we obtain

$$\Omega\mathring{E}(\varepsilon)\Omega^+ = \Omega\mathring{E}(\varepsilon)P\Omega^+ = E(\varepsilon)Q \tag{4.3.4}$$

(since $E(\varepsilon)$ commutes with Q, $E(\varepsilon)Q$ is clearly a spectral family!). Thus, from (4.3.4), we obtain the following theorem:

Th. 4.3.1. *The spectral families $\mathring{E}(\varepsilon)P$ in $P\mathscr{H}$ and $E(\varepsilon)Q$ in $Q\mathscr{H}$ are unitarily equivalent.*

According to IV, §2.5 and IX, §§2 and 3 $E(\varepsilon)$ defines a "spectral measure" on the set of real numbers which divides the set of real numbers into the set k_d of the discrete spectrum, k_{cc} the absolutely continuous spectrum, k_{sc} the singular continuous spectrum and k_0 where the latter is the complement of $k_d \cup k_{cc} \cup k_{sc} = \mathrm{Sp}(H)$ (the sets k_d, k_{cc}, k_{sc} are disjoint). We shall denote the projectors $E(k_d)$, $E(k_{cc})$, $E(k_{sc})$ by P_d, P_{cc}, P_{sc}, respectively. Clearly $P_{cc} + P_{sc} + P_d = \mathbf{1}$.

We shall denote the corresponding quantities for the spectral family $\mathring{E}(\varepsilon)$ by an additional index $\circ$ as follows: $\mathring{P}_d$, $\mathring{P}_{cc}$, $\mathring{P}_{sc}$, respectively.

Each spectral measure $E(k)$ may be decomposed into an absolutely continuous part $E_{cc}(k) = E(k \cap k_{cc}) = P_{cc}E(k)$, a singular continuous part $E_{sc}(k) = E(k \cap k_{sc}) = P_{sc}E(k)$ and a discrete part $E_d(k) = E(k \cap k_d) = P_d E(k)$ as follows:

$$E(k) = E(k \cap k_{cc}) + E(k \cap k_{sc}) + E(k \cap k_d).$$

The projections P_{cc}, P_{sc}, P_d are uniquely determined (see IV, §2.5) because $P_{cc}\mathscr{H}$ consists of all vectors φ in $\mathscr{H}$ for which $\|E(k)\varphi\|^2$ is absolutely continuous with respect to Lebesgue measure; $P_d\mathscr{H}$ consists of all vectors φ in $\mathscr{H}$ for which the function $\|E(\varepsilon)\varphi\|^2$ is constant except for jump discontinuities; $P_{sc}\mathscr{H}$ consists of all vectors φ in $\mathscr{H}$ for which $\|E(\varepsilon)\varphi\|^2$ is continuous but $\|E(k)\varphi\|^2$ is singular with respect to Lebesgue measure.

We shall now use the following abbreviated notation:

$$P_1 = \mathring{P}_{cc}P, \qquad P_2 = \mathring{P}_{sc}P, \qquad P_3 = \mathring{P}_dP$$

and

$$Q_1 = QP_1\Omega^+, \qquad Q_2 = \Omega P_2\Omega^+, \qquad Q_3 = \Omega P_3\Omega^+.$$

Then, since $\mathring{P}_{cc} + \mathring{P}_{sc} + \mathring{P}_d = \mathbf{1}$, we obtain

$$P_1 + P_2 + P_3 = P, \qquad Q_1 + Q_2 + Q_3 = Q.$$

Thus it follows that

$$\Omega\mathring{E}_{cc}(k)\Omega^+ = \Omega\mathring{E}_{cc}(k)P_1\Omega^+ = E(k)Q_1$$

and therefore that $E(k)Q_1$ is absoluely continuous (because for each $\varphi \in Q_1\mathscr{H}$, $E(k)Q_1\varphi = \Omega\mathring{E}_{cc}(k)\Omega^+\varphi$ is absolutely continuous), that is, $Q_1 \geq P_{cc}$. Therefore since

$$E(k)Q_1 = E(k)P_{cc}Q_1 = E_{cc}(k)Q_1$$

it follows that

$$\Omega\mathring{E}_{cc}(k)\Omega^+ = E_{cc}(k)Q_1.$$

In a similar manner we obtain $Q_2 \leq P_{sc}$ and $Q_3 \leq P_d$. Thus, multiplying $Q = Q_1 + Q_2 + Q_3$ by P_{cc} it follows that

$$Q_1P_{cc} = Q_1 = P_{cc}Q.$$

Therefore we may also write

$$\Omega\mathring{E}_{cc}(k)\Omega^+ = E_{cc}(k)Q. \tag{4.3.5}$$

D 4.3.1. A wave operator $\Omega(H, \mathring{H})$ is said to be normal if $\mathring{P}_{cc} \leq P$, that is, $\Omega(H, \mathring{H})$ is defined at least on $\mathring{P}_{cc}$.

In Th. 4.2.5 we may therefore consider the special case in which $P_1 = \mathring{P}_{cc}$ and $Q_1 = P_{cc}Q$. From (4.3.5) it then follows that

Th. 4.3.2. *If Ω is normal, then the spectrum of H in $P_{cc}Q$ (that is, the absolutely continuous part of the spectrum of H which lies in $Q\mathscr{H}$) is unitarily equivalent to the absolutely continuous spectrum of $\mathring{H}$.*

D 4.3.2. If $\Omega(H, \mathring{H})$ and $\Omega(\mathring{H}, H)$ are normal, that is, $P \geq \mathring{P}_{cc}$ and $Q \geq P_{cc}$ then we say that $\Omega(H, \mathring{H})$ is complete.

For a complete operator Ω, equation (4.3.5) therefore holds in the special form

$$\Omega\mathring{E}_{cc}(k)\Omega^+ = E_{cc}(k), \tag{4.3.6}$$

that is,

Th. 4.3.3. *If Ω is complete, then the absolutely continuous parts of the spectrum of H and $\mathring{H}$ are unitarily equivalent.*

Th. 4.2.6 can easily be formulated for the special case of normal operators as follows:

Th. 4.3.4. *If $\Omega(H, \mathring{H})$ and $\Omega(H_1, H)$ are normal operators, then the following equation is satisfied:*

$$\Omega(H_1, H)\Omega(H, \mathring{H})\mathring{P}_{cc} = \Omega(H_1, \mathring{H})\mathring{P}_{cc}.$$

PROOF. Since $\Omega(H, \mathring{H})$ is normal, then $\mathring{P}_{cc} \leq P$, and we may therefore set $P_1 = \mathring{P}_{cc}$ in Th. 4.2.6. Then $Q_1 = \Omega P_1 \Omega^+ \leq P_{cc}$. Since $\Omega(H_1, H)$ is normal, $\tilde{P} \geq P_{cc} \geq Q_1$ and therefore $\tilde{P}$ commutes with Q_1. Thus we obtain $Q_2 = \tilde{P}Q = Q_1$ and we finally obtain $P_2 = P_1 = \mathring{P}_{cc}$.

Th. 4.3.5. *If $\Omega(H, \mathring{H})$ and $\Omega(H_1, H)$ are complete, then $\Omega(H_1, \mathring{H})$ is also complete and satisfies the equation*

$$\Omega(H_1, \mathring{H})\mathring{P}_{cc} = \Omega(H_1, H)\Omega(H, \mathring{H})\mathring{P}_{cc}.$$

PROOF. Since $\Omega(H, \mathring{H})$ and $\Omega(H_1, H)$ are complete, $\Omega(\mathring{H}, H)$ and $\Omega(H, H_1)$ are normal, and we may use Th. 4.3.4. We obtain

$$\Omega^+(H_1, \mathring{H})P_{1cc} = \Omega(\mathring{H}, H_1)P_{1cc} = \Omega(\mathring{H}, H)\Omega(H, H_1)P_{1cc}$$

thus the domain of $\Omega(\mathring{H}, H_1)$ is larger than $P_{1cc}\mathscr{H}$ and $\Omega(\mathring{H}, H_1)$ is therefore complete.

4.4 The *S* Operator

Here we shall only consider the S operator for the case in which both Ω_+ and Ω_- are complete wave operators. According to Th. 4.2.2 for each W satisfying $\mathring{P}_{cc} W \mathring{P}_{cc} = W$ it follows that $\Omega(t)W\Omega(t) \to \tilde{W} = \Omega_\pm W \Omega_\pm^+$ for $t \to \pm\infty$ where $\tilde{W} = P_{cc}\tilde{W}P_{cc}$. The same situation also holds for each $\tilde{W}$ for which $P_{cc}\tilde{W}P_{cc} = \tilde{W}$: $\Omega^+(t)\tilde{W}\Omega(t) \to \Omega_\pm^+ \tilde{W}\Omega_\pm$ for $t \to \pm\infty$. According to (4.1.2) we may replace W by $W_0^{f,i}$ (and $\tilde{W}$ by W_0) in the above relations. For $W_0^{f,i}$ satisfying $\mathring{P}_{cc} W_0^{f,i} \mathring{P}_{cc} = W_0^{f,i}$ we therefore obtain

$$\Omega_\pm W_0^{f,i} \Omega_\pm^+ = W_0 \tag{4.4.1a}$$

and

$$\Omega_\pm^+ W_0 \Omega_\pm = W_0^{f,i}. \tag{4.4.1b}$$

At the end of §2 we noted that the "jump from W_0^i to W_0^f" is of crucial importance in scattering theory. From (4.4.1b) it follows that

$$W_0^f = \Omega_+^+ W_0 \Omega_+.$$

If we substitute the following relation (obtained from (4.4.1a))

$$W_0 = \Omega_- W_0^i \Omega_-^+$$

into the above, we obtain

$$W_0^f = \Omega_+^+ \Omega_- W_0^i \Omega_-^+ \Omega_+ .$$

D 4.4.1. The operator $S = \Omega_+^+ \Omega_- \mathring{P}_{cc}$ is called the "Scattering Operator" or the S operator.

Since we have assumed that $\Omega_\pm$ is complete, S is therefore a unitary operator which maps $\mathring{P}_{cc}\mathscr{H}$ into itself. If $\mathring{P}_{cc} = \mathbf{1}$ then S would be a unitary operator in all of $\mathscr{H}$. If $\Omega_\pm$ is not complete, then $Q_\pm$ could be different and we must carry out investigations which are similar to those carried out in Th. 4.2.6 in order to describe the behavior of the operator $\Omega_+^+ \Omega_-$. We therefore see that it is an important problem to determine the conditions under which complete wave operators exist.

For the case of a complete wave operator we may, with the help of the S operator, write

$$W_0^f = S W_0^i S^+. \tag{4.4.2}$$

According to the discussions at the end of §2 we may therefore describe the scattering process as follows: Before time T_i (which of course, depends upon W_0^i), that is, for $t < T_i$ we may replace W_t in the Schrödinger picture by

$$W_t \sim e^{-i\mathring{H}t} W_0^i e^{i\mathring{H}t}. \tag{4.4.3}$$

For $t > T_i$ the expression (4.4.3) for W_t is not correct. Since, however, W_t is prepared for $t < T_i$, the knowledge of the expression (4.4.3) is sufficient for the description of the preparation.

The "error" caused by the use of $e^{i\mathring{H}t}$ instead of e^{iHt} in the time evolution of W_t is corrected by assuming that "at time $t = 0$" W makes a sudden transition from W_0^i to W_0^f according to (4.4.2), and that afterwards, the time evolution is computed with $e^{i\mathring{H}t}$, whereupon for $t > T_f$ we obtain the correct W_t as follows:

$$W_t \sim e^{-i\mathring{H}t} W_0^f e^{i\mathring{H}t} = e^{-i\mathring{H}t} S W_0^i S^+ e^{i\mathring{H}t}. \tag{4.4.4}$$

Using (4.4.4) we may compute the probabilities of "effects F registered after the scattering" as follows:

$$\operatorname{tr}(e^{-i\mathring{H}t} S W_0^i S^+ e^{i\mathring{H}t} F) \tag{4.4.5a}$$

(4.4.5a) holds in the Schrödinger picture, where F is assumed to be constant in time. In the Heisenberg picture F changes under the time translation of the effect apparatus by time τ according to the expression

$$F_\tau = e^{i\mathring{H}\tau} F e^{-i\mathring{H}\tau}$$

and the probability is computed according to the equation

$$\operatorname{tr}(SW_0^i S^+ F_\tau), \tag{4.4.5b}$$

where this result is valid only for values of t for which the approximation of the "free Heisenberg picture" is valid.

In the special case in which F_τ is an effect of system (1) (or (2)) then we replace F_τ by $F_\tau \times \mathbf{1}$ (or $\mathbf{1} \times F_\tau$) in (4.4.5b) where the product representation of $\mathscr{H} = \mathscr{H}_1 \times \mathscr{H}_2$ is the same as that before the scattering (see §1)! The fact that this situation is not changed is obtained from the transformation property of W_0^i in $SW_0^i S^+$. If we fix the ensemble, that is, if we rewrite (4.4.5) in the form

$$\operatorname{tr}(W_0^i \tilde{F}_\tau) \quad \text{with} \quad \tilde{F}_\tau = S^+ F_\tau S \tag{4.4.5c}$$

then, in the sense of VIII, §2, the product representation is transformed by S, that is, the maps $T_1^{(f)}$, $T_2^{(f)}$ defined in VIII, §2 are obtained from $T_1^{(i)}$, $T_2^{(i)}$ as follows:

$$T_1^{(f)} g_1 = S^+(T_1^{(i)} g_1)S, \qquad T_2^{(f)} g_2 = S^+(T_2^{(i)} g_2)S,$$

respectively, where

$$g_1 \in L_1, \qquad g_2 \in L_2.$$

Th. 4.4.1. *S commutes with $e^{i\mathring{H}t}$.*

PROOF. From Th. 4.2.3 it follows that

$$\begin{aligned} Se^{i\mathring{H}t} &= \Omega_+^+ \Omega_- \mathring{P}_{cc} e^{i\mathring{H}t} = \Omega_+^+ \Omega_- e^{i\mathring{H}t} \mathring{P}_{cc} \\ &= \Omega_+^+ e^{iHt} \Omega_- \mathring{P}_{cc} = e^{i\mathring{H}t} \Omega_+^+ \Omega_- \mathring{P}_{cc} = e^{i\mathring{H}t} S. \end{aligned}$$

According to Th. 4.4.1 equation (4.4.5) can be rewritten as follows:

$$\operatorname{tr}(Se^{-i\mathring{H}t} W_0^i e^{i\mathring{H}t} S^+ F) = \operatorname{tr}(e^{-i\mathring{H}t} W_0^i e^{i\mathring{H}t} S^+ F S). \tag{4.4.6}$$

From Th. 4.4.1 it follows that the jump from W_0^i to W_0^f can take place, with the aid of the S operator, not only "at time $t = 0$" but also at *any* time τ, because, according to (4.4.4), for $t > T_f$, we obtain

$$\begin{aligned} W_t \sim e^{-i\mathring{H}t} SW_0^i S^+ e^{i\mathring{H}t} &= e^{-i\mathring{H}(t-\tau)} Se^{-i\mathring{H}\tau} W_0^i e^{i\mathring{H}\tau} S^+ e^{i\mathring{H}(t-\tau)} \\ &= e^{-i\mathring{H}(t-\tau)} SW_\tau^i S^+ e^{-i\mathring{H}(t-\tau)}. \end{aligned}$$

Thus there is no particular time point "zero" for the application of the scattering operator, but only the two times T_i and T_f (where T_i and T_f depend upon W_0^i), so that (4.4.3) can be used for $t < T_i$ and (4.4.4) can be used for $t > T_f$. Therefore it is not physically meaningful to speak of a "time point of scattering".

4.5 A Sufficient Condition for the Existence of Normal Wave Operators

According to the discussion of the previous section, it is possible to determine the structure of a "scattering theory" providing that the existence of the S operator is assured. This, in turn, depends upon the existence of complete wave operators. According to D4.3.2 a wave operator $\Omega(H, \mathring{H})$ is complete if both $\Omega(H, \mathring{H})$ and $\Omega(\mathring{H}, H)$ are normal. Thus, all of the above points to the question of finding conditions which insure that a wave operator $\Omega(H, \mathring{H})$ (or $\Omega(\mathring{H}, H)$) is normal. This is therefore the central question of scattering theory, a problem to be discussed following the discussion of the "formal" properties of the wave operator.

In this section we shall provide sufficient conditions for the existence of normal wave operators and we shall provide a number of interesting conclusions. The following theorems will be formulated somewhat generally; they will, for the most part, only be used to show that a wave operator is normal.

Th. 4.5.1. *Let P, as before, denote the projection onto the domain of definition of $\Omega(H, \mathring{H})$. To a projection P_1 there may exist a fundamental set X from $P_1\mathscr{H}$ (that is, the closed subspace $[X]$ generated by X is $P_1\mathscr{H}$; see AIV, §2) which satisfies the following properties: For each $\varphi \in X$ there exists a τ for which $e^{-i\mathring{H}t}\varphi \in \mathscr{D}(\mathring{H}) \cap \mathscr{D}(H)$ for $t \geq \tau$ (here $\mathscr{D}(H)$ is the domain of definition for the Hermitian operator H—see AIV, §10) and $Ve^{-i\mathring{H}t}$ is continuous in t for $t \geq \tau$ and $\|Ve^{-i\mathring{H}t}\varphi\|$ is integrable over t from τ to ∞ (here we have used the abbreviation $H - \mathring{H} = V$ where $V = H_J$ was defined in §3). We then obtain $P_1 \leq P$. If the assumptions hold for $P_1 = \mathring{P}_{cc}$, then $\Omega(H, \mathring{H})$ is normal; if the same is also true for $P_1 = P_{cc}$ and if we exchange H and $\mathring{H}$ (and replace V by $(-V)$), then $\Omega(H, \mathring{H})$ will be complete.*

PROOF. From $\varphi \in X$ and from $e^{-i\mathring{H}t}\varphi \in \mathscr{D}(\mathring{H}) \cap \mathscr{D}(H)$ and $\Omega(t)\varphi = e^{iHt}e^{-i\mathring{H}t}\varphi$, according to §4.3 it follows that

$$\frac{d}{dt}\Omega(t)\varphi = ie^{iHt}(H - \mathring{H})e^{-i\mathring{H}t}\varphi = ie^{iHt}Ve^{-i\mathring{H}t}\varphi. \tag{4.5.1}$$

Since $Ve^{-i\mathring{H}t}$ was continuous in t, this is also the case for the right-hand side of (4.5.1); we therefore obtain

$$\Omega(t_2)\varphi - \Omega(t_1)\varphi = i\int_{t_1}^{t_2} e^{iHt}Ve^{-i\mathring{H}t}\varphi\, dt. \tag{4.5.2}$$

Thus it follows that

$$\|\Omega(t_2)\varphi - \Omega(t_1)\varphi\| \leq \int_{t_1}^{t_2} \|Ve^{-i\mathring{H}t}\varphi\|\, dt.$$

Since, by hypothesis, $\|Ve^{-i\mathring{H}t}\varphi\|$ is integrable from τ to ∞, the right-hand side will be arbitrarily small providing that t_1, t_2 are chosen sufficiently large. Therefore $\Omega(t)\varphi$ converges for $t \to \infty$:

$$\Omega(t)\varphi \to \Omega\varphi,$$

that is $\varphi \in P\mathscr{H}$. Since this is true for all $\varphi \in X$, $\Omega(t)\varphi$ converges for all φ in the space spanned by X (see §4.1) and therefore for all $\varphi \in P_1\mathscr{H}$. Therefore $P \geq P_1$.

With the aid of Th. 4.5.1 for certain "interactions" V it may be proven that $\Omega(H, \mathring{H})$ is normal. Corresponding to §3, we set $\mathring{H} = \mathring{H}_i$ and $H = H_i$ according to (3.9) and (3.10).

It is easy to see that the operator $\mathring{H}_i = (1/2m)\vec{P}_r^2$ has an absolutely continuous spectrum for $\varepsilon \geq 0$, and we obtain $\mathring{P}_{cc} = \mathbf{1}$. A fundamental set X for $\mathscr{H}$ is given by the set of all vectors (with different $\vec{a}$):

$$\varphi(\vec{r} - \vec{a}) = e^{-(|\vec{r}-\vec{a}|^2)/2}. \tag{4.5.3}$$

In order to eliminate unnecessary numerical factors, we shall set $m = \frac{1}{2}$, in order that $\mathring{H} = -\Delta$. Then $e^{-i\mathring{H}t}\varphi(\vec{r} - \vec{a})$ describes the "force free" motion of a wave packet, the position of which at time $t = 0$ is given by $\varphi(\vec{r} - \vec{a})$. According to the discussion at the end of X, §1

$$e^{-i\mathring{H}t}\varphi(\vec{r} - \vec{a}) = (1 + 2it)^{-3/2}e^{-(|\vec{r}-\vec{a}|^2)/[2(1+2it)]}. \tag{4.5.4}$$

We may also consider the case in which the particles have spin. The Hilbert space $\mathscr{H}_i$ for the relative motion is, according to VIII (2.23), given by $\mathscr{H}_i = \mathscr{H}_{rb} \times \mathscr{H}_s$ where $\mathscr{H}_s = \imath_{1s} \times \imath_{2s}$. Here $\mathscr{H}_{rb}$ is the "relative orbit space" for which the vectors (4.5.3) form a fundamental set and $\mathscr{H}_s$ is the finite-dimensional (!) spin space of the particle pair. A fundamental set for $\mathscr{H}_i$ is obtained from the set of all vectors of the form

$$\varphi_\nu(\vec{r} - \vec{a}) = e^{-(|\vec{r}-a|2)/2}u_\nu, \tag{4.5.5}$$

where the u_ν ($\nu = 1, \ldots, n$) form a basis for $\mathscr{H}_s$. Thus (4.5.4) is simply multiplied by the u_ν.

The interaction operator V can be written as a matrix operator $V_{\nu\mu}$ where the $V_{\nu\mu}$ act as operators in $\mathscr{H}_{rb}$ as follows:

$$Ve^{-i\mathring{H}t}\varphi_\nu(\vec{r} - \vec{a}) = \sum_\mu (V_{\mu\nu}e^{-i\mathring{H}t}\varphi(\vec{r} - \vec{a}))u_\mu. \tag{4.5.6}$$

We shall at first assume that (as is already implicit in (4.5.6)) that wave packets of the form (4.5.4) belong to the domain of definition of the operators $V_{\mu\nu}$. Thus they are elements of $\mathscr{D}(\mathring{H}) \cap \mathscr{D}(H)$.

From (4.5.6) it follows that (assuming that the u_μ are orthonormal):

$$\|Ve^{-i\mathring{H}t}\varphi_\nu(\vec{r} - \vec{a})\|^2 = \sum_\mu \int |V_{\mu\nu}e^{-i\mathring{H}t}\varphi(\vec{r} - \vec{a})|^2\, d^3\vec{r}. \tag{4.5.7}$$

Since, for arbitrary $b_\mu > 0$ $(\sum_\mu b_\mu)^2 \geq \sum_\mu b_\mu^2$, we find that $\|Ve^{-\mathring{H}t}\varphi_\nu(\vec{r} - \vec{a})\|$ is integrable over all t from $t = -\infty$ to ∞ provided that the

$$\|V_{\mu\nu}e^{-i\mathring{H}t}\varphi(\vec{r} - \vec{a})\| = \left[\int |V_{\mu\nu}e^{-i\mathring{H}t}\varphi(\vec{r} - \vec{a})|^2\, d^3\vec{r}\right]^{1/2} \tag{4.5.8}$$

are integrable over all t. Now that we have seen that, due to the finite dimensionality of $\mathscr{H}_s$, that it is "mathematically" sufficient to consider only the case without spin, we shall omit the indices in (4.5.8) in order to simplify the notation.

We shall now assume that the operator V is of the form

$$V = v_0(\vec{r}) + \sum_{\nu=1}^{3} v_\nu(\vec{r})\frac{\partial}{\partial x_\nu}. \tag{4.5.9}$$

Then, for

$$\psi_t(\vec{r}) = V e^{-i\mathring{H}t} \varphi(\vec{r} - \vec{a})$$

and from (4.5.4) it follows that

$$\psi_t(\vec{r}) = \left[v_0(\vec{r})(1 + 2it)^{-3/2} - (1 + 2it)^{-5/2} \sum_{\nu=1}^{3} v_\nu(\vec{r})(x_\nu - a_\nu) \right] \cdot e^{-(|\vec{r}-\vec{a}|^2)/[2(1+2it)]}.$$

From

$$\psi_t^{(1)} = (1 + 2it)^{-3/2} v_0(\vec{r}) e^{-(|\vec{r}-\vec{a}|^2)/[2(1+2it)]},$$

$$\psi_t^{(2)} = (1 + 2it)^{-5/2} \sum_\nu v_\nu(\vec{r})(x_\nu - a_\nu) e^{-(|\vec{r}-\vec{a}|^2)/[2(1+2it)]},$$

it follows that

$$\|\psi_t\| \leq \|\psi_t^{(1)}\| + \|\psi_t^{(2)}\|.$$

For arbitrary ε, where $0 < \varepsilon < 1$ it follows that

$$\psi_t^{(1)} = (1 + 2it)^{-1-\varepsilon/2} \frac{v_0(\vec{r})}{|\vec{r} - \vec{a}|^{(1-\varepsilon)/2}} \frac{|\vec{r} - \vec{a}|^{(1-\varepsilon)/2}}{(1 + 2it)^{(1-\varepsilon)/2}} e^{-(|\vec{r}-\vec{a}|^2)/[2(1+2it)]},$$

$$\psi_t^{(2)} = (1 + 2it)^{-1-\varepsilon/2} \frac{\sum_\nu v_\nu(\vec{r})(x_\nu - a_\nu)}{|\vec{r} - \vec{a}|^{(3-\varepsilon)/2}} \frac{|\vec{r} - \vec{a}|^{(3-\varepsilon)/2}}{(1 + 2it)^{(3-\varepsilon)/2}} e^{-(|\vec{r}-\vec{a}|)/[2(1+2it)]}.$$

Since $y^{(1-\varepsilon)/2} e^{-y^2/2} < c_1$ and $y^{(3-\varepsilon)/2} e^{-y^2/2} < c_2$ (for real $y \geq 0$) where c_1 and c_2 are suitable constants, we may estimate (where $A > 0$ may be chosen arbitrarily):

(1) for $|\vec{r} - \vec{a}| \geq A$

$$|\psi_t^{(1)}(\vec{r})| \leq (1 + 4t^2)^{-1/2-\varepsilon/4} c_1 \frac{|v_0(\vec{r})|}{|\vec{r} - \vec{a}|^{(1-\varepsilon)/2}},$$

$$|\psi_t^{(2)}(\vec{r})| \leq (1 + 4t^2)^{-1/2-\varepsilon/4} c_2 \frac{\sqrt{\sum_\nu |v_\nu(\vec{r})|^2}}{|\vec{r} - \vec{a}|^{(1-\varepsilon)/2}};$$

(2) for $|\vec{r} - \vec{a}| \leq A$ we simply obtain

$$|\psi_t^{(1)}(\vec{r})| \leq (1 + 4t^2)^{-3/4} |v_0(\vec{r})|,$$

$$|\psi_t^{(2)}(\vec{r})| \leq (1 + 4t^2)^{-5/4} |\vec{r} - \vec{a}| \sqrt{\sum_\nu |v_\nu(\vec{r})|^2}.$$

Thus it follows that

$$\int_{\tau}^{\infty} \|\psi_t\| \, dt$$

exists if there exists an ε satisfying $0 < \varepsilon < 1$ for which the integrals

$$\int_{|\vec{r}-\vec{a}|\geq A} \frac{|v_0(\vec{r})|^2}{|\vec{r}-\vec{a}|^{1-\varepsilon}} \, d^3\vec{r}; \qquad \int_{|\vec{r}-\vec{a}|\geq A} \frac{\sum_{\nu=1}^{3} |v_\nu(\vec{r})|^2}{|\vec{r}-\vec{a}|^{1-\varepsilon}} \, d^3\vec{r};$$
$$\int_{|\vec{r}-\vec{a}|\leq A} |v_0(\vec{r})|^2 \, d^3\vec{r}; \qquad \int_{|\vec{r}-\vec{a}|\leq A} |\vec{r}-\vec{a}|^2 \sum_{\nu=1}^{3} |v_\nu(\vec{r})|^2 \, d^3\vec{r} \tag{4.5.10}$$

exist. The first conditions from (4.5.10) state that the functions $v_0(\vec{r})$, $v_\nu(\vec{r})$ must decrease more quickly, as $|\vec{r}| \to \infty$ than $|\vec{r}|^{-1}$. The Coulomb potential does not fall within this case, a fact which is not mathematically pleasant. This fact is, however, not of particular physical importance since there is no experiment for which we encounter arbitrarily large values of $|\vec{r}|$, that is, distances between the scattered particles. It is frequently useful to multiply the functions $v_0(\vec{r})$ and $v_\nu(\vec{r})$ by a factor $e^{-\alpha|\vec{r}|}$ where α^{-1} corresponds to the maximum distance of the microsystems in actual experiments. α^{-1} is therefore an experimentally determined macroscopic distance. For the mathematical problem of the limits $t \to \pm\infty$ in the case $|v(\vec{r})| \sim |r|^{-1}$ we refer readers to [18].

Th. 4.5.2. *If V is bounded, then for each $\varphi \in P\mathscr{H}$ we obtain*

$$\|\Omega\varphi - \Omega(t)\varphi\|^2 = 2 \operatorname{Im} \int_t^{\infty} \langle \varphi, e^{i\mathring{H}\tau}\Omega + Ve^{-i\mathring{H}\tau}\varphi \rangle \, d\tau.$$

PROOF. Since V is bounded, $\mathscr{D}(H) = \mathscr{D}(\mathring{H})$ and, therefore for $\varphi \in \mathscr{D}(\mathring{H})$ we obtain $e^{-i\mathring{H}t}\varphi \in \mathscr{D}(H) \cap \mathscr{D}(\mathring{H})$. In addition $Ve^{-i\mathring{H}t}\varphi$ is continuous in t, and (4.5.2) therefore holds for each $\varphi \in \mathscr{D}(\mathring{H})$. We obtain

$$\Omega(t_2)\varphi - \Omega(t_1)\varphi = i \int_{t_1}^{t_2} e^{iHt} V e^{-i\mathring{H}t} \varphi \, dt.$$

Since $\mathscr{D}(\mathring{H})$ is dense in $\mathscr{H}$, and V is bounded, this equation holds for all $\varphi \in \mathscr{H}$. In particular, for $\varphi \in P\mathscr{H}$, for $t_2 \to \infty$ we obtain

$$\Omega\varphi - \Omega(t)\varphi = i \int_t^{\infty} e^{iH\tau} V e^{-i\mathring{H}\tau} \varphi \, d\tau.$$

Thus from

$$\begin{aligned} \|\Omega\varphi - \Omega(t)\varphi\|^2 &= \|\Omega\varphi\|^2 + \|\Omega(t)\varphi\|^2 \\ &\quad - 2 \operatorname{Re}\langle \Omega\varphi, \Omega(t)\varphi \rangle = 2\|\varphi\|^2 - 2 \operatorname{Re}\langle \Omega\varphi, \Omega(t)\varphi \rangle \\ &= 2 \operatorname{Re}\langle \Omega\varphi, \Omega\varphi - \Omega(t)\varphi \rangle \end{aligned}$$

it follows that

$$\|\Omega\varphi - \Omega(t)\varphi\|^2 = 2\,\mathrm{Re}\left\{i \int_t^\infty \langle \varphi, \Omega^+ e^{iH\tau} V e^{-i\mathring{H}\tau}\varphi\rangle\, dt\right\}$$

$$= -2\,\mathrm{Im} \int_t^\infty \langle \varphi, e^{i\mathring{H}\tau}\Omega^+ V e^{-i\mathring{H}\tau}\varphi\rangle\, d\tau,$$

where we have used Th. 4.2.3.

4.6 The Existence of Complete Wave Operators

In this section we shall seek to find sufficient conditions for V in order to insure completeness of $\Omega(H, \mathring{H})$. To this purpose we shall now prove the following theorem:

Th. 4.6.1. *If $V = cP_f$ ($\|f\| = 1$), then $\Omega_\pm(H, \mathring{H})$ will be complete.*

We shall carry out the proof of this theorem in several steps, in which we first prove a special case of Th. 4.6.1.

Case (a).

$$\mathscr{H} = \mathscr{L}^2(-\infty, \infty), \qquad \mathring{H}\varphi(x) = x\varphi(x).$$

Here $P = \mathbf{1}$.

Proof. Clearly $\mathscr{D}(H) = \mathscr{D}(\mathring{H})$. We therefore use Th. 4.5.1. From

$$Ve^{-i\mathring{H}t}\varphi(x) = cf(x) \int_{-\infty}^{\infty} \overline{f(x')} e^{-ix't}\varphi(x')\, dx'$$

it follows that

$$\|Ve^{-i\mathring{H}t}\varphi(x)\| = |c| \left| \int_{-\infty}^{\infty} \overline{f(x)} e^{-ixt}\varphi(x)\, dx\right|.$$

We now consider only functions $\varphi(x)$ and $f(x)$ which are twice differentiable and for $x \to \pm\infty$ decrease towards zero sufficiently fast that the Fourier integral can be integrated by parts twice

$$\int_{-\infty}^{\infty} e^{-ixt} g(x)\, dx = \frac{1}{it} \int_{-\infty}^{\infty} e^{ixt} g'(x)\, dx = \frac{1}{(it)^2} \int_{-\infty}^{\infty} e^{ixt} g''(x)\, dx.$$

Then it follows that $\|Ve^{-i\mathring{H}t}\varphi\|$ is integrable for both $t \to \infty$ and $t \to -\infty$. Such functions $\varphi(x)$ are dense in $\mathscr{H}$ and therefore for $\Omega(H, \mathring{H})$, $P = 1$.

Since Ω exists, according to Th. 4.5.2, for all $\varphi \in \mathscr{H}$ we obtain

$$\|\Omega\varphi - \Omega(t)\varphi\|^2 = -2\,\mathrm{Im} \int_t^\infty \langle \varphi, e^{i\mathring{H}\tau}\Omega^+ V e^{-i\mathring{H}\tau}\varphi\rangle\, d\tau$$

$$= 2c\,\mathrm{Im} \int_t^\infty \langle \varphi, e^{i\mathring{H}\tau}\Omega^+ f\rangle\langle f, e^{-i\mathring{H}\tau}\varphi\rangle\, d\tau \tag{4.6.1}$$

$$\leq 2|c| \left[\int_t^\infty |\langle \varphi, e^{i\mathring{H}\tau}\Omega^+ f\rangle|^2\, d\tau\right]^{1/2} \left[\int_t^\infty |\langle f, e^{-i\mathring{H}\tau}\varphi\rangle|^2\, d\tau\right]^{1/2}.$$

This estimate is meaningful only if the integrals over τ in the last row exist. To show this we may, in general, estimate

$$\int_{-\infty}^{\infty} |\langle g, e^{i\mathring{H}\tau}h\rangle|^2 \, d\tau = \int_{-\infty}^{\infty} \left| \int_{-\infty}^{\infty} e^{ix\tau}h(x)\overline{g(x)} \, dx \right|^2 d\tau$$

$$= 2\pi \int_{-\infty}^{\infty} |h(x)g(x)|^2 \, dx \leq 2\pi g_s^2 \|h\|^2, \qquad (4.6.2)$$

where, for the Fourier integral we have used the following general result: From

$$u(k) = \frac{1}{\sqrt{2\pi}} \int_{-\infty}^{\infty} e^{ikx}v(x) \, dx$$

it follows that

$$\int_{-\infty}^{\infty} |u(k)|^2 \, dk = \int_{-\infty}^{\infty} |v(x)|^2 \, dx.$$

Here g_s is the effective supremum of $g(x)$. In (4.6.2) we may exchange the estimates of g and h. The integrals over τ therefore exist if one of the two functions g and h (not only belongs to $\mathscr{L}^2(-\infty, \infty)$) but is essentially bounded.

From (4.6.1) for an essentially bounded $\varphi(x)$ with φ_s (since $\|\Omega^+ f\| \leq \|f\| = 1$!) it follows that

$$\|\Omega\varphi - \Omega(t)\varphi\| \leq 2|c|(2\pi)^{1/2}\varphi_s \left[\int_t^{\infty} \langle f, e^{-i\mathring{H}\tau}\varphi\rangle|^2 \, d\tau \right]^{1/2}.$$

From which it follows that

$$\|\Omega(t_2)\varphi - \Omega(t_1)\varphi\| \leq \|\Omega(t_2)\varphi - \Omega\varphi\| + \|\Omega\varphi - \Omega(t_1)\varphi\|$$

$$\leq (8\pi)^{1/4}|c|^{1/2}\varphi_s^{1/2} \left\{ \left[\int_{t_2}^{\infty} |\langle f, e^{-i\mathring{H}\tau}\varphi\rangle|^2 \, d\tau \right]^{1/4} \right.$$

$$\left. + \left[\int_{t_1}^{\infty} |\langle f, e^{-iH\tau}\varphi\rangle|^2 \, d\tau \right]^{1/4} \right\}. \qquad (4.6.3)$$

This inequality holds, according to Th. 4.5.2, for all $\varphi \in \mathscr{H}$ (since $P = 1$!) with finite φ_s; it is only proven for such $f(x)$ which were twice differentiable and vanish sufficiently quickly for $x \to \pm\infty$. The right side of (4.6.3) is defined for all $f \in \mathscr{H}$, a result which follows from (4.6.2) for finite φ_s. We are therefore led to the conjecture that (4.6.3) holds for all $f \in \mathscr{H}$—here we note that $\Omega(t)$ also depends upon f!

If $f \in \mathscr{H}$, then there exists a sequence f_n ($\|f_n\| = 1$) which satisfies $\|f_n - f\| \to 0$ for which (4.6.3) holds for each f_n. We set $H_n = \mathring{H} + cP_{fn}$ and $\Omega_n(t) = e^{iH_nt}e^{-i\mathring{H}t}$. We then obtain

$$\|H_n - H\| = |c| \, \|P_{f_n} - P_f\| \leq 2|c| \, \|f_n - f\|.$$

We will now show that $\|\Omega_n(t) - \Omega(t)\| \to 0$:

We obtain

$$\|e^{iH_nt}e^{-i\mathring{H}t} - e^{iHt}e^{-i\mathring{H}t}\| = \|e^{iH_nt} - e^{iHt}\| = \|e^{iH_nt}e^{-iHt} - \mathbf{1}\|.$$

For a $\psi \in \mathscr{D}(H_n) = \mathscr{D}(H)$ we obtain

$$\frac{d}{dt}(e^{iH_n t}e^{-iHt}\psi) = ie^{iH_n t}(H_n - H)e^{-iHt}\psi$$

and we therefore find that

$$e^{iH_n t}e^{-iHt}\psi - \psi = i\int_0^t e^{iH_n\tau}(H_n - H)e^{-iH\tau}\psi\, d\tau.$$

This equation holds for all $\psi \in \mathscr{H}$ since the operators which are found both on the left and right side are bounded. Therefore for $\|\psi\| = 1$ we obtain

$$\|e^{iH_n t}e^{-iHt}\psi - \psi\| \leq \int_0^t \|(H_n - H)e^{-iH\tau}\psi\|\, d\tau \leq \int_0^t \|H_n - H\|\, d\tau = t\|H_n - H\|.$$

Thus from

$$\begin{aligned}\Omega(t_2)\varphi - \Omega(t_1)\varphi = (\Omega(t_2)\varphi &- \Omega_n(t_2)\varphi) \\ &+ (\Omega_n(t_2)\varphi - \Omega_n(t_1)\varphi) + (\Omega_n(t_1)\varphi - \Omega(t_1)\varphi)\end{aligned}$$

we obtain

$$\|\Omega_n(t_2)\varphi - \Omega_n(t_1)\varphi\| \underset{n}{\rightarrow} \|\Omega(t_2)\varphi - \Omega(t_1)\varphi\|.$$

For an integral on the right side of (4.6.3) it follows from the triangle inequality that

$$\begin{aligned}&\left|\left[\int_t^\infty |\langle f_n, e^{-i\mathring{H}\tau}\varphi\rangle|^2\, d\tau\right]^{1/2} - \left[\int_t^\infty |\langle f, e^{-i\mathring{H}\tau}\varphi\rangle|^2\, d\tau\right]^{1/2}\right| \\ &\qquad \leq \left[\int_t^\infty \langle f_n - f, e^{-i\mathring{H}\tau}\varphi\rangle|^2\, d\tau\right]^{1/2} \leq (2\pi)^{1/2}\|f_n - f\|\varphi_s.\end{aligned}$$

Thus (4.6.3) is proven for all $f \in \mathscr{H}$ ($\|f\| = 1$). From (4.6.3) and (4.6.2) it follows that $\Omega(t)\varphi$ converges for all $\varphi \in \mathscr{H}$ with finite φ_s. Since these φ are dense in $\mathscr{H}$, we have proven that $P = \mathbf{1}$ in Case (a).

Case (b). *$\mathring{H}_{cc}$ is unitarily equivalent to the operator $\mathring{H}$ defined in Case* (a). *$V = cP_f$ for arbitrary $f \in \mathscr{H}$. We obtain $P \geq \mathring{P}_{cc}$.*

Proof. We set $f = g + h$ where $g = \mathring{P}_{cc}f$. Then for a $\varphi \in \mathring{P}_{cc}\mathscr{H}$: we obtain

$$\|Ve^{-i\mathring{H}t}\varphi\| = |c|\,|\langle f, e^{-i\mathring{H}t}\mathring{P}_{cc}\varphi\rangle| = |c|\,|\langle \mathring{P}_{cc}f, e^{-i\mathring{H}t}\varphi\rangle| = |c\,|\langle g, e^{-i\mathring{H}t}\varphi\rangle|.$$

Since $\mathring{H}$ in $\mathring{P}_{cc}\mathscr{H}$ is unitarily equivalent to the $\mathring{H}$ in (a), equation (4.6.3) holds as before with g instead of f and for $\varphi \in \mathring{P}_{cc}\mathscr{H}$ with finite φ_s. Thus it follows that $P \geq \mathring{P}_{cc}$.

Case (c). *Let S be a Lebesgue measurable set on the reals. Let $\mathscr{H} = \mathscr{L}^2(S)$; $\mathring{H}\varphi(x) = x\varphi(x)$, $H = \mathring{H} + cP_f$. Then $P = \mathbf{1}$.*

Proof. We will shortly find that $\mathring{H}$ has only an absolutely continuous spectrum. $\mathscr{H}$ can be considered to be a subspace of $\mathscr{L}^2(-\infty, \infty)$ as follows:

$$\mathscr{H} = \mathscr{L}^2(S) = P_S\mathscr{L}^2(-\infty, \infty)$$

where $P_S\chi(x) = \{\chi(x) \text{ for } x \in S; 0 \text{ for } x \notin S\}$. Let $\mathring{H}'\chi(x) = x\chi(x)$ for $\chi(x) \in \mathscr{L}^2(-\infty, \infty)$. Therefore $\mathring{H} = P_S \mathring{H}' P_S$. We set $H' = \mathring{H}' + cP'_f$ as an operator in $\mathscr{L}^2(-\infty, \infty)$ for an $f \in P_S \mathscr{L}^2(-\infty, \infty)$. Then the limit

$$\Omega\varphi = \lim \Omega(t)\varphi = \lim e^{iH't} e^{-i\mathring{H}'t}\varphi$$

exists for all $\varphi \in \mathscr{L}^2(-\infty, \infty)$, and therefore for all $\varphi \in \mathscr{H}$; For $\varphi = P_S\varphi$ it follows that

$$\Omega\varphi = \lim e^{iH't - i\mathring{H}'t} P_S \varphi = \lim e^{iH't} P_S e^{-i\mathring{H}t} P_S \varphi$$

since $e^{-i\mathring{H}'t}$ commutes with P_S and $e^{i\mathring{H}'t}$ is identical to $e^{i\mathring{H}t}$ in $\mathscr{H}$. Since $P_S f = f$ the operator $e^{iH't}$ does not lead us out of $\mathscr{H}$. Therefore we obtain

$$e^{iH't} P_S = P_S e^{iHt} P_S.$$

Therefore in $\mathscr{H}$ we obtain

$$\Omega\varphi = \lim e^{iHt} e^{-i\mathring{H}t}\varphi.$$

Case (d). *$\mathring{H}_{cc}$ is unitarily equivalent with an operator $\mathring{H}$, as we have assumed in* (c). *$H = \mathring{H} + cP_f$. Then $P \geq \mathring{P}_{cc}$.*

PROOF. $\mathscr{H} = \mathring{P}_{cc}\mathscr{H} \oplus \mathscr{H}_r$; $\mathring{H}\mathring{P}_{cc} = \mathring{H}_{cc}$. As in (c) we extend the space $\mathring{P}_{cc}\mathscr{H}$ to the space $\mathscr{L}^2(-\infty, \infty)$ and consider the operators $\mathring{H}'$ and $H' = \mathring{H}' + cP_f$. As in (b) we decompose $f = g + h$ where $h \in \mathscr{H}_r$ and $g \in \mathscr{L}^2(-\infty, \infty)$. Then g is an element of the subspace $\mathring{P}_{cc}\mathscr{H}$ of $\mathscr{L}^2(-\infty, \infty)$. According to (b) the limit of $\Omega(t)$ exists in all of $\mathscr{L}^2(-\infty, \infty)$ whereby, according to (c) $\Omega(t)$ transforms the subspace $\mathring{P}_{cc}\mathscr{H}$ of $\mathscr{L}^2(-\infty, \infty)$ into itself since $g \in \mathring{P}_{cc}\mathscr{H}$.

Case (e). *H_0 is arbitrary, $H = \mathring{H} + cP_f$. Then $P \geq \mathring{P}_{cc}$.*

PROOF. Let $\mathscr{H}_0$ denote the subspace of $\mathscr{H}$ spanned by all the $\mathring{E}(\varepsilon)f$ $(-\infty < \varepsilon < \infty)$ (see IX, §1). Thus $f \in \mathscr{H}_0$; for each "spectral measurable" function $\phi(\varepsilon)$ for which

$$\int |\phi(\varepsilon)|^2 \, d\|E(\varepsilon)f\|^2 < \infty$$

it follows that

$$g = \int \phi(\varepsilon) \, d\mathring{E}(\varepsilon) f \in \mathscr{H}_0$$

$\mathscr{H}_0$ is therefore an invariant subspace of $e^{i\mathring{H}t}$, but also of e^{iHt} since $f \in \mathscr{H}_0$.

We set $\mathscr{H} = \mathscr{H}_0 \oplus \mathscr{H}_1$. Then $\mathscr{H}_1$ is therefore also an invariant subspace of $e^{i\mathring{H}t}$ and e^{iHt} and therefore of $\Omega(t)$. Since $f \in \mathscr{H}_0$, it follows that (see above (c)) $e^{iHt}\chi = e^{i\mathring{H}t}\chi$ for all $\chi \in \mathscr{H}_1$ and we therefore obtain $\Omega(t)\chi = \chi$. For P_1 as the projection onto $\mathscr{H}_1$ P_1 commutes with $\mathring{P}_{cc}$ and we therefore obtain

$$\lim \Omega(t) P_1 \mathring{P}_{cc} = \Omega \mathring{P}_{cc} P_1 = \mathring{P}_{cc} P_1.$$

In order to prove that $\lim \Omega(t)\mathring{P}_{cc}$ exists, it is therefore only necessary to show that the following limit exists:

$$\Omega(t)\mathring{P}_{cc}(\mathbf{1} - P_1),$$

that is, that $\lim \Omega(t)\mathring{P}_{cc}\varphi$ exists for all $\varphi \in \mathscr{H}_0$.

We therefore need only consider the operators e^{iHt} and $e^{i\mathring{H}t}$ in $\mathscr{H}_0$. We shall denote these operators by $e^{iH't}$ and $e^{i\mathring{H}'t}$. $\mathring{P}_{cc}(1 - P_1)\mathscr{H} = \mathring{P}_{cc}\mathscr{H}_0$ is then the subspace of the absolutely continuous part of $\mathring{H}'$, since $\mathring{P}_{cc}$ and $(1 - P_1)$ commute. We will show that $\mathring{H}'_{cc}$ is unitarily equivalent to the operator $\mathring{H}$ given in (c), and, according to (d) $\lim \Omega(t)\mathring{P}_{cc}\varphi$ exists for all $\varphi \in \mathscr{H}_0$.

$\mathscr{H}_0$ is spanned by the vectors $\mathring{E}(\varepsilon)f$. Let $f = g + h$ where $g = \mathring{P}_{cc}f$. Then for $\mathring{E}(\varepsilon)g$ we obtain

$$\mathring{P}_{cc}\mathring{E}(\varepsilon)g = \mathring{E}(\varepsilon)\mathring{P}_{cc}g = \mathring{E}(\varepsilon)g.$$

The $\mathring{E}(\varepsilon)g$ and $\mathring{E}(\varepsilon)h$ span two mutually orthogonal subspaces of $\mathscr{H}_0$ which together span all of $\mathscr{H}_0$. Each vector of the subspace spanned by $E_0(\varepsilon)g$ is absolutely continuous, while the vectors from the subspace spanned by the $E_0(\varepsilon)h$ are orthogonal to $\mathring{P}_{cc}\mathscr{H}$. The vectors $\mathring{E}(\varepsilon)g$ therefore span $\mathring{P}_{cc}\mathscr{H}_0$.

All the vectors $\varphi \in \mathring{P}_{cc}\mathscr{H}_0$ may therefore be represented (see IX, §2) by functions $\phi(\varepsilon)$ which are defined on a Lebesgue measurable set S from $-\infty < \varepsilon < \infty$ for which

$$\int_S |\phi(\varepsilon)|^2 \, d\varepsilon < \infty,$$

where $H_0\phi(\varepsilon) = \varepsilon\phi(\varepsilon)$. Therefore the problem posed above corresponds to case (d) above.

We now prove Th. 4.6.1 as follows:

For $\Omega(H, \mathring{H})$ we therefore obtain $P \geq \mathring{P}_{cc}$. Since $\Omega^+ = \Omega(\mathring{H}, H)$ we may apply (e) and we obtain $Q \geq P_{cc}$. Thus $\Omega(H, \mathring{H})$ is complete.

Th. 4.6.2. *If $V = \sum_{\nu=1}^{n} c_\nu P_{\varphi_\nu}$ then $\Omega(H, \mathring{H})$ is complete.*

Proof. Let $H_\mu = \mathring{H} + \sum_{\nu=1}^{\mu} c_\nu P_{\varphi_\nu}$. Then $H = H_n$ and $H_{\mu+1} = H_\mu + c_{\mu+1}P_{\varphi_{\mu+1}}$. It therefore follows that all the $\Omega(H_{\mu+1}, H_\mu)$ are complete.

By finitely many applications of Th. 4.3.5 it follows directly that $\Omega(H, \mathring{H})$ is complete.

Th. 4.6.3. *Let $H = \int \varepsilon \, dE(\varepsilon)$ and $\varphi \in P_{cc}\mathscr{H}$. For*

$$\|\varphi\|_s^2 = \text{ess sup}\left\{\frac{d\|E(\varepsilon)\varphi\|^2}{d\varepsilon}\middle|\varepsilon\right\}.$$

it follows that for all $\psi \in \mathscr{H}$

$$\int_{-\infty}^{\infty} |\langle\psi, e^{-iHt}\varphi\rangle|^2 \, dt \leq 2\pi\|\varphi\|_s^2\|\psi\|^2.$$

Proof. We begin with

$$\langle\psi, e^{-iHt}\varphi\rangle = \int_{-\infty}^{\infty} e^{-i\varepsilon t} \, d\langle\psi, E(\varepsilon)\varphi\rangle.$$

Since φ is absolutely continuous, $\langle\psi, E(\varepsilon)\varphi\rangle$ is absolutely continuous; we therefore obtain

$$d\langle\psi, E(\varepsilon)\varphi\rangle = \frac{d\langle\psi, E(\varepsilon)\varphi\rangle}{d\varepsilon} \, d\varepsilon$$

from which it follows that

$$\langle \psi, e^{-iHt}\varphi \rangle = \int_{-\infty}^{\infty} e^{-i\varepsilon t} \frac{d\langle \psi, E(\varepsilon)\varphi \rangle}{d\varepsilon} d\varepsilon.$$

If $d\langle \psi, E(\varepsilon)\varphi \rangle / d\varepsilon$ is quadratically integrable with respect to ε, it follows directly (as in (4.6.1)) that

$$\int_{-\infty}^{\infty} |\langle \psi, e^{-iHt}\varphi \rangle|^2 \, dt = 2\pi \int_{-\infty}^{\infty} \left| \frac{d\langle \psi, E(\varepsilon)\varphi \rangle}{d\varepsilon} \right|^2 d\varepsilon.$$

For $\psi' = P_{cc}\psi$ (since φ is absolutely continuous) it follows that

$$\langle \psi, E(\varepsilon)\varphi \rangle = \langle \psi', E(\varepsilon)\varphi \rangle = \langle E(\varepsilon)\psi', E(\varepsilon)\varphi \rangle.$$

Thus, for each measurable set k we obtain

$$\langle \psi, E(k)\varphi \rangle = \langle E(k)\psi', E(k)\varphi \rangle$$

it therefore follows that

$$|\langle \psi, E(k)\varphi \rangle|^2 \leq \|E(k)\psi'\|^2 \, \|E(k)\varphi\|^2.$$

Since $\|E(k)\psi'\|^2$ is also absolutely continuous, it therefore follows that

$$\left| \int_k \frac{d\langle \psi, E(\varepsilon)\varphi \rangle}{d\varepsilon} d\varepsilon \right|^2 \leq \int_k \frac{d\|E(\varepsilon)\psi'\|^2}{d\varepsilon} d\varepsilon \int_k \frac{d\|E(\varepsilon)\varphi\|^2}{d\varepsilon} d\varepsilon.$$

Since the above result is valid for every measurable set k, it follows that (almost everywhere)

$$\left| \frac{d\langle \psi, E(\varepsilon)\varphi \rangle}{d\varepsilon} \right|^2 \leq \frac{d\|E(\varepsilon)\psi'\|^2}{d\varepsilon} \frac{d\|E(\varepsilon)\varphi\|^2}{d\varepsilon}.$$

From which we conclude that

$$\int \left| \frac{d\langle \psi, E(\varepsilon)\varphi \rangle}{d\varepsilon} \right|^2 d\varepsilon \leq \|\varphi\|_s^2 \int \frac{d\|E(\varepsilon)\psi'\|^2}{d\varepsilon} d\varepsilon$$
$$= \|\varphi\|_s^2 \, \|\psi'\|^2 \leq \|\varphi\|_s^2 \, \|\psi\|^2.$$

Th. 4.6.4. *If V is a trace class operator, then $\Omega(H, \mathring{H})$ is complete.*
(From Th. 4.6.4 it follows, according to Th. 4.3.3, that the absolute continuous parts of H and $\mathring{H}$ are unitarily equivalent.)

Proof. We need only show that $\Omega(H, \mathring{H})$ is normal, then the same is true if we replace H with $\mathring{H}$ for $\Omega(\mathring{H}, H) = \Omega^+(H, \mathring{H})$ and it immediately follows that $\Omega(H, \mathring{H})$ is complete.

We have $V = \sum_{\nu=1}^{\infty} c_\nu P_{\varphi_\nu}$ where $\sum_{\nu=1}^{\infty} |c_\nu| < \infty$. Let $V_n = \sum_{\nu=1}^{n} c_\nu P_{\varphi_\nu}$, $H_n = \mathring{H} + V_n$, $\Omega_n(t) = e^{iH_n t} e^{-i\mathring{H}t}$ and let Ω_n denote the limit of $\Omega_n(t)$.

According to Th. 4.6.2 the Ω_n are normal. According to Th. 4.5.2, for all $\varphi \in \mathring{P}_{cc}\mathscr{H}$ we obtain

$$\|\Omega_n \varphi - \Omega_n(t)\varphi\|^2 = -2 \, \mathrm{Im} \int_t^{\infty} \langle \varphi, e^{i\mathring{H}\tau} \Omega_n^+ V_n e^{-i\mathring{H}\tau} \varphi \rangle \, d\tau.$$

For

$$\langle \varphi, e^{i\mathring{H}\tau}\Omega_n^+ V_n e^{-i\mathring{H}\tau}\varphi\rangle = \sum_{\nu=1}^{n} c_\nu \langle \varphi, e^{i\mathring{H}\tau}\Omega_n^+ \varphi_\nu\rangle\langle \varphi_{\bar{k}}, e^{-i\mathring{H}\tau}\varphi\rangle$$

it follows that (using Th. 4.6.3)

$$\|\Omega_n\varphi - \varphi_n(t)\varphi\|^2 \leq 2\left[\sum_{\nu=1}^{\infty} |c_\nu| \int_t^\infty |\langle \varphi, e^{i\mathring{H}\tau}\Omega_n^+\varphi\rangle|^2\, d\tau\right]^{1/2}$$
$$\cdot\left[\sum_{\nu=1}^{\infty} |c_\nu| \int_t^\infty |\langle \varphi_\nu, e^{-i\mathring{H}\tau}\varphi\rangle|^2\, d\tau\right]^{1/2}$$
$$\leq 2(2\pi)^{1/2}\|\varphi\|_s\left(\sum_{\nu=1}^{\infty} |c_\nu|\right)^{1/2}$$
$$\cdot\left[\sum_{\nu=1}^{\infty} |c_\nu| \int_t^\infty |\langle \varphi_\nu, e^{-i\mathring{H}\tau}\varphi\rangle|^2\, d\tau\right]^{1/2},$$

Thus we obtain the following inequality

$$\|\Omega_n(t_2)\varphi - \Omega_n(t_1)\varphi\| \leq (8\pi)^{1/4}\|\varphi\|_s^{1/2}\|V\|_s^{1/4}[r(t_2, \varphi)^{1/4} + r(t_1, \varphi)^{1/4}]$$

with trace norm

$$\|V\|_s = \sum_\nu |c_\nu|,$$

where we used the following abbreviated notation:

$$r(t, \varphi) = \sum_{\nu=1}^{\infty} |c_\nu| \int_t^\infty |\langle \varphi_\nu, e^{-i\mathring{H}\tau}\varphi\rangle|^2\, d\tau.$$

Since $H = H_n + (V - V_n)$ and $\|V - V_n\| \to 0$ it follows that (as shown above) $\|e^{iHt} - e^{iH_nt}\| \underset{n}{\to} 0$ and $\|\Omega(t) - \Omega_n(t)\| \underset{n}{\to} 0$. Thus it follows that

$$\|\Omega(t_2)\varphi - \Omega(t_1)\varphi\| \leq (8\pi)^{1/4}\|\varphi\|_s^{1/2}\|V\|_s^{1/4}[r(t_2, \varphi)^{1/4} + r(t_1, \varphi)^{1/4}].$$

Since

$$r(t, \varphi) \leq 2\pi\|\varphi\|_s\|V\|_s$$

it therefore follows that $r(t, \varphi) \to 0$ for $t \to \infty$. Thus $\lim \Omega(t)$ exists for all $\varphi \in \mathring{P}_{cc}\mathscr{H}$ having finite $\|\varphi\|_s$. It only remains to show that the set of all φ with finite $\|\varphi\|_s$ is dense in $\mathring{P}_{cc}\mathscr{H}$.

For $\rho(\varepsilon) = d\|E(\varepsilon)\varphi\|^2/d\varepsilon$ we obtain $\int_{-\infty}^{\infty} \rho(\varepsilon)\, d\varepsilon = \|\varphi\|^2$ thus for $k_n = \{\varepsilon | \rho(\varepsilon) > n\}$: $E(k_n)\varphi \to 0$. Thus it follows that $\varphi_n = (\mathbf{1} - E(k_n))\varphi \underset{n}{\to} \varphi$ and $\|\varphi_n\|_s^2 \leq n$.

Th. 4.6.5. *Let $H = \mathring{H} + V$, let V be a trace class operator and let $\phi(\varepsilon)$ be a function which satisfies the following conditions: There exists a partition of $-\infty \cdots \infty$ into finitely many intervals such that $\phi(\varepsilon)$ is differentiable in each interval, $\phi'(\varepsilon)$ is continuous, of finite variation in each finite sub-interval and $\phi'(\varepsilon) > 0$ (therefore $\phi'(\varepsilon) \neq 0$!). Then $\Omega(\phi(H), \phi(\mathring{H}))$ is a complete wave operator, and for all $\varphi \in \mathring{P}_{cc}\mathscr{H}$ the following equation is satisfied:*

$$\Omega(\phi(H), \phi(\mathring{H}))\varphi = \Omega(H, \mathring{H})\varphi$$

PROOF. We must show that

$$\lim e^{i\phi(H)t}e^{-i\phi(\mathring{H})t}\mathring{P}_{cc}$$

exists and is equal to $\Omega(H, \mathring{H})\mathring{P}_{cc}$. Since $e^{i\phi(H)t}$ is (uniformly in t) bounded, it suffices to show that

$$e^{-i\phi(\mathring{H})t}\mathring{P}_{cc} - e^{-i\phi(H)t}\Omega(H, \mathring{H})\mathring{P}_{cc} \to 0.$$

From (4.3.6) it follows that

$$\Omega(H, \mathring{H}) \int e^{-i\phi(\varepsilon)t}\, d\mathring{E}(\varepsilon)\mathring{P}_{cc} = \int e^{-i\phi(\varepsilon)t}\, dE(\varepsilon)\Omega(H, \mathring{H})$$

and we therefore obtain

$$e^{-i\phi(H)t}\Omega(H, \mathring{H}) = \Omega(H, \mathring{H})e^{-i\phi(\mathring{H})t}.$$

Thus it remains to prove that

$$[\mathbf{1} - \Omega(H, \mathring{H})]e^{-i\phi(\mathring{H})t}\mathring{P}_{cc} \to 0.$$

It suffices to show that

$$[\mathbf{1} - \Omega(H, \mathring{H})]e^{-i\phi(\mathring{H})t}\varphi \to 0$$

holds for a dense set of vectors φ in $\mathring{P}_{cc}\mathscr{H}$.

Using the last estimate from the proof of Th. 4.6.4, for $t_2 \to \infty$ it follows that:

$$\|\Omega\varphi - \Omega(t)\varphi\| \le \|\varphi\|_s^{1/2}(8\pi\|V\|_s)^{1/4}r(t, \varphi)^{1/4}.$$

For the particular case in which $\Omega(0)\varphi = \varphi$ we obtain:

$$\|(\Omega - \mathbf{1})\varphi\| \le \|\varphi\|_s^{1/2}(8\pi\|V\|_s)^{1/4}r(0, \varphi)^{1/4}.$$

Since

$$\|\mathring{E}(\varepsilon)\varphi\|^2 = \|e^{-i\phi(\mathring{H})t}\mathring{E}(\varepsilon)\varphi\|^2$$

we obtain

$$\|\varphi\|_s = \|e^{-i\phi(\mathring{H})t}\mathring{E}(\varepsilon)\varphi\|_s$$

and we therefore obtain

$$\|[\mathbf{1} - \Omega(H, \mathring{H})]e^{-i\phi(\mathring{H})t}\varphi\| \le \|\varphi\|_s^{1/2}(8\pi\|V\|_s)^{1/4}r(0, e^{-i\phi(\mathring{H})t}\varphi)^{1/4}.$$

We therefore need only show that

$$r(0, e^{-i\phi(\mathring{H})t}\varphi) \to 0$$

for $t \to \infty$. We obtain:

$$r(0, e^{-i\phi(\mathring{H})t}\varphi) = \sum_{\nu=1}^{\infty} |c_\nu| \int_0^\infty |\langle \varphi_\nu, e^{-i\mathring{H}\tau - i\phi(\mathring{H})t}\varphi\rangle|^2\, d\tau.$$

As in the proof of Th. 4.6.3 we obtain

$$\langle \varphi_\nu, e^{-i\mathring{H}\tau - i\phi(\mathring{H})t}\varphi\rangle = \int_{-\infty}^{\infty} e^{-i\varepsilon\tau - i\phi(\varepsilon)t} \frac{d\langle \varphi_\nu, \mathring{E}(\varepsilon)\varphi\rangle}{d\varepsilon}\, d\varepsilon,$$

where $d\langle\varphi_\nu, \mathring{E}(\varepsilon)\varphi\rangle/d\varepsilon$ is a quadratically integrable function satisfying (since $\|\varphi_\nu\| = 1$):

$$\int \left| \frac{d\langle\varphi_\nu, \mathring{E}(\varepsilon)\varphi\rangle}{d\varepsilon} \right|^2 d\varepsilon \leq \|\varphi\|_s^2.$$

If we show, that as $t \to \infty$ each integral

$$\int_0^\infty |\langle\varphi_\nu, e^{-i\mathring{H}\tau - i\phi(\mathring{H})t}\varphi\rangle|^2 \, d\tau \to 0,$$

then we also obtain

$$r(0, e^{-i\phi(\mathring{H})t}\varphi) \to 0$$

since (using Th. 4.6.3)

$$r(0, e^{-i\phi(\mathring{H})t}) \leq \sum_{\nu=N+1}^{\infty} |c_\nu| \, 2\pi \, \|\varphi\|_s^2 + \sum_{\nu=1}^{N} |c_\nu| \int_0^\infty |\langle\varphi_\nu, e^{-i\mathring{H}\tau - i\phi(\mathring{H})t}\varphi\rangle|^2 \, d\tau.$$

Therefore, it only remains to show that, for

$$f(t, \tau) = \int_{-\infty}^{\infty} e^{-i\varepsilon\tau - i\phi(\varepsilon)t} g(\varepsilon) \, d\varepsilon$$

and a quadratic integrable function $g(\varepsilon)$

$$\int_0^\infty |f(t, \tau)|^2 \, d\tau \to 0 \quad \text{for } t \to \infty.$$

$f(t, \tau)$ is, as a function of τ, the Fourier transform of the function $e^{-i\phi(\varepsilon)t} g(\varepsilon)$. If R is the projection

$$R\chi(\tau) = \begin{cases} 0 & \text{for } \tau < 0, \\ \chi(\tau) & \text{for } \tau \geq 0, \end{cases}$$

then $\int_0^\infty |f(t, \tau)|^2 d\tau \to 0$ is equivalent to the statement that $Re^{-i\phi(\mathring{H})t} g \to 0$. If this is the case for a basis set of vectors $g(\varepsilon)$ in the Hilbert space of quadratically integrable functions of ε, then it holds for all quadratically integrable functions $g(\varepsilon)$. It therefore suffices to choose $g(\varepsilon)$ to be characteristic functions for arbitrarily small intervals. Since the $\phi(\varepsilon)$ may be decomposed into finitely many intervals in which it is continuously differentiable, the characteristic functions may be so chosen that $\phi(\varepsilon)$ is continuously differentiable in the intervals of the characteristic functions. For such a characteristic function we obtain (for $\tau > 0$ and $t > 0$):

$$\begin{aligned} f(t, \tau) &= \int_a^b e^{-i\varepsilon\tau - i\phi(\varepsilon)t} \, d\varepsilon = i \int_a^b (\tau + t\phi'(\varepsilon))^{-1} \frac{d}{d\varepsilon} (e^{-i\varepsilon\tau - i\phi(\varepsilon)t}) \, d\varepsilon \\ &= i \left[\frac{e^{-i\varepsilon\tau - i\phi(\varepsilon)t}}{\tau + t\phi'(\varepsilon)} \right]_a^b - i \int_a^b e^{-i\varepsilon\tau - i\phi(\varepsilon)t} \frac{d}{d\varepsilon} (\tau + t\phi'(\varepsilon))^{-1} \, d\varepsilon; \end{aligned}$$

$$|f(t, \tau)| \leq (\tau + t\phi'(a))^{-1} + (\tau + t\phi'(b))^{-1} + \int_a^b |d(\tau + t\phi'(\varepsilon))^{-1}|.$$

We obtain

$$\begin{aligned} |(\tau + t\phi'(\varepsilon_1))^{-1} - (\tau + t\phi'(\varepsilon_2))^{-1}| &= |t(\phi'(\varepsilon_2) - \phi'(\varepsilon_1))(\tau + t\phi'(\varepsilon_1))^{-1}(\tau + t\phi'(\varepsilon_2))^{-1}| \\ &\leq t|\phi'(\varepsilon_2) - \phi'(\varepsilon_1)|(\tau + ct)^2, \end{aligned}$$

where c (>0)!) is the minimum of $\phi'(\varepsilon)$ in the closed interval $[a, b]$. Therefore we obtain

$$\int_a^b |d(\tau + t\phi'(\varepsilon))^{-1}| \leq \frac{Mt}{(\tau + ct)^2},$$

where M is the total variation of $\phi'(\varepsilon)$ in the interval $[a, b]$. Thus it follows that

$$\begin{aligned} |f(t, \tau)|^2 &\leq 2(\tau + ct)^{-1} + Mt(\tau + ct)^2 \\ &\leq 2(\tau + ct)^{-1} + Mc^{-1}(\tau + ct)^{-1} = \frac{2c + M}{c(\tau + ct)} \end{aligned}$$

and

$$\int |f(t, \tau)|^2 \, d\tau \leq \frac{(2c + M)^2}{c^2} \frac{1}{ct} \to 0$$

for $t \to \infty$.

We shall begin by choosing $\mathring{H} = (1/2m)\vec{P}^2$ and $H = \mathring{H} + V$ where V is such that H is bounded from below. We see, that

$$\Omega(H, \mathring{H}) = \Omega(H + \alpha\mathbf{1}, \mathring{H} + \alpha\mathbf{1})$$

and that the statement "$\Omega(H, \mathring{H})$ is complete" is equivalent to the statement "$\Omega(H + \alpha\mathbf{1}, \mathring{H} + \alpha\mathbf{1})$ is complete". We can choose α such that, to $H_\alpha = H + \alpha\mathbf{1}$, $\mathring{H}_\alpha = \mathring{H} + \alpha\mathbf{1}$ there exists a $\delta > 0$ such that $H_\alpha \geq \delta\mathbf{1}$ and $\mathring{H}_\alpha > \delta\mathbf{1}$. Thus it follows that H_α^{-1} and $\mathring{H}_\alpha^{-1}$ are bounded operators.

Th. 4.6.6. *If $H_\alpha^{-1} - \mathring{H}_\alpha^{-1}$ is a trace class operator then $\Omega(H, H_0)$ exists and is complete.*

Proof. According to Th. 4.6.4, for $H' = H_\alpha^{-1}$ and $\mathring{H}' = \mathring{H}_\alpha^{-1}$, $\Omega(H', \mathring{H}')$ exists and is complete. $\phi(\varepsilon) = -\varepsilon^{-1}$ for $\varepsilon \geq \delta$ (with the above δ) and $\phi(\varepsilon) = \varepsilon$ for $\varepsilon < \delta$ satisfies the conditions from Th. 4.6.5. Therefore $\Omega(\phi(H_\alpha^{-1}), \phi(\mathring{H}_\alpha^{-1})) = \Omega(-H_\alpha, -\mathring{H}_\alpha)$ is complete. Since $\Omega_\pm(-H_\alpha, -\mathring{H}_\alpha) = \Omega_\mp(H_\alpha, \mathring{H}_\alpha)$ it follows that $\Omega(H_\alpha, \mathring{H}_\alpha) = \Omega(H, H_0)$ is complete.

Now let V have the form assumed in §4.5 where $v_v(\vec{r}) = 0$ in (4.5.9). For the sake of simplicity we will now assume that V is bounded. Then, $H = \mathring{H} + V$ is (trivially) bounded from below. We may therefore apply Th. 4.6.6 providing that $H_\alpha^{-1} - \mathring{H}_\alpha^{-1}$ is a trace class operator.

From $H_\alpha = \mathring{H}_\alpha + V$ it follows that $H_\alpha^{-1}\mathring{H}_\alpha = 1 - H_\alpha^{-1}V$. Since V and H_α^{-1} are bounded operators, $H_\alpha^{-1}\mathring{H}_\alpha$ is also a bounded operator. Furthermore it follows that

$$H_\alpha^{-1} - \mathring{H}_\alpha^{-1} = -H_\alpha^{-1}V\mathring{H}_\alpha^{-1} = -H_\alpha^{-1}\mathring{H}_\alpha\mathring{H}_\alpha^{-1}V\mathring{H}_\alpha^{-1}.$$

Since $H_\alpha^{-1}\mathring{H}_\alpha$ is bounded, $H_\alpha^{-1} - \mathring{H}_\alpha^{-1}$ is a trace class operator if $\mathring{H}_\alpha^{-1}V\mathring{H}_\alpha^{-1}$ is a trace class operator. Let V_+ be the positive and let $(-V_-)$ be the negative part of V. Since $\mathring{H}_\alpha^{-1}$ is positive, $\mathring{H}_\alpha^{-1} V_+ H_\alpha^{-1}$ is the positive and $\mathring{H}_\alpha^{-1}(-V_-)\mathring{H}_\alpha^{-1}$ is the negative part of $\mathring{H}_\alpha^{-1} V \mathring{H}_\alpha^{-1}$. It is therefore sufficient if

$$\operatorname{tr}(\mathring{H}_\alpha^{-1} V_+ \mathring{H}_\alpha^{-1}) < \infty$$

and $\operatorname{tr}(\mathring{H}_\alpha^{-1} V_- \mathring{H}_\alpha^{-1}) < \infty$. This is the case if the condition

$$\int \operatorname{tr}_s(V_\pm(\vec{r}))\, d^3\vec{r} < \infty$$

is satisfied where tr_s is the trace in spin space (see §4.5).

Then we find that

$$\operatorname{tr}(\mathring{H}_\alpha^{-1} V_\pm \mathring{H}_\alpha^{-1}) = \int \operatorname{tr}_s(V_\pm(\vec{r}))\, d^3\vec{r} \int \frac{d^3\vec{k}}{(\vec{k}^2 + \alpha)^2}.$$

Here it is possible (and of considerable *mathematical* interest) to improve the results described above (see [10] and [18]). However, with respect to the physics described here, no new essential results are obtained, because the interaction operators between two elementary systems (satisfying VIII (5.8)) do not satisfy all of the conditions for the existence of a complete wave operator. In §§4.5 and 5.2 we have easily obtained such an operator by means of a "trick" in which we multiply the interaction operator V by $e^{-\alpha r}$ (where $\alpha > 0$ is very small). If there is a singularity at $r = 0$, we may modify V for $r < \delta(\delta$ very small) such that V is bounded. Such modifications must not change (in physical approximation) the spectrum of H nor the wave operators; otherwise, the Hamiltonian operator defined by VIII (5.8) *must be considered to be physically unusable*. Naturally, the mathematical problem which reflects the above approximation remains to be formulated.

4.7 Stationary Scattering Theory

Earlier, in §§4.1 to 4.6, we have outlined the problem of the wave operator and scattering operators in terms of "time limits". We shall now refer to a number of results of the so-called stationary scattering theory without giving precise descriptions of the mathematical problems to be found there. This section shall only serve to establish the connection between this method and other methods.

We shall begin with (4.5.2). Here we shall assume that the limit $\Omega(t) \underset{t\to-\infty}{\rightarrow} \Omega_-$ exists in all of $\mathscr{H}$, that is, that $P = \mathring{P}_{cc} = \mathbf{1}$. Then in the limit we obtain

$$\Omega_- = \mathbf{1} + i \int_0^{-\infty} e^{iHt} V e^{-i\mathring{H}t}\, dt. \tag{4.7.1}$$

Similarly, by exchanging H and $\mathring{H}$, we obtain

$$\Omega_-^+ = Q - i \int_0^{-\infty} e^{i\mathring{H}t} V e^{-iHt} Q\, dt, \tag{4.7.2}$$

where Q is defined in §4.2. If Ω is complete, then $Q = P_{cc}$. From $\Omega_-^+\Omega_- = 1$ and $Q\Omega_- = \Omega_-$, from (4.7.2) it follows that, by multiplication with Ω_-

$$\Omega_- = \mathbf{1} + i\int_0^{-\infty} e^{i\mathring{H}t}Ve^{iHt}\Omega_-\,dt.$$

From (4.2.8) it follows that

$$\Omega_- = \mathbf{1} + i\int_0^{-\infty} e^{i\mathring{H}t}V\Omega_-\,e^{-i\mathring{H}t}\,dt. \tag{4.7.3}$$

Completely analogous equations may be derived for Ω_+. Equation (4.7.3) for Ω_- has the form of a stationary integral equation because t is only an integration variable. It is useful to multiply the integrands in (4.7.3) by a factor $e^{\alpha t}$ and then to take the limit $0 < \alpha \to 0$. We obtain

$$\Omega_- = \mathbf{1} + \lim_{\alpha\to 0} i\int_0^{-\infty} e^{i\mathring{H}t}V\Omega_-e^{-i\mathring{H}t+\alpha t}\,dt. \tag{4.7.4}$$

Equation (4.7.4) has, as a consequence, that it often remains meaningful when applied to improper eigenvectors of $\mathring{H}$.

Beginning with the improper eigenvector

$$\langle\vec{r}\,|\,\vec{k}\rangle u = \frac{1}{(2\pi)^{3/2}}\,e^{i\vec{k}\cdot\vec{r}}u$$

in the position representation where $u \in \mathscr{H}_s$, and using the notation

$$\psi_{\vec{k}}(\vec{r}) = \Omega_-\langle\vec{r}\,|\,\vec{k}\rangle u \tag{4.7.5}$$

then from (4.7.4) it follows that

$$\psi_{\vec{k}}(\vec{r}) = \langle\vec{r}\,|\,\vec{k}\rangle u + \lim_{\alpha\to 0} i\int_0^{-\infty} e^{(i\mathring{H}-i(k^2/2m)+\alpha)t}\,dt\,V\psi_{\vec{k}}(\vec{r})$$

and we therefore obtain

$$\psi_{\vec{k}}(\vec{r}) = \langle\vec{r}\,|\,\vec{k}\rangle u - \lim_{\alpha\to 0}\left(\mathring{H} - \frac{k^2}{2m} - i\alpha\right)V\psi_{\vec{k}}(\vec{r}). \tag{4.7.6}$$

Equation (4.7.6) is known as the Lippmann–Schwinger equation. In the position representation it follows that

$$\lim_{\alpha\to 0}\left\langle\vec{r}\,\middle|\left(\mathring{H} - \frac{k^2}{2m} - i\alpha\right)^{-1}\middle|\,\vec{r}'\right\rangle = \lim_{\alpha\to 0}\frac{1}{(2\pi)^3}\int\frac{e^{i\vec{k}'(\vec{r}'-\vec{r})}}{k'^2/2m - k^2/2m - i\alpha}\,d^3\vec{k}'.$$

This is precisely a Green's function of the differential operator $\Delta - k^2\mathbf{1}$:

$$\lim_{\alpha\to 0}\left\langle\vec{r}\,\middle|\left(\mathring{H} - \frac{k^2}{2m} - i\alpha\right)^{-1}\middle|\,\vec{r}'\right\rangle = \frac{m}{2\pi}\frac{e^{ik|\vec{r}-\vec{r}'|}}{|\vec{r}-\vec{r}'|}. \tag{4.7.7}$$

Thus (4.7.6) is transformed into

$$\psi_{\vec{k}}(\vec{r}) = \frac{1}{(2\pi)^{3/2}}\,e^{i\vec{k}\cdot\vec{r}}u - \frac{m}{2\pi}\int\frac{e^{ik|\vec{r}-\vec{r}'|}}{|\vec{r}-\vec{r}'|}V(\vec{r}')\psi_{\vec{k}}(\vec{r}')\,d^3\vec{r}', \tag{4.7.8}$$

where V as an operator also acts in spin space because $\psi_{\vec{k}}(\vec{r})$ must, according to (4.7.5), be considered as a vector in spin space:

$$\psi_{\vec{k}}(\vec{r}) = \sum_{\nu} \psi_{\vec{k}\nu}(\vec{r}) u_{\nu},$$

where $\psi_{\vec{k}\nu}(\vec{r}) \in \mathscr{H}_{rb}$ and the u_ν is a basis in $\mathscr{H}_s$ (see §4.5).

From (4.7.8) it follows that $\psi_{\vec{k}}(\vec{r})$ is an eigenvector of H with eigenvalue $k^2/2m$:

$$H\psi_{\vec{k}}(\vec{r}) = (\mathring{H} + V)\psi_{\vec{k}}(\vec{r}) = \frac{k^2}{2m} \psi_{\vec{k}}(\vec{r}). \tag{4.7.9}$$

The above equation is only an alternative description of the relationship (4.3.6) making use of improper eigenvectors.

The solution (4.7.8) of (4.7.9) has the following form: For large $r = |\vec{r}|$ we obtain the asymptotic solution:

$$\psi_{\vec{k}}(\vec{r}) \approx \frac{1}{(2\pi)^{3/2}} e^{i\vec{k}\cdot\vec{r}} u + \frac{1}{(2\pi)^{3/2}} g_{\vec{k}}\left(\frac{\vec{r}}{r}\right) \frac{e^{ikr}}{r}. \tag{4.7.10}$$

Equation (4.7.10) is called the "radiation condition" for the solution $\psi_{\vec{k}}(\vec{r})$ of (4.7.9). This condition states that $\psi_{\vec{k}}(\vec{r})$ differs from a plane wave by the addition of an out-going spherical wave. The intuitive meaning of (4.7.10) is very clear: If we consider, in the Schrödinger picture (4.2.4) an ensemble W_0^i in $\mathscr{H}_i$ of the form $W_0^i = P_\psi$ where

$$\psi(\vec{r}) = \frac{1}{(2\pi)^{3/2}} \int e^{i\vec{k}\cdot\vec{r}} \sum_{\nu} u_\nu \chi_\nu(\vec{k})\, d^3\vec{k}$$

then W_t will, according to (4.2.4), be equal to P_{ψ_t} where

$$\psi_t(\vec{r}) = \int \psi_{\vec{k}\nu}(\vec{r}) \sum_{\nu} u_\nu \chi_\nu(\vec{k}) e^{-i(k^2/2m)t}\, d^3\vec{k}. \tag{4.7.11}$$

For $t < T_i$ (T_i is defined in §1) we approximately obtain

$$\psi_t(\vec{r}) \approx \frac{1}{(2\pi)^{3/2}} \int e^{i(\vec{k}\cdot\vec{r} - (k^2/2m)t)} \sum_{\nu} u_\nu \chi_\nu(\vec{k})\, d^3\vec{k},$$

that is, a free moving wave packet. Upon reaching the region of interaction in the vicinity of $|r| = 0$ spherical waves are produced. For $t > T_f$ we again obtain free motion, given by

$$\begin{aligned}\psi_t(\vec{r}) \approx{}& \frac{1}{(2\pi)^{3/2}} \int e^{i(\vec{k}\cdot\vec{r} - (k^2/2m)t)} \sum_{\nu} u_\nu \chi_\nu(\vec{k})\, d^3\vec{k} \\ &+ \frac{1}{(2\pi)^{3/2}} \int \frac{e^{i(kr - (k^2/2m)t)}}{r} \sum_{\nu} g_{\vec{k}\nu}\left(\frac{\vec{r}}{r}\right) \chi_\nu(\vec{k})\, d^3\vec{k}.\end{aligned} \tag{4.7.12}$$

The solution of (4.7.9) is uniquely determined by the condition (4.7.10), that is, we obtain (4.7.8), where the latter is nothing other than (4.7.3) since Ω_- is determined if $\Omega_- e^{i\vec{k}\cdot\vec{r}}$ is known for all $\vec{k}$.

The problem of solving (4.7.9) subject to the condition (4.7.10) is often called the problem of stationary scattering theory. In §6 we will discover the practical value of the solutions $\psi_{\vec{k}}(\vec{r})$.

The fact that for many practical purposes the knowledge of the operator Ω_- (instead of $S = \Omega_+^+\Omega_-$) is sufficient is a direct consequence of (4.7.10). In the same way we obtain the following result for $\Omega_+ : \Omega_+$ applied to a "free" ψ produces a wave packet which has the same outgoing spherical wave as ψ. $\Omega_+^+\psi_{\vec{k}}(r) = \eta_{\vec{k}}(\vec{r})$ is therefore a solution of the free Schrödinger equation (that is, of (4.7.9) where $V = 0$) which is identical to $\psi_{\vec{k}}(\vec{r})$ with respect to the outgoing spherical wave. Therefore, for $\eta_{\vec{k}}(\vec{r})$, for large $|r|$ we obtain:

$$\eta_{\vec{k}}(\vec{r}) \approx \frac{1}{(2\pi)^{3/2}} e^{i\vec{k}\cdot\vec{r}}u + \frac{1}{(2\pi)^{3/2}} g_{\vec{k}}\left(\frac{\vec{r}}{r}\right)\frac{e^{ikr}}{r} + \frac{1}{(2\pi)^{3/2}} h_{\vec{k}}\left(\frac{\vec{r}}{r}\right)\frac{e^{-ikr}}{r}, \tag{4.7.13}$$

where $h_{\vec{k}}(\vec{r}/r)$ is a yet-to-be-determined amplitude of the incoming spherical wave. The scattering operator S is determined by

$$\eta_{\vec{k}\nu}(\vec{r}) = S\left(\frac{1}{(2\pi)^{3/2}} e^{i\vec{k}\cdot\vec{r}}u_\nu\right). \tag{4.7.14}$$

For the effects to be measured after the scattering it is only necessary to consider the outgoing spherical wave portion of the wave packet

$$\int \sum_\nu \eta_{\vec{k}\nu}(\vec{r}) e^{-(k^2/2m)t}\chi_\nu(\vec{k})\, d^3\vec{k}.$$

This outgoing spherical wave portion is identical to the last term on the right-hand side of (4.7.12). For the application of scattering theory it is sufficient to know the amplitude $g_{\vec{k}}(\vec{r}/r)$ of the outgoing spherical wave in (4.7.10), the knowledge of which is somewhat less than that of Ω_- (see §6).

4.8 Scattering of a Pair of Identical Elementary Systems

Earlier, for the sake of conceptual simplicity, we considered only the scattering of two different elementary systems. In fact most of the considerations of the previous section did not depend upon this special assumption. We need only to make a few minimal changes in order to apply the above methods to the case of two identical systems.

Equation (1.15) from Coll. 1 is to be replaced by an operator W_0^i from $\{\mathscr{H}_1^2\}_\pm$ as follows

$$W_0^i = \alpha[W_{10}^i \times W_{20}^i + W_{20}^i \times W_{10}^i], \tag{4.8.1}$$

where α is a normalization factor, so that $\mathrm{tr}(W_0^i) = 1$ where tr is the trace in $\{\mathscr{H}_1^2\}_\pm$.

Here it is important to choose such W_{10}^i and W_{20}^i which are mutually orthogonal (that is $W_{10}^i W_{20}^i = 0$), since both preparation procedures a_1

(where $\varphi(a_1) = W^i_{10}$) and a_2 (where $\varphi(a_2) = W^i_{20}$) must be combined to a single preparation procedure, as described in §1. We have to assume that $(a_1, a_2) \in \pi_c$ implies that $\varphi(a_1)$ and $\varphi(a_2)$ are mutually orthogonal. Many pseudoproblems will arise if this condition is ignored.

Otherwise, all other considerations remain the same except for the fact that for the space $\mathscr{H}_i$ this subspace of $\mathscr{H}_{rb} \times \imath_s^2$ must be chosen for which the operator $R \times \mathbf{P}_s$ from VIII (3.12) takes on a $(+1)$ eigenvalue for Bose systems and a (-1) eigenvalue for Fermi systems. Then the operators $\Omega_\pm$ are defined in $\mathscr{H}_i$ as in the previous sections.

In §6.3 we shall provide an example of a scattering process with identical systems.

4.9 Multiple-Channel Scattering Theory

Up until now we have only considered the mathematical structure associated with scattering only for the case of elementary systems because in that case it is easy to obtain a general overview. With respect to the totality of experiments, the cases of scattering of two composite systems or of an elementary system with a composite system play a much more encompassing role.

The complication of the scattering theory of composite systems rests upon the fact that there are multiple "channels", that is, after the scattering, a system which is different than the incoming systems may be present.

A system composed of N elementary systems can often be decomposed into noninteracting parts in several different ways if these parts are widely separated. The entire system consisting of N elementary systems may, for example, be decomposed into f parts in which the ith subsystem is such that it consists of n_i elementary systems which are "bound" together. Here we would therefore obtain $N = \sum_{i=1}^{n} n_i$.

A system composed of n_i elementary systems is said to be bound if the ensemble W satisfies the condition

$$W = (\mathbf{1} \times E_d)W(\mathbf{1} \times E_d), \tag{4.9.1}$$

where $\mathbf{1} \times E_d$ refers to the product decomposition VIII (1.1) and the projection E_d is the projection onto the subspace of the discrete spectrum of the Hamiltonian operator H_i in VIII (1.2).

Here we have used E_d as a characterization for the definition of the "bound" state W according to (4.9.1). There are alternative characterizations (see, for example, [18]) which are, under certain assumptions about the interaction, equivalent to (4.9.1).

In order to demonstrate a decomposition of a system consisting of N elementary systems into "channels" we shall choose $N = 3$ and assume that all three systems are of different types. We shall write the Hamiltonian operator in the following form

$$H = H_{01} + H_{02} + H_{03} + V_{12} + V_{23} + V_{13}, \tag{4.9.2}$$

where H_{0i} is the kinetic energy of the ith subsystem and V_{ik} is the interaction operator between the ith and kth system.

Let E_{dik} denote the projection onto the appropriate subspace of $\mathscr{H}_i \times \mathscr{H}_k$ into which (after separating out the motion of the center of mass of the two systems) the Hamiltonian operator $H_{ik} = H_{0i} + H_{0k} + V_{ik}$ has a point spectrum. The subspace of the continuous spectrum of H in the space $\mathscr{H} = \mathscr{H}_1 \times \mathscr{H}_2 \times \mathscr{H}_3$ can be split into "channels", that is, into the following subspaces as follows:

Channel f_-: All three systems are free.
$\mathscr{H}_{f-}$ is the subspace of all the φ for which the limit

$$\psi = \lim_{t \to -\infty} e^{iH_0 t} e^{-iHt} \varphi, \tag{4.9.3}$$

where $H_0 = H_{01} + H_{02} + H_{03}$ exists.

Channel $(12)_-$: (1, 2) are bound, 3 is free.

$\mathscr{H}_{(12)-}$ is the subspace of all φ for which the limit

$$\psi = \lim_{t \to -\infty} e^{i(H_0 + V_{12})t} e^{-iHt} \varphi \tag{4.9.4}$$

exists and $(E_{d12} \times \mathbf{1})\psi = \psi$.

The channels $(13)_-$ and $(23)_-$ may be defined in a completely similar way. Similar definitions can be made for the limits $t \to \infty$.

The subspaces $\mathscr{H}_{f-}$, $\mathscr{H}_{(12)-}$, $\mathscr{H}_{(13)-}$, $\mathscr{H}_{(23)-}$ are pairwise orthogonal. In each of the channels an operator Ω_{f-}^+, $\Omega_{(12)-}^+$, etc. is defined by (4.9.3), (4.9.4), etc. The domains of $\Omega_{f-}^+ \mathscr{H}_{f-}$, $\Omega_{(12)-}^+ \mathscr{H}_{(12)-}$, etc., are not mutually orthogonal (by choosing suitable assumptions about the interaction, the domains may all be equal to $\mathscr{H}$). Thus, in this case, a unique operator $\Omega_- W \Omega_-^+$ as was the case in §4.2 does not exist. How do we select the "correct" W_0 in the case of multiple channel theory?

As in the simpler cases which we have seen in §1, this is not a mathematical question. It is a question of a new structure in which several preparation procedures are combined to form a new preparation procedure. As in §1, the individual preparation procedures must be such that, if the corresponding ensembles are extrapolated backwards in time, that there is, for all practical purposes, no interaction between the individually prepared systems. Thus, by analogy with §1, we are led to introduce the following extension of Coll. 1:

If three preparation procedures a_1, a_2, a_3 (where a_k is preparing elementary systems of type k with $\varphi(a_k) = W_{0k} \in \mathscr{B}(\mathscr{H}_k)$) such that the systems are "shot" at each other, then we set

$$W_0 = \Omega_{f-}(W_{01} \times W_{02} \times W_{03})\Omega_{f-}^+.$$

If two preparation procedures a_{12} and a_3 where a_3 is preparing elementary systems of type 3 with $\varphi(a_3) = W_{03} \in \mathscr{B}(\mathscr{H}_3)$ and a_{12} is preparing bound states of a pair of elementary systems 1, 2 according to

$$\varphi(a_{12}) = W_{012} \in \mathscr{B}(\mathscr{H}_1 \times \mathscr{H}_2)$$

such that the two systems are "shot" at each other, then we set

$$W_0 = \Omega_{(1,2)-}(W_{012} \times W_{03})\Omega^+_{(1,2)-}.$$

The asymptotic measurement of effects after scattering can already be defined with the aid of W_0 if we know the asymptotic behavior of W_0. In general this problem, of course, cannot be simply analyzed, as, for example, in §4.7. Later we shall only give an example (see §5.3) because, we cannot give an overview of the most important applications (for multi-channel theory see, for example, [19]).

5 Examples of Wave Operators and Scattering Operators

We shall now illustrate the general theory presented in §§1–4 by the use of a few examples.

5.1 Scattering of an Elementary System of Spin $\frac{1}{2}$ by an Elementary System of Spin 0

In §§3 and 4 we have not taken into account an important symmetry—the rotational symmetry of the interaction V. In $\mathscr{H}_i = \mathscr{H}_{rb} \times \imath_{1/2}$ we therefore find it necessary to make use of the reduction of the rotation group which was carried out in XI, §10. In XI (10.21) and XI (10.22) we obtained vectors which belong to an irreducible representation $\mathscr{D}_j$ of the rotation group. These two vectors differ in their symmetry under reflection $\vec{r} \to -\vec{r}$. The scattering operator S must therefore commute with the operators of the representation $\mathscr{D}_j$ and the reflection operator.

Since S also commutes with $\mathring{H}$, we use the (improper) eigenvectors of $\mathring{H}$. Let $g_{kl}(r)$ denote the solution of the equation (see XI (3.4)):

$$-\frac{1}{r^2}\frac{d}{dr}\left(r^2\frac{dg_{kl}}{dr}\right) + \frac{l(l+1)}{r^2}g_{kl} = k^2 g_{kl}. \tag{5.1.1}$$

Then the functions φ^{lj}_{kM} given by

$$\varphi^{l\,l+1/2}_{k\,m+1/2} = g_{kl}(r)\left(\sqrt{\frac{l+m+1}{2l+1}}\,Y^l_m u_+ + \sqrt{\frac{l-m}{2l+1}}\,Y^l_{m+1}u_-\right)$$

and

$$\varphi^{l+1\,l+1/2}_{k\,m+1/2} = g_{kl+1}(r)\left(-\sqrt{\frac{l-m+1}{2l+3}}\,Y^{l+1}_m u_+ + \sqrt{\frac{l+m+2}{2l+3}}\,Y^{l+1}_{m+1}u_-\right)$$

are improper eigenvectors of S, where the eigenvalue does not depend upon M; we obtain

$$S\varphi^{lj}_{kM} = e^{i2\delta_{lj}(k)}\varphi^{lj}_{kM}, \tag{5.1.2}$$

where the quantity $\delta_{lj}(k)$ is called the phase shift, where the latter is, of course, only determined up to integer multiples of π. The factor 2 in the exponent is chosen by convention.

The $g_{kl}(r)$ may be directly expressed in terms of Bessel functions. Later we shall see that they may also be obtained by using other methods than solving the differential equation (5.1.1) directly.

If the interaction V (approximately) does not depend upon spin, then $\delta_{lj}(k)$ will be independent of j. In that case, we write $\delta_{lj}(k) = \delta_l(k)$.

In order to actually compute S we must obtain a method of computing the $\delta_{lj}(k)$. We will now carry out this computation for the case in which the interaction V arises from a potential function $V(r)$.

In order to determine Ω_- we now seek a solution $\psi_{\vec{k}}(\vec{r})$ of equation (4.7.9) which satisfies (4.7.10). First we seek a solution of (4.7.9) of the form $\eta_{kl}(r)Y^l_m$. Then it follows that (see the derivation of XI (3.4))

$$-\frac{1}{r^2}\frac{d}{dr}\left(r^2\frac{d\eta_{kl}}{dr}\right) + \frac{l(l+1)}{r^2}\eta_{kl} + 2mV(r)\eta_{kl} = k^2\eta_{kl}. \tag{5.1.3}$$

Setting $u_{kl} = r\eta_{kl}$ it follows that

$$-u''_{kl} + \frac{l(l+1)}{r^2}u_{kl} + 2mV(r)u_{kl} = k^2 u_{kl}. \tag{5.1.4}$$

Thus it follows that if $V(r)$ decreases more rapidly than $1/r$ as $r \to \infty$ then u_{kl} will behave asymptotically like

$$u_{kl}(r) \sim c \sin\left(kr - \frac{l}{2}\pi + \tilde{\delta}_l(k)\right). \tag{5.1.5}$$

Later we shall show that $\tilde{\delta}_l(k)$ has been defined in such a way that $\tilde{\delta}_l = 0$ for $V = 0$.

The $\tilde{\delta}_l(k)$ will be determined by (5.1.5) up to integer multiples of π because only solutions u_{kl} of (5.1.4) may be used which are regular for $r = 0$ (see also XI, §3).

The desired solution $\psi_{\vec{k}}(\vec{r})$ must therefore have the form

$$\sum_{lm} a_{lm}\frac{u_{kl}(r)}{r}Y^l_m(\theta, \varphi),$$

where for the sake of simplicity the polar axis for θ, φ is chosen to be in the direction of $\vec{k}$. For large r, according to (4.7.10) we must have the asymptotic behavior

$$\sum_{l,m} a_{lm}\frac{u_{kl}(r)}{r}Y^l_m \approx \frac{1}{(2\pi)^{3/2}}e^{i\vec{k}\cdot\vec{r}} + \frac{1}{(2\pi)^{3/2}}f_k(\theta, \varphi)\frac{e^{ikr}}{r}. \tag{5.1.6}$$

In order to determine the a_{lm}, we need to expand the plane wave in terms of spherical harmonics

$$e^{i\vec{k}\cdot\vec{r}} = e^{ikr\cos\theta} = \sum_{l=0}^{\infty} g_{kl}(r)Y^l_0(\theta). \tag{5.1.7}$$

From (5.1.7) it directly follows that $g_{kl}(r)$ must be the solutions of (5.1.1). We may therefore obtain these solutions in the following form

$$g_{kl}(r) = 2\pi \int_{-1}^{1} e^{ikr\xi} Y_0^l(\xi)\, d\xi, \tag{5.1.8}$$

where $\xi = \cos\theta$. For Y_0^l given by VII (3.55) it follows that

$$g_{kl}(r) = \sqrt{2\pi}\sqrt{\frac{2l+1}{2}}\frac{1}{2^l l!}\int_{-1}^{1} e^{ikr\xi}\frac{d^l}{d\xi^l}(1-\xi^2)^l\, d\xi. \tag{5.1.9}$$

From which we may obtain g_{kl} by repeated integration by parts. Here we shall only carry out the first integration by parts; we obtain

$$\begin{aligned} g_{kl}(r) &= \sqrt{2\pi}\sqrt{\frac{2l+1}{2}}\frac{1}{2^l l!}\frac{1}{ikr}\int_{-1}^{1}\left(\frac{d}{d\xi}e^{ikr\xi}\right)\frac{d^l}{d\xi^l}(1-\xi^2)^l\, d\xi \\ &= \sqrt{2\pi}\frac{\sqrt{2l+1}}{2^l l!}\frac{1}{ikr}\left[e^{ikr\xi}\frac{d^l}{d\xi^l}(1-\xi^2)^l\right]\Bigg|_{-1}^{1} + O\left(\frac{1}{r^2}\right), \end{aligned}$$

where $O(1/r^2)$ means that additional terms decrease at least as rapidly as $1/r^2$ as $r \to \infty$. We therefore obtain the following asymptotic behavior:

$$g_{kl}(r) \sim 2\sqrt{\pi}\sqrt{2l+1}\,(-i)^l \frac{1}{kr}\sin\left(kr - \frac{l}{2}\pi\right).$$

We may choose the yet undefined constants in η_{kl} such that η_{kl} has, according to (5.1.5), the asymptotic behavior

$$\eta_{kl}(r) \sim \frac{1}{r}\sin\left(kr - \frac{l}{2}\pi + \tilde{\delta}_l(k)\right).$$

In this way the constants a_{lm} in (5.1.6) are uniquely determined. It follows that

$$\begin{aligned} \Omega_-\left(\frac{1}{(2\pi)^{3/2}}e^{i\vec{k}\cdot\vec{r}}\right) &= \psi_{\vec{k}}(\vec{r}) \\ &= \frac{1}{(2\pi)^{3/2}}\sum_{l=0}^{\infty}\frac{2\sqrt{\pi}}{k}\sqrt{2l+1}\,(-i)^l e^{i\tilde{\delta}_l(k)}\eta_{kl}(r)Y_0^l(\theta). \end{aligned} \tag{5.1.10}$$

Thus $f_k(\theta)$ is also determined by (5.1.6); we obtain

$$f_k(\theta) = \sum_{l=0}^{\infty} Y_0^l(\theta)\frac{1}{k}\sqrt{\pi}\sqrt{2l+1}\,(-i)^{l+1}e^{-i(l\pi/2)}(e^{2i\tilde{\delta}_l(k)} - 1). \tag{5.1.11}$$

We shall now determine S. Since $\Omega_+^+\psi_{\vec{k}}(\vec{r})$ must be a solution of (4.7.9) we may now write

$$\Omega_+^+\psi_{\vec{k}}(\vec{r}) = \frac{1}{(2\pi)^{3/2}}\sum_{l=0}^{\infty} b_l g_{kl}(r)Y_0^l(\theta),$$

where b_l are undetermined coefficients. Asymtotically it therefore follows that

$$\Omega_+^+\psi_{\vec{k}}(\vec{r}) \sim \frac{1}{(2\pi)^{3/2}} \sum_{l=0}^{\infty} b_l Y_0^l(\theta) 2\sqrt{\pi}\sqrt{2l+1}(-i)^l \frac{1}{kr} \sin\left(kr - \frac{l\pi}{2}\right).$$

The outgoing spherical waves of $\Omega_+^+\psi_{\vec{k}}(\vec{r})$ and $\psi_{\vec{k}}(\vec{r})$ must, however be identical from which it follows that

$$b_l = e^{2i\tilde{\delta}_l(k)}. \tag{5.1.12}$$

Therefore we obtain

$$S\left(\frac{1}{(2\pi)^{3/2}} e^{i\vec{k}\cdot\vec{r}}\right) = \frac{1}{(2\pi)^{3/2}} \sum_{l=0}^{\infty} e^{2i\tilde{\delta}_l(k)} g_{kl}(r) Y_0^l(\theta) \tag{5.1.13}$$

from which it follows that

$$S g_{kl}(r) Y_0^l(\theta) = e^{2i\tilde{\delta}_l(k)} g_{kl}(r) Y_0^l(\theta), \tag{5.1.14}$$

that is, $\tilde{\delta}_l(k)$ is identical to the quantity $\delta_{lj}(k) = \delta_l(k)$ defined in (5.12).

Therefore the scattering operator S can be obtained by solving the equation (5.1.4) and by computing the phase shifts determined by equation (5.1.5).

5.2 The Born Approximation

The approximate solution of (4.7.8) obtained by an iteration scheme is known as the Born approximation. Here we shall only consider the first step for the case in which $V(r)$ is a potential. The first iteration step consists of replacing $\psi_{\vec{k}}(\vec{r})$ on the right-hand side of (4.7.8) by $(1/(2\pi)^{3/2})e^{i\vec{k}\cdot\vec{r}}$. It then follows that

$$\psi_{\vec{k}}(\vec{r}) = \frac{e^{i\vec{k}\cdot\vec{r}}}{(2\pi)^{3/2}} - \frac{m}{(2\pi)^{5/2}} \int \frac{e^{ik|\vec{r}-\vec{r}'|}}{|\vec{r}-\vec{r}'|} V(\vec{r}')e^{i\vec{k}\cdot\vec{r}'}\, d^3\vec{r}'.$$

For $\vec{e} = \vec{r}/|\vec{r}|$, for large $|\vec{r}|$ we obtain the asymptotic behavior:

$$e^{ik|\vec{r}-\vec{r}'|} \sim e^{ik|\vec{r}| - ik\vec{r}'\cdot\vec{e}}.$$

In this we obtain the $f_k(\theta)$ defined in (5.1.6):

$$f_k(\theta) = -\frac{m}{2\pi} \int V(r')e^{ik\vec{r}\cdot(\vec{e}_0 - \vec{e})} d^3\vec{r}',$$

where $\vec{e}_0 = \vec{k}/|\vec{k}|$. Since θ is the angle between $\vec{k}$ and $\vec{r}$ it follows that

$$\begin{aligned} f_k(\theta) &= -2m \frac{1}{k|\vec{e}_0 - \vec{e}|} \int_0^{\infty} r'V(r') \sin(kr'|\vec{e}_0 - \vec{e}|)\, dr' \\ &= \frac{m}{k \sin(\theta/2)} \int_0^{\infty} r'V(r') \sin\left(2kr' \sin\frac{\theta}{2}\right) dr'. \end{aligned} \tag{5.2.1}$$

We shall now consider the example of Rutherford scattering; in particular, we shall consider the scattering of an electron by a nucleus of charge Z, that is, $V(r) = -Ze^2/r$. In order to make it possible to apply scattering theory, and in order to obtain better descriptions of real experiments (see §4.5), we shall replace $V(r)$ by $V(r)\,e^{-\alpha r}$ and let $\alpha \to 0$. Then we obtain

$$f_k(\theta) = \frac{mZe^2}{2k^2 \sin^2(\theta/2)} \tag{5.2.2}$$

which should, except for small θ, be a good approximation. For θ close to zero it is, however, extremely difficult to distinguish the scattered systems from nonscattered systems (see §6).

5.3 Scattering of an Electron by a Hydrogen Atom

Here we will now give an example of a multiple-channel scattering experiment which will not only illustrate the general discussions of §4.9 but will also serve as a demonstration for the discussion to follow in XVII, §§6.2 and 6.4.

The problem of the scattering of an electron by a hydrogen atom will be simplified in that we shall consider the nucleus as a fixed stationary field of force. The fact that this is permissible in the case of scattering experiments because of the large mass of the nucleus (this is also true for the case of light emission; see XI to XIV) will be justified in more detail in §6.2.

The Hamiltonian operator has the form XII (2.1) where $Z = 1$. We will neglect the effect of spin. In stationary scattering theory (that is, in the determination of Ω_-) we shall seek a solution ϕ of

$$H\phi(\vec{r}_1, \vec{r}_2) = \left(\frac{k^2}{2m} + E_1\right)\phi(\vec{r}_1, \vec{r}_2) \tag{5.3.1}$$

of the form

$$\phi_\pm(\vec{r}_1, \vec{r}_2) = \frac{1}{(2\pi)^{3/2}}\,[e^{i\vec{k}\cdot\vec{r}_1}\varphi_1(r_2) \pm e^{i\vec{k}\cdot r_2}\varphi_1(\vec{r}_1) + \chi_{\vec{k}}^{\pm}(\vec{r}_1, \vec{r}_2)], \tag{5.3.2}$$

where $\chi_{\vec{k}}^{\pm}$ describes the outgoing spherical wave. Here $\varphi_1(\vec{r})$ is the eigenfunction for the ground state of the hydrogen atom where the energy eigenvalue is E_1.

We shall now determine the form of W_0^i according to (4.8.1). To this purpose we shall make assertions about W_{10}^i and W_{20}^i, that is, assertions about how the experiment is to be carried out.

We shall assume that the colliding electrons and the hydrogen atoms are "unpolarized", that is, the following condition with respect to the product representation $\mathscr{H}_b \times \imath_s$ of the Hilbert space of an electron holds:

$$W_{10}^i = \tfrac{1}{2}W_{10b}^i \times \mathbf{1}, \qquad W_{20}^i = \tfrac{1}{2}P_{\varphi_1(\vec{r}_2)} \times \mathbf{1}.$$

Since we may write

$$W^i_{10b} = \sum_\nu w_\nu P_{\psi_\nu},$$

where

$$\psi_\nu(\vec{r}_1) = \frac{1}{(2\pi)^{3/2}} \int e^{i\vec{k}\cdot\vec{r}_1} \chi_\nu(\vec{k})\, d^3\vec{k}$$

it suffices to only consider the case $W^i_{10b} = P_\psi$. According to (4.8.1) we obtain

$$W^i_0 = \alpha[(P_\psi \times P_{\varphi_1}) \times (\mathbf{1} \times \mathbf{1}) + (P_{\varphi_1} \times P_\psi) \times (\mathbf{1} \times \mathbf{1})] \tag{5.3.3}$$

with respect to the product representation $(\mathscr{H}^2_b \times \imath^2_s)$ using the normalization factor α. As we explained in §4.8, we may assume that ψ is orthogonal to φ_1. Then, according to (5.3.3) W^i_0 as an operator in

$$\{\mathscr{H}^2_b \times \imath^2_s\}_- = \{\mathscr{H}^2_b\}_+ \times \{\imath^2_s\}_- \oplus \{\mathscr{H}^2_b\}_- \times \{\imath^2_s\}_+,$$

will have the form

$$W_0 = \tfrac{1}{4}[P_{\psi_+} \times P_0 + P_{\psi_-} \times P_1], \tag{5.3.4}$$

where

$$\psi_+(\vec{r}_1, \vec{r}_2) = \frac{1}{\sqrt{2}}(\psi(\vec{r}_1)\varphi_1(\vec{r}_2) + \varphi_1(\vec{r}_1)\psi(\vec{r}_2)),$$

$$\psi_-(\vec{r}_1, \vec{r}_2) = \frac{1}{\sqrt{2}}(\psi(\vec{r}_1)\varphi_1(\vec{r}_2) - \varphi_1(\vec{r}_1)\psi(\vec{r}_2))$$

where P_0 is an operator in $\imath^2_s$ and is the projection onto the subspace $\{\imath^2_s\}_-$ with total spin 0 and P_1 is the projection onto $\{\imath^2_s\}_+$ with total spin 1.

From the form (5.3.4) we suggest that it is reasonable to seek solutions of the form (5.3.2) because it will then be possible to determine the operator

$$W = \Omega_- W^i_0 \Omega^+_-.$$

In order to obtain solutions of (5.3.1) of the form (5.3.2) it suffices to first find solutions ϕ of (5.3.1) of the form

$$\phi(\vec{r}_1, \vec{r}_2) = \frac{1}{(2\pi)^{3/2}} e^{i\vec{k}\cdot\vec{r}} \varphi_1(r_2) + \cdots \tag{5.3.5}$$

[here $+ \cdots$ refers to outgoing spherical wave terms] because $\phi_\pm(\vec{r}_1, \vec{r}_2) = \phi(\vec{r}_1, \vec{r}_2) \pm \phi(\vec{r}_2, \vec{r}_1)$ are solutions of (5.3.1) which satisfy the conditions (5.3.2).

In order to find solutions of (5.3.1) of the form (5.3.5), we decompose H in the following way:

$$H = H_{01} + H_{A2} - \frac{Ze^2}{r_1} + \frac{e^2}{r_{12}} \tag{5.3.6}$$

where

$$H_{01} = \tfrac{1}{2}\vec{P}_1^2, \qquad H_{A2} = \frac{1}{2m}\vec{P}_2^2 - \frac{Ze^2}{r_2}. \tag{5.3.7}$$

H_A is the Hamiltonian operator of the hydrogen atom.

In terms of the notation used in XI, §§4 and 5, we may write $\phi(\vec{r}_1, \vec{r}_2)$ in the $(H_A, \vec{L}^2, L_3, \vec{Q}_1)$ representation as follows:

$$\phi(\vec{r}_1, \vec{r}_2) \to \langle \varepsilon, l, m; \vec{r}_1 | \phi \rangle. \tag{5.3.8}$$

Here ε runs through the discrete and continuous spectrum of H_A. In this representation (5.3.1) becomes

$$\left(-\frac{1}{2m_1}\Delta_1 - \frac{Ze^2}{r_1}\right)\langle \varepsilon, l, m; \vec{r}_1 | \phi \rangle + e^2 \sum_{l', m'} \int_0^\infty \left\langle \varepsilon, l, m \left| \frac{1}{r_{12}} \right| \varepsilon', l', m' \right\rangle$$
$$\cdot \langle \varepsilon', l', m'; \vec{r}_1 | \phi \rangle \, d\varepsilon' + \sum_{l', m', n} \left\langle \varepsilon, l, m \left| \frac{1}{r_{12}} \right| E_n, l', m' \right\rangle$$
$$\cdot \langle E_n, l', m'; \vec{r}_1 | \phi \rangle = \left(\frac{k^2}{2m} + E_1 - \varepsilon\right) \langle \varepsilon, l, m; \vec{r}_1 | \phi \rangle. \tag{5.3.9}$$

In this way $\langle \varepsilon, l, m | 1/r_{12} | \varepsilon', l', m' \rangle$ is a function of r_1, namely the matrix which represents $|\vec{r}_1 - \vec{r}_2|^{-1}$ as an operator in $\mathscr{H}_{b_2}$ which depends upon the parameter $\vec{r}_1$.

Of course, it is not possible to obtain an exact solution of (5.3.9). Thus we shall consider two approximations.

For the case of low excitation probabilities (that is, for sufficiently slow electrons) we consider the equation (5.3.9) for $\varepsilon = E_1$, where we neglect all nondiagonal terms of $\langle E_1, 0, 0 | 1/r_{12} | \varepsilon', l' \, m' \rangle$. We therefore obtain

$$\left[-\frac{1}{2m_1}\Delta_1 + V(r_1)\right] \langle E_1, 0, 0; \vec{r}_1 | \phi \rangle = \frac{k^2}{2m} \langle E_1, 0, 0; \vec{r}_1 | \phi \rangle, \tag{5.3.10}$$

where

$$V(r_1) = -\frac{Ze^2}{r_1} + e^2 \left\langle E_1, 0, 0 \left| \frac{1}{r^{12}} \right| E_1, 0, 0 \right\rangle. \tag{5.3.11}$$

$V(r_1)$ is, according to (5.3.11), the Coulomb potential for a nucleus shielded by the charge density $-e|\varphi_1(\vec{r})|^2$. Condition (5.3.5) may then be expressed as follows

$$\langle E_1, 0, 0; \vec{r}_1 | \phi \rangle = \frac{1}{(2\pi)^{3/2}} e^{i\vec{k}\cdot\vec{r}_1} + \cdots, \tag{5.3.12}$$

where $+ \cdots$ refers to outgoing spherical wave terms. Here (5.3.10) together with (5.3.11) represent nothing other than a scattering problem with potential function $V(r)$. For slow electrons the atom behaves like an elementary system with the interaction potential $V(r)$ between the atom and the colliding electron given by (5.3.11). This example illustrates the importance of scattering problems (other than the Rutherford scattering in §5.2) which are described by a potential function.

The quantity $f_k(\theta)$ is determined by the solution of (5.3.10) and the condition (5.3.12) [its asymptotic behavior] as follows:

$$\langle E_1, 0, 0; \vec{r}|\phi\rangle \sim \frac{1}{(2\pi)^{3/2}} e^{i\vec{k}\cdot\vec{r}} + \frac{1}{(2\pi)^{3/2}} f_k(\theta) \frac{e^{ikr}}{r} \tag{5.3.13}$$

a result which is of great importance for the computation of the interaction cross section in §6.3.

We shall now assume that we have solved (5.3.10). Then, in the next approximation step of equation (5.3.9) for $\varepsilon = E_n$ where $n \neq 1$ we only retain the sum of two terms:

$$\begin{aligned}\left[-\frac{1}{2m}\Delta_1 - \frac{Ze_2}{r_1} - \left(\frac{k^2}{2m} + E_1 - E_n\right)\right] \langle E_n, l, m; \vec{r}_1|\phi\rangle \\ + e^2 \sum_{l', m'} \left\langle E_n, l, m \left| \frac{1}{r_{12}} \right| E_n, l', m' \right\rangle \langle E_n, l', m'; \vec{r}_1|\phi\rangle \\ = -e^2 \left\langle E_n, l, m \left| \frac{1}{r_{12}} \right| E_1, 0, 0 \right\rangle \langle E_1, 0, 0; \vec{r}_1|\phi\rangle.\end{aligned} \tag{5.3.14}$$

Condition (5.3.5) takes on the following form for $\langle E_n, l, m; \vec{r}|\phi\rangle$ $(n \neq 1)$

$$\langle E_n, l, m; \vec{r}_1|\phi\rangle = \text{outgoing spherical waves only.} \tag{5.3.15}$$

This condition may be satisfied because (6.3.14) represents an inhomogeneous equation for $\langle E_n, l, m; \vec{r}_1|\phi\rangle$. For its solution we need only choose the Green's function for the operator on the left side of (6.3.14) which satisfies the "radiation condition", that is, for large distances, only contains outgoing spherical waves. In order that such a Green's function exist, it is necessary that

$$\frac{k^2}{2m} + E_1 - E_n > 0,$$

that is, the kinetic energy of the colliding electrons must be greater than the excitation energy $E_n - E_1$. If, for example, $k^2/2m < E_2 - E_1$, then there is no excitation, that is, in the sense of the collision process, for $t \to \infty$ only a scattered electron and an atom in the ground state can remain. For elastic scattering the solution of (5.3.10) with (5.3.11) is only an approximation, albeit, a very good one.

In the case in which $k^2/2m < E_2 - E_1$ there are, according to (5.3.9) additional possibilities for an elastic collision, namely a term in (5.3.5) which, for large r_2, takes on the form

$$\cdots \varphi_1(r_1) \frac{e^{ikr_2}}{r_2},$$

that is, the case in which electron #1 remains in the ground state and electron #2 is distant from the nucleus. Here (5.3.9) is ill-suited for an approximate computation of this effect because it would be necessary to work with a complicated eigenfunction $\langle \vec{r}_2 | \varepsilon, l, m \rangle$ of H_{A2}. Instead, we return to (5.3.1) and make the following approximation assumption:

$$\phi(\vec{r}_1, \vec{r}_2) = \chi(\vec{r}_1)\varphi_1(\vec{r}_2) + \varphi_1(\vec{r}_1)\eta(\vec{r}_2), \tag{5.3.16}$$

where $\chi(\vec{r}_1)$ denotes the solution of (5.3.10) subject to the condition (5.3.12).

From (5.3.1) it follows that with

$$\tilde{V}(r) = \int \frac{|\varphi_1(\vec{r}')|^2}{|\vec{r} - \vec{r}'|} d^3\vec{r}' \tag{5.3.17}$$

we obtain the following equation for $\eta(\vec{r}_2)$:

$$\begin{aligned} &\left[-\frac{1}{2m} \Delta_2 - \frac{Ze^2}{r_2} + e^2 \tilde{V}(r_2) - \frac{k^2}{2m} \right] \eta(\vec{r}_2) \\ &\quad = e^2 \varphi(\vec{r}_2) \int \left(\tilde{V}(r_1) - \frac{1}{|\vec{r}_1 - \vec{r}_2|} \right) \varphi_1(\vec{r}_1)\chi(\vec{r}_1) \, d^3\vec{r}_1. \end{aligned} \tag{5.3.18}$$

This equation may be solved in the same way as (5.3.14) with the help of the Green's function which satisfies the radiation condition.

The asymptotic form of the solution of (5.3.18) is given by

$$\eta(\vec{r}_2) \sim \frac{1}{(2\pi)^{3/2}} h_k(\theta) \frac{e^{ikr_2}}{r_2} \tag{5.3.19}$$

and determines the important amplitude function $h_k(\theta)$.

We may also seek to obtain another approximation for the solutions of (5.3.10) and (5.3.18) by using the Born approximation, as in §5.2. For $\chi(\vec{r}_1)$ from (5.3.10) and (5.3.13) it follows that

$$f_k(\theta) = -\frac{m}{2\pi} \int V(\vec{r}') e^{ik\vec{r}' \cdot (\vec{e}_0 - \vec{e})} \, d^3\vec{r}', \tag{5.3.20}$$

where $V(\vec{r})$ is given by (5.3.11). In order to compute η from (5.3.18) we simplify (5.3.18) as follows:

$$\left(-\frac{1}{2m} \Delta_2 - \frac{k^2}{2m} \right) \eta(\vec{r}_2) = \frac{e^2}{(2\pi)^{3/2}} g(\vec{r}_2), \tag{5.3.21}$$

where

$$g(\vec{r}_2) = \varphi_1(r_2) \int \left(\tilde{V}(\vec{r}_1) - \frac{1}{|\vec{r}_1 - \vec{r}_2|} \right) \varphi_1(\vec{r}_1) e^{i\vec{k} \cdot \vec{r}_1} \, d^3\vec{r}_1. \tag{5.3.22}$$

Thus it follows that

$$\eta(\vec{r}_2) = -\frac{me^2}{(2\pi)^{5/2}} \int \frac{e^{ik|\vec{r}_2 - \vec{r}'|}}{|\vec{r}_2 - \vec{r}'|} g(\vec{r}') \, d^3\vec{r}'$$

from the asymptotic form (5.3.19) we obtain

$$h_k(\theta) = -\frac{me^2}{2\pi}\int e^{-ik\vec{r}'\cdot\vec{e}}g(\vec{r}')\,d^3\vec{r}', \tag{5.3.23}$$

where $\vec{e}$ is the direction of $\vec{r}_2$.

In the same approximation we may solve (5.3.13). For $k_n^2 = k^2 + 2m(E_1 - E_n)$ it follows that

$$\langle E_n, l, m; \vec{r}_1|\phi\rangle = -\frac{me^2}{(2\pi)^{5/2}}\int\frac{e^{ik_n|\vec{r}_1-\vec{r}'|}}{|\vec{r}_1-\vec{r}'|}\left\langle E_n, l, m\left|\frac{1}{|\vec{r}'-\vec{r}_2|}\right|E_1, 0, 0\right\rangle \cdot e^{i\vec{k}\cdot\vec{r}'}\,d^3\vec{r}'$$

and, in asymptotic form, we obtain

$$\langle E_n, l, m; \vec{r}_1|\phi\rangle \sim -\frac{me^2}{(2\pi)^{5/2}}\frac{e^{ik_n|\vec{r}_1|}}{|\vec{r}_1|} \cdot \int e^{i(k\vec{e}_0-k_n\vec{e})\cdot\vec{r}'}\left\langle E_n, l, m\left|\frac{1}{|\vec{r}'-\vec{r}_2|}\right|E_1, 0, 0\right\rangle d^3\vec{r}'. \tag{5.3.24}$$

We will not attempt to evaluate the integrals in (5.3.20), (5.3.23) and (5.3.24). The derived expressions were obtained more to demonstrate the methods than to show (in approximation) the exact behavior of the individual scattering channels (after the scattering).

The system of equations (5.3.9) or the equation (5.3.1) also permit the possibility of describing the ionization of the hydrogen atom by means of collisions with electrons. For approximation methods for such computations we refer readers to the literature [20].

6 Examples of Registrations in Scattering Experiments

Now that we have considered (in §§4 and 5) the mathematical questions concerning the existence and the computational methods for the Wave Operators and Scattering Operator, we now turn to physical questions concerning the registration of systems after the scattering has occurred. In §2 we have only sought to obtain a mathematical formulation of what we mean by the expression "after" the scattering process. In order to obtain experimentally verifiable results we must know which effects are to be measured in the individual experiments. Since we have not yet developed a theory which permits us, on the basis of the construction of the registration apparatus, to draw conclusions about the nature of the corresponding effects, it remains for us to discover such F with the aid of theoretical considerations and experimental experience, and then to use the above information in order to establish axioms together with the corresponding mapping principles. Our procedures will be completely analogous to the procedures described and used in XI, §1. We will also show that the registration effects for scattering experiments are not, in general, decision effects.

6.1 The Effects of the "Impact" of a Microsystem on a Surface

The apparatuses by which microsystems are registered in scattering experiments have, in general, a wide variety of forms. Experimental physicists are always inventing new registration methods. Here it is usually necessary that the operator $F = \psi(b_0, b)$ which corresponds to a given effect procedure (b_0, b) must be determined from a mixture of theoretical and experimental considerations. There should be no doubt that working experimental physicists will take great care in this direction even though they may not express that care by using such an abstract expression as "seeking the operator $\psi(b_0, b)$". In their considerations, the intuitive desire for a special operator $F = \psi(b_0, b)$ plays an important role: They seek to construct a "surface" $\mathscr{F}$ such that "every microsystem reaching the surface" is registered. What is the meaning of these "intuitive" words?

We return to the special case of an elementary microsystem. In the Schrödinger picture, and in the position representation an ensemble of the form P_{ψ_t} can be described by a ψ_t which satisfies the Schrödinger equation

$$-\frac{1}{i}\frac{\partial}{\partial t}\langle\vec{r}\,|\psi_t\rangle = -\frac{1}{2m}\Delta\langle\vec{r}\,|\psi_t\rangle. \tag{6.1.1}$$

The position probability density is given by

$$w(\vec{r}, t) = |\langle\vec{r}\,|\psi_t\rangle|^2. \tag{6.1.2}$$

A direct consequence of (6.1.1) is the equation of continuity

$$\frac{\partial}{\partial t}w(\vec{r}, t) + \operatorname{div}\vec{s}(\vec{r}, t) = 0, \tag{6.1.3}$$

where

$$\vec{s}(\vec{r}, t) = \frac{1}{2im}[\overline{\langle\vec{r}\,|\psi_t\rangle}\operatorname{grad}\langle\vec{r}\,|\psi_t\rangle - \langle\vec{r}\,|\psi_t\rangle\operatorname{grad}\overline{\langle\vec{r}\,|\psi_t\rangle}]. \tag{6.1.4}$$

Here $\vec{s}(\vec{r}, t)$ is called the probability current density because of its role in equation (6.1.3). Such terminology does not, however, specify if and how the named quantities may be measured. From Gauss' theorem it follows that by integrating (6.1.3) over a volume $\mathscr{V}$ bounded by the surface $\mathscr{F}$ we obtain

$$\frac{d}{dt}\int_{\mathscr{V}} w(\vec{r}, t)\, d^3\vec{r} + \int_{\mathscr{F}} \vec{s}(\vec{r}, t)\cdot\vec{n}\, df = 0, \tag{6.1.5}$$

where $\vec{n}$ is the (outwardly pointed) unit normal vector to the surface element.

$$\int_{\mathscr{V}} w(\vec{r}, t)\, d^3\vec{r} \tag{6.1.6}$$

is the probability of finding a microsystem in the volume $\mathscr{V}$ for a measurement of the "position at time t". This probability may depend upon time. Since the position at different times are not commensurable, the time dependence of the probability (6.1.6) *cannot be determined* by measuring the position at various times on the same systems; instead, it must be determined in the following way: We first measure the "position at time t_1" on systems prepared by a procedure a for which $\varphi(a) = P_\psi$. Then, in another experiment we measure the "position at time t_2" on *other microsystems* which are prepared by the *same procedure a.* The time dependence of (6.1.6) (which is measurable only in the above way) satisfies (6.1.5). This fact suggests the following *intuitive* interpretation of $\vec{s}(\vec{r}, t)$: $\vec{s}(\vec{r}, t)$ represents something like the net flow of particles through the surface area. With this intuitive interpretation, it is easy to be deceived about an actual physical problem: Here it is often believed that it has been proven that

$$\vec{s}(\vec{r}, t) \cdot \vec{n} \Delta f \Delta t \tag{6.1.7}$$

describes the "probability" that a microsystem has passed through the surface element Δf in the time interval Δt. Underlying this description there are physical problems because we do not know how to measure a quantity like (6.1.7). We will now show that (6.1.7) has the form $\operatorname{tr}(WA)$ where $W = P_\psi$. Instead of (6.1.7) we shall use the mathematically more precise expression

$$\int_{t_1}^{t_2} dt \int_{\mathscr{F}} \vec{s}(\vec{r}, t) \cdot \vec{n}\, df, \tag{6.1.8}$$

where $\mathscr{F}$ may be an arbitrary nonclosed surface. Using the formula

$$\langle \vec{r} | \psi_t \rangle = \frac{1}{(2\pi)^{3/2}} \int \langle \vec{k} | \psi_0 \rangle e^{i(\vec{k}\cdot\vec{r} - (k^2/2m)t)} d^3\vec{k} \tag{6.1.9}$$

for the transition from the position to the momentum representation, it follows that the expression (6.1.8) is equal to

$$\iint \overline{\langle \vec{k} | \psi_0 \rangle} \frac{1}{(2\pi)^3} \int_{t_1}^{t_2} e^{(i/2m)(k^2 - k'^2)t}\, dt \int_{\mathscr{F}} e^{i(\vec{k}' - \vec{k})\cdot\vec{r}} \frac{1}{2m} (\vec{k} + \vec{k}') \cdot \vec{n}\, df_{\vec{r}}$$
$$\cdot \langle \vec{k}' | \psi_0 \rangle\, d^3\vec{k}\, d^3\vec{k}'.$$

We now define a self-adjoint operator A as follows:

$$A\langle \vec{k} | \psi \rangle = \int \langle \vec{k} | A | \vec{k}' \rangle \langle \vec{k}' | \psi \rangle\, d^3\vec{k}', \tag{6.1.10}$$

where

$$\langle \vec{k} | A | \vec{k}' \rangle = \frac{1}{(2\pi)^3} \int_{t_1}^{t_2} e^{(i/2m)(k^2 - k'^2)t}\, dt \int_{\mathscr{F}} e^{i(\vec{k}' - \vec{k})\cdot\vec{r}} \frac{1}{2m} (\vec{k} + \vec{k}') \cdot \vec{n}\, df_{\vec{r}}$$

then (6.1.8) takes on the form

$$\operatorname{tr}(WA), \quad \text{where } W = P_{\psi_0}. \tag{6.1.11}$$

Here (6.1.11) has the form of the Heisenberg picture because P_{ψ_0} is time independent, while the operator A, under displacements of t_1, t_2 by γ changes, corresponding to the Heisenberg picture. For A given by (6.1.10) $\operatorname{tr}(WA)$ is determined for arbitrary W. What is, however, the physical interpretation of A in the formula $\operatorname{tr}(WA)$? Is A an observable or an effect? Since A is obviously not a positive definite operator, A cannot be an effect.

If we interpret A as a decision observable, then from (6.1.10) it follows that A^2 is not a projection operator, that is, A has eigenvalues (and therefore measurement values) other than 1, 0 and -1. This is hardly what we would expect "intuitively" by a probability current. From the expression (6.1.11) alone we cannot make any conclusions concerning the physical interpretation of the operator A! Failure to take heed of this general result often leads to errors.

The quantity A obtained from (6.1.4) should, however, have something to do with the passage of a microsystem through a surface $\mathscr{F}$. Since, however, there are no objective properties for microsystems described by quantum mechanics (see IV, §8), it therefore makes no sense to say whether or not a microsystem, as such, passes through the surface $\mathscr{F}$ in the time interval t_1 to t_2. A "registerable" effect can only be produced or not produced at a registration apparatus. We must therefore have a registration apparatus which is capable (at least in principle) to register this "passage".

As our first attempt it seems that it would be reasonable to indirectly define an observable (by analogy with the indirect definition of the position observable) which registers the times for which a system passes from one side to the other of the surface $\mathscr{F}$. Here it is reasonable to construct the following Boolean ring Σ:

We begin by selecting intervals on the time scale. To each such interval $\mathscr{I}$ we assign a pair of indices $(+)$ and $(-)$: $\mathscr{I}_+$ and $\mathscr{I}_-$. We define Σ to be the Boolean ring generated by these intervals $\mathscr{I}_+$ and $\mathscr{I}_-$, where for each pair of intervals $\mathscr{I}_+$ and $\mathscr{I}_-$ we require that $\mathscr{I}_+ \wedge \mathscr{I}_- = 0$.

We now seek to obtain observables $\Sigma \overset{F}{\rightarrow} L$ such that $F(\mathscr{I}_+)$, $F(\mathscr{I}_-)$ have the following intuitive meaning: the microsystem passes through the surface $\mathscr{I}$ in the time interval t_1 to t_2 from left to right and from right to left, respectively. In order to justify this intuitive description we shall, among other things, assert that $F(\mathscr{I}'_+) = e^{iH\gamma}F(\mathscr{I}_+)e^{-H\gamma}$ where $\mathscr{I}'$ is obtained from $\mathscr{I}$ by a time displacement γ. This requirement cannot be satisfied by a decision observable! For nondecision observables we have no intuitive basis by which we may *uniquely* define the map F. On the contrary, we expect that real registration procedures should be described by effects which are inseparable(!) from the actual details of the construction of the registration apparatus, even in the case in which they appear as *intuitively* natural as a registration whether a microsystem "passes" or "strikes" the surface $\mathscr{F}$ or not.

Nevertheless, experimental physicists look for effects after scattering which, for all practical purposes, do not depend upon the particular construction of the registration procedures; they measure, for example, the so-called "interaction cross sections" which depend only on the interaction between the colliding microsystems. How is this possible?

This possibility essentially depends upon two facts: First, experimental physicists use only "collection surfaces" which are large compared to $|\vec{k}|^{-1}$ where the $\vec{k}$ are the various values of the momentum where $\langle \vec{k} | \psi_0 \rangle$ is, for all practical purposes, nonzero. Second, in scattering experiments experimental physicists do not use ensembles for which differences between various registration methods are readily detectable. These two different viewpoints are, of course, not mutually independent because the form of the surface $\mathscr{F}$ and the ensemble have to be adapted to each other.

Such situations are not uncommon in physics, but are disturbing to mathematicians because they are not aesthetically pleasing. Physics does not always present aesthetically pleasing proofs. In fact, the situation described above could be considered to be a type of asymptotic approximate description, which we actually have already introduced from the beginning of our description of the scattering process. The operator A in (6.1.10) has already been introduced in order to obtain an asymptotic approximate description of the effect of a system striking a surface area $\mathscr{F}$.

We now consider a plane E where $\mathscr{F}$ will refer to different portions of this plane. We shall denote the operator defined in (6.1.10) by $A(\mathscr{I}, \mathscr{F})$ where $\mathscr{I}$ is the time interval between t_1 and t_2.

$A(\mathscr{I}, \mathscr{F})$ defines an additive measure on the Boolean ring $\Sigma_{t,E}$ generated by the intervals $\mathscr{I}$ and the surfaces $\mathscr{F}$, a result which follows directly from (6.1.10). This measure $\Sigma_{t,E} \xrightarrow{A} \mathscr{B}'(\mathscr{H})$ is, however, not positive. Here it may be suggested splitting A into positive and negative parts. For scalar measures such a decomposition is unique (see IV, §2.1). We will not discuss this question for measures for which the domain is a subset of a vector space $\mathscr{B}'(\mathscr{H})$, because it is first necessary to examine the physical situation more closely.

The actual situation of registration is not identical with any intuitive description of the passage of microsystems through a surface $\mathscr{F}$. On the contrary, experimentally we seek to construct "detection surfaces" in which the systems can only "strike" the surface from one side in order to produce the registration effect (see Figure 39). This diagram is to be understood only in a very symbolic way. The technical details of the construction are not specified in Figure 39, but apart from all details the construction shall be

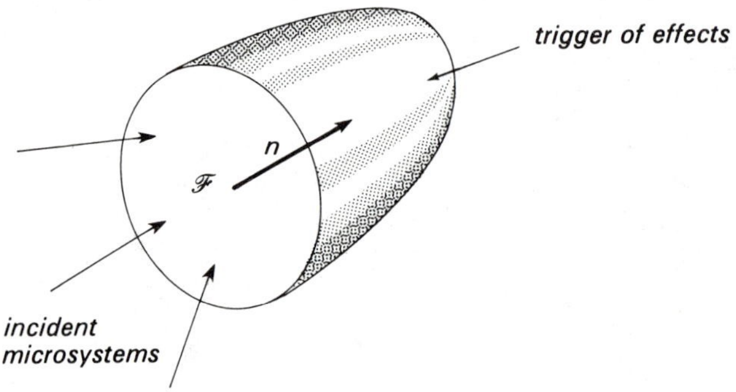

Figure 39

made in such a way that a "sensitive" detecting surface $\mathcal{F}$ is defined. The microsystems interact with the material of the detecting surface in such a way that a macroscopic effect is triggered at time t.

From the theory of diffraction (in particular, diffraction of light) it is well known that it is possible to describe the radiation falling on a surface $\mathcal{F}$ to very good approximation as if the radiation from behind (right in Figure 39) has been shielded, and the radiation from the front is unhindered. The diffraction phenomena at the apparatus can therefore be neglected, providing that we are only interested in the radiation falling upon the surface $\mathcal{F}$. We shall now seek to formulate this approximation more precisely.

For simplicity we shall choose the coordinate system such that the plane E is spanned by the (2) and (3) axis, and the (1) axis points towards the interior of the apparatus (see Figure 39). Here it is reasonable to characterize the incoming radiation on the surface $\mathcal{F}$ by requiring that the operator A in (6.1.10) be replaced by the null operator whenever $\vec{k}\cdot\vec{n}$ or $\vec{k}'\cdot\vec{n}$ are negative. We now define an operator F (using the coordinate system introduced above) as follows:

$$F(\mathcal{I}, \mathcal{F}')\langle\vec{k}|\psi\rangle = \int\langle\vec{k}|F|\vec{k}'\rangle\langle\vec{k}'|\psi\rangle\, d^3\vec{k}',$$

where

$$\langle\vec{k}|F|\vec{k}'\rangle = \begin{cases} \dfrac{1}{(2\pi)^3}\displaystyle\int_{t_1}^{t_2} e^{(i/2m)(\vec{k}^2-\vec{k}'^2)}\,dt \int e^{i[(k_2'-k_2)x_2+(k_3'-k_3)x_3]} \\ \quad\cdot\dfrac{1}{2m}(k_1+k_1')\,dx_2\,dx_3 \quad \text{for } k_1>0,\ k_1'>0, \\ 0 \quad \text{otherwise.}\end{cases} \tag{6.1.12}$$

$F(\mathcal{I}, \mathcal{F})$ defines, like $A(\mathcal{I}, \mathcal{F})$, an additive measure on the Boolean ring $\Sigma_{t,E}$. The conjecture that $\Sigma_{t,E} \xrightarrow{F} L(\mathcal{H})$ is, unfortunately false, although in (6.1.12) $k_1 > 0$ and $k_2 > 0$, because the operator defined by (6.1.12) is not positive definite. Therefore (6.1.12) does not represent an exact observable, that is, it cannot precisely define the production of effects.

Equation (6.1.12) should, however, approximate the measure $\mathcal{R}(b_0)\to L(\mathcal{H})$ for a registration method b_0. On the basis of the actual characteristics of an apparatus, an experimental physicist knows that the occurrence of a detection signal cannot be localized to arbitrarily small time intervals $(t_2 - t_1)$. In addition, the surface $\mathcal{F}$ may not be too small in order that diffraction should not play a role, as we have described earlier. In addition, a real apparatus cannot respond to arbitrarily small k_1. The approximation of $\mathcal{R}(b_0) \to L(\mathcal{H})$ by $F(\mathcal{I}, \mathcal{F})$ is therefore meaningful only if $\Delta t = t_2 - t_1$ is not too small, $\mathcal{F}$ is not too small and the ensemble does not contain too small components k_1 of momentum.

If Δt and $\mathcal{I}$ is not too small, then

$$\int_{t_1}^{t_2} e^{(i/2m)(k^2-k'^2)}\,dt \int_{\mathcal{F}} e^{i[(k'_2-k_2)x_2+(k'_3-k_3)x_3]}\,dx_2\,dx_3$$

decreases so rapidly in $k_1 - k_1'$, that in the slowly variable factor $(k_1 + k_1')$, for all practical purposes k_1 may be set equal to k_1'. Therefore it is reasonable to choose the following operator $\bar{F}(\mathscr{I}, \mathscr{F})$

$$\langle \vec{k} | \bar{F} | \vec{k}' \rangle = \begin{cases} \dfrac{1}{(2\pi)^3} \displaystyle\int_{t_1}^{t_2} e^{(i/2m)(k^2 - k'^2)t}\, dt \int_{\mathscr{F}} e^{i[(k_2' - k_2)x_2 + (k_3' - k_3)x_3} \\ \cdot \dfrac{1}{m} \sqrt{k_1}\sqrt{k_1'}\, dx_2\, dx_3 & \text{for } k_1 > 0 \text{ and } k_1' > 0, \\ 0 & \text{otherwise} \end{cases} \qquad (6.1.13)$$

instead of $F(\mathscr{I}, \mathscr{F})$.

Here we shall not attempt to estimate the "error"

$$\left| \iint \langle \psi | \vec{k} \rangle (\langle \vec{k} | \bar{F} | \vec{k}' \rangle - \langle \vec{k} | F | \vec{k}' \rangle) \langle k' | \psi \rangle\, d^3\vec{k}\, d^3\vec{k}' \right|$$

and we will not show that this "error" decreases with increasing Δt and increasing $\mathscr{F}$. This "error" has no physical meaning since we already know that (6.1.12) is only an approximation to an actual registration procedure. Therefore it is meaningless to consider equation (6.1.12) to be the "correct" formula and (6.1.13) to be the approximation. We may only hope that *real* experiments (where $\Delta t = t_2 - t_1$ and $\mathscr{F}$ cannot be chosen arbitrarily small) can be described "in good approximation" by (6.1.13); by making test experiments, we may determine whether our hopes are justified. An experimental error can only be defined as the difference between (6.1.13) and the effect of the *actual registration procedure.*

We will now show that (6.1.13) actually determines an additive measure $\Sigma_{t,E} \xrightarrow{\bar{F}} L(\mathscr{H})$. The additivity of the measure follows directly from (6.1.13). From

$$\iint \langle \psi | \vec{k} \rangle \langle \vec{k} | \bar{F} | \vec{k}' \rangle \langle \vec{k}' | \psi \rangle\, d^3\vec{k}\, d^3\vec{k}'$$
$$= \frac{1}{(2\pi)^3} \frac{1}{m} \int_{t_1}^{t_2} dt \int_{\mathscr{F}} dx_1\, dx_2 \left| \int_{k_1 > 0} e^{(i/2m)k^2 t} e^{-i(k_2 x_2 + k_3 x_3)} \sqrt{k_1} \langle \vec{k} | \psi \rangle\, d^3\vec{k} \right|^2$$

it follows that $\bar{F}(\mathscr{I}, \mathscr{F})$ is a positive operator. If we show that for $t_1 \to -\infty$ and $t_2 \to \infty$, and for $\mathscr{F} = E$ (the entire plane) then $\bar{F} \leq \mathbf{1}$ we will have proven that $\sum_{t,E} \xrightarrow{\bar{F}} L(\mathscr{H})$. For $t_1 \to -\infty$ and $t_2 \to \infty$, and $\mathscr{F} = E$, from (6.1.13) it follows that

$$\langle \vec{k} | \bar{F} | \vec{k}' \rangle \to \frac{1}{m} \sqrt{k_1}\sqrt{k_1'}\, \delta\!\left(\frac{1}{2m}(k^2 - k'^2)\right) \delta(k_2' - k_2) \delta(k_3' - k_3)$$
$$\text{for } k_1 > 0 \text{ and } k_1' > 0$$

and we therefore obtain

$$\langle \psi, \bar{F}\psi \rangle \to \int_{k_1 > 0} |\langle k | \psi \rangle|^2\, d^3\vec{k} \leq 1.$$

For real scattering experiments only such ensembles are used (because they are simpler to produce) for which $W = \Sigma_\nu w_\nu P_{\psi_\nu}$ where the individual ψ_ν have small variance in momentum (W itself can have a larger variance!). If $\vec{k}^{(\nu)}$ is the average momentum for the ensemble P_{ψ_ν}, then we approximately obtain

$$\operatorname{tr}(P_{\psi_\nu}\bar{F}) = \iint \langle \psi_\nu | \vec{k} \rangle \langle \vec{k} | \bar{F} | \vec{k}' \rangle \langle k' | \psi_\nu \rangle \, d^3\vec{k}\, d^3\vec{k}' \tag{6.1.14}$$
$$\approx \frac{k_1^{(\nu)}}{m(2\pi)^3} \int_{t_1}^{t_2} dt \int_{\mathscr{F}} dx_1\, dx_2 \left| \int_{k_1>0} e^{i[k_2x_2 - k_3x_3 - (k^2/2m)t]} \langle \vec{k} | \psi_\nu \rangle \, d^3\vec{k} \right|^2 .$$

The result (6.1.14) is very intuitive, because from

$$\langle \vec{r} | \psi_{\nu t} \rangle = \frac{1}{(2\pi)^{3/2}} \int e^{i[\vec{k}\cdot\vec{r} - (k^2/2m)t]} \langle \vec{k} | \psi_\nu \rangle \, d^3\vec{k}$$

(6.1.14) becomes

$$\operatorname{tr}(P_{\psi_\nu}\bar{F}) \approx \frac{k_1^{(\nu)}}{m} \int_{t_1}^{t_2} dt \int_{\mathscr{F}} |\langle 0, x_2, x_3 | \psi_{\nu t} \rangle|^2 \, dx_1\, dx_2 \tag{6.1.15}$$

where $k_1^{(\nu)} m^{-1}$ is the mean velocity in the direction of the normal to the surface and $|\langle 0, x_1, x_2 | \psi_{\nu t} \rangle|$ is the position probability function at time t on the position (x_2, x_3) on the plane E.

We shall now provide a summary of our considerations: Experimental evidence shows that it is possible to construct registration methods b_0 with a surface $\mathscr{F}$ in the sense of Figure 39 in such a way that for the registration procedure b which produces a signal in the time interval $\mathscr{I}$ the effect $\psi(b_0, b)$ is represented to a high degree of accuracy by $\bar{F}(\mathscr{I}, \mathscr{F})$ providing that no extremely small values of k_1 are present in the ensemble. We call $\bar{F}(\mathscr{I}, \mathscr{F})$ given by (6.1.13) the "effect of the impact of a microsystem on the surface $\mathscr{F}$ in the time interval $\mathscr{I}$".

The intuitiveness of the expression (6.1.15) is the basic reason why it is usually not necessary to examine the real difficulties associated with the registration process (and thereby avoid making serious errors), because the ensembles used in scattering experiments are already so constituted, and the surfaces $\mathscr{F}$ so selected that the conditions for the application of (6.1.15) are satisfied.

Since we are concerned in this book with a realistic and sober conceptual analysis of quantum mechanics, including the experiments used for measurement, we have found it necessary to examine the conditions for the application of (6.1.13) and (6.1.15) more closely.

6.2 Counting Microsystems Scattered into a Solid Angle

Now that we have discussed the fundamentals about the effects $\bar{F}(\mathscr{I}, \mathscr{F})$ which describe the impact of microsystems on a surface $\mathscr{F}$ in the time interval $\mathscr{I}$, we can now proceed to discuss additional specialization and applications. We shall not carry out all computations and estimates explicitly.

Scattering experiments are carried out in such a way that the surfaces $\Delta\mathscr{F}$ are placed at great distances from the scattering region, and the number of those systems which strike the surface $\Delta\mathscr{F}$ after scattering in a given time interval $\mathscr{I}$ are measured. The experimental situation described in the previous sentence will now be expressed in mathematical form, that is, in terms of assertions about the structure of the prepared ensembles and registered effects. These assertions will not be laws of nature in a new axiomatic formulation associated with the theory. Instead they will be brief formulations of the structure of the experiments which can be verified before carrying out the "proper scattering experiments". Here experimental physicists will either speak of test experiments, calibration of the apparatus or of verification of the technical specifications of the apparatus. These yet-to-be-formulated structure assertions are to be deduced in the theory from those relations which are nothing other than the mathematical formulation of the *results of the test experiments.*

In the momentum representation we require that $\langle\vec{k}_1|W^i_{10}|\vec{k}_1\rangle$ and frequently introduced concerning the experiments without much discussion in such words as the following: "We consider experiments in which . . . is true". Here we note such a statement is a substitute for the more precise statement— additional structures are introduced based upon experimental facts which have been experimentally verified. Such situations are distasteful to mathematicians because, by admitting new formal axioms based upon experimental results, the mathematical part of a physical theory will never be "axiomatically closed". Readers who are interested in a more precise conceptual formulation of this situation are referred to [8] and especially to the notion of a "real hypothesis" in [8], §10.4.

In this sense we now "consider experiments for which the following is true": Before the scattering the ensembles satisfy the requirements made in §1 where W^i_{10} and W^i_{20} are such that the position probability densities $\langle\vec{r}_1|W^i_{10}|\vec{r}_1\rangle$ and $\langle\vec{r}_2|W^i_{20}|\vec{r}_2\rangle$ are nonzero only in a macroscopic neighborhood of the origin of the laboratory reference system. Here we refer to this region as the collision region for the experiment.

In the momentum representation we require that $\langle\vec{k}_1|W^i_{10}|\vec{k}_1\rangle$ and $\langle\vec{k}_2|W^i_{20}|\vec{k}_2\rangle$ is nonzero only in a (as small as possible) region about $\vec{k}^0_1$ (resp. $\vec{k}^0_2$). Here the experimental physicist will seek the smallest possible dispersion in the momenta for scattering experiments.

Later we shall impose additional requirements on W^i_{10} and W^i_{20}; at present these requirements will not be well understood:

The position of the collection surface $\Delta\mathscr{F}$ is adjusted relative to the collision area such that no system will be counted if no scattering is taking place (that is, when experiments involving only systems of type 1 from ensemble W^i_{10} or type 2 with ensemble W^i_{20} are taking place). We say concisely that the surface $\Delta\mathscr{F}$ is not allowed to be placed in the direct beams of either system 1 or 2.

In addition we require that $\Delta\mathscr{F}$ is placed at a great distance from the collision domain in order that the normal vector for $\Delta\mathscr{F}$ shall be parallel to

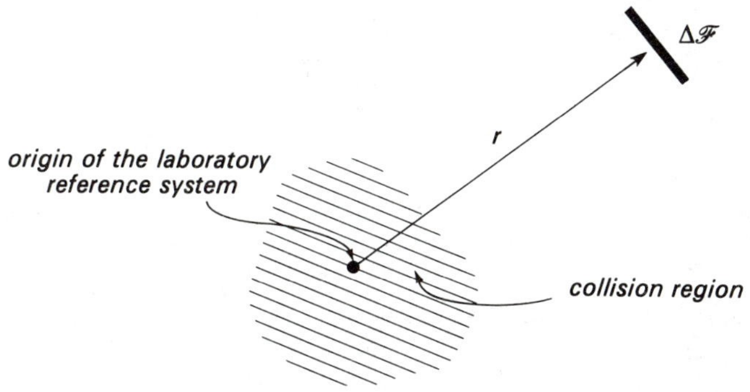

Figure 40

the direction of the position vector of the surface $\vec{r}$ (see Figure 40). By this we mean that $|\vec{r}|$ is large compared to the size of the scattering region. The vectors $\vec{k}_1^0$, $\vec{k}_2^0$ and $\vec{r}$ (towards $\Delta\mathscr{F}$) determine three directions of considerable experimental importance.

We are now interested in the case in which the surface denoted by $\Delta\mathscr{F}$ responds only to systems of type 1. Our purpose is to compute the probability

$$\operatorname{tr}(S(W_{10}^i \times W_{20}^i)S^+(\bar{F}_1(\mathscr{I}, \Delta\mathscr{F}) \times \mathbf{1})) \tag{6.2.1}$$

under the assumptions made above. The $\times$ product in (6.2.1) refers to the representation $\mathscr{H} = \mathscr{H}_1 \times \mathscr{H}_2$ and $\bar{F}_1(\mathscr{I}, \Delta\mathscr{I})$ is the effect introduced in §6.1 for the detection of systems of type 1.

In order to use the assumptions previously introduced for W_{10}^i, W_{20}^i, we rewrite S in the form

$$S = \mathbf{1} - 2\pi i T \tag{6.2.2}$$

by which we define the operator T. In the absence of the interaction, T would be the null operator.

The assumption that the undisturbed beam does not hit $\Delta\mathscr{F}$ means that

$$\operatorname{tr}(W_{10}^i \bar{F}_1(\mathscr{I}, \Delta\mathscr{F})) \approx 0, \tag{6.2.3}$$

from which it follows that $W_{10}^i \bar{F}_1 \approx 0$. Thus, from the relation $\operatorname{tr}(AB) = \operatorname{tr}(BA)$ it follows that (6.2.1) is transformed into

$$(2\pi)^2 \operatorname{tr}(T(W_{10}^i \times W_{20}^i)T^+[\bar{F}_1(\mathscr{I}, \Delta\mathscr{F}) \times \mathbf{1}]). \tag{6.2.4}$$

The operator T has a close relationship with the function $g_{\vec{k}}(\vec{r}/|\vec{r}|)$ introduced in (4.7.10). As we have stated in §5.1, $\Omega_- e^{i\vec{k}\cdot\vec{r}} = \psi_{\vec{k}}(\vec{r})$ and $Se^{i\vec{k}\cdot\vec{r}}$ are identical with respect to the outgoing spherical wave. This means that, for large $|\vec{r}|$

$$-2\pi i T e^{i\vec{k}\cdot\vec{r}} \approx g_{\vec{k}}\left(\frac{\vec{r}}{|\vec{r}|}\right)\frac{e^{ikr}}{r} + \cdots \frac{e^{-ikr}}{r}. \tag{6.2.5}$$

In (6.2.5) T is to be understood as an operator in $\mathscr{H}_i$ where $\mathscr{H}_i$ occurs in the product representation VIII (1.1) of $\mathscr{H}$. Since S commutes with $\mathring{H}$, this is also the case for T. In the momentum representation we prefer to express T in the form

$$\langle \vec{k}' | T | \vec{k} \rangle = \delta\left(\frac{k'^2}{2m} - \frac{k^2}{2m}\right) T(\vec{k}', \vec{k}), \tag{6.2.6}$$

where $T(\vec{k}', \vec{k})$ is defined only on the energy surface $k' = k$. Introducing polar coordinates ($\vec{r}$ is the polar axis) we obtain

$$\begin{aligned} Te^{i\vec{k}\vec{r}} &= \int e^{i\vec{k}'\cdot\vec{r}} \langle \vec{k}' | T | \vec{k} \rangle \, d^3\vec{k}' \\ &= \int_0^{2\pi} d\varphi \int_0^{\infty} k'^2 \, dk' \int_{-1}^{1} d\xi e^{ik'r\xi} \delta\left(\frac{k'^2}{2m} - \frac{k^2}{2m}\right) T(\vec{k}', \vec{k}), \end{aligned}$$

where we note that $\vec{k}'$ is to be expressed in polar coordinates as follows:

$$k', \xi = \cos\theta, \varphi.$$

Integrating by parts over ξ, for large $|\vec{r}|$ it follows that

$$Te^{i\vec{k}\cdot\vec{r}} \approx 2\pi \int_0^{\infty} k'^2 \, dk' \frac{e^{ik'r}}{ik'r} \delta\left(\frac{k'^2}{2m} - \frac{k^2}{2m}\right) T\left(k' \frac{\vec{r}}{|\vec{r}|}, \vec{k}\right) + \cdots \frac{e^{-ik'r}}{r}.$$

By comparison with (6.2.5) we therefore find that

$$\begin{aligned} g_{\vec{k}}\left(\frac{\vec{r}}{|\vec{r}|}\right) \frac{e^{ikr}}{r} &= -(2\pi)^2 \int_0^{\infty} k' \, dk' \frac{e^{ik'r}}{r} \delta\left(\frac{k'^2}{2m} - \frac{k^2}{2m}\right) T\left(k' \frac{\vec{r}}{|\vec{r}|}, \vec{k}\right) \\ &= -(2\pi)^2 m \frac{e^{ikr}}{r} T\left(k \frac{\vec{r}}{|\vec{r}|}, \vec{k}\right). \end{aligned}$$

We therefore obtain

$$g_{\vec{k}}\left(\frac{\vec{r}}{|\vec{r}|}\right) = -(2\pi)^2 m T\left(k \frac{\vec{r}}{|\vec{r}|}, \vec{k}\right) \tag{6.2.7a}$$

and

$$\int e^{i\vec{k}'\cdot\vec{r}} \delta\left(\frac{k'^2}{2m} - \frac{k^2}{2m}\right) T(\vec{k}', \vec{k}) \, d^3\vec{k}' \approx \frac{2\pi}{i} m \frac{e^{ikr}}{r} T\left(k \frac{\vec{r}}{|\vec{r}|}, \vec{k}\right). \tag{6.2.7b}$$

Since T has, with respect to $\mathscr{H} = \mathscr{H}_b \times \mathscr{H}_i$, the form $\mathbf{1} \times T$ (where the latter does not correspond to the $\times$ product in (6.2.4)), it is preferable to express T in terms of the momentum representation with respect to $\vec{P}_1, \vec{P}_2$, and to use the separation of the center of mass, as described in §3. Using $\vec{k}_1, \vec{k}_2$ as the variables of the $(\vec{P}_1, \vec{P}_2)$ representation, and $\vec{K}, \vec{k}$ defined by (3.4) and (3.5), we obtain

$$\begin{aligned} \vec{K} &= \vec{k}_1 + \vec{k}_2, \\ \vec{k} &= \frac{m_2}{M} \vec{k}_1 - \frac{m_1}{M} \vec{k}_2 = m\left(\frac{1}{m_1} \vec{k}_1 - \frac{1}{m_2} \vec{k}_2\right) \end{aligned} \tag{6.2.8}$$

then, from the form $\mathbf{1} \times T$ with respect to $\mathscr{H} = \mathscr{H}_b \times \mathscr{H}_i$ we obtain

$$\langle \vec{k}_1', \vec{k}_2' | T | \vec{k}_1, k_2 \rangle = \delta(\vec{K}' - \vec{K})\delta\left(\frac{k'^2}{2m} - \frac{k^2}{2m}\right) T(\vec{k}', \vec{k}). \tag{6.2.9}$$

Here (6.2.4) takes on the form (using $\bar{a}$ as the complex conjugate of a):

$$\begin{aligned}(2\pi)^2 \int &\langle \vec{k}_1', \vec{k}_2' | T | \vec{k}_1, \vec{k}_2 \rangle \langle \vec{k}_1 | W_{10}^i | \vec{k}_1'' \rangle \\ &\cdot \langle \vec{k}_2 | W_{20}^i | \vec{k}_2'' \rangle \overline{\langle \vec{k}_1''', \vec{k}_2' | T | \vec{k}_1'', \vec{k}_2'' \rangle} \\ &\cdot \langle \vec{k}_1''' | F_1 | \vec{k}_1' \rangle \delta(\vec{k}_2' - \vec{k}_2''') \\ &\cdot d^3\vec{k}_1\, d^3\vec{k}_2\, d^3\vec{k}_1'\, d^3\vec{k}_2'\, d^3\vec{k}_1''\, d^3\vec{k}_2''\, d^3\vec{k}_1'''\, d^3\vec{k}_2''',\end{aligned} \tag{6.2.10a}$$

where, according to §6.1

$$\begin{aligned}\langle \vec{k}_1''' | F_1 | \vec{k}_1' \rangle = \frac{1}{m_1(2\pi)^3} &\int_{t_1}^{t_2} dt e^{(i/2m_1)(k_1'''^2 - k_1'^2)t} \\ &\cdot \int_{\Delta\mathscr{F}} e^{i(\vec{k}_1' - \vec{k}_1''')\cdot\vec{r}} \sqrt{\vec{k}_1' \cdot \vec{n}} \sqrt{\vec{k}_1''' \cdot \vec{n}}\, df\end{aligned} \tag{6.2.10b}$$

(where $\vec{n} = \vec{r}/|\vec{r}|$) or is equal to 0 if $\vec{k}_1' \cdot \vec{n}$ or $\vec{k}_1''' \cdot \vec{n}$ is negative.

Using (6.2.9), (6.2.10) takes on the form

$$\begin{aligned}(2\pi)^2 \int &\delta(\vec{K}' - \vec{K})\delta(\vec{K}''' - \vec{K}'')\delta\left(\frac{k'^2}{2m} - \frac{k^2}{2m}\right)\delta\left(\frac{k'''^2}{2m} - \frac{k''^2}{2m}\right) \\ &\cdot T(\vec{k}', \vec{k})\overline{T(\vec{k}''', \vec{k}'')}\langle \vec{k}_1 | W_{10}^i | \vec{k}_1'' \rangle \langle \vec{k}_2 | W_{20}^i | \vec{k}_2'' \rangle \\ &\cdot \langle \vec{k}_1''' | F_1 | \vec{k}_1' \rangle \delta(\vec{k}_2' - \vec{k}_2''')\, d^3\vec{K}'\, d^3\vec{k}' d^3\vec{K}''' \\ &\cdot d^3\vec{k}'''\, d^3\vec{k}_1\, d^3\vec{k}_2\, d^3\vec{k}_1''\, d^3\vec{k}_2'' \\ = \frac{1}{2\pi} \int &\delta\left(\frac{k'^2}{2m} - \frac{k^2}{2m}\right) T(\vec{k}', \vec{k})\delta\left(\frac{k'''^2}{2m} - \frac{k''^2}{2m}\right) \\ &\overline{T(\vec{k}''', k'')}\langle \vec{k}_1 | W_{10}^2 | \vec{k}_1'' \rangle \langle \vec{k}_2 | W_{20}^2 | \vec{k}_2'' \rangle \\ &\cdot \langle \vec{k}''' | F_1 | \vec{k}_1' \rangle e^{i(\vec{k}_2' - \vec{k}_2'')\cdot\vec{r}_2}\, d^3\vec{k}'\, d^3\vec{k}''' \\ &\cdot d^3\vec{k}_1\, d^3\vec{k}_2\, d^3\vec{k}_1''\, d^3\vec{k}_2''\, d^3\vec{r}_2,\end{aligned} \tag{6.2.11a}$$

where, on the right-hand side of the last equation, $\vec{K}'$ is to be replaced by $\vec{K}$ and $\vec{K}'''$ by $\vec{K}''$. Since the function $\delta(\vec{k}_2' - \vec{k}_2''')$ occurs in (6.2.10a), we may replace

$$\frac{1}{2m_1}(k_1'''^2 - k_1'^2)$$

in (6.2.10b) by

$$\frac{1}{2m_1}(k_1'''^2 - k_1'^2) + \frac{1}{2m_2}(k_2'''^2 - k_2'^2) = \frac{1}{2m}(k'''^2 - k'^2) + \frac{1}{2M}(\vec{K}'''^2 - \vec{K}'^2),$$

that is, in (6.2.11a) we may replace $\langle \vec{k}_1''' | \bar{F}_1 | \vec{k}_1' \rangle$ by the value

$$\frac{1}{m_1(2\pi)^3} \int_{t_1}^{t_2} dt\, e^{(i/2m)(k'''^2 - k'^2)t} e^{(i/2M)(\vec{K}'''^2 - \vec{K}'^2)t}$$

$$\cdot \int_{\Delta \mathcal{F}} e^{i(\vec{k}_1' - \vec{k}_1''') \cdot \vec{r}} \sqrt{\vec{k}_1' \cdot \vec{n}} \sqrt{\vec{k}_1''' \cdot \vec{n}}\, df, \tag{6.2.11b}$$

where we have to set $\vec{K}''' = \vec{K}''$, $\vec{K}' = \vec{K}$.

From (6.2.8) it follows that

$$\vec{k}_1' \cdot \vec{r} = \vec{k}' \cdot \vec{r} + \frac{m}{m_2} \vec{K} \cdot \vec{r},$$

$$\vec{k}_2' \cdot \vec{r}_2 = -\vec{k}' \cdot \vec{r}_2 + \frac{m}{m_1} \vec{K} \cdot \vec{r}_2.$$

Therefore we obtain

$$\vec{k}_1' \cdot \vec{r} + \vec{k}_2' \cdot \vec{r}_2 = \vec{k}' \cdot (\vec{r} - \vec{r}_2) + \vec{K} \cdot \left(\frac{m}{m_2} \vec{r} + \frac{m}{m_1} \vec{r}_2 \right),$$

and similarly,

$$\vec{k}_1''' \cdot \vec{r} + \vec{k}_2''' \cdot \vec{r}_2 = \vec{k}''' \cdot (\vec{r} - \vec{r}_2) + \vec{K}'' \cdot \left(\frac{m}{m_2} \vec{r} + \frac{m}{m_1} \vec{r}_2 \right).$$

Thus it follows that the integration over $\vec{k}'$, $\vec{k}'''$ can be carried out separately. Using (6.2.7b), for large $|\vec{r} - \vec{r}_2|$ we obtain for (6.2.11a)

$$\frac{m^2}{m_1(2\pi)^2} \int_{t_1}^{t_2} dt \int_{\Delta \mathcal{F}} df \int e^{-(i/2m)(k^2 - k''^2)t - (i/2M)(K^2 - K''^2)t}$$

$$\cdot \sqrt{\vec{k}_1' \cdot \vec{n}} \sqrt{\vec{k}_1''' \cdot \vec{n}}\, \frac{e^{i(k - k'')|\vec{r} - \vec{r}_2|}}{|\vec{r} - \vec{r}_2|}$$

$$\cdot T\left(k \frac{\vec{r} - \vec{r}_2}{|\vec{r} - \vec{r}_2|}, \vec{k}\right) \overline{T\left(k'' \frac{\vec{r} - \vec{r}_2}{|\vec{r} - \vec{r}_2|}, \vec{k}''\right)}$$

$$\cdot \langle \vec{k}_1 | W_{10}^i | \vec{k}_1'' \rangle \langle \vec{k}_2 | W_{20}^i | \vec{k}_2'' \rangle$$

$$\cdot e^{i(\vec{K} - \vec{K}'') \cdot (m/m_2 \vec{r} + m/m_1 \vec{r}_2)}$$

$$\cdot d^3\vec{k}_1\, d^2\vec{k}_2\, d^3\vec{k}_1''\, d^3\vec{k}_2''\, d^3\vec{r}_2, \tag{6.2.12}$$

where $\vec{k}_1'$, $\vec{k}_1'''$, k, k'' are defined as follows:

$$\vec{k}_1' = k \frac{\vec{r} - \vec{r}_2}{|\vec{r} - \vec{r}_2|} + \frac{m}{m_2} \vec{K}; \qquad \vec{k}_1''' = k'' \frac{\vec{r} - \vec{r}_2}{|\vec{r} - \vec{r}_2|} + \frac{m}{m_2} \vec{K}'';$$

$$k = \left| \frac{m}{m_1} \vec{k}_1 - \frac{m}{m_2} \vec{k}_2 \right|; \qquad k'' = \left| \frac{m}{m_1} \vec{k}_1'' - \frac{m}{m_2} \vec{k}_2'' \right|. \tag{6.2.13}$$

The time dependent exponent can be replaced by the original expression

$$e^{-(i/2m_1)(k_1'^2 - k_1''^2)t}. \tag{6.2.14}$$

Equation (6.2.12) is the general formula for the computation of the probability for the impact of a system of type 1 on the surface $\Delta\mathscr{F}$ (at the location $\vec{r}$) in the time interval t_1 to t_2. This probability depends, in general, upon the "wave packets" $\langle\vec{k}_1|W^i_{10}|\vec{k}''_1\rangle$ and $\langle\vec{k}_2|W^i_{20}|\vec{k}''_2\rangle$. Here precise control measurements are required to determine the structure of these wave packets, that is, of the particle beams for systems 1 and 2. It is clear that without such control measurements for the incoming beams scattering experiments will not work. Generally we seek to keep these to a minimum *by choosing special forms* of W^i_{10} and W^i_{20} in order that we may be able to investigate the interaction between the systems undergoing the scattering process. We will consider only two examples to demonstrate how we may choose special forms of the experimentally prepared ensembles W^i_{10}, W^i_{20}:

(1) $m_2 \gg m_1$ and we choose the average momentum $\vec{k}^0_2 = 0$.
(2) the interval $t_1 - t_2$ is chosen so large (that is, we consider the mathematical limits $t_1 \to -\infty$, $t_2 \to \infty$) that all the systems striking the surface $\Delta\mathscr{F}$ are counted (independent of when they strike). We shall now begin with case (1).

In addition to $m_1/m_2 \ll 1$ we require that $k_2/m_2 \ll k_1/m_1$ for all $\vec{k}_1, \vec{k}_2$ in the wave packets. This means not only that the average of the velocities of system 2 is, for all practical purpose, zero but also that these velocities have very small dispersions (compared to velocities of system 1). Under these assumptions, the velocity of the center of mass $\vec{K}/m$ is, for all practical purposes, equal to zero: $\vec{K}/M \ll k_1/m_1$.

From these assumptions it follows directly that

$$\vec{k} \approx \vec{k}_1, \qquad \vec{k}'' \approx \vec{k}''_1, \qquad \vec{k}'_1 \approx k_1 \frac{\vec{r} - \vec{r}_2}{|\vec{r} - \vec{r}_2|}, \qquad \vec{k}'''_1 \approx k''_1 \frac{\vec{r} - \vec{r}_2}{|\vec{r} - \vec{r}_2|}.$$

Using (6.2.14) in (6.2.12), the latter takes on the form

$$\begin{aligned}
&\frac{m}{(2\pi)^2} \int_{t_1}^{t_2} dt \int_{\Delta\mathscr{F}} df \int e^{-(i/2m_1)(k_1^2 - k_1''^2)t}\, k_1^0 \vec{n} \cdot \frac{\vec{r} - \vec{r}_2}{|\vec{r} - \vec{r}_2|} \\
&\cdot \frac{e^{i(k-k'')|\vec{r}-\vec{r}_2|}}{|\vec{r} - \vec{r}_2|^2}\, T\!\left(k_1 \frac{\vec{r} - \vec{r}_2}{|\vec{r} - \vec{r}_2|}, \vec{k}_1\right) \overline{T\!\left(k''_1 \frac{\vec{r} - \vec{r}_2}{|\vec{r} - \vec{r}_2|}, \vec{k}''_1\right)} \\
&\cdot \langle\vec{k}_1|W^i_{10}|\vec{k}''_1\rangle e^{i(\vec{k}_1 + \vec{k}_2 - \vec{k}''_1 - \vec{k}''_2)\cdot((m_1/m_2)\vec{r} + \vec{r}_2)} \\
&\cdot \langle\vec{k}_2|W^i_{20}|\vec{k}''_2\rangle\, d^3\vec{k}_1\, d^3\vec{k}_2\, d^3\vec{k}''_1\, d^3\vec{k}''_2\, d^3\vec{r}_2.
\end{aligned} \tag{6.2.15}$$

From (6.2.15) it follows that $\vec{r}_2$ need only be integrated over a region for which $|\vec{r}_2| \ll |\vec{r}|$, as we shall soon see. In order to evaluate (6.2.15) more precisely, we shall, for $|\vec{r}_2| \ll |\vec{r}|$ use the following approximation

$$|\vec{r} - \vec{r}_2| \approx |\vec{r}| - \vec{r}_2 \cdot \frac{\vec{r}}{r}.$$

From (6.2.13) we obtain the approximation

$$k - k'' = \left|\frac{m}{m_1}\vec{k}_1 - \frac{m}{m_2}\vec{k}_2\right| - \left|\frac{m}{m_1}\vec{k}_1'' - \frac{m}{m_2}\vec{k}_2''\right|$$
$$\approx (k_1 - k_1'') - \frac{m_1}{m_2}\vec{k}_2\cdot\vec{e}_0 + \frac{m_1}{m_2}\vec{k}_2''\cdot\vec{e}_0,$$

where $\vec{e}_0 = \vec{k}_1^0/k_1^0$. Thus, in the exponent we finally obtain

$$(k - k'')|\vec{r} - \vec{r}_2| = (k_1 - k_1'')r - (k_1 - k_1'')\vec{r}_2\cdot\vec{e} - \frac{m_1}{m_2}(\vec{k}_2 - \vec{k}_2'')\cdot\vec{e}_0 r, \tag{6.2.16}$$

where $\vec{e} = \vec{r}/r$.

This better estimate is required for the exponent because r and k_1 can be relatively large. If we use this approximation, then we can easily perform the integration over $\vec{k}_2$ and $\vec{k}_2''$ by using the formula

$$\langle\vec{r}_2'|W_{20}^i|\vec{r}_2''\rangle = \frac{1}{(2\pi)^3}\int e^{i\vec{k}_2\cdot\vec{r}_2}\langle\vec{k}_2|W_{20}^i|\vec{k}_2''\rangle e^{-i\vec{k}_2''\cdot\vec{r}_2''}d^3\vec{k}_2\,d^3\vec{k}_2''$$

for the transition from the momentum representation to the position representation. Thus from (6.2.15) and using (6.2.16) we obtain

$$2\pi m_1\int_{t_1}^{t_2}dt\int_{\Delta\mathscr{F}}df\int e^{-(i/2m_1)(k_1^2 - k_1''^2)t}k_1^0\frac{e^{i(k_1 - k_1'')r}}{r^2}$$
$$\cdot T(k_1\vec{e},\vec{k}_1)\overline{T(k_1''\vec{e},\vec{k}_1'')}\langle\vec{k}_1|W_{10}^i|\vec{k}_1''\rangle$$
$$\cdot e^{i(\vec{k}_1 - \vec{k}_1'')\cdot(\vec{r}_2 + m_1/m_2\vec{r}) - i(k_1 - k_1'')\vec{e}\cdot\vec{r}_2}$$
$$\cdot\left\langle\frac{m_1}{m_2}\vec{r} + \vec{r}_2 - \frac{m_1}{m_2}\vec{e}_0 r\,\middle|\,W_{20}^i\,\middle|\,\frac{m_1}{m_2}\vec{r} + \vec{r}_2 - \frac{m_1}{m_2}\vec{e}_0 r\right\rangle$$
$$\cdot d^3\vec{k}_1\,d^3\vec{k}_1''\,d^3\vec{r}_2. \tag{6.2.17}$$

We now introduce the new integration variable

$$\vec{r}_2' = \vec{r}_2 + \frac{m_1}{m_2}\vec{r} - \frac{m_1}{m_2}\vec{e}_0 r \tag{6.2.18}$$

and set $df = r^2\,d\omega(\vec{e})$, then (6.2.17) is transformed into

$$2\pi m_1\int_{t_1}^{t_2}dt\int_{\Delta\Omega}d\omega(\vec{e})\int e^{-i[(k_1 - k_1'')r - (1/2m_1)(k_1^2 - k_1''^2)t]}$$
$$\cdot k_1^0 T(k_1\vec{e},\vec{k}_1)\overline{T(k_1''\vec{e},\vec{k}_1'')}\langle\vec{k}_1|W_{10}^i|\vec{k}_1''\rangle$$
$$\cdot e^{i(\vec{k}_1 - \vec{k}_1'')\cdot(\vec{r}_2' + (m_1/m_2)\vec{e}_0 r)}\cdot e^{-i(k_1 - k_1'')(\vec{e}\cdot\vec{r}_2' - (m_1/m_2)\vec{r} + (m_1/m_2)\vec{e}\cdot\vec{e}_0 r)}$$
$$\cdot\langle\vec{r}_2'|W_{20}^i|\vec{r}_2'\rangle\,d^3k_1\,d^3\vec{k}_1''\,d\vec{r}_2', \tag{6.2.19}$$

where $\Delta\Omega$ is the solid angle corresponding to the surface $\Delta\mathscr{F}$.

$\langle \vec{r}_2' | W_{20}^i | \vec{r}_2' \rangle$ is the position probability density of the system 2 at "time $t = 0$" if there is no scattering. According to the instructions for the experiment $\langle \vec{r}_2 | W_{20}^i | \vec{r}_2' \rangle$ is only noticeably different from zero in the collision region. In addition r must be large compared to the collision region, that is, $|\vec{r}_2'| \ll r$. Thus from (6.2.17) it also follows that $|\vec{r}_2| \ll r$, from which it follows that the above assumption about $\vec{r}_2$ and therefore our approximation (6.2.18) are justified.

Since the momentum in W_{10}^i has small dispersion, we may approximately set

$$(\vec{k}_1 - \vec{k}_1'') = (k_1 - k_1'')\vec{e}_0 + \vec{k}_{1\perp} - \vec{k}_{1\perp}'',$$

where $\vec{k}_{1\perp}$, $\vec{k}_{1\perp}''$ are the components of $\vec{k}_1$ and $\vec{k}_1''$ which are orthogonal to $\vec{e}_0$. In the case in which $T(\cdots)$ does not vary too quickly with the direction of $\vec{e}$ to $\vec{k}$, we may set

$$T(k_1\vec{e}, \vec{k}_1) = T(k_1\vec{e}, k_1\vec{e}_0), \qquad T(k_1''\vec{e}, \vec{k}_1'') = T(k_1''\vec{e}, k_1''\vec{e}_0)$$

(6.2.19) is then transformed into

$$\begin{aligned}
2\pi m_1 k_1^0 \int_{t_1}^{t_2} dt \int_{\Delta\Omega} d\omega(\vec{e}) \int & e^{i(k_1 - k_1'')(r + 2(m_1/m_2)r - (m_1/m_2)\vec{e}\cdot\vec{e}_0 r + \vec{e}_0\cdot\vec{r}_2' - \vec{e}\cdot\vec{r}_2')} \\
& \cdot e^{-(i/2m_1)(k_1^2 - k_1''^2)t}\, T(k_1\vec{e}, k_1\vec{e}_0)\overline{T(k_1''\vec{e}, k_1''\vec{e}_0)} \\
& \cdot \langle \vec{k}_1 | W_{10}^i | \vec{k}_1'' \rangle e^{i(\vec{k}_{1\perp} - \vec{k}''_{1\perp})\cdot\vec{r}_2'} \langle \vec{r}_2' | W_{20}^i | \vec{r}_2' \rangle \\
& \cdot d^2\vec{k}_{1\perp}\, d^2\vec{k}_{1\perp}''\, d^3\vec{r}_2'.
\end{aligned} \tag{6.2.20}$$

Here we may integrate over $\vec{k}_{1\perp}$, $\vec{k}_{1\perp}''$. To simplify the notation we set

$$\begin{aligned}
\frac{1}{(2\pi)^2} \int & e^{i\vec{k}_{1\perp}\cdot\vec{r}_1} \langle \vec{k}_1 | W_{10}^i | \vec{k}_1'' \rangle e^{-i\vec{k}_{1\perp}''\cdot\vec{r}_1''} d^2\vec{k}_{1\perp}\, d^2\vec{k}_{1\perp}'' \\
&= \langle k_1, \vec{r}_{1\perp} | W_{10}^i | k_1'', \vec{r}_{1\perp}'' \rangle,
\end{aligned} \tag{6.2.21}$$

where $\vec{r}_{1\perp}$ and $\vec{r}_{1\perp}''$ are the components of $\vec{r}_1$ or $\vec{r}_1''$ which are orthogonal to $\vec{e}_0$. Here (6.2.20) is transformed into

$$\begin{aligned}
(2\pi)^3 m_1 k_1^0 \int_{t_1}^{t_2} dt \int_{\Delta\Omega} d\omega(\vec{e}) \int & e^{i(k_1 - k_1'')(r + 2(m_1/m_2)r - (m_1/m_2)\vec{e}\cdot\vec{e}_0 r + \vec{e}_0\cdot\vec{r}_2' - \vec{e}\cdot\vec{r}_2')} \\
& \cdot e^{-(i/2m)(k_1^2 - k_1''^2)}\, T(k_1\vec{e}, k_1\vec{e}_0)\, \overline{T(k_1''\vec{e}, k_1''\vec{e}_0)} \\
& \cdot \langle k_1, \vec{r}_{2\perp}' | W_{10}' | k_1'', \vec{r}_{2\perp}' \rangle \langle \vec{r}_2' | W_{20}^i | \vec{r}_2' \rangle\, dk_1\, dk_1''\, d^3\vec{r}_2'.
\end{aligned} \tag{6.2.22}$$

We obtain a further simplification under the following assumption: If $(t_2 - t_1)(k_1/m_1)$ is large compared to the size of the collision region and compared to $(m_1/m_2)\vec{r}$ (that is, r is not "too large"), then we may neglect in the exponent in (6.2.22) $(\vec{e}_0 - \vec{e})\cdot\vec{r}_2'$ and the terms in $(m_1/m_2)\vec{r}$. If the wave packet $\langle \vec{r}_1 | W_{10}'^i | \vec{r}_1' \rangle$ has the property that, in the "target region" (that is, where $\langle \vec{r}_2 | W_{20}'^i | \vec{r}_2 \rangle$ is essentially different from zero) the quantity

$$\langle k_1, \vec{r}_{2\perp}' | W_{10}' | k_1'', \vec{r}_{2\perp}' \rangle$$

does not, for all practical purposes, depend upon $\vec{r}_{2\perp}$, then we may integrate over $\vec{r}'_2$ and, using

$$\int \langle \vec{r}'_2 | W^i_{20} | \vec{r}'_2 \rangle \, d^3\vec{r}'_2 = \operatorname{tr}(W^i_{20}) = 1$$

obtain the following expression for the probability of a system striking $\Delta\mathscr{F}$ in the time interval t_1 to t_2:

$$\begin{aligned}(2\pi)^3 m_1 k_1^0 \int_{t_1}^{t_2} dt \int_{\Delta\Omega} d\omega(\vec{e}) \int e^{i[(k_1 - k''_1)r - (1/2m_1)(k_1^2 - k_1''^2)t]} \\ \cdot T(k_1\vec{e}, k_1\vec{e}_0)\overline{T(k''_1\vec{e}, k''_1\vec{e}_0)} \\ \cdot \langle k_1, 0 | W^i_{10} | k''_1, 0 \rangle \, dk_1 \, dk''_1. \end{aligned} \tag{6.2.23}$$

The assumption that $\langle k_1, \vec{r}'_{2\perp} | W'_{10} | k''_1, \vec{r}'_{2\perp} \rangle$ does not, for all practical purposes, depend upon $r'_{2\perp}$ in the target region is the so-called postulate of "homogeneous beam density for system 1". To see what the notion of a homogeneous beam density means we now compute the probability for the impact of particles of type 1 on a surface $\mathscr{F}$ perpendicular to $\vec{e}_0$ at the location $\vec{r} \cdot \vec{e}_0 = x$ (that is, for the case in which there is no interaction with particles of type 2). Then, from (6.1.13) we obtain the probability (using the same approximation)

$$\begin{aligned}&\frac{k_1^0}{m_1(2\pi)^3} \int_{t_1}^{t_2} dt \int_{\mathscr{F}} d^2\vec{r}\bot \int e^{i[(\vec{k}_{1\perp} - \vec{k}''_{1\perp}) \cdot \vec{r} + (k_1 - k''_1)x - (1/2m)(k_1^2 - k_1''^2)t]} \\ &\quad \cdot \langle \vec{k}_1 | W^i_{10} | \vec{k}''_1 \rangle \, d^3\vec{k}_1 \, d^3\vec{k}''_1 \\ &\approx \frac{k_1^0}{m_1 2\pi} \int_{t_1}^{t_2} \int_{\mathscr{F}} d^2\vec{r}\bot \int e^{i[(k_1 - k''_1)x - (1/2m_1)(k_1^2 - k_1''^2)t]} \\ &\quad < k_1, \vec{r}_\perp | W^i_{10} | k''_1, \vec{r}_\perp \rangle \, dk_1 \, dk''_1. \end{aligned} \tag{6.2.24}$$

If, for the location of the surface $\mathscr{F}$, we set $x = 0$ and consider (6.2.24) for different small surfaces $\Delta\mathscr{F}$ in the target region, from (6.2.24) we obtain

$$\frac{k_1^0 \Delta F}{2\pi m_1} \int_{t_1}^{t_2} dt \int e^{-(i/2m_1)(k_1^2 - k_1''^2)t} \langle k_1, \vec{r}_\perp | W^i_{10} | k''_1, \vec{r}_\perp \rangle \, dk_1 \, dk''_1, \tag{6.2.25}$$

where ΔF is the surface area of $\Delta\mathscr{F}$, and $\vec{r}_\perp$ is the location of $\Delta\mathscr{F}$. The probability (6.2.25) is constant if $\langle k_1, \vec{r}_\perp | W^i_{10} | k''_1, \vec{r}_\perp \rangle$ does not significantly vary with $\vec{r}_\perp$ within the target region, that is, if we may replace

$$\langle k_1, \vec{r}_{1\perp} | W^i_{10} | k''_1, \vec{r}_{1\perp} \rangle$$

by $\langle k_1, 0 | W^i_{10} | k''_1, 0 \rangle$. This is the meaning of a homogeneous beam. From (6.2.24), in the same approximation for $x = r$ we obtain

$$\frac{k_1^0 \Delta F}{2\pi m_1} \int_{t_1}^{t_2} dt \int e^{i[(k_1 - k''_1)r - (1/2m_1)(k_1^2 - k_1''^2)t} \langle k_1 \, 0 | W^i_{10} | k''_1, 0 \rangle \, dk_1 \, dk''_1. \tag{6.2.26}$$

The equation (6.2.26) describes a comparison experiment to (6.2.23) in which we place the surface $\Delta\mathscr{F}$ directly in the beam path of system 1 in the absence of the system 2 beam, that is, without scattering. Here it is interesting to determine whether the probability for time intervals in the case (6.2.23) is essentially delayed in time compared with the case (6.2.26); see, for example XVII, §6.5.

For $t_1 \to -\infty$, $t_2 \to \infty$, (6.2.22) is transformed into

$$(2\pi)^4 m_1 k_1^0 \int_{\Delta\Omega} d\omega(\vec{e}) \int dk_1 \, |T(k_1\vec{e}, k_1\vec{e}_0)|^2 \int \delta\left(\frac{k_1^2}{2m_1} - \frac{k_1''^2}{2m_1}\right) \qquad (6.2.27)$$
$$\cdot \langle k_1, 0 | W^i_{10} | k_1'', 0 \rangle \, dk_1''$$

and from (6.2.25) we obtain

$$\frac{k_1^0 \Delta F}{m_1} \int \delta\left(\frac{k_1^2}{2m_1} - \frac{k_1''^2}{2m_1}\right) \langle k_1, 0 | W^i_{10} | k_1'', 0 \rangle \, dk_1''. \qquad (6.2.28)$$

The formulas (6.2.27), (6.2.28) are the basis for the computation of the experimentally important notion of the "interaction cross section" (see §6.3).

We shall now consider case (2): We further assume that $T(k\vec{e}, k\vec{e}_0)$ not only varies slowly with $\vec{e}$ and $\vec{e}_0$, but does not vary rapidly with k. This is not the case when there are sharp "resonances", that is, when there exist "decaying states" (see XVII, §6.5) with long lifetimes. For $t_1 \to -\infty$ and $t_2 \to \infty$, (6.2.12) takes on the form

$$\frac{m^2}{(2\pi)^2 m_1} \int_{\Delta\mathscr{F}} df \int \left[\int_{-\infty}^{\infty} e^{-i[(1/2m)(k^2 - k''^2) + (1/2M)(K^2 - K''^2)]t} \, dt \right]$$
$$\cdot (\vec{k}_1'^0 \cdot \vec{n}) \frac{e^{i(k - k'')|\vec{r} - \vec{r}_2|}}{|\vec{r} - \vec{r}_2|} \left| T\left(k^0 \frac{\vec{r} - \vec{r}_2}{|\vec{r} - \vec{r}_2|}, \vec{k}^0\right) \right|^2$$
$$\cdot \langle \vec{k}_1 | W^i_{10} | \vec{k}_1'' \rangle e^{i(\vec{k}_1 + \vec{k}_2 - \vec{k}_1'' - \vec{k}_2'') \cdot ((m/m_2)\vec{r} + (m/m_1)\vec{r}_2)}$$
$$\cdot \langle \vec{k}_2 | W^i_{20} | \vec{k}_2'' \rangle \, d^3\vec{k}_1 \, d^3\vec{k}_2 \, d^3\vec{k}_1'' \, d^3\vec{k}_2'' \, d^3\vec{r}_2, \qquad (6.2.29)$$

where

$$\vec{k}_1^0 = \frac{m}{m_1} k_1^0 \vec{e}_0 - \frac{m}{m_2} \vec{k}_2^0,$$
$$\vec{k}_1'^0 = k^0 \frac{\vec{r} - \vec{r}_2}{|\vec{r} - \vec{r}_2|} + \frac{m}{m_2}(k_1^0 \vec{e}_0 + \vec{k}_2^0). \qquad (6.2.30)$$

We will evaluate (6.2.29) for two subcases:

(α) $k_2^0 = -k_1^0$, that is, the center of mass is, on average, at rest. This case can be realized, for example, by using colliding beams.

(β) $\vec{k}_2 = 0$, that is, system 2 is at rest (on average) before the collision; system 2 as target for the system 1; this case is called the collision in the laboratory system.

For the case (β) we have already found an expression (6.2.27) using the assumption $m_1/m_2 \ll 1$.

In case (α) from $\vec{k}_2^0 = -\vec{k}_1^0$ and (6.2.30) it follows that

$$\vec{K}^0 = 0, \qquad \vec{k}_1'^0 = k_1^0 \frac{\vec{r} - \vec{r}_2}{|\vec{r} - \vec{r}_2|} = \vec{k}'^0. \tag{6.2.31}$$

As long as the momentum distributions of systems 1 and 2 do not have high dispersion, we may expand the expressions in the exponent in (6.2.29) into linear terms in the differences $\vec{k}_1 - \vec{k}_1^0, \vec{k}_2 - \vec{k}_2^0$, etc. From (6.2.29) it follows that since $K^2 - K''^2$ is quadratic in the deviations:

$$\frac{m^2}{(2\pi)^2 m_1} \int_{\Delta\mathcal{F}} df \int \left[\int_{-\infty}^{\infty} e^{-(i/2m)(k^2 - k'^2)t}\, dt\right] \frac{\vec{k}_1'^0 \cdot \vec{n}}{|\vec{r} - \vec{r}_2|^2}$$
$$\cdot \left|T\left(k^0 \frac{\vec{r} - \vec{r}_2}{|\vec{r} - \vec{r}_2|}, \vec{k}^0\right)\right|^2 \langle \vec{k}_1 | W_{10}^i | \vec{k}_1'' \rangle \langle \vec{k}_2 | W_{20}^i | \vec{k}_2'' \rangle$$
$$\cdot e^{i(\vec{k}_1 + \vec{k}_2 - \vec{k}_1'' - \vec{k}_2'') \cdot ((m/m_2)\vec{r} + (m/m_1)\vec{r}_2)}\, d^3\vec{k}_1\, d^3\vec{k}_2\, d^3\vec{k}_1''\, d^3\vec{k}_2''\, d^3\vec{r}_2. \tag{6.2.32}$$

Since the integral over the time is proportional to $\delta(k - k'')$, we have set the factor $e^{i(k-k'')|\vec{r}-\vec{r}_2|} = 1$.

In accord with our assumptions, we may approximately set

$$k^2 - k''^2 = 2k_1^0 \left[\frac{m}{m_1} (\vec{k}_1 - \vec{k}_1'') \cdot \vec{e}_0 - \frac{m}{m_2} (\vec{k}_2 - \vec{k}_2'') \cdot \vec{e}_0\right]. \tag{6.2.33}$$

Thus, in (6.2.32) we may integrate over $d^3\vec{k}_1\, d^3\vec{k}_2\, d^3\vec{k}_1''\, d^3k_2''$; we obtain

$$\frac{m^2(2\pi)^4}{m_1} \int_{\Delta\Omega} df \int \frac{\vec{k}_1'^0 \cdot \vec{n}}{|\vec{r} - \vec{r}_2|^2} \left|T\left(k^0 \frac{\vec{r} - \vec{r}_2}{|\vec{r} - \vec{r}_2|}, k^0\vec{e}_0\right)\right|^2$$
$$\cdot \left\langle \frac{m}{m_2}\vec{r} + \frac{m}{m_1}\vec{r}_2 - \frac{k_1^0}{m_1}\vec{e}_0 t \right| W_{10}^i \left| \frac{m}{m_2}\vec{r} + \frac{m}{m_1}\vec{r}_2 - \frac{k_1^0}{m_1}\vec{e}_0 t \right\rangle$$
$$\cdot \left\langle \frac{m}{m_2}\vec{r} + \frac{m}{m_1}\vec{r}_2 + \frac{k_1^0}{m_2}\vec{e}_0 t \right| W_{20}^i \left| \frac{m}{m_2}\vec{r} + \frac{m}{m_1}\vec{r}_2 + \frac{k_1^0}{m_1}\vec{e}_0 t \right\rangle d^3\vec{r}_2\, dt.$$

According to the assumptions the quantities

$$\vec{r}_1' = \frac{m}{m_2}\vec{r} + \frac{m}{m_1}\vec{r}_2 - \frac{k_1^0}{m_1}\vec{e}_0 t$$

and

$$\vec{r}_2' = \frac{m}{m_2}\vec{r} + \frac{m}{m_1}\vec{r}_2 + \frac{k_1^0}{m_1}\vec{e}_0 t$$

are of the order of magnitude of the size of the collision region in order that neither $\langle \vec{r}_1' | W_{10}^i | \vec{r}_1' \rangle$ or $\langle \vec{r}_2' | W_{20}^i | \vec{r}_2' \rangle$ are equal to zero. Thus it follows that

$$\frac{k_1^0}{m_1}\vec{e}_0 t, \quad \frac{k_1^0}{m_2}\vec{e}_0 t \quad \text{and} \quad \frac{m}{m_2}\vec{r} + \frac{m}{m_1}\vec{r}_2$$

must also be of the same order of magnitude. For r large compared to the size of the collision region we therefore find

$$\vec{r}_2 \sim -\frac{m_1}{m_2}\vec{r}$$

so that we may replace

$$\frac{1}{|\vec{r}-\vec{r}_2|} \text{ by } \frac{1}{(1+m_1/m_2)^2 r^2}$$

and

$$\frac{\vec{r}-\vec{r}_2}{|\vec{r}-\vec{r}_2|} \text{ by } \frac{\vec{r}}{r} \overset{\text{def}}{=} \vec{e}.$$

On the basis of these considerations we may, instead of (6.2.32), also write:

$$\frac{m^2 k_1^0}{(1+m_1/m_2)^2 m_1 (2\pi)^2} \int_{\Delta\Omega} d\omega(\vec{e}) \int \left[\int_{-\infty}^{\infty} e^{-(i/2m)(k^2-k''^2)t}\,dt\right] \tag{6.2.34}$$

$$\cdot |T(k_1^0\vec{e}, k_1^0\vec{e}_0)|^2 \langle \vec{k}_1 | W^i_{10} | \vec{k}''_1\rangle\langle \vec{k}_2 | W^i_{20} | \vec{k}''_2\rangle$$

$$\cdot\, e^{i(\vec{k}_1+\vec{k}_2-\vec{k}''_1-\vec{k}''_2)\cdot((m/m_2)\vec{r}+(m/m_1)\vec{r}_2)} d^3\vec{k}_1\, d^3\vec{k}_2\, d^3\vec{k}''_1\, d^3\vec{k}''_2\, d^3\vec{r}_2.$$

Integration over $\vec{r}_2$ yields a δ function. For this reason we may use the following approximation instead of (6.2.33)

$$k^2 - k''^2 = 2k_1^0(\vec{k}_1 - \vec{k}''_1)\cdot\vec{e}_0 = k_1^2 - k_1''^2.$$

From (6.2.34) we therefore obtain the new expression:

$$\frac{m^2 k_1^0}{(1+m_1/m_2)^2 m_1 2\pi} \int_{\Delta\Omega} d\omega(\vec{e}) \int \delta\left(\frac{1}{2m}(k_1^2 - k_1''^2)\right) |T(k_1^0\vec{e}, k_1^0\vec{e}_0)|^2$$

$$\cdot \langle \vec{k}_1 | W^i_{10} | \vec{k}''_1\rangle\langle \vec{k}_2 | W^i_{20} | \vec{k}''_2\rangle$$

$$\cdot\, e^{i(\vec{k}_1+\vec{k}_2-\vec{k}''_1-\vec{k}''_2)\cdot((m/m_2)\vec{r}+(m/m_1)\vec{r}_2)}\, d^3\vec{k}_1\, d^3\vec{k}_2\, d^3\vec{k}''_1\, d^3\vec{k}''_2\, d^3\vec{r}_2.$$

We may now carry out the integrations over $\vec{k}_2$ and $\vec{k}''_2$. Together with

$$\delta\left(\frac{1}{2m}(k_1^2 - k_1''^2)\right) = \frac{m}{m_1}\,\delta\left(\frac{1}{2m_1}(k_1^2 - k_1''^2)\right)$$

we finally obtain

$$\frac{m^5 k_1^0 (2\pi)^2}{m_1^4} \int_{\Delta\Omega} d\omega(\vec{e}) \int \delta\left(\frac{1}{2m_1}(k_1^2 - k_1''^2)\right) |Tk_1^0\vec{e}, k_1^0\vec{e}_0|^2 \langle \vec{k}_1 | W^i_{10} | \vec{k}''_1\rangle$$

$$e^{i(\vec{k}_1-\vec{k}''_1)\cdot((m/m_2)\vec{r}+(m/m_1)\vec{r}_2)} \left\langle \frac{m}{m_2}\vec{r} + \frac{m}{m_1}\vec{r}_2 \middle| W^i_{20} \middle| \frac{m}{m_2}\vec{r} + \frac{m}{m_1}\vec{r}_2 \right\rangle d^3\vec{k}_1\, d^3\vec{k}''_1\, d^3\vec{r}_2.$$

Similarly as in the evaluation of (6.2.20) it follows that the impact probability will be given by

$$\frac{m^2 k_1^0 (2\pi)^4}{m_1} \int_{\Delta\Omega} d\omega(\vec{e}) \int \delta\left(\frac{1}{2m_1}(k_1^2 - k_1''^2)\right) |T(k_1^0 \vec{e}, k_1^0 \vec{e}_0)|^2$$
$$\cdot \langle k_1, \vec{r}_{2\perp} | W_{10}^i | k_1, \vec{r}'_{2\perp} \rangle \langle \vec{r}_2 | W_{20}^i | \vec{r}'_2 \rangle \, dk_1 \, dk_1'' \, d^3\vec{r}'_2$$

and, under the same assumptions, since $\langle k_1, \vec{r}_{2\perp} | W_{10}^i | k_1 \vec{r}'_{2\perp} \rangle$ does not, for all practical purposes, significantly vary with $\vec{r}'_{2\perp}$ in the target region, we may integrate over $\vec{r}'_2$. We obtain

$$\frac{m^2 k_1^0 (2\pi)^4}{m_1} \int_{\Delta\Omega} d\omega(\vec{e}) |T(k_1^0 \vec{e}, k_1^0 \vec{e}_0)|^2$$
$$\cdot \int \delta\left(\frac{1}{2m_1} k_1^2 - \frac{1}{2m_1} k_1''^2\right) \langle k_1, 0 | W_{10}^i | k_1'', 0 \rangle \, dk_1 \, dk_1''. \quad (6.2.35)$$

We now consider case (β): Since $\vec{k}_2^0 = 0$, from (6.2.30) it follows that

$$\vec{k}^0 = \frac{m}{m_1} k_1^0 \vec{e}_0, \qquad \vec{k}_1'^0 = k^0 \frac{\vec{r} - \vec{r}_2}{|\vec{r} - \vec{r}_2|} + \frac{m}{m_2} k_1^0 \vec{e}_0. \quad (6.2.36)$$

In the time dependent exponent in (6.2.29) we may write the energy difference in the form

$$\frac{1}{2m_1}(k_1^2 - k_1''^2) + \frac{1}{2m_2}(k_2^2 - k_2''^2).$$

Since $\vec{k}_2^0 = 0$, and since we consider only linear terms in $\vec{k}_2$; we may neglect $k_2^2 - k_2''^2$. Equation (6.2.29) then can be written as follows

$$\frac{m^2}{2\pi m_1} \int_{\Delta\mathscr{F}} df \int \delta\left(\frac{1}{2m_1} k_1^2 - \frac{1}{2m_1} k_1''^2\right) \vec{k}_1'^0 \cdot \vec{n} \, \frac{e^{i(k - k'')|\vec{r} - \vec{r}_2|}}{|\vec{r} - \vec{r}_2|}$$
$$\cdot \left| T\left(k^0 \frac{\vec{r} - \vec{r}_2}{|\vec{r} - \vec{r}_2|}, \vec{k}^0\right)\right|^2 \langle k_1 | W_{10}^i | k_1'' \rangle$$
$$\cdot \langle \vec{k}_2 | W_{20}^i | k_2'' \rangle$$
$$\cdot \, e^{i(\vec{k}_1 + \vec{k}_2 - \vec{k}_1'' - \vec{k}''_2) \cdot ((m/m_2)\vec{r} + (m/m_1)\vec{r}_2)} \, d^3\vec{k}_1 \, d^3\vec{k}_2 \, d^3k_1'' \, d^3k_2'' \, d^3\vec{r}_2. \quad (6.2.37)$$

Including linear terms in $\vec{k}_2, \vec{k}_2''$, it follows that

$$k - k'' = -\frac{m}{m_2} \vec{e}_0 \cdot (\vec{k}_2 - \vec{k}_2''),$$

where we have set $k_1'' = k_1$ because of the δ function. Thus (6.2.37) is transformed into

$$\frac{m^2}{2\pi m_1}\int_{\Delta\mathscr{F}} df \int \delta\left(\frac{1}{2m_1}(k_1^2 - k_1''^2)\right)\left|T\left(k^0 \frac{\vec{r}-\vec{r}_2}{|\vec{r}-\vec{r}_2|}, \vec{k}^0\right)\right|^2 \frac{\vec{k}_1'^0\cdot\vec{n}}{|\vec{r}-\vec{r}_2|^2}$$

$$\cdot\, e^{i(\vec{k}_1-\vec{k}_1'')\cdot((m/m_2)\vec{r}+(m/m_1)\vec{r}_2)}\, e^{i(\vec{k}_2-\vec{k}_2'')\cdot((m/m_2)\vec{r}+(m/m_1)\vec{r}_2-(m/m_2)e_0|\vec{r}-\vec{r}_2|)}$$

$$\cdot\,\langle\vec{k}_1|W_{10}^i|\vec{k}_1''\rangle\langle\vec{k}_2|W_{20}^i|\vec{k}_2''\rangle\, d^3\vec{k}_1\, d^3\vec{k}_2\, d^3\vec{k}_1''\, d^3\vec{k}_2''\, d^3\vec{r}_2.$$

The integrations over $\vec{k}_2, \vec{k}_2''$ may be carried out as follows:

$$\frac{m^2(2\pi)^2}{m_1}\int_{\Delta\mathscr{F}} df \int \delta\left(\frac{1}{2m_1}(k_1^2 - k_1''^2)\right)\left|T\left(k^0 \frac{\vec{r}-\vec{r}_2}{|\vec{r}-\vec{r}_2|}, \vec{k}^0\right)\right|^2$$

$$\frac{\vec{k}_1'^0\cdot\vec{n}}{|\vec{r}-\vec{r}_2|^2}\, e^{i(\vec{k}_1-\vec{k}_1'')\cdot((m/m_2)\vec{r}+(m/m_1)\vec{r}_2)}\langle\vec{k}_1|W_{10}^i|\vec{k}_1''\rangle$$

$$\cdot\left\langle\frac{m}{m_2}\vec{r}+\frac{m}{m_1}\vec{r}_2-\frac{m}{m_2}\vec{e}_0|\vec{r}-\vec{r}_2|\,\middle|\,W_{20}^i\,\middle|\,\frac{m}{m_2}\vec{r}+\frac{m}{m_1}\vec{r}_2-\frac{m}{m_2}\vec{e}_0|\vec{r}-\vec{r}_2|\right\rangle$$

$$\cdot\, d^3\vec{k}_1\, d^3\vec{k}_1''\, d^3\vec{r}_2. \tag{6.2.38}$$

Instead of $\vec{r}_2$ we introduce the following new integration variable

$$\vec{r}_2' = \frac{m}{m_2}\vec{r}+\frac{m}{m_1}\vec{r}_2-\frac{m}{m_2}\vec{e}_0|\vec{r}-\vec{r}_2|. \tag{6.2.39}$$

Thus (6.2.28) is transformed into

$$\frac{m^2(2\pi)^2}{m_1}\int_{\Delta\mathscr{F}} df \int \delta\left(\frac{1}{2m_1}(k_1^2 - k_1''^2)\right)\left|T\left(k^0 \frac{\vec{r}-\vec{r}_2}{|\vec{r}-\vec{r}_2|}, \vec{k}^0\right)\right|^2$$

$$\cdot\,\frac{\vec{k}_1'^0\cdot\vec{n}}{|\vec{r}-\vec{r}_2|^2}\, e^{i(\vec{k}_1-\vec{k}_1'')\cdot(\vec{r}_2'+(m/m_2)e_0|\vec{r}-\vec{r}_2|)}$$

$$\cdot\,\langle\vec{k}_1|W_{10}^i|\vec{k}_1''\rangle\langle\vec{r}_2'|W_{20}^i|\vec{r}_2'\rangle\left|\frac{\partial\vec{r}_2'}{\partial\vec{r}_2}\right|^{-1} d^3\vec{k}_1\, d^3\vec{k}_1''\, d^3\vec{r}_2', \tag{6.2.40}$$

where $\vec{r}_2$ is considered to be a function of $\vec{r}_2'$ and $|\partial\vec{r}_2'/\partial r_2|$ is the functional determinant of the transformation (6.2.39). Since $\vec{r}_2'$ is only of the order of magnitude of the size of the target region, from (6.2.39) we obtain the following approximation

$$\frac{m}{m_2}\vec{r}+\frac{m}{m_1}\vec{r}_2-\frac{m}{m_2}\vec{e}_0|\vec{r}-\vec{r}_2|\approx 0 \tag{6.2.41}$$

which suffices as an approximation for $\vec{r}_2$ in $|\vec{r}-\vec{r}_2|^{-2}$. In the same way we may compute $\vec{k}_1'^0$ with this approximation according to (6.2.30). We will show that $\vec{k}_1'^0$ has the same direction as $\vec{r}$. For simplification we introduce the following notation

$$\vec{e} = \frac{\vec{r} - \vec{r}_2}{|\vec{r} - \vec{r}_2|} \quad \text{and} \quad \vec{e}_1 = \frac{\vec{r}}{|\vec{r}|}.$$

From (6.2.36) it follows that

$$\vec{k}_1'^0 = k_1^0\left(\frac{m}{m_1}\,\vec{e} + \frac{m}{m_2}\,\vec{e}_0\right). \tag{6.2.42}$$

From (6.2.41) it follows that

$$r\vec{e}_1 = |\vec{r}_1 - \vec{r}_2|\left(\frac{m}{m_1}\,\vec{e} + \frac{m}{m_2}\,\vec{e}_0\right). \tag{6.2.43}$$

Thus it follows that $\vec{k}_1'^0$ has the direction of $\vec{e}_1 = \vec{n}$, so that in (6.2.40) we may replace $\vec{k}_1'^0 \cdot \vec{n}$ by $k_1'^0$. The functional determinant is given by

$$\left|\frac{\partial \vec{r}_2'}{\partial \vec{r}_2}\right| = \left(\frac{m}{m_1}\right)^3\left(1 + \frac{m_1}{m_2}\,\vec{e}_0 \cdot \vec{e}\right). \tag{6.2.44}$$

As earlier, we may forego integration over $\vec{k}_1, \vec{k}_1''$ in (6.2.40), in which we use the δ function to obtain

$$(\vec{k}_1 - \vec{k}_1'')\cdot\left(\vec{r}_2' + \frac{m}{m_2}\,\vec{e}_0|\vec{r} - \vec{r}_2|\right) = (\vec{k}_{1\perp} - \vec{k}_{1\perp}'')\cdot\vec{r}_{2\perp}'.$$

Thus, under the same assumptions about $\langle k_1', \vec{r}_{2\perp}|W_{20}^i|k_1'\vec{r}_{2\perp}'\rangle$ we obtain the impact probability

$$\frac{m^2(2\pi)^4}{m_1}\int_{\Delta\mathscr{F}} df\,\frac{k_1'^0}{|\vec{r} - \vec{r}_2|^2}\,\frac{\left|T\left(\frac{m}{m_1}\,k_1^0\vec{e}, \frac{m}{m_1}\,k_1^0\vec{e}_0\right)\right|^2}{\left(\frac{m}{m_1}\right)^3\left(1 + \frac{m_1}{m_2}\,\vec{e}_0 \cdot \vec{e}\right)}$$
$$\cdot \int \delta\left(\frac{1}{2m_1}\,(k_1^2 - k_1''^2)\right)\langle k_1, 0|W_{10}^i|k_1'', 0\rangle\, dk_1\, dk_1''. \tag{6.2.45}$$

From (6.2.42) and (6.2.43) it follows that

$$r^2 = |\vec{r} - \vec{r}_2|^2\left|\frac{m}{m_1}\,\vec{e} + \frac{m}{m_2}\,\vec{e}_0\right|^2, \qquad k_1'^0 = k_1^0\left|\frac{m}{m_1}\,\vec{e} + \frac{m}{m_2}\,\vec{e}_0\right|.$$

(6.2.45) therefore takes on the following form:

$$\frac{m^2k_1^0(2\pi)^4}{m_1}\int_{\Delta\Omega} d\omega(\vec{e}_1)\,\frac{\left|\frac{m}{m_1}\,\vec{e}_1 + \frac{m}{m_2}\,\vec{e}_0\right|^2}{\left(\frac{m}{m_1}\right)^3\left(1 + \frac{m_1}{m_2}\,\vec{e}_0 \cdot \vec{e}\right)}\left|T\left(\frac{m}{m_1}\,k_1^0\vec{e}, \frac{m}{m_1}\,k_1^0\vec{e}_0\right)\right|^2$$
$$\cdot \int \delta\left(\frac{1}{2m_1}\,(k_1^2 - k_1''^2)\right)\langle k_1, 0|W_{10}^i|k_1'', 0\rangle\, dk_1\, dk_1''. \tag{6.2.46}$$

Equation (6.2.43) fixes the "auxiliary" direction $\vec{e}$ in terms of the experimentally determined measurement directions $\vec{e}_1$ and $\vec{e}_0$; using the abbreviation $\lambda = r/|\vec{r} - \vec{r}_2|$ we obtain

$$\lambda \vec{e}_1 = \frac{m}{m_1}\vec{e} + \frac{m}{m_2}\vec{e}_0 \quad \text{and} \quad \lambda^2 = \left(\frac{m}{m_1}\vec{e} + \frac{m}{m_2}\vec{e}_0\right)^2. \tag{6.2.47}$$

Let θ denote the experimental value of the scattering angle. We therefore obtain $\vec{e}_1 \cdot \vec{e}_0 = \cos\theta$. The auxiliary angle ϑ is introduced as the angle between $\vec{e}$ and $\vec{e}_0$, that is, $\vec{e}\cdot\vec{e}_0 = \cos\vartheta$. Thus from (6.2.47) it follows that

$$\left|\frac{m}{m_1}\vec{e} + \frac{m}{m_2}\vec{e}_0\right| = \frac{m}{m_1}\left[1 + 2\frac{m_1}{m_2}\cos\vartheta + \left(\frac{m_1}{m_2}\right)^2\right]^{1/2}.$$

Therefore (6.2.46) can also be written in the form

$$\frac{m^2 k_1^0 (2\pi)^4}{m_1}\int_{\Delta\Omega} d\omega(\vec{e}_1)\frac{\left[1 + 2\frac{m_1}{m_2}\cos\vartheta + \left(\frac{m_1}{m_2}\right)^2\right]^{3/2}}{1 + \frac{m_1}{m_2}\cos\vartheta}\left|T\left(\frac{m}{m_1}k_1^0\vec{e}, \frac{m}{m_1}k_1^0\vec{e}_0\right)\right|^2$$
$$\cdot \int \delta\left(\frac{1}{2m_1}(k_1^2 - k_1''^2)\right)\langle k_1, 0|W_{10}^i|k_1'', 0\rangle\, dk_1\, dk_1''. \tag{6.2.48}$$

According to (6.2.47) the scattering angle θ is obtained from the auxiliary angle ϑ as follows

$$\tan\theta = \frac{\sin\vartheta}{m_1/m_2 + \cos\vartheta}. \tag{6.2.49}$$

Here we have provided detailed derivations of several formulas for the impact probability of systems on surfaces for scattering experiments in order to illustrate the following problems.

First, it is important to realize that the mathematically idealized limiting process described in §§2 to 5 for the comparison between theory and experiment must be reduced to its actual physical meaning, that is, to finite times and finite distances. The necessary comparisons between the idealized approximation of the scattering operator and the exact behavior of the wave packet cannot be carried out because of the level of difficulty. Because of the small range of the interaction, the idealized approximation obtained by using the scattering operator is very good.

Second, it is important to realize that it is not sufficient to carry out an arbitrarily designed scattering experiment, because implicit in the formulas is the structure of the ensembles W_{10}^i, W_{20}^i. We must, in addition, carry out additional experiments in order to obtain information concerning the nature of the beams W_{10}^i and W_{20}^i of systems 1 and 2, respectively. We must therefore always assume that experimentally determined facts must be included in the mathematical form of the theory, as we have discussed at the beginning of this section and expressed as the so-called assumptions about W_{10}^i, W_{20}^i.

A third point is evident in the derivations: In order to use simple formulas such as (6.2.24), (6.2.27), (6.2.35) and (6.2.48), additional approximation steps are necessary in which, again, because of the level of difficulty, we must omit even rough estimates of the accuracy of the approximation. In particular, it is evident that the ratio of the distances of the detection surfaces to the size of the collision region is more important than the range of the interaction. However, the concept of the size of the collision region is already an "approximation", because every wave packet is, for almost all times, mathematically extended to infinity. In this relationship, the dispersion of the wave packet (see X, §§1 and 2) is such that some of the above formulas cannot be correct for "arbitrarily large" distances.

The previously mentioned estimates of the approximation are described in experimental physics by the use of the catchword "error estimate". For conceptial clarity, we emphasize the fact that experiments have, in themselves, no "errors" since they are the real facts. "Errors" refer exclusively to the comparison between experiments and mathematically deduced formulas. Such "errors" may have three different origins. First, every mathematical theory, that is, every $\mathscr{MT}$ from a $\mathscr{PT}$ (see I) represents reality imperfectly. Second, real experiments are, in general, not the experiments which one "wanted" to perform in order to test a mathematical relation, which was derived in $\mathscr{MT}$. Third, the relationships derived in $\mathscr{MT}$ are often approximations to an exact relation which is too difficult to derive. The first two points cannot be easily distinguished, since the mapping principles ($\mathscr{MAP}$) (see I) for experimental facts which cannot be represented in digital terms require the use of imprecision sets (see [8], §6 and both [6], II and [6], III, §5). The third point is not concerned with the conceptual structure of a $\mathscr{PT}$, but only an often used mathematical approximation procedure, which just permits us to apply a $\mathscr{MT}$ to a reality domain $\mathscr{W}$.

If the complex structure of the question of the comparison between theory and experiment were made clear, then we would understand, that we would prefer, for the most part, to apply methods which are not very precise but would lead more quickly to the results which we shall obtain in the following section.

In closing this section we shall extend the results obtained above by showing how the formulas for the "impact of the microsystems after scattering" have to be changed for the case of two "identical" systems. First, we must consider how (6.2.1) is to be changed since the apparatus symbolized by $\Delta\mathscr{F}$ cannot distinguish between the two "identical" systems 1 and 2.

The purely "symbolic" introduction of the map T_0 in VIII (3.5), VIII (3.7) is of little help because we have to consider the question of the response of a concrete apparatus.

Since two systems can occur, we consider the more general case of an apparatus with two detection surfaces $\Delta\mathscr{F}_a$ and $\Delta\mathscr{F}_b$. We then obtain the following coexistent effect processes: In the time interval $\mathscr{I}_a$ a system strikes $\Delta\mathscr{F}_a$; in the time interval $\mathscr{I}_b$ a system strikes $\Delta\mathscr{F}_b$, and all logical combinations of the latter.

We now consider the following logical combination: In the time interval $\mathscr{I}_a$ a system strikes $\Delta\mathscr{F}_a$ AND in the time interval $\mathscr{I}_b$ a system strikes $\Delta\mathscr{I}_b$, where $\Delta\mathscr{I}_a$ and $\Delta\mathscr{I}_b$ are two nonoverlapping portions of a spherical surface centered on the scattering region. To this effect we assign the effect

$$F(\mathscr{I}_a, \Delta\mathscr{F}_a) \times F(\mathscr{I}_b, \Delta\mathscr{F}_b) + F(\mathscr{I}_b, \Delta\mathscr{F}_b) \times F(\mathscr{I}_a, \Delta\mathscr{F}_a). \quad (6.2.50)$$

(6.2.50) is an effect because $F(\mathscr{I}_a, \Delta\mathscr{F}_a)$ and $F(\mathscr{I}_b, \Delta\mathscr{F}_b)$ cannot simultaneously occur, that is, since

$$F(\mathscr{I}_a, \Delta\mathscr{F}_a) + F(\mathscr{I}_b, \Delta\mathscr{F}_b) \le \mathbf{1}.$$

It is somewhat more difficult to discuss the case in which $\Delta\mathscr{F}_a = \Delta\mathscr{F}_b$, that is, only one detction surface is present $\Delta\mathscr{F} = \Delta\mathscr{F}_a = \Delta\mathscr{F}_b$. If we assume that the apparatus characterized by $\Delta\mathscr{F}$ has sufficiently high time resolution that both systems can be separately counted, then from $\mathscr{I}_a \cap \mathscr{I}_b = \phi$ we will obtain the same formula as (6.2.50) where $\Delta\mathscr{F} = \Delta\mathscr{F}_a = \Delta\mathscr{F}_b$. The case that both systems strike the surface $\Delta\mathscr{F}$ simultaneously can, for sufficiently small $\Delta\mathscr{F}$, for all practical purposes, be excluded. Thus we do not have to consider the form of such an effect for such a simultaneous impact on a small surface $\Delta\mathscr{F}$.

For the case of multi-channel scattering theory (see, for example, §5.3) in which one of the two systems is bound on the nucleus after scattering, the probability for (6.2.50) will be zero. Thus only one of the two systems will strike the surface $\Delta\mathscr{F}$. What is the effect operator to be used to describe this impact?

In complete analogy to (6.2.50) we shall select

$$F(\mathscr{I}, \Delta\mathscr{F}) \times \mathbf{1} + \mathbf{1} \times F(\mathscr{I}, \Delta\mathscr{F}) \quad (6.2.51)$$

as the effect of the impact of a system on the surface $\Delta\mathscr{F}$ in the time interval $\mathscr{I}$. For the ensemble used above (one system remains bound) the operator (6.2.51) acts like an effect because $F(\mathscr{I}, \Delta\mathscr{F}) \times F(\mathscr{I}, \Delta\mathscr{F})$ has expectation value 0. If we would like an effect operator which can also be used for the case in which both systems are not bound after scattering we may (for sufficiently small $\Delta\mathscr{F}$) select

$$\begin{aligned} &F(\mathscr{I}, \Delta\mathscr{F}) \times (\mathbf{1} - F(\mathscr{I}, \Delta\mathscr{F})) + (\mathbf{1} - F(\mathscr{I}, \Delta\mathscr{F})) \times F(\mathscr{I}, \Delta\mathscr{F}) \\ &\qquad + F(\mathscr{I}, \Delta\mathscr{F}) \times F(\mathscr{I}, \Delta\mathscr{F}) \\ &\quad = F(\mathscr{I}, \Delta\mathscr{F}) \times \mathbf{1} + \mathbf{1} \times F(\mathscr{I}, \Delta\mathscr{F}) - F(\mathscr{I}, \Delta\mathscr{F}) \times F(\mathscr{I}, \Delta\mathscr{F}), \quad (6.2.52) \end{aligned}$$

which can be replaced by (6.2.51) if the expectation value of

$$F(\mathscr{I}, \Delta\mathscr{F}) \times F(\mathscr{I}, \Delta\mathscr{F})$$

is zero.

6.3 The Scattering Cross Section

One of the most important experimentally measurable quantities in scattering theory is the so-called scattering cross section. This quantity is appropriate for experiments which are configured in such a way that the application of the formulas (6.2.27) or (6.2.35) or (6.2.48) is permissible. Then the application of the formula (6.2.28) for test experiments with respect to W^i_{10} will be appropriate.

The following expression follows directly from (6.2.28):

$$\frac{k_1^0}{m_1}\int\delta\left(\frac{1}{2m_1}(k_1^2 - k_1''^2)\right)\langle k_1, 0|W^i_{10}|k_1'', 0\rangle\, dk_1\, dk_1'' \tag{6.3.1}$$

and is called the "beam density" for the ensemble. (6.3.1) is the probability density that a system of type 1 will impinge on the surface $\mathscr{F}$, normalized with respect to a unit surface area.

We now consider case (1), that is, $m_2 \gg m_1$. The differential cross section $d\sigma$ is the ratio of the quantities defined by (6.2.27) and (6.3.1), that is,

$$d\sigma(\vec{e}) = (2\pi)^4 m_1^2 |T(k_1^0\vec{e}, k_1^0\vec{e}_0)|^2\, d\omega(\vec{e}), \tag{6.3.2}$$

where in (6.2.27) we have assumed that $|T(k_1^0\vec{e}, k_1^0\vec{e}_0)|$ is so weakly dependent on k_1 and that $\langle k_1, 0|W^i_{20}|k_1, 0\rangle$ has such small dispersion in k_1, that $|T(k_1\vec{e}_1, k_1\vec{e}_0)|^2$, upon integration over k_1 can be replaced by $|T(k_1^0\vec{e}, k_1^0\vec{e}_0)|^2$. Then, using (6.2.7a) we may rewrite (6.3.2) as follows:

$$d\sigma(\vec{e}) = |g_{k_1^0\vec{e}_0}(\vec{e})|^2\, d\omega(\vec{e}). \tag{6.3.3}$$

Using the special notation $f_k(\theta)$ for $g_{\vec{k}}(\vec{e})$ in (5.1.6), (6.3.3) is transformed into

$$d\sigma(\vec{e}) = |f_{k_1^0}(\theta)|^2\, d\omega(\vec{e}), \tag{6.3.4}$$

where θ is the "scattering angle"—the angle between $\vec{e}$ and $\vec{e}_0$.

For (6.2.35), that is, for "collisions in the center of mass system ($\vec{K}^0 = 0$)" we obtain the same result:

$$d\sigma(\vec{e}) = m^2(2\pi)^4 |T(k_1^0\vec{e}, k_1^0\vec{e}_0)|^2\, d\omega(\vec{e}).$$

Using (6.2.7a) we may rewrite this expression as follows

$$d\sigma(\vec{e}) = |g_{k^0,\vec{e}_0}(\vec{e})|^2\, d\omega(\vec{e}) \tag{6.3.5}$$

or (6.3.4). The reduced mass m no longer appears in $d\sigma(\vec{e})$!

For the case in which systems 2 are at rest before the collision (6.2.27) is a good approximation for (6.2.48) only for the case in which $m_2 \gg m_1$. If this is not the case, then from (6.2.48) it follows that

$$d\sigma(\vec{e}_1) = m^2(2\pi)^4 \frac{\left[1 + 2\frac{m_1}{m_2}\cos\vartheta + \left(\frac{m_1}{m_2}\right)^2\right]^{3/2}}{1 + \frac{m_1}{m_2}\cos\vartheta}$$

$$\times \left|T\left(\frac{m}{m_1}k_1^0\vec{e}, \frac{m}{m_1}k_1^0\vec{e}_0\right)\right|^2 d\omega(e_1)$$

from which, using (6.2.7a) and (5.1.6), we obtain

$$d\sigma(\vec{e}_1) = \frac{\left[1 + 2\frac{m_1}{m_2}\cos\vartheta + \left(\frac{m_1}{m_2}\right)^2\right]^{3/2}}{1 + \frac{m_1}{m_2}\cos\vartheta} \, | f_{(m/m_1)k_1^0}(\vartheta)|^2 \, d\omega(\vec{e}_1), \quad (6.3.6)$$

where the angle ϑ is related to the scattering angle $\theta(\theta$ is the angle between $\vec{e}_0$ and $\vec{e}_1$!) by the formula (6.2.49).

These formulas may easily be applied to the examples in §5; for this reason we shall not consider (6.3.6) any further, except to note that it is of great importance for experiments. We have derived it here because of its principle importance in illuminating the conceptual foundations and the approximation schemes which are required in order to obtain (6.3.6). We now return to (6.3.4) in which we need only determine $| f_k(\theta)|^2$.

From (5.1.11) it follows that

$$| f_k(\theta)|^2 = \sum_{l=0,\, l'=0}^{\infty} Y_0^l(\theta) Y_0^{l'}(\theta) \frac{1}{k^2} \pi \sqrt{2l+1}\sqrt{2l'+1}$$

$$\cdot (-i)^{l-l'} e^{-i(\pi/2)(l-l')} (e^{2i\delta_l(k)} - 1)(e^{-2i\delta_l(k)} - 1). \quad (6.3.7)$$

This formula is practical only in cases in which it suffices to consider only a small number of terms in l and l', that is, $\delta_l(k) \approx 0$ for large values of l. The l value for which $\delta_{l'}(k) \approx 0$ for $l' \geq l$ can be obtained with the aid of the range of the interaction R as follows: For

$$l' \gg kR \qquad \text{we obtain} \quad \delta_{l'}(k) \sim 0. \quad (6.3.8)$$

For the special case in which $kR \ll 1$, then one term $l = 0$ suffices and we obtain

$$| f_k(\theta)|^2 = \frac{1}{k^2} \sin^2 \delta_0(k). \quad (6.3.9)$$

In this case the differential cross section is isotropic.

The "total cross section" is defined as follows:

$$\delta = \int d\sigma(\vec{e}). \quad (6.3.10)$$

From (6.3.7) it follows that, from the orthogonality of the spherical harmonics:

$$\sigma = \frac{4\pi}{k^2} \sum_{l=0}^{\infty} (2l+1) \sin^2 \delta_l(k). \quad (6.3.11)$$

From (5.2.2) it follows that the differential cross section for the scattering of an electron by an atomic nucleus is given by

$$d\sigma = \left(\frac{mZe^2}{2k^2}\right)^2 \frac{1}{\sin^4(\theta/2)} \, d\omega. \quad (6.3.12)$$

Here it is important to note that the Rutherford scattering formula (6.3.12) is not an approximation, even though (5.2.2) was obtained by the use of the Born approximation. It is also possible to obtain (6.3.12) by using the exact eigenfunctions of the continuous spectrum given in XI, §5 (see, for example, [21]). The same result may also be obtained from classical mechanics (see, for example, [2], XI, §1.1). These results are a consequence of the special properties of the Coulomb potential.

From (6.3.12) it follows that the total scattering cross section $\sigma = \infty$. We note, however, that (6.3.12) is not meaningful for small θ in actual experiments, because the actual experiments are carried out by using apparatuses which have finite dimensions. In order to test (6.3.12) for very small angles, we would need a completely empty(!) space of thousands of kilometers between the preparation and registration apparatus. For the case of a Coulomb field the total cross section is not a characteristic of the Coulomb field but of the structure of the actual experimental targets.

In closing we now pay attention to the results of §5.3. From (5.3.4) it follows that for $W = \Omega_- W_0^i \Omega_-^+$:

$$W = \tfrac{1}{4}[P_{\Omega_-\psi_+} \times P_0 + P_{\Omega_-\psi_+} \times P_1]. \tag{6.3.13}$$

Thus, using $\phi_{\vec{k}}(\vec{r}_1, \vec{r}_2)$ defined according to (5.3.5), where we use $\vec{k}$ as an index, we obtain

$$\Omega_-\psi_+ = \frac{1}{(2\pi)^{3/2}} \int \chi(\vec{k}) \frac{1}{\sqrt{2}} [\phi_{\vec{k}}(\vec{r}_1, \vec{r}_2) + \phi_{\vec{k}}(\vec{r}_2, \vec{r}_1)]\, d^3\vec{k}$$

and

$$\Omega_-\psi_- = \frac{1}{(2\pi)^{3/2}} \int \chi(\vec{k}) \frac{1}{\sqrt{2}} [\phi_{\vec{k}}(\vec{r}_1, \vec{r}_2) - \phi_{\vec{k}}(\vec{r}_2, \vec{r}_1)]\, d^3\vec{k}. \tag{6.3.14}$$

Let $\bar{F}_1$ denote the effect of the impact of a microsystem on a surface given in §6.2. Then, from (6.2.52) and (6.3.13) it follows that

$$\begin{aligned} &\operatorname{tr}(W[(\bar{F}_1 \times \mathbf{1}) + (\mathbf{1} \times \bar{F}_2) - \bar{F}_1 \times \bar{F}_2] \times \mathbf{1}) \\ &\quad = \tfrac{1}{2}[\langle \Omega_-\psi_+, (\bar{F}_1 \times \mathbf{1})\Omega_-\psi_+ \rangle + 3\langle \Omega_-\psi_-, (\bar{F}_1 \times \mathbf{1})\Omega_-\psi_- \rangle]. \end{aligned} \tag{6.3.15}$$

In (6.3.15) we have used the fact that $\bar{F}_1 \varphi_1(\vec{r}_1) = 0$ (see also (6.2.51)) since the effect $\bar{F}_1$ is measured for large $|\vec{r}_1|$. From (6.3.15), using the approximation (5.3.16) together with the asymptotic forms (5.3.13) for

$$\chi(\vec{r}) = \langle E_1, 0, 0; \vec{r} \,|\, \phi \rangle,$$

and similarly, (5.3.19) for $\eta(\vec{r})$ we obtain the differential cross section for elastic scattering:

$$d\sigma(\vec{e}) = [|f_k(\theta)|^2 + |h_k(\theta)|^2 - \tfrac{1}{2}(\overline{f_k(\theta)}h_k(\theta) + f_k(\theta)\overline{h_k(\theta)})]\, d\omega(\vec{e}). \tag{6.3.16}$$

Using (5.3.20), (5.3.23) we may compute an approximation of (6.3.16).

The amplitude $h_k(\theta)$ describes an "exchange effect" for the pair of electrons. If $|h_k(\theta)| \ll |f_k(\theta)|$ then this effect can be neglected, and we obtain the approximation

$$d\sigma(\vec{e}) = |f_k(\theta)|^2 \, d\omega(\vec{e}), \tag{6.3.17}$$

that is, the same result we have obtained in (6.3.4). The cross section for the scattering of "slow" electrons on atoms (a similar derivation to that in §5.3 holds also for atoms other than hydrogen atoms) is therefore similar to a potential scattering. From the formula (6.3.9) it follows that the total cross section for very slow electrons can "accidently" be very small for certain energy values, that is, if $\sin \delta_0(k) \approx 0$ (see, for example, [22]).

From the asymptotic form of the function $\langle E_n, l, m; \vec{r} | \phi \rangle$ defined in §5.3

$$\langle E_n, l, m; \vec{r} | \phi \rangle \approx f_{k, nlm}(\vec{e}) \frac{e^{ik_n r}}{r}, \tag{6.3.18}$$

noting that (5.3.24) represents an approximation of (6.3.18), we may compute the total cross section σ_n for the excitation of the energy level E_n as follows:

$$\sigma_n = \frac{k_n}{k} \int \sum_{l=0}^{n-1} \sum_{m=-l}^{l} |f_{k, nlm}(\vec{e})|^2 \, d\omega(\vec{e}). \tag{6.3.19}$$

In order to measure this cross section we need to be able to experimentally distinguish between the different energies $k^2/2m$, $k_n^2/2m$ ($n = 2, 3, \ldots$), a topic which we shall consider in the following section and in XVII, §6.2.

7 Survey of Other Problems in Scattering Theory

In this section we shall only present a brief overview of other problems in scattering theory. The variety of different scattering experiments for which quantum mechanics provides a precise description (providing the corresponding mathematical problems can be solved to reasonable accuracy) is vast. All possible atoms and molecules can be scattered against each other. In addition to two-body problems it is also possible to consider three-body problems.

While it was possible to present a brief overview of the structure of "bound" atoms and molecules in XI–XV, such an overview of scattering problems is not possible without the need for an additional book, the scope of which would be far larger than the present book. Why is there a disparity between the ability, on the one hand, to describe the structure of atoms and molecules, and on the other, to describe the scattering process?

The difference lies primarily in the fact that the structure of atoms and molecules is, for the most part, determined by the discrete spectrum of the Hamiltonian operators for the different system types; in particular, the structure of a molecule is often determined by the behavior of the ground states of the Hamiltonians. The discrete character of the energy levels makes it possible to make qualitative statements about their structure based upon symmetry considerations and by performing an imaginary "continuous variation of parameters." We have already considered these in XI to XV.

For the case of the continuous spectrum the situation is completely different. The spectrum is continuous, and, as such, has no structure (except for degree of degeneracy [which is always infinite] and the decomposition of the spectrum into absolutely continuous and singular-continuous components). The physical significance of the continuous spectrum has to be determined from considerations other than the "possible" energy eigenvalues. The physical significance is associated with questions which arise in scattering theory. It is, of course, possible to obtain general statements about the structure of the wave operator and the scattering operator on the basis of the symmetry of the Hamiltonian operator; in §5.1 we have considered examples of such. Readers who are interested in the implications of symmetry for the case of scattering theory are referred to the specialized literature [23].

Symmetry considerations can be very useful for scattering theory because they often reduce the work necessary in solving problems and, for many applications, there are important symmetries associated with cross sections. In addition to rotational symmetry, reflection and time-reversal symmetries play an important role (see, for example, [23]).

Symmetry considerations, on the other hand, provide no insights into the structure of the scattering process, neither in principle or quantitatively. This has been shown in the general statement of the problem in §4 and the survey of particular scattering problems in §5.

We now come to an additional distinction—the measurement methods needed in scattering experiments. The discrete terms of the Hamiltonian operator determine the discrete frequencies of the emitted radiation in the simplest way—and already provide experimentally provable results without making it necessary to provide a precise analysis of the function of a spectral apparatus. With the aid of simple arguments it is possible to obtain theoretical expressions for the relative intensities of the spectral lines (see XI, §1). In the case of the scattering process we must give the measurement process a great deal of thought, as we have done in §§6.1 and 6.2. In fact, the registration processes of "striking a surface at time t" discussed are not the only ones used by experimental physicists. For example, the surface considered in §6.1 can be experimentally achieved by constructing a hole, through which the system enters into a complicated apparatus. This apparatus may be such that it is possible to measure the energy of the incoming systems with, for example, the aid of fields (see XVII, §6.2). For such a registration there is the problem that there does not exist a coexistent registration procedure which can resolve both times of incidence and energies to arbitrary accuracy. Often this fact is expressed in terms of a Heisenberg uncertainty relation for energy and time:

$$\Delta t \, \Delta E \gtrsim \tfrac{1}{2}. \tag{7.1}$$

Although (7.1) appears to have the same form as the uncertainty relation between position and momentum (see IV, (8.3.18)), there are essential differences.

First, we note that time is not a scale observable like position and momentum, a fact which is physically reasonable but is often (at first) disturbing to many people (see VII). Second, the relation IV (8.3.18) refers to the preparation, because it represents the dispersion of position and momentum in all possible ensembles, as we have discussed in IV, §8.3. Of course, a relationship which is similar to IV (8.3.18) must hold also for "measurement errors", and in IV, §8.3 we have interpreted them in this way.

We may also interpret (7.1) as an assertion about the preparation process, that is, about the ensemble. Then we must define what we mean by Δt and ΔE more precisely. The probability for striking a surface at time t can only be meaningfully defined with the help of (6.1.13) for free moving systems. For W in the $\vec{K}$ representation we will therefore define

$$E = \int \frac{k^2}{2m} \langle \vec{k} | W | \vec{k} \rangle \, d^3\vec{k} \quad \text{and} \quad (\Delta E)^2 = \int \left(\frac{k^2}{2m} - E \right)^2 \langle \vec{k} | W | \vec{k} \rangle \, d^3\vec{k}. \tag{7.2}$$

Similarly, for $\bar{F}$ defined by (6.1.13), we may introduce Δt. In order to be able to describe these simply, we introduce the following projection operators. First, in the $\vec{K}$ representation we define

$$P_+ \langle \vec{k} | \psi \rangle = \begin{cases} \langle \vec{k} | \psi \rangle & \text{if } k_1 \geq 0, \\ 0 & \text{if } k_1 < 0. \end{cases} \tag{7.3}$$

We define a mapping of $P_+ \langle \vec{k} | \psi \rangle$ on $\eta(t, x_2, x_3)$ by

$$\eta(t, x_2, x_3) = \frac{1}{\sqrt{m(2\pi)^3}} \int_0^\infty dk_1 \int_{-\infty}^\infty dk_2 \int_{-\infty}^\infty dk_3 \sqrt{k_1}\, e^{i(k_2 x_2 + k_3 x_3)(i/2m)k^2 t} \langle \vec{k} | \psi \rangle. \tag{7.4}$$

It then follows that

$$\int_{-\infty}^\infty dt \int_{-\infty}^\infty dx_2 \int_{-\infty}^\infty dx_3 |\eta(t, x_2, x_3)|^2 = \int_0^\infty dk_1 \int_{-\infty}^\infty dk_2 \int_{-\infty}^\infty dk_3 |\langle \vec{k} | \psi \rangle|^2. \tag{7.5}$$

We define the projection operator $Q_{\mathscr{F}}$ as follows

$$Q_{\mathscr{F}} \eta(t, x_2, x_3) = \begin{cases} \eta(t, x_2, x_3) & \text{if } (x_2, x_3) \in \mathscr{F}, \\ 0 & \text{otherwise.} \end{cases} \tag{7.6}$$

Let A be the multiplication operator (with t) applied to η according to (7.4). We define the following, on the basis of (6.1.13) in the following obvious way

$$\bar{t} = a^{-1} \operatorname{tr}(Q_{\mathscr{F}} P_+ W P_+ Q_{\mathscr{F}} A), \tag{7.7}$$

$$(\Delta t)^2 = a^{-1} \operatorname{tr}(Q_{\mathscr{F}} P_+ W P_+ Q_{\mathscr{F}} (A - \bar{t}1)^2), \tag{7.8}$$

where the normalization factor a is given by

$$a = \operatorname{tr}(Q_{\mathscr{F}} P_+ W P_+ Q_{\mathscr{F}}). \tag{7.9}$$

A relationship of the form (7.1) can be anticipated only if, in the definition of ΔE in (7.2) we replace the ensemble W by the ensemble

$$\tilde{W} = a^{-1} Q_{\mathscr{F}} P_+ W P_+ Q_{\mathscr{F}} \tag{7.10}$$

because not all of W (but only $\tilde{W}$) contributes to the measurement of the incidence of systems on the surface $\mathscr{F}$.

If we apply the operator $\tilde{W}$ and the operator $B = i(\partial/\partial t)$ to η using (7.2) we obtain

$$\begin{aligned} \bar{E} &= \operatorname{tr}(\tilde{\mathscr{W}} B), \\ (\Delta E)^2 &= \operatorname{tr}(\tilde{W}(B - \bar{E}1)^2). \end{aligned} \tag{7.11}$$

From (7.7), (7.8) we obtain

$$\begin{aligned} \bar{t} &= \operatorname{tr}(\tilde{W} A), \\ (\Delta t)^2 &= \operatorname{tr}(\tilde{W}(A - \bar{t} \cdot 1)^2). \end{aligned} \tag{7.12}$$

On the basis of the commutation relation $AB - BA = (1/i)\mathbf{1}$, from IV (8.3.17) we obtain (7.1).

In order to avoid errors, we observe that the mapping defined by (7.4) does not establish an isomorphism between the Hilbert spaces $\mathscr{L}^2(R^3, d^3\vec{k})$ and $\mathscr{L}^2(R^3, dt\, dx_2\, dx_3)$. In order to derive relation (7.1) for ensembles of the form $\tilde{W}$ (7.10) it suffices to prove that the map (7.4) is well defined.

The relation (7.1) implies that it is only interesting to experimentally analyze the time incidence after the scattering, that is, to make experimental comparisons between the formulas (6.2.23) and (6.2.26)—if Δt in (7.12) (for $\tilde{W}$ before scattering) is not greater than the theoretically expected difference in the time structure between (6.2.23) and (6.2.26). In particular, it does not make sense to prepare decaying states in the sense of XVII, §6.5 if, before the scattering $\Delta t > \tau$ (where τ is the decay time; see XVII, §6.5), that is, $\Delta E < \tau^{-1}$. On the contrary, we must choose $\Delta E \gg \tau^{-1}$. Experimentally, for the usual values of τ, there is, of course, no difficulty, because it is experimentally difficult to make ΔE small.

For the registration of scattering processes, not only impacts on surfaces but also completely different methods (such as cloud chambers, bubble chambers, etc.) are used, the effect procedures are not as simple as those described in §§6.1 and 6.2. The "evaluation" of such experiments is somewhat more difficult. Here by "evaluation" we mean nothing other than obtaining a good (more or less) determination of the effect operator F corresponding to the procedure. What theory should we use to carry out such an evaluation? With the help of quantum mechanics, or with classical theories which precede quantum mechanics? In order to obtain another step in the answer to this question, in the following chapter we shall investigate certain partial aspects of the physical processes of preparation and registration. For this purpose we shall require, as a fundamental assumption, the description of the scattering process presented in this chapter.

Scattering theory raises another question concerning the possibilities of preparing and registering—a question which is of great importance for our view of quantum mechanics. According to the description of scattering experiments presented in §1, it appears that it should be a simple matter to prepare experimentally an ensemble of composite systems in such a manner that it has the form (1.15) before the scattering, that is, it has the form $W = \Omega_-(W_{10}^i \times W_{20}^i)\Omega_-^+$ where Ω_- is the wave operator introduced in §4.

Our question is as follows: Is it possible to prepare the ensemble $\tilde{W} = C\Omega_-(W_{10}^i \times W_{20}^i)\Omega_-^+C$? Here C is the time reversal operator introduced in X, §4. What is the structure of $\tilde{W}$? From $C\Omega_-C^{-1} = \Omega_+$ it follows that $\tilde{W} = \Omega_+[CW_{10}^iC^{-1} \times CW_{20}^iC^{-1}]\Omega_+^+$. $\tilde{W}$ is therefore an ensemble which "after" the scattering is equal to $CW_{10}^iC^{-1} \times CW_{20}^iC^{-1}$.

If, for example, W_{10}^i, W_{20}^i describe a pair of colliding beams of systems 1 and 2, respectively, then we must seek to "prepare" W such that, "after" scattering, two outgoing beams $CW_{10}^iC^{-1}$ and $CW_{20}^iC^{-1}$ arise. This would mean that, before the scattering, "infalling spherical waves" must be prepared which have two strongly correlated spherical waves of systems 1 and 2, respectively, in such a way that after the scattering two uncorrelated separated beams $CW_{10}^iC^{-1}$, $CW_{20}^iC^{-1}$ are created. To satisfy this problem will certainly be considered hopeless by any experimental physicist. Have we encountered a principle impossibility in the preparation process?

If yes, then quantum mechanics would not be (at least for composite microsystems) a closed theory (g.G. closed theory in the sense of [8], §10.3). Then, we need to seek a more comprehensive theory which describes what really can be prepared.

Where is the place where we have introduced requirements about the preparation possibilities which are too strong?

It does not appear to be difficult to prepare the two diverging beams $CW_{10}^iC^{-1}$ and $CW_{20}^iC^{-1}$ as described earlier (as illustrated in Figure 38). How does the preparation according to Figure 38 differ from the preparation of $\tilde{W}$ by preparation apparatuses before the scattering?

Let $a_1 \in \mathcal{Q}'$ be the preparation which corresponds to Figure 38, then $\varphi(a_1) = \tilde{W}$ where φ is given in III, D 1.3. In very good approximation $\varphi(a_1) \sim CW_{10}^iC^{-1} \times CW_{20}^iC^{-1}$, since a_1 can only be combined with those registration methodes $b_0 \in \mathcal{R}_0$ (II, D 4.3.1) which register *after* the preparation, where we may replace $\tilde{W}$ by $CW_{10}^iC^{-1} \times CW_{20}^iC^{-1}$.

We seek a preparation procedure a_2 which is equivalent to a_1 (that is, $\varphi(a_2) = \varphi(a_1) = \tilde{W}$) and which prepares before scattering, and may be combined with more registration methods b_0 than a_1. Our reason to doubt that such a possible a_2 exists has something to do with Axioms APS 5.1.3 and APS 5.1.4 from III, §1. If we examine the arguments of III, §1, which lead to the introduction of APS 5.1.4, we will find that our doubt about the existence of such an a_2 merely raises questions concerning that axiom. Then if b_0 is a registration method which cannot be combined with a_1 because b_0 already registers before a_1 prepares, then the equivalence class $\tilde{W}$ to which a_1 belongs does not necessarily contain an a_2 which may be combined with b_0.

The above problem is closely related to the question of the time translation $(1, 0, 0, \gamma)$ of registration procedures. If b_{01} is a registration method which can be combined with a_1, then for negative γ there does not necessarily exist a $a_2 \in \varphi(a_1)$ which can be combined with the translated registration method $b_{02} = (1, 0, 0, \gamma)b_{01}$.

The form of a more comprehensive theory than quantum mechanics is, as yet, unknown. In [7] there is a proposed solution for systems composed of "many" elementary systems. For the problem of smaller composite systems we can only refer readers to [7].

CHAPTER XVII

The Measurement Process and the Preparation Process

In our terminology the above chapter heading probably should be called "Registration Procedures and Preparation Procedures" because it is concerned with a theoretical treatment of the phenomena which we have described in this book beginning with II, §4. Nevertheless we have used the expression "Measurement Process" because the literature commonly refers to the problems which are treated here as the "measurement process in quantum mechanics". Actually, the usual quantum mechanics viewpoint is too one-sided, emphasizing the measurement process (in our terminology—the registration process). This onesidedness is related to the philosophical meaning and interpretation of quantum mechanics, where the measurement process is, of course, a central issue in the problem of meaning.

Because of size limitations, in this book, we find it impossible to provide an extensive survey of the possible meanings of quantum mechanics and their corresponding viewpoints concerning the measurement process (see, for example, [24]). In order to remedy an underlying shortcoming in our presentation, we will now briefly describe the measurement process in terms which do not correspond to the principles described in this book.

According to this viewpoint, the problem of measurement is that of measuring one or more unknown "properties of an object", that is, determine whether or not the object under consideration has the given property. If the measurement is the "determination of a given property" then the objects must "have" these properties (at least at the time of measurement). Of course these properties can be changed shortly after the measurement by perturbing the object (by the measurement apparatus). By an ideal measurement process we mean a measurement process in which a post-measurement perturbation of the measured quantity does not occur. In this viewpoint the projection operators are considered to mathematically characterize these properties.

This viewpoint contradicts the principles underlying our formulation of quantum mechanics and its interpretation in II, §4. In our viewpoint the preparation and registration procedures are the basis for the interpretation of quantum mechanics and are not a means for the determination of the properties of micro-objects (registration procedures) or of the production of micro-objects having the desired properties (preparation procedures). On the contrary, we have seriously questioned whether it is possible to attribute objective properties to microsystems (III, §4 and IV, §8). Therefore the structure of preparation and registration procedures is not a means by which we may experimentally confirm or refute statements about previously interpreted "micro-objects having properties". Instead, concepts such as "object", "property", "pseudoproperty" can only be given meaning with the help of preparation and registration procedures. Therefore, in this connection, the expression "theory of the measurement process" does not mean to explain how we may measure properties, but how we may make a transition to a more comprehensive theory (in the sense of [8], §8) than that described earlier in this book. Here we would like to obtain a theory which will permit us to use information about the technical construction of the preparation apparatus and registration apparatus in order to determine the operators $W = \varphi(a)$ and $F = \psi(b_0, b)$. The fact that we cannot do this in this book reflects the current state of affairs. At present we must "guess" the operators W and F, and then axiomatically assert them for certain procedures (see, for example, XI and XVI). The fundamental difficulty for a complete solution lies in the fact that we do not know how such a desired theory would be related to quantum mechanics. The first section of this chapter will provide a brief survey of these difficulties, the solution of which can be found in [7].

1 The Problem of Consistency

In I we have already described a difficulty in the problem of formulating an axiomatic basis of quantum mechanics. For microsystems nothing can be directly measured. A more precise formulation of the notions of direct and indirect measurements can be found in [8]. Here we shall only intuitively motivate the difference between these notions, without referring to [8].

By a direct measurement we mean the determination of facts without use of the $\mathscr{PT}$ under consideration—here without the use of quantum mechanics. Of course this does not mean that we cannot use other theories (so-called pre-theories). Thus, with respect to quantum mechanics we may use, for example, classical electrodynamics as a pre-theory, in order, for example, to speak of a specific electric current.

The so-called classical theories are characterized by the fact they are theories of "physical objects" (in the sense of III, §4.1 and [8], §12) whose objective properties can be directly measured. The same is not true for the case of quantum mechanics. Indeed, the microsystems themselves are only indirectly determinable physically real facts. We have not taken the last fact into account in our formulation of an axiomatic basis from Chapter II onward, since we have introduced the set M of microsystems as our basis set. We have permitted this inconsistency in order not to make our representation more difficult. In [6], XVI we briefly outline and in [7] we provide more detail how this inconsistency can be alleviated. The measurement problem in quantum mechanics would not exist if it were possible to consider microsystems and their properties to be directly measurable.

By indirectly measurable we mean those facts which may be obtained with the aid of the theory under consideration and can be considered to be real physical facts, where the determination of indirectly measured facts are obtained on the basis of direct measurement and by making deductions within the framework of the theory.

The questions concerning the indirectly determinable facts is not, however, a problem of measurement, as we shall see later in this chapter. The facts obtained from indirect measurements (on the basis of direct determinable preparation and registration procedures) has been the subject of Chapters III–XVI of this book, beginning with the introduction of ensembles, effects, decision effects in III, continuing with the discussion of observables and preparators, the objective properties and pseudoproperties in IV until the special structure investigations of atoms, molecules and the scattering processes in XI to XVI.

What more shall we expect from the discussions of this chapter? We shall develop a theory which permits the theoretical description of the interaction process between the preparation apparatus and the microsystems on the one hand and the interaction process between the registration apparatus and the microsystems on the other hand and therefore develop a more comprehensive theory of the maps φ and ψ from III, D 1.3. But with the help of what theory?

For obvious reasons we shall use the formulation of quantum mechanics developed in II–XVI, because this theory, according to VIII, permits us to combine systems to form new systems. Thus it is possible to combine microsystems in order to form larger systems such as preparation and registration apparatuses. This means nothing other than introducing the apparatus themselves as elements of M. In this way we encounter the problem whether macrosystems themselves (such as preparation and registration apparatuses) can actually belong to the fundamental domain of quantum mechanics, as the concept of fundamental domain is introduced in [8] and [6], III. The fundamental domain is, however, the domain of experience upon which $\mathscr{PT}$ is applicable. This problem is not trivial. Adherents to the "universality" doctrine—those who believe that quantum mechanics is universally valid—claim that quantum mechanics remains valid for macrosystems, while sober-minded individuals claim that it is necessary to seek a theory of macrosystems which is more comprehensive than the quantum mechanics of many particles. Such a theory of macrosystems should be compatible (in a yet to be described way) with the theory of many-particle quantum mechanics. Since we encounter the problem of the relationship between quantum mechanics extrapolated to many particles and a theory of macrosystems by asking questions concerning the measurement process, our goal of obtaining a deeper theoretical description of the preparation and registration process is very remote. An outline of the problem of the relationship between quantum mechanics and macrophysics can be found in [8], XV, XVI and [7], X, XI. We shall again return to the general problem of the compatibility of quantum mechanics (as developed in II–XIV) with a theory of macrosystems in XVIII. Here we shall restrict our considerations in order to attain a simpler goal. We will only show that quantum mechanics is self-consistent, that is, the description of the interactions of microsystems does not lead us out of the realm of quantum mechanics as described by the scheme presented in II. This "special consistency problem" will be discussed in the following sections. It will lead us to considerable progress in the direction towards a theory of the preparation and registration, not only in terms of epistomological questions but also with respect to the actions of experimental physicists.

An enrichment by additional structures of the theory described earlier (up to XVI) is made possible by the observation already made in I, §2 that it is possible to divide many experiments into preparation and registration portions in many *different* ways. In this way we obtain relationships (between the maps φ, ψ in III, D 1.3) which are determined in a purely quantum

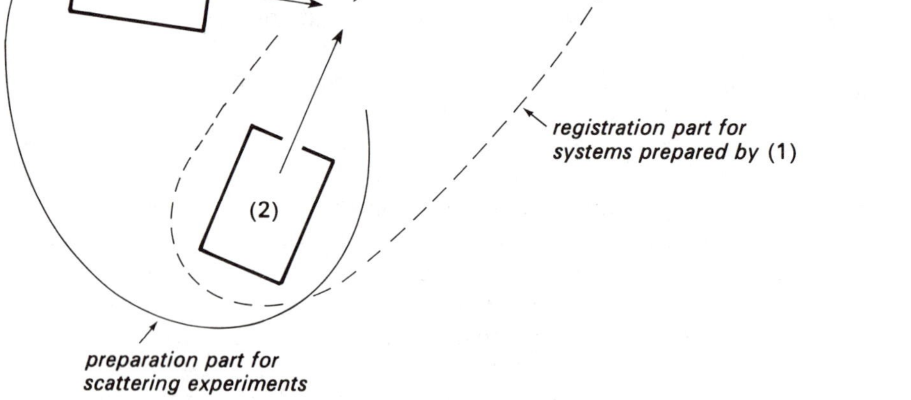

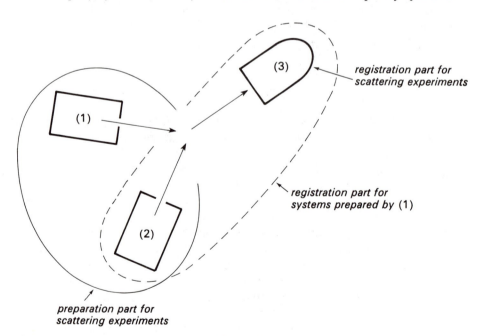

Figure 41

mechanical way by the Hamiltonian operator. We will not examine the general problem of obtaining a mathematical description of the different possible ways of dividing an experiment into preparation and registration parts, but only present a few typical examples. For our first example we shall consider the scattering experiment described in XVI consisting of two preparation apparatuses and a registration apparatus. In XVI we have considered the two preparation parts (1) and (2) as a single composite preparation apparatus for scattering experiments (see Figure 41). We may also consider (1) to be the preparation part and both (2) and (3) to be a composite registration apparatus (see Figure 41).

The *special consistency* problem consists of introducing by axioms additional structures for the different possibilities of dividing experiments into preparation and registration parts and that these axioms do not contradict previously introduced axioms (especially in II and III). The fact that this is not completely trivial we shall see in the following section when we try to strengthen the axioms in III.

2 Measurement Scattering Processes

In our terminology the expression "measurement scattering process" refers to the use of the scattering process for the purpose of registration. Such processes are not only of considerable importance for the clarification of the measurement process but are also frequently used in experimental physics. Before we formulate the general notion of measurement scattering processes we will first consider a simple example in which it appears. In order to make the example as simple as possible, we shall choose a "gedanken" experiment.

2.1 Measurement with a Microscope

Our "gedanken" experiment will consist of the determination of the position of an electron with the aid of a light quantum (photon), a microscope and a photographic plate (see Figure 42). The only imaginable possible (strongly

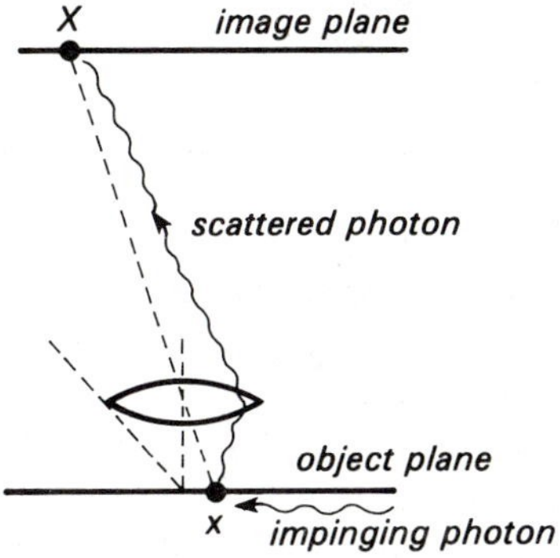

Figure 42

idealized) version of such an experiment requires that the electrons to be measured are confined to the object plane of a microscope, but they are free to move within the plane, that is, we are considering two-dimensional electrons.

We wish to determine the statistics associated with the following experiment: A light quantum with an approximate experimentally determined momentum value p impinges onto the object plane from an approximate well-determined direction. It is then scattered by an electron, passes through the microscope and blackens a silver grain near X (see Figure 42) on the photographic plate which lies in the image plane of the microscope. The position X of the blackened silver grain will, of course, be different for different trials. Here we wish to make it clear that the statistics for the position X of the blackened silver grains permits us to determine the statistics for the position of the electrons in the ensemble W which was used in the experiment.

Now it is necessary that we experimentally fix a time t_1 (at least approximately) at which we wish to measure the position—or more correctly—that we arrange the experimental system in such a way that it corresponds (at least approximately) to a measurement of $Q(t_1)$. Here we may think of a situation in which the photons are generated by light flashes which illuminate the field of the microscope during a time interval $t_1 - \tau/2$ to $t_1 + \tau/2$. Here we note that τ will become larger as the momentum p of the light quanta is more precisely defined. Here the velocity of light (in the units chosen here) is taken to be 1; according to the Heisenberg uncertainty relation the position uncertainty τ of the photons corresponds to a momentum uncertainty $\Delta p > 1/\tau$ for the photons.

In the $Q(t_1)$ representation $w(r) = \langle\vec{r}|W|\vec{r}\rangle$ is the position probability density for the ensemble W of electrons. We now wish to establish the relationship between this probability density and the probability density for the blackened silver grains in the photographic plate. This is usually done as follows: In Figure 42 the dashed line from X of the blackened grain which passes through the center of the lens corresponds to a "measured" position x of the electron. This inference from X to x is a consequence of geometrical optics; every bundle of light rays emanating from x which passes through the lens crosses the photographic plate at X. This geometric construction establishes a correspondence between a surface area F in the image plane with a surface area F_0 in the object plane. For this reason we say that the occurrence of a blackened silver grain in F corresponds to the determination of a position measurement in F_0.

If b_0 denotes the registration method described above consisting of a light source (which produces the photons), a microscope and a photographic plate, and if b_F denotes the registration procedure in which a silver grain is blackened in F, then we obtain $\psi(b_0, b_F) \approx E(F_0)$ where $E(F_0)$ is the decision effect corresponding to the situation that the position (two-dimensional) of an electron at time t_1 lies in F_0.

Of course every experimental physicist knows that $\psi(b_0, b_F) = E(F_0)$ cannot be an exact expression because the position of an electron cannot be measured without a finite error associated with the finite resolution of the microscope, and because the scattering does not occur precisely at time t_1 but during the time interval $t_1 - \tau/2$ and $t_1 + \tau/2$. It is interesting that we speak of errors here even though an apparatus (as an object of nature) has a specific function and is not faulty. The expression "error" therefore cannot refer to the apparatus as such, but to the relationship between the apparatus and the idealization which we have in mind. We shall discuss this situation later.

We shall now discuss this gedanken experiment in terms of the registration procedures which were presented in III, §4.2 and in terms of the formulation of quantum mechanics presented in this book, especially those in XVI!

We have already spoken of the blackening of a silver grain in the surface area F as a registration procedure b_F. This expression already contains the following "idealization" which does not completely correspond to the experimental situation: A surface area F is a mathematical object, that is, an idealized concept; therefore, in an exact sense, there is no such "registration" in which a silver grain located in "F" is blackened. In order to characterize the actual registration process we must describe the evaluation procedure more precisely with the aid of a physical grid on the photographic plate. The blackened silver grains within each cell of the grid are then counted. Since such an evaluation procedure will, however, be of no significant value if the magnification of the microscope is so large that the imprecision due to lack of resolution is larger than the dimensions of the silver grain, we often use the idealization that the surface F_p of the entire photographic plate is decomposed into surface area elements F. We have already considered such idealizations in IV where we have discussed the transition from the "real" Boolean ring $\mathscr{R}(b_0)$ to the abstract Boolean ring Σ and its completion. In this sense the b_F are elements of the "idealized" Boolean ring of surface elements F of the photographic plate. After we have clarified this step of idealization we shall make use of it in the following.

If b_0 is the registration method (in the sense of II, §4.2) for the above gedanken experiment, then the b_F are registration procedures for which $b_F \subset b_0$. Let F_p denote the surface of the entire photographic plate. Clearly $b_{F_p} = b_0$ is incorrect since, in an individual trial a single silver grain need not necessarily be blackened. Therefore $b_0 \backslash b_{F_p} \neq \varnothing$. According to the axioms of quantum mechanics in III, to a surface area F of the photographic plate there corresponds an effect $\psi(b_0, b_F) \in L$. In the sense of the framework presented in II, III there is, therefore, no "error" in the registration method b_0. There are, therefore, several physical reasons (for example, finite resolution) why $\psi(b_0, b_F) \neq E(F_0)$ where F_0 is the surface area in the object plane corresponding to (according to geometrical optics) F. The wish (!) of the experimental physicist that $\psi(b_0, b_F) = E(F)$ is not satisfied by the registration method described above. Here the word "error" is therefore used by experimental physicists to describe the discrepancy between the actual registration method and the "desired" registration method.

We now have obtained a very important result for quantum mechanics.

A measurement error for microsystems does not represent the difference between a "measured value" and a "real or intrinsic value" of the measured quantity. According to IV, §8 for microsystems it is impossible to attribute meaning to the expression "real or intrinsic value" because microsystems have no objective (that is, intrinsic) properties (with the exception of those associated with the elements of the center). The "error" associated with the finite resolution of the microscope is therefore not a difference between the measurement value x corresponding to the location of a blackened silver grain X and the "real" position x but is only definable as the difference between the actually measured effect $\psi(b_0, b_F)$ and the desired effect $E(F_0)$ of an idealized position measurement procedure. This difference may be computed, in principle, in the above example (and for many other real measurements) with the help of quantum mechanics alone. Thus quantum mechanics itself can provide important contributions for the measurement of microsystems.

The possibility of using quantum mechanics for the above example depends upon the following structure of the registration methods. The measurement process described above may evidently be divided into three consecutive processes:

1 The scattering of a photon on an electron, in which the presence of the microscope and photographic plate plays no role.
2 The deflection of the photon after scattering into the lens (microscope objective).
3 After passing through the microscope, this photon strikes the photographic plate, and activates a silver bromide grain, where the latter turns black when the plate is developed.

The process 1 corresponds to the scattering problem which we have examined in XVI. We have certainly not considered the scattering between photons and electrons there because we do not yet have a mathematically "clean" formulation of quantum electrodynamics. However, if we replace the optical microscope by an electron microscope and replace the electron to be measured by a proton, then we may consider process 1 as the scattering of an electron by a proton, a process which may be described by quantum mechanics.

The process 2 may also be described by quantum mechanics if we replace the lenses by a refracting medium for photons. Again, this is more meaningful for the case of an electron microscope where the lenses are obtained by the use of magnetic fields. Here the deflection of electrons by a field can be described in terms of the methods described in VIII, §6.

The process 3 cannot be described only in terms of quantum mechanical processes because the system encounters the macrosystems—the silver bromide crystals. Here we may, however, make use of the "effect of the microsystems striking a surface F" described in XVI, §6.1. If the sensitivity of the photographic plate was 100 percent, then the expression given in XVI (6.1.13) would be used (for $t_1 \to -\infty$, $t_2 \to +\infty$) to describe the blackening of a silver grain in the surface F. Otherwise, it is necessary to obtain an efficiency factor λ where $0 < \lambda < 1$ which is to multiply expression XVI (6.1.13), in order to obtain the "effect of blackening" from the effect of "striking the surface".

Our gedanken experiment clearly illustrates why we wish to refer to process (1) as a measurement scattering process.

2.2 Measurement Scattering Morphisms

Measurement scattering refer to the processes investigated in XVI but considered from a special viewpoint, namely, the viewpoint of their application to registration procedures. We now return to Figure 41. The apparatus (1) represents a preparation procedure a_1 which may be used to produce systems of type 1. We may therefore simply write: $\varphi(a_1) = W_1$ where W_1 is an operator in the Hilbert space $\mathscr{H}_1$ for systems of type 1. Similarly apparatus (2) in Figure 41 represents a preparation procedure a_2 which may be used in order to produce systems of type 2. We therefore write $\varphi(a_2) = W_2$ where W_2 is an operator in the Hilbert space for systems of type 2. The registration apparatus (3) in Figure 41 will register only "after the scattering".

In the case of scattering experiments we have introduced a new structure in XVI, §1—a new preparation procedure $a = \gamma(a_1, a_2)$ corresponding to the composite system. We now consider (and this is our new viewpoint) the apparatuses (2) and (3) together as a registration apparatus for the systems produced by (1) (see Figure 41). We will now express this viewpoint mathematically in terms of a new structure which identifies certain pairs (a_2, b_0) and (a_2, b) (where $a_2 \in Q'$, $b_0 \in \mathscr{R}'_0$, $b \in \mathscr{R}$ and $b \subset b_0$) with a registration method $\tilde{b}_0$ and a registration procedure $\tilde{b}$ as follows:

$$\begin{aligned} (a_2, b_0) &\to \tilde{b}_0, \\ (a_2, b) &\to \tilde{b}, \end{aligned} \tag{2.2.1}$$

what, for fixed a_2, represents an isomorphic mapping of $\mathscr{R}(b_0)$ onto $\mathscr{R}(\tilde{b}_0)$. Of particular interest are the corresponding effects $\psi(\tilde{b}_0, \tilde{b})$ if $\varphi(a_2)$ and $\psi(b_0, b)$ are known. This problem can be solved by purely quantum mechanical methods.

Here we shall only consider the case in which the scattering of systems 1 onto systems 2 can be described with the aid of a scattering operator. This is the case when both system types 1 and 2 are elementary system types. Under these assumptions we have to use the ensemble

$$W = S(W_1 \times W_2)S^+ \tag{2.2.2}$$

for the calculation of the registration probabilities for the apparatus (3) in Figure 41 (see, for example, XIV, (4.4.5b)) where $W_1 = \varphi(a_1)$, $W_2 = \varphi(a_2)$.

According to the meaning of process 1 in the example in §2.1 we will discuss the case that the system 2 is used for the "measurement of system 1", that is, after the collision only system 2 is registered. We therefore assume that the registration apparatus (3) which is described by the registration method $b_0 \in \mathscr{R}_0$ registers effects of the form $\mathbf{1} \times F_2$, that is,

$$\psi(b_0, b) = \mathbf{1} \times F_2. \tag{2.2.3}$$

From (2.2.2) and (2.2.3) we find that the probability that b responds is given by

$$\lambda(a \cap b_0, a \cap b) = \operatorname{tr}(S(W_1 \times W_2)S^+(\mathbf{1} \times F_2)), \tag{2.2.4}$$

where a is the preparation procedure $\gamma(a_1, a_2)$. On the other hand, on the basis of the correspondence (2.2.1) we find that

$$\begin{aligned} \lambda(a \cap b_0, a \cap b) &= \lambda(a_1 \cap \tilde{b}_0, a_1 \cap \tilde{b}) \\ &= \operatorname{tr}_1(\varphi(a_1)\psi(\tilde{b}_0, \tilde{b})) = \operatorname{tr}_1(W_1 F_1), \end{aligned} \tag{2.2.5}$$

where $\psi(\tilde{b}_0, \tilde{b}) = F_1$. The correspondence (2.2.1) is a structure which corresponds to a different division of the experiment into preparation and registration parts. Therefore, it is consistent with the formulation of quantum mechanics presented earlier in this book (up to XVI) only if (2.2.5) is satisfied. On the basis of quantum mechanics this is the case since the right-hand side of (2.2.4) is, for fixed W_1 and F_2, a bounded linear form over $W_1 \in K(\mathscr{H}_1)$ and therefore over $\mathscr{B}(\mathscr{H}_1)$. Thus there exists a $F_1 \in \mathscr{B}'(\mathscr{H}_1)$ such that

$$\operatorname{tr}(S(W_1 \times W_2)S^+(\mathbf{1} \times F_2)) = \operatorname{tr}_1(W_1 F_1). \tag{2.2.6}$$

Since the left-hand side of (2.2.6) takes on values between 0 and 1 for $W_1 \in K(\mathscr{H}_1)$, it follows that $F_1 \in L(\mathscr{H}_1)$, from which we have proven consistency.

From (2.2.6) it is evident that, for fixed W_2, a $\mathscr{B}$ continuous effect morphism $L(\mathscr{H}_2) \to L(\mathscr{H}_1)$ is defined by

$$\mathbf{T}(1, 2; W_2)F_2 = F_1. \tag{2.2.7}$$

Here it is easy to discover examples for which $\mathbf{T}(1, 2; W_2)$ does not, in general, map the set of decision effects $G(\mathscr{H}_2)$ into the set of decision effects $G(\mathscr{H}_1)$. Indeed, to the author's knowledge, there is no realistic scattering operator S for which, for suitably chosen W_2, the relation

$$\mathbf{T}(1, 2; W_2)G(\mathscr{H}_2) \subset G(\mathscr{H}_1)$$

is satisfied. If we introduced additional axioms in Chapter III which lead to a strengthening of the map ψ in the form $\mathscr{F} \xrightarrow{\psi} G$, then (2.2.1) would no longer be consistent with the previously introduced axioms. The consistency problem (that is, freedom from contradiction) for further development of the theory arising from the addition of new structure is nontrivial. The consistency shown above shows us, conversely, that the axioms introduced in III were realistic, because the exclusion of the situation described by (2.2.1) would be unrealistic. Therefore it is particularly important that $\psi(b_0, b) \in G$ does not hold in general.

In order to make evident the fact that (2.2.7) describes the relationship between the maps $\mathscr{Q}' \xrightarrow{\varphi} K$ and $\mathscr{F} \xrightarrow{\psi} L$ and the correspondence (2.2.1) we rewrite (2.2.7) in the form

$$\psi(\tilde{b}_0, \tilde{b}) = \mathbf{T}(1, 2; \varphi(a_2))\psi(b_0, b). \qquad (2.2.8)$$

The effect morphism defined by (2.2.6) will be called the *measurement scattering morphism*. The use of a scattering of systems of type 1 on systems of type 2 for the purpose of measurement (that is, registration) of systems of type 1 will therefore be completely described by a measurement scattering morphism $\mathbf{T}(1, 2, W_2)$.

We have introduced the maps φ and ψ in (2.2.8) for the purpose of clarification. Later we shall often describe different partitions of experiments into preparation and registration parts only with words, without providing maps in the form (2.2.1). We shall only give the corresponding formulas for the ensembles and effects. We hope that in these cases the physical meaning of these formulas will be clear, namely, the relation between the different preparation and registration procedures. In the example from §2.1 the apparatuses described by b_0 and a_2 are the "lens and photographic plate" and "light source", respectively. The apparatus described by $\tilde{b}_0$ is therefore the "light source + lens + photographic plate". The arrangement of the apparatuses (lens and photographic plate) characterized by b_0—in particular, corresponding to the analysis of the processes 2 in §2.1—have not been considered here; this will be done in §3.1.

2.3 Properties of Measurement Scattering Morphisms

In XVI we have seen that it does not make sense to speak of a precise "time point" for the scattering. For this reason the "discontinuous nature" of the transition from $W_1 \times W_2$ to $S(W_1 \times W_2)S^+$ is not "physically real", but instead is only a symbolic discontinuity which enables us to ignore the detailed evolution of the systems during the scattering process. In the Schrödinger picture we may, in principle (as we have seen in XVI, §4.4) attribute the jump from $W_1 \times W_2$ to $S(W_1 \times W_2)S^+$ to "any" time. Again, this description makes evident the fact that there is no precise "time point" associated with the measurement in applications of the measurement scattering process. Indeed this is the case even if the time interval of the interaction is relatively short, we may not, in advance, consider this particular time t as the time at which an observable is measured, since the observable which we measure for system 1 with the help of the scattering of system 2 will strongly depend on what we measure for system 2 after the scattering.

If $\Sigma_2 \xrightarrow{F_2} L(\mathscr{H}_2)$ is an observable of system type 2, then, from (2.2.7) and from the diagram

$$\begin{array}{ccc} & \overset{F_2}{\nearrow} & L(\mathscr{H}_2) \\ \Sigma_2 & & \big\downarrow \mathbf{T}(1,2;W_2) \\ & \underset{F_1}{\searrow} & L(\mathscr{H}_1) \end{array}$$

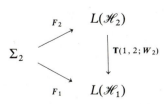

we obtain an observable $\Sigma_2 \xrightarrow{F_1} L(\mathscr{H}_1)$. According to (2.2.7) the measurement of an observable $\Sigma_2 \xrightarrow{F_2} L(\mathscr{H}_2)$ on system 2 corresponds to the measurement of an observable $\Sigma_2 \xrightarrow{F_1} L(\mathscr{H}_1)$ on system 1. In particular it follows that coexistent effects for systems of type 2 will be coexistent effects for systems of type 1. This situation can be illustrated by the realization (see IV, §4) $\mathscr{R}(b_0) \xrightarrow{\psi} L(\mathscr{H}_2)$ of an observable $\Sigma_2 \xrightarrow{F_2} L(\mathscr{H}_2)$ and the corresponding map $\mathscr{R}(\tilde{b}_0) \xrightarrow{\psi} L(\mathscr{H}_1)$ since on account of (2.2.1), $\mathscr{R}(\tilde{b}_0)$ and $\mathscr{R}(b_0)$ are isomorphic. The reader may mind that noncoexistent effects F_1 can be obtained from coexistent effects F_2 if we alter W_2, that is, the preparation procedure a_2!

The choice of what we may measure for systems of type 2 after the scattering on system 1 remains completely open. The measurement scattering process does not uniquely determine the observables of system 1 which can be measured by the use of the scattering process. At first this situation is irritating—because we wish to, with the aid of the measurement process, determine what is actually registered. The measurement scattering process itself cannot clarify this final registration because we have to assume that the final registration is already given by a registration method b_0 for system 2.

The set $\mathbf{T}(1, 2; W_2)L(\mathscr{H}_2)$ is decisive for the possibility to make conclusions about systems 1 from the registration of systems 2. This set is a σ compact subset of $L(\mathscr{H}_2)$. We therefore call the set $\mathbf{T}(1, 2; W_2)L(\mathscr{H}_2)$ the "set of information possibilities" for system 1 on the basis of the measurement scattering on systems 2 in the ensemble W_2.

The extreme case, in which "no" information possibilities may be obtained from the scattering experiment does not mean that $\mathbf{T}(1, 2; W_2)L(\mathscr{H}_2)$ will be empty, because $\mathbf{T}(1, 2; W_2)$ contains all F_1 of the form $F_1 = \lambda\mathbf{1}$ where $0 \leq \lambda \leq 1$, a result which is easily shown by setting $F_2 = \lambda\mathbf{1}$. We shall call the case in which $\mathbf{T}(1, 2; W_2)L(\mathscr{H}_2) = \{\lambda\mathbf{1} \mid 0 \leq \lambda \leq 1\}$ the case of "no" information possibilities since effects of the form $F_1 = \lambda\mathbf{1}$ do not permit us to distinguish between the W_1.

The latter case occurs when, for example, there is no interaction between systems 1 and 2, that is, when $S = \mathbf{1}$. However, $S = \mathbf{1}$ is not necessary for this case, as we shall see from the "examples" following below.

Of particular interest is the case in which

$$\mathbf{T}(1, 2; W_2)L(\mathscr{H}_2) = \overline{\mathrm{co}}(F\Sigma) \tag{2.3.1}$$

holds for a particular observable $\Sigma \xrightarrow{F} L(\mathscr{H}_1)$. Then we say that the information possibilities in scattering are restricted to the "measurement of the observable $\Sigma \xrightarrow{F} L(\mathscr{H}_1)$". For the case in which

$$\mathbf{T}(1, 2; W_2)L(\mathscr{H}_2) = L(\mathscr{H}_1) \tag{2.3.2}$$

we say that by the scattering "no information possibilities are lost".

A systematic examination of the relation between S and the sets $\mathbf{T}(1, 2; W_2)L(\mathscr{H}_2)$ has not yet been done. Here we shall only consider certain "principal" examples, without knowing whether they can be achieved for real interactions (that is, for realistic Hamiltonian operators). In these examples we shall set $W_2 = P_\psi$ where $\psi \in \mathscr{H}_2$. Furthermore, for a suitably chosen complete orthonormal basis $\{\varphi_\nu\}$ in $\mathscr{H}_1$ we assume that

$$S\varphi_\nu\psi = \chi_\nu\eta_\nu, \tag{2.3.3}$$

where $\chi_\nu \in \mathscr{H}_1$, $\eta_\nu \in \mathscr{H}_2$. Since the $\varphi_\nu\psi$ are pairwise orthogonal and S is unitary, the $\chi_\nu\eta_\nu$ must be pairwise orthogonal:

$$\langle \chi_\nu\eta_\nu, \chi_\mu\eta_\mu\rangle = \langle \chi_\nu, \chi_\mu\rangle\langle \eta_\nu, \eta_\mu\rangle = \delta_{\nu\mu}. \tag{2.3.4}$$

Case 1. The χ_ν are pairwise orthogonal.

For $W_1 = P_\varphi$ and $S(W_1 \times W_2)S^+ = S(P_\varphi \times P_\psi)S^+ = SP_{\varphi\psi}S^+ = P_{S_{\varphi\psi}}$ and $\varphi = \Sigma_\nu\, \varphi_\nu\alpha_\nu$ it follows that

$$\begin{aligned}\operatorname{tr}(S(W_1 \times W_2)S^+(1 \times F_2)) &= \sum_\nu |\alpha_\nu|^2\langle \eta_\nu, F_2\eta_\nu\rangle \\ &= \operatorname{tr}_1(W_1F_1) = \sum_{\nu\mu} \bar{\alpha}_\nu\alpha_\mu\langle \varphi_\nu, F_1\varphi_\mu\rangle.\end{aligned}$$

Since this holds for all vectors φ, that is, for all possible α_ν satisfying

$$\sum_\nu |\alpha_\nu|^2 = 1,$$

it follows that

$$\langle \varphi_\nu, F_1\varphi_\mu\rangle = \delta_{\nu\mu}\langle \eta_\nu, F_2\eta_\nu\rangle,$$

that is,

$$F_1 = \sum_\nu \langle \eta_\nu, F_2\eta_\nu\rangle P_{\varphi_\nu}. \tag{2.3.5}$$

The complete (atomic) Boolean subring Σ_1 of $G(\mathscr{H}_1)$ generated by the P_{φ_ν} is a maximal decision observable. We will show that the relationship

$$\mathbf{T}(1, 2; P_\psi)L(\mathscr{H}_2) \subset \overline{\text{co}}(\Sigma_1) \tag{2.3.6}$$

follows directly from (2.3.5).

Since $F_2 \in L(\mathscr{H}_2)$ it follows that $0 \le \langle \eta_\nu, F_2\eta_\nu\rangle \le 1$. If λ_ν is a sequence of real numbers for which $0 \le \lambda_\nu \le 1$, then in the σ topology it follows that

$$\sum_{\nu=1}^{N} \lambda_\nu P_{\varphi_\nu} \to \sum_{\nu=1}^{\infty} \lambda_\nu P_{\varphi_\nu}$$

since $\operatorname{tr}(W\sum_{\nu=1}^N \lambda_\nu P_{\varphi_\nu}) \to \operatorname{tr}(W\sum_{\nu=1}^\infty \lambda_\nu P_{\varphi_\nu})$ holds for all $W \in K$, a result which is easily proven with the aid of $\operatorname{tr}(W) = \sum_\nu \langle \varphi_\nu, W\varphi_\nu\rangle = 1$. In order to prove (2.3.6) we need only show that, for any finite sum

$$\sum_{\nu=1}^{N} \lambda_\nu P_{\varphi_\nu} \in \text{co}(\Sigma_1)$$

holds. For this purpose we shall renumber the finite sum in order that $\lambda_1 \le \lambda_2 \le \lambda_3 \le \cdots \le \lambda_1$, holds. Then we may write

$$\begin{aligned}\sum_{\nu=1}^{N} \lambda_\nu P_{\varphi_\nu} = \lambda_1 \sum_{\nu=1}^{n} P_{\varphi_\nu} &+ (\lambda_2 - \lambda_1) \sum_{\nu=2}^{N} P_{\varphi_\nu} \\ &+ (\lambda_3 - \lambda_2) \sum_{\nu=3}^{N} P_{\varphi_\nu} + (\lambda_N - \lambda_{N-1})P_{\varphi_N} \\ &+ (1 - \lambda_N)\,\mathbf{0}.\end{aligned}$$

Since

$$\lambda_1 + (\lambda_2 - \lambda_1) + (\lambda_3 - \lambda_2) + \cdots + (1 - \lambda_N) = 1,$$

it follows that $\sum_{\nu=1}^{N} \lambda_\nu P_{\varphi_\nu}$ is a convex combination of elements of Σ_1.

If the η_ν are also pairwise orthogonal, then, for a given choice of F_2 we find that $\Sigma_1 \subset \mathbf{T}(1, 2; P_\psi)L(\mathscr{H}_2)$ and we obtain

$$\mathbf{T}(1, 2, P_\psi)L(\mathscr{H}_2) = \overline{\text{co}}(\Sigma_1) \tag{2.3.7}$$

and we obtain an example for the case (2.3.1). We may also easily given an example of an observable for system 2, for which the measurement of system 2 is equivalent to the measurement of the decision observable Σ_1 for system 1. To this purpose we define Σ_2 to be the complete Boolean subring of $G(\mathscr{H}_2)$ which is generated by the P_{η_ν}. Then $\mathbf{T}(1, 2; P_\psi)$ is the isomorphic map of Σ_2 onto Σ_1 for which $\mathbf{T}(1, 2; P_\psi)P_{\eta_\nu} = P_{\varphi_\nu}$. A "measurement of P_{η_ν}" on the system 2 therefore corresponds to a measurement of P_{φ_ν} on system 1.

If all the η_ν are identical, that is, $\eta_\nu = \eta$ then we obtain $F_1 = \lambda\mathbf{1}$ where $\lambda = \langle \eta, F_2 \eta \rangle$ and $\mathbf{T}(1, 2; P_\psi)L(\mathscr{H}_2) = \{\lambda\mathbf{1} \,|\, 0 \leq \lambda \leq 1\}$.

Case 2. The η_ν are pairwise orthogonal.

As in Case 1, for $\varphi = \Sigma_\nu\, \varphi_\nu \alpha_\nu$ we obtain

$$\sum_{\nu\mu} \bar{\alpha}_\nu \alpha_\mu \langle \varphi_\nu, F_1 \varphi_\mu \rangle = \sum_{\nu\mu} \bar{\alpha}_\nu \alpha_\mu \langle \chi_\nu, \chi_\mu \rangle \langle \eta_\nu, F_2 \eta_\mu \rangle$$

and we therefore obtain

$$\langle \varphi_\nu, F_1 \varphi_\mu \rangle = \langle \chi_\nu, \chi_\mu \rangle \langle \eta_\nu, F_2 \eta_\nu \rangle. \tag{2.3.8}$$

For the special case in which all χ_ν are identical (that is $\chi_\nu = \chi$) it follows that

$$\langle \varphi_\nu, F_1 \varphi_\mu \rangle = \langle \eta_\nu, F_2 \eta_\mu \rangle$$

and we therefore obtain $\langle \varphi, F_1 \varphi \rangle = \langle \eta, F_2 \eta \rangle$ where $\eta = \sum_\nu \eta_\nu \alpha_\nu$. From $\langle \varphi, F_1 \varphi \rangle = \langle \eta, F_2 \eta \rangle$ we therefore obtain (2.3.2). The case $S\varphi_\nu \psi = \chi \eta_\nu$ is therefore an example for which no information possibilities are lost. In particular, we may measure the observable $\Sigma_1 \subset G(\mathscr{H}_1)$ described earlier in Case 1 on system 1 by measuring the observable Σ_2 defined above on system 2. In the same way we may, however, measure *every* observable on system 1 by measuring a corresponding observable on system 2.

3 Measurement Transformations

Process 2 in the example from §2.1, namely, the reflection of a photon by a lens is an example of a measurement transformation. In general, by a measurement transformation we mean the application of interactions of other systems upon the system to be registered in order to register the system by "previously known" methods after the interaction. Measurement transformations are used in great variety by experimental physicists. This topic will be discussed in great detail in §6.2.

3.1 Measurement Transformation Morphisms

We shall return to the considerations of §2.2 with the single difference that we shall measure the system 1 after the scattering, that is, we replace (2.2.3) by

$$\psi(b_0, b) = F_1^f \times \mathbf{1}. \tag{3.1.1}$$

Here we have introduced the index f for F_1 in order to indicate that this F_1^f is to be registered after the scattering. System 2 will therefore only be used as a registration tool, because, after the scattering we shall only register systems of type 1. As in §2.2, from

$$\operatorname{tr}(S(W_1 \times W_2)S^+(F_1^f \times \mathbf{1})) = \operatorname{tr}_1(W_1 F_1^i) \tag{3.1.2}$$

it follows that there exists a $\mathscr{B}$ continuous effect morphism $\mathbf{T}(1, 1; W_2)$ of $L(\mathscr{H}_1)$ into itself such that

$$\psi(\tilde{b}_0, \tilde{b}) = F_1^i = \mathbf{T}(1, 1; W_2)F_1^f. \tag{3.1.3}$$

$\mathbf{T}(1, 1; W_2)$ is called a *measurement transformation morphism.* Here it is evident that the measurement transformations are very useful tools which permit us to obtain "new" effect procedures $(\tilde{b}_0, \tilde{b})$ from "previously known" effect procedures (b_0, b).

Measurement transformation morphisms may lead to interesting results even in cases in which no information is transferred from systems 1 to systems 2 (see §2.3). Such a special case occurs if we consider an external field as system 2 (see VIII, §6). Then the entire measurement transformation can be considered in terms of the Hilbert space $\mathscr{H}_1$ of system 1. The registration method $\tilde{b}_0$ is obtained from b_0 by "switching on" the external field before b_0 registers.

Let S denote the scattering operator for the external field. In this case we replace (3.1.2) by

$$\operatorname{tr}_1(SW_1S^+F_1^f) = \operatorname{tr}_1(W_1F_1^i). \tag{3.1.4}$$

Thus it follows that

$$F_1^i = S^+F_1^fS. \tag{3.1.5}$$

Therefore the measurement transformation morphism is defined by

$$\mathbf{T}(1, 1)F_1^f = S^+F_1^fS. \tag{3.1.6}$$

$\mathbf{T}(1, 1)$ is therefore an automorphism of $L(\mathscr{H}_1)$ which is determined by the unitary transformation S.

3.2 Properties of Measurement Transformation Morphisms

From the considerations of §2.3 it simply follows that the measurement transformation morphisms have the same properties as the measurement scattering morphisms. They transform coexistent effects into coexistent effects and observables into observables.

Here it is important to perform a systematic analysis of the relationship between S and the set $\mathbf{T}(1, 1; W_2)L(\mathscr{H}_1)$ which describes the change of information possibilities which takes place as the result of the measurement transformation. Since such an analysis has not yet been done, we will consider the same examples of operators S as we have demonstrated in §2.3, that is, we assume (2.3.3).

In Case 1 (in which the χ_ν are pairwise orthogonal), for $W_2 = P_\psi$ and $W_1 = P_\varphi$, from

$$\operatorname{tr}(S(W_1 \times W_2)S^+(F_1^f \times \mathbf{1})) = \sum_{\nu, \mu} \bar{\alpha}_\nu \alpha_\mu \langle \chi_\nu, F_1^f \chi_\mu \rangle \langle \eta_\nu, \eta_\mu \rangle$$

$$= \operatorname{tr}(W_1 F_1^i) = \sum_{\nu, \mu} \bar{\alpha}_\nu \alpha_\mu \langle \varphi_\nu, F_1^i \varphi_\mu \rangle$$

we obtain the relation

$$\langle \varphi_\nu, F_1^i \varphi_\mu \rangle = \langle \chi_\nu, F_1^f \chi_\mu \rangle \langle \eta_\nu, \eta_\mu \rangle. \tag{3.2.1}$$

If the η_ν are pairwise orthogonal, it follows that

$$F_1^i = \sum_\nu \langle \chi_\nu, F_1^f \chi_\nu \rangle P_{\varphi_\nu}. \tag{3.2.2}$$

Thus, as in §2.3 it follows that

$$\mathbf{T}(1, 1; P_\psi)L(\mathscr{H}_1) = \overline{\text{co}}(\Sigma_1^i), \tag{3.2.3}$$

where Σ_1^i is identical to Σ_1 in §2.3. If we define Σ_1^f as the decision observable defined by the P_{χ_ν}, then we find that in this case $\mathbf{T}(1, 1; P_\psi)$ transforms the observable Σ_1^f into the observable Σ_1^i.

If the η_ν are all equal, that is, $\eta_\nu = \eta$, then from (3.2.1) it follows that $\langle \varphi_\nu, F_1^i \varphi_\mu \rangle = \langle \chi_\nu, F_1^f \chi_\mu \rangle$ and we therefore obtain

$$\mathbf{T}(1, 1; P_\psi)L(\mathscr{H}_1) = L(\mathscr{H}_1). \tag{3.2.4}$$

from which we cannot conclude that $\mathbf{T}(1, 1; P_\psi)$ is an automorphism of $L(\mathscr{H}_1)$ because the χ_ν, for example, need not form a complete orthonormal basis in

In Case 2 (the η_ν are pairwise orthogonal) from

$$\operatorname{tr}(S(P_\varphi \times P_\psi)S^+(F_1^f \times \mathbf{1})) = \sum_\nu |\alpha_\nu|^2 \langle \chi_\nu, F_1^f \chi_\nu \rangle$$

$$= \operatorname{tr}(P_\varphi F_1^i) = \sum_{\nu\mu} \bar{\alpha}_\nu \alpha_\mu \langle \varphi_\nu, F_1^i \varphi_\mu \rangle$$

it follows that

$$\langle \varphi_\nu, F_1^i \varphi_\mu \rangle = \delta_{\nu\mu} \langle \chi_\nu, F_1^f \chi_\nu \rangle$$

and we therefore obtain

$$F_1^i = \sum_\nu \langle \chi_\nu, F_1^f \chi_\nu \rangle P_{\varphi_\nu}. \tag{3.2.5}$$

Thus it follows that

$$\mathbf{T}(1, 1; P_\psi)L(\mathscr{H}_1) \subset \overline{\text{co}}(\Sigma_1^i) \tag{3.2.6}$$

with the previously defined Σ_1^i. The case in which the χ_ν are pairwise orthogonal we have treated earlier.

If all χ_ν are identical, that is, $\chi_\nu = \chi$, then it follows that $F_1^i = \lambda\mathbf{1}$ where $\lambda = \langle\chi, F_1^f\chi\rangle$ and we obtain

$$\mathbf{T}(1, 1; P_\psi)L(\mathscr{H}_1) = \{\lambda\mathbf{1} \,|\, 0 \leq \lambda \leq 1\}. \tag{3.2.7}$$

For our examples, together with the results of §2.3 we obtain: If a scattering process is extremely unsuitable as a measurement scattering process, then we will lose no information possibilities from its use as a measurement transformation. Conversely, if a scattering process is extremely unsuitable as a measurement transformation, then no information possibilities are lost for its use as a measurement scattering process.

The special case of a scattering by an external field is extremely ill-suited for a measurement scattering process. In this case $\mathbf{T}(1, 1)$ is an automorphism of $L(\mathscr{H}_1)$, and is therefore extremely well-suited as a measurement transformation.

4 Transpreparations

We now return to the measurement situation described by Figure 41 in which we replace the composite apparatus consisting of the two apparatuses (1) and (2) by an arbitrary preparation apparatus (1, 2) for systems composed of systems of type (1) and (2) (see Figure 43). In this way the Hilbert space $\mathscr{H}$ of the composite systems can be written (without additional symmetrization) in the form $\mathscr{H} = \mathscr{H}_1 \times \mathscr{H}_2$ where $F_1 \times \mathbf{1}$ and $\mathbf{1} \times F_2$ are the effects which

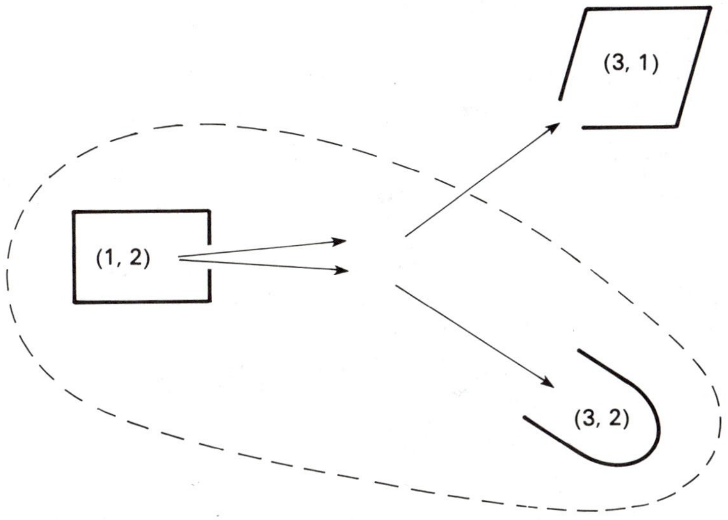

Figure 43

are produced by subsystems 1 and 2, respectively. We may consider the registration apparatus (3) in Figure 41 to be composed of two parts (3, 1) and (3, 2) (see Figure 43) where (3, 1) will respond only to systems of type 1 and (3, 2) will only respond to systems of type 2.

Let a denote the preparation procedure corresponding to the apparatus (1, 2). We therefore obtain (using somewhat simplified notation—see §2.2) $\varphi(a) = W$ where W is an operator in $\mathscr{H} = \mathscr{H}_1 \times \mathscr{H}_2$. The composite apparatus consisting of (3, 1) and (3, 2) corresponds to a registration method b_0 and a registration procedure b where $\psi(b_0, b) \in L(\mathscr{H})$.

It is now necessary to develop an axiomatic formulation of the statement that apparatus (3) is obtained by combining the two apparatuses (3, 1) and (3, 2). Suppose we are given an $\mathscr{R}(b_{01})$ and an $\mathscr{R}(b_{02})$ such that

$$\psi(b_{01}, b_1) \in L(\mathscr{H}_1)$$

and $\psi(b_{02}, b_2) \in L(\mathscr{H}_2)$ where $\mathscr{R}(b_{01})$ and $\mathscr{R}(b_{02})$ correspond to the apparatuses (3, 1) and (3, 2), respectively.

Let Σ_{12} be the free Boolean ring generated by the Boolean rings $\mathscr{R}(b_{01})$ and $\mathscr{R}(b_{02})$. This new Boolean ring can be easily constructed from the two Boolean rings $\mathscr{R}(b_{01})$ and $\mathscr{R}(b_{02})$ by constructing the appropriate "logical switching network". We then require that there exists an isomorphic map h of Σ_{12} onto $\mathscr{R}(b_0)$. The two rings $\mathscr{R}(b_{01})$, $\mathscr{R}(b_{02})$ may then be identified with subrings of Σ_{12}, that is, $\mathscr{R}(b_{01})$ may be identified with the ring of all $b_1 \wedge b_{02}$ where $\wedge$ is defined in the abstract Boolean ring Σ_{12}. In this sense hb_1 and hb_2 are defined for all $b_1 \in \mathscr{R}(b_{01})$ and $b_2 \in \mathscr{R}(b_{02})$, respectively. We then require that

$$\psi(b_0, hb_1 \cap hb_2) = \psi(b_{01}, b_1) \times \psi(b_{02}, b_2). \tag{4.1}$$

The requirement (4.1) is equivalent to the somewhat more physically intuitive requirement

$$\mathrm{tr}((W_1 \times W_2)\psi(b_0, hb_1 \cap hb_2)) = \mathrm{tr}_1(W_1\psi(b_{01}, b_1))\mathrm{tr}(W_2\psi(bo_2, b_2)) \tag{4.2}$$

for all W_1, W_2: the proof of this statement is left to the reader (see VIII, §2). Similarly, we suggest that the reader prove that $\psi(b_0, b)$ is well defined by (4.1) for all $b \in \mathscr{R}(b_0)$.

In the same way from two observables $\Sigma_1 \overset{F_1}{\to} L(\mathscr{H}_1)$ and $\Sigma_2 \overset{F_2}{\to} L(\mathscr{H}_2)$ we obtain a uniquely determined observable $\hat{\Sigma}_{12} \overset{F}{\to} L(\mathscr{H}_1 \times \mathscr{H}_2)$ by taking the completion $\hat{\Sigma}_{12}$ (in the sense of IV, §1.4) of the free Boolean ring Σ_{12} generated by Σ_1 and Σ_2 using the following condition

$$F(\sigma_1 \wedge \sigma_2) = F_1(\sigma_1) \times F_2(\sigma_2) \quad \text{for } \sigma_1 \in \Sigma_1, \sigma_2 \in \Sigma_2$$

which is analogous to (4.1).

In the next sections 4.1 to 4.4 we shall examine the structure obtained by considering the apparatus (1, 2) together with the apparatus (3, 2) as a preparation apparatus of systems of type (1) (see Figure 43).

4.1 Reduction of a Preparation Procedure by Means of a Registration Procedure

We may represent the preparation apparatus obtained by combining the apparatuses (1, 2) and (3, 2) (see Figure 43) by a mapping k of a subset of M into M, that is, by mapping systems of type (1, 2) (that is, pairs 1, 2) into systems of type 1 where kx is "partner 1" in the pair x. This formulation suggests that there is a mapping principle concerned with further experiments with partner 1.

Here k maps the set $a \cap hb_2$ (using the above map h) onto a set $\tilde{a} = k(a \cap hb_2)$. This map which assigns a set $\tilde{a}$ to each pair (a, b_2) with $b_2 \in \mathscr{R}(b_{02})$ will, for simplicity, also be denoted by k. We now require, as an axiom, that $\tilde{a} \in \mathscr{Q}$:

$$(a, b_2) \xrightarrow{k} \tilde{a} \in \mathscr{Q}. \tag{4.1.1}$$

For fixed a this map is a homomorphism of the Boolean ring $\mathscr{R}(b_{02})$ onto a Boolean subring of $\mathscr{Q}$ for which the probabilities satisfy:

$$\lambda(k(a, b_2) \cap b_{01}, k(a, b_2) \cap b_1) = \lambda(a \cap hb_2, a \cap hb_2 \cap hb_1) \tag{4.1.2}$$

(where the latter is an additional axiomatic requirement for k!), since both sides experimentally yield the same probability where $k(a, b_2)$ is considered to be the preparation procedure on the left-hand side and a is considered to be the preparation procedure on the right-hand side of (4.1.2).

According to quantum mechanics from (4.1.2) it follows that $\varphi(k(a, b_2))$ is known whenever $\psi(b_{02}, b_2)$ and $\varphi(a)$ are known. In order to obtain the expression for $\varphi(k(a, b_2))$ we begin with

$$\lambda(a \cap b_0, a \cap hb_1 \cap hb_2) = \lambda(a \cap b_0, a \cap hb_2)\lambda(a \cap hb_2, a \cap hb_1 \cap hb_2)$$

from which it follows that

$$\operatorname{tr}(\varphi(a)\psi(b_0, hb_1 \cap hb_2)) = \operatorname{tr}(\varphi(a)\psi(b_0, hb_2))\lambda(a \cap hb_2, a \cap hb_1 \cap hb_2).$$

Using the abbreviated notation $\varphi(a) = W$, $\psi(b_{01}, b_1) = F_1$, $\psi(b_{02}, b_2) = F_2$, from (4.1) we obtain

$$\operatorname{tr}(W(F_1 \times F_2)) = \operatorname{tr}(W(\mathbf{1} \times F_2))\lambda(a \cap hb_2, a \cap hb_1 \cap hb_2).$$

From (4.1.2) we finally obtain

$$\operatorname{tr}(W(\mathbf{1} \times F_2))\lambda(k(a_1 b_2) \cap b_{01}, k(a, b_2) \cap b_1) = \operatorname{tr}(W(F_1 \times F_2)). \tag{4.1.3}$$

According to (4.1.3) (as we shall shortly find) $\varphi(k(a, b_2))$ is well defined providing that $\operatorname{tr}(W(\mathbf{1} \times F_2)) \neq 0$. This result demonstrates the consistency of the new structure with the previously developed quantum mechanics. Here $\operatorname{tr}(W(\mathbf{1} \times F_2)) = 0$ means that there are no systems of type 1 the partner 2 of which will produce the effect $\psi(b_{02}, b_2)$. In order to calculate $\varphi(k(a, b_2))$ we rewrite the right-hand side of (4.1.3) as follows:

$$\begin{aligned}\operatorname{tr}(W(F_1 \times F_2)) &= \operatorname{tr}(W(\mathbf{1} \times F_2^{1/2})(F_1 \times \mathbf{1})(\mathbf{1} \times F_2^{1/2}))\\ &= \operatorname{tr}((\mathbf{1} \times F_2^{1/2})W(\mathbf{1} \times F_2^{1/2})(F_1 \times \mathbf{1}))\\ &= \operatorname{tr}_1(R_1[(\mathbf{1} \times F_2^{1/2})W(\mathbf{1} \times F_2^{1/2})]F_1),\end{aligned}$$

where R_1 is the reduction operator onto subsystem 1 (see XVI, §1). Thus it follows that

$$\lambda(k(a, b_2) \cap b_{01}, k(a, b_2) \cap b_1) = \mathrm{tr}_1(\varphi(k(a, b_2))F_1),$$

where

$$\varphi(k(a, b_2)) = \frac{R_1[(\mathbf{1} \times F_2^{1/2})W(\mathbf{1} \times F_2^{1/2})]}{\mathrm{tr}(W(\mathbf{1} \times F_2))}. \tag{4.1.4}$$

Instead of the map $\varphi(a) \to \varphi(k(a, b_2))$ we shall consider the map

$$\varphi(a) \to \mathrm{tr}(W(\mathbf{1} \times F_2))\varphi(k(a, b)),$$

that is, the map $\mathbf{S}(1, F_2)$ defined by

$$\mathbf{S}(1, F_2)W = R_1[(\mathbf{1} \times F_2^{1/2})W(\mathbf{1} \times F_2^{1/2})]. \tag{4.1.5}$$

$\breve{K} \xrightarrow{\mathbf{S}(1, F_2)} \breve{K}_1$ is, as it is easy to show, an operation (see V, D 4.1.2). $\mathbf{S}(1; F_2)$ is also defined by means of the following identity:

$$\mathrm{tr}(W(F_1 \times F_2)) = \mathrm{tr}_1([\mathbf{S}(1; F_2)W]F_1). \tag{4.1.6}$$

An operation measure χ is defined by $b_2 \xrightarrow{\chi} \mathbf{S}(1; \psi(b_{02}, b_2))$ on the Boolean ring $\mathscr{R}(b_{02})$ as follows:

$$\mathscr{R}(b_{02}) \xrightarrow{\chi} \Pi \tag{4.1.7}$$

(where Π is given in V, D 4.3.1), a result which follows from (4.1.6). By completion of the Boolean ring $\mathscr{R}(b_{02})$ to a Boolean ring Σ_2 we obtain from (4.1.7) a transpreparator $\Sigma_2 \xrightarrow{\chi} \Pi$ in the sense of V, D 4.3.3.

This transpreparator, for fixed a, and therefore for fixed $W = \varphi(a)$, defines a preparator $\Sigma_2 \xrightarrow{\eta} \breve{K}(\mathscr{H}_1)$ (in the sense of IV, D 5.4) of $R_1 W$ as follows ($\sigma \in \Sigma$):

$$\eta(\sigma) = \chi(\sigma)W. \tag{4.1.8}$$

For the observable $\Sigma_2 \xrightarrow{F_2} L(\mathscr{H}_2)$ corresponding to $\mathscr{R}(b_{02})$ we therefore find that

$$\eta(\sigma) = \mathbf{S}(1, F_2(\sigma))W. \tag{4.1.9}$$

Σ_2 may be considered to be an idealized completion of the Boolean ring which is the image of $\mathscr{R}(b_{02})$ under the map k, where $\Sigma_2 \xrightarrow{\eta} K(\mathscr{H}_1)$ describes the preparator which corresponds to the formation of a preparation apparatus by means of putting the apparatuses (1, 2) and (3, 2) together.

In our considerations we have chosen a fixed preparation procedure a. The apparatus (1, 2) itself may, however, correspond to a preparator. It is not difficult to extend the above considerations to this case. Let $\Sigma_{12} \xrightarrow{\eta_{12}} \vec{K}(\mathscr{H})$ where $\mathscr{H} = \mathscr{H}_1 \times \mathscr{H}_2$ be the preparator which corresponds to the apparatus (1, 2). Let $\Sigma_2 \xrightarrow{F_2} L(\mathscr{H}_2)$ be the observable corresponding to the apparatus (3, 2). Let Σ denote the free (complete) Boolean ring generated by Σ_{12} and Σ_2.

For $\sigma_{12} \in \Sigma_{12}$, $\sigma_2 \in \Sigma_2$ and therefore for $\sigma_{12} \wedge \sigma_2 \in \Sigma$ the expression

$$\eta(\sigma_{12} \wedge \sigma_2) = \mathbf{S}(1, F_2(\sigma_2))\eta_{12}(\sigma_{12}) \tag{4.1.10}$$

defines an additive measure $\Sigma \xrightarrow{\eta} \vec{K}(\mathscr{H}_1)$ which represents a preparator of the ensemble $R_1\eta_{12}(\varepsilon_{12})$ (where ε_{12} is the unit element of Σ_{12}).

The "combination" of the preparator $\Sigma_{12} \xrightarrow{\eta_{12}} \breve{K}(\mathscr{H})$ with the observable $\Sigma_2 \xrightarrow{F_2} L(\mathscr{H}_2)$ in the manner described above leads, therefore, to a new preparator $\Sigma \xrightarrow{\eta} \breve{K}(\mathscr{H}_1)$ which corresponds to the combination of the apparatuses (1, 2) and (3, 2) to form a preparation apparatus for systems of type 1.

All the above is consistent with quantum mechanics. However, this "combination" can be disturbing, as we shall find in §4.4. First we shall carefully study an often used special case of the above.

4.2 Transpreparation by Means of Scattering

We shall consider the special case of the situation described in §4.1 and illustrated in Figure 43 in which the apparatus (1, 2) is a combination of two apparatuses (1) and (2) in the sense of a scattering experiment, as we have described in Figure 41. In this case we do not vary the ensemble produced by apparatus (2). The apparatuses (3, 1), (3, 2) will register only after the scattering.

In the considerations of §4.1 we need only replace W by the special choice $S(W_1 \times W_2)S^+$. By analogy with (4.1.5) from:

$$\mathbf{S}(1, 2; W_2, F_2)W_1 = R_1[(\mathbf{1} \times F_2^{1/2})S(W_1 \times W_2)S^+(\mathbf{1} \times F_2^{1/2})] \tag{4.2.1}$$

we define an operation (for fixed W_2) as follows:

$$\breve{K}(\mathscr{H}_1) \xrightarrow{\mathbf{S}(1,2;W_2,F_2)} \breve{K}(\mathscr{H}_1). \tag{4.2.2}$$

If we consider the (idealized) observable $\Sigma_2 \xrightarrow{F_2} L(\mathscr{H}_2)$ corresponding to the apparatus (3.2), then we define a transpreparator

$$\Sigma_2 \xrightarrow{\chi} \Pi, \tag{4.2.3}$$

where

$$\chi(\sigma) = \mathbf{S}(1, 2; W_2, F_2(\sigma)). \tag{4.2.4}$$

It describes the transpreparation of an ensemble W_1 with the aid of the scattering on systems 2 from the ensemble W_2 and with the help of the observable $\Sigma_2 \xrightarrow{F_2} L(\mathscr{H}_2)$.

For fixed W_1 the transpreparator defines a preparator

$$\Sigma_2 \xrightarrow{\eta} \breve{K}(\mathscr{H}_1),$$

where

$$\eta(\sigma) = \mathbf{S}(1, 2; W_2, F_2(\sigma))W_1, \tag{4.2.5}$$

where

$$\eta(\varepsilon) = R_1[S(W_1 \times W_2)S^+].$$

Here we abbreviate $\eta(\varepsilon) = W_1^f$.

This preparator leads to coexistent decompositions of W_1^f (see IV, §5). If, for example, $\varepsilon = \sum_i \sigma_i$ is a partition of unity ε of Σ_2 (that is, $\sigma_i \wedge \sigma_j = 0$ for $i \neq j$) then

$$W_1^f = \sum_i \eta(\sigma_i)W_1 \tag{4.2.6}$$

is a decomposition of the W_1^f. All such decompositions of W_1^f obtained from the same observable $\Sigma_2 \overset{F_2}{\to} L(\mathscr{H}_2)$ are coexistent decompositions (W_2 fixed). If we change the observable $\Sigma_2 \overset{F_2}{\to} L(\mathscr{H}_2)$, however, then we may obtain noncoexistent decompositions of the same ensemble W_1^f.

We will illustrate this situation by means of two examples. First, we shall consider the "theoretical" example of the scattering operator S defined by (2.3.3). Here we shall only consider the case in which both the η_ν and the χ_ν are pairwise orthogonal We then obtain (2.3.7). Therefore, since all effects $\mathbf{T}(1, 2; P_\psi)L(\mathscr{H}_2)$ are coexistent, we may make the conjecture that all possible decompositions with the aid of observables $\Sigma_2 \overset{F_2}{\to} L(\mathscr{H}_2)$ are coexistent. This is, however, not the case.

In order to show this we consider, as an observable, $\Sigma_2 \overset{F_2}{\to} L(\mathscr{H}_2)$ the decision observable generated by the P_{γ_ν} where $\gamma_\nu \in \mathscr{H}_2$ is a complete orthonormal basis. In order to compute $\mathbf{S}(1, 2; P_\psi, P_\gamma)$ for a $\gamma \in \mathscr{H}_2$, we set $W_1 = P_\varphi$ where $\varphi = \sum_\nu \varphi_\nu \alpha_\nu$. Then for $\eta(P_\gamma)$, according to (4.2.5) it follows that

$$\operatorname{tr}(S(P_\varphi \times P_\psi)S^+(F_1 \times P_\gamma)) = \operatorname{tr}_1(\eta(P_\gamma)F_1).$$

Since $S(P_\varphi \times P_\psi)S^+ = P_\phi$ where $\phi = \sum_\nu \chi_\nu \eta_\nu \alpha_\nu$ it follows that

$$\begin{aligned}\operatorname{tr}_1(\eta(P_\gamma)F_1) &= \langle \phi, (F_1 \times P_\gamma)\phi \rangle \\ &= \sum_{\nu\mu} \bar{\alpha}_\nu \alpha_\mu \langle \chi_\nu \eta_\nu, (F_1 \times P_\gamma)\chi_\mu \eta_\mu \rangle \\ &= \sum_{\nu\mu} \bar{\alpha}_\nu \alpha_\mu \langle \chi_\nu, F_1 \chi_\mu \rangle \langle \eta_\nu, P_\gamma \eta_\mu \rangle \\ &= \sum_{\nu\mu} \bar{\alpha}_\nu \alpha_\mu \langle \chi_\nu, F_1 \chi_\mu \rangle \langle \eta_\nu, \gamma \rangle \langle \gamma, \eta_\mu \rangle.\end{aligned}$$

For

$$\delta = \sum_\mu \alpha_\mu \langle \gamma, \eta_\mu \rangle \chi_\mu$$

it follows that

$$\eta(P_\gamma) = |\delta\rangle\langle\delta|$$

and therefore

$$\mathbf{S}(1, 2; P_\psi, P_{\gamma_\lambda})P_\varphi = |\delta_\lambda\rangle\langle\delta_\lambda|, \tag{4.2.7}$$

where

$$\delta_\lambda = \sum_\mu \langle \varphi_\mu, \varphi \rangle \langle \gamma_\lambda, \eta_\mu \rangle \chi_\mu. \tag{4.2.8}$$

From

$$\phi = \sum_{\nu} \chi_\nu \eta_\nu \alpha_\nu$$

we easily obtain

$$W_1^f = R_1 P_\phi = \sum_{\nu} |\langle \varphi_\nu, \varphi \rangle|^2 P_{\chi_\nu}. \tag{4.2.9}$$

The decomposition of W_1^f after the registration of the P_{γ_λ} is therefore given by

$$W_1^f = \sum_{\lambda} |\delta_\lambda\rangle\langle\delta_\lambda| = \sum_{\lambda} \|\delta_\lambda\|^2 P_{\delta_\lambda/\|\delta_\lambda\|}. \tag{4.2.10}$$

The decompositions (4.2.10) are not coexistent for arbitrary choices of the γ_ν. We shall show this for an example:

$$\varphi = \frac{1}{\sqrt{2}}(\varphi_1 + \varphi_2), \qquad \gamma_1 = \beta_1 \eta_1 + \beta_2 \eta_2, \qquad \gamma_2 = \bar{\beta}_2 \eta_1 - \bar{\beta}_1 \eta_2$$

and additional vectors which are orthogonal to γ_1, γ_2. Since $\|\gamma_1\| = 1$, $|\beta_1|^2 + |\beta_2|^2 = 1$. Thus it follows that

$$\delta_1 = \frac{1}{\sqrt{2}}(\bar{\beta}_1 \chi_1 + \bar{\beta}_2 \chi_2),$$

$$\delta_2 = \frac{1}{\sqrt{2}}(\beta_2 \chi_1 - \beta_1 \chi_2)$$

and $\delta_\nu = 0$ for $\nu > 2$.

Thus we obtain

$$W_1^f = \tfrac{1}{2} P_{\chi_1} + \tfrac{1}{2} P_{\chi_2} \tag{4.2.11}$$

with the decomposition corresponding to the effects P_{γ_1}, P_{γ_2} given by

$$\begin{aligned} W_1^f &= |\delta_1\rangle\langle\delta_1| + |\delta_2\rangle\langle\delta_2| \\ &= P_{\delta_1/\|\delta_1\|} + \tfrac{1}{2} P_{\delta_2/\|\delta_2\|}. \end{aligned} \tag{4.2.12}$$

$\delta_1 \|\delta_1\|^{-1}$ and $\delta_2 \|\delta_2\|^{-1}$ are two orthonormal vectors in the plane spanned by χ_1, χ_2. The mathematically possible different decompositions (4.2.12) with "any two" orthogonal vectors in the plane spanned by χ_1, χ_2 can also be "experimentally" produced by choice of the observables $\{P_{\gamma_\nu}\}$. If δ_1 is oblique to χ_1, χ_2, then the decompositions (4.2.11) and (4.2.12) are NOT coexistent, as we have already seen in IV, §6.

For our second example we shall use the gedanken experiment described in §2.1, which we shall again encounter in a different context in the next section.

The operation $\mathbf{S}(1, 2; W_2, F_2)$ defined in (4.2.1) must be closely related with the measurement scattering morphism described in (2.2.7). From (4.1.6) it follows that, for the special case in which $W = S(W_1 \times W_2)S^+$

$$\operatorname{tr}(S(W_1 \times W_2)S^+(F_1 \times F_2)) = \operatorname{tr}_1([\mathbf{S}(1, 2; W_2 F_2)W_1]F_1). \tag{4.2.13}$$

Using the mapping $\mathbf{S}'(1, 2; W_2, F_2)$ which is adjoint to $\mathbf{S}(1,2; W_2, F_2)$ it follows that

$$\operatorname{tr}(S(W_1 \times W_2)S^+(F_1 \times F_2)) = \operatorname{tr}_1(W_1[\mathbf{S}'(1, 2; W_2, F_2)F_1]). \quad (4.2.14)$$

For the special case in which $F_1 = \mathbf{1}$, if we compare (4.2.14) with (2.2.6) and (2.2.7) we obtain the relation:

$$\mathbf{T}(1, 2; W_2)F_2 = \mathbf{S}'(1, 2; W_2, F_2)\mathbf{1}. \quad (4.2.15)$$

The observable corresponding to the transpreparator (4.2.4) (the associated observable in the sense of V, D 4.3.4) is therefore given by $\Sigma_2 \xrightarrow{F_1} L(\mathscr{H}_1)$ (where $F_1(\sigma) = \mathbf{T}(1, 2; W_2)F_2(\sigma)$) and was introduced in the diagram in §2.3. The effect transformer which corresponds to the transpreparator (4.2.4) in the sense of V, D 4.3.4 is given by

$$\Sigma \xrightarrow{\chi'} P,$$

where

$$\chi'(\sigma) = \mathbf{S}(1, 2; W_2, F_2(\sigma)).$$

Similarly, as described near the end of §4.1, these considerations can be extended to the case in which the apparatus (1) is not described by a single preparation procedure a_1 nor by a single ensemble W_1 but by a preparator $\Sigma_1 \xrightarrow{\eta_1} \breve{K}(\mathscr{H}_1)$. We shall now require that W_2 be fixed! Let Σ be the free (complete) Boolean ring generated by Σ_1 and Σ_2. For $\sigma_1 \in \Sigma_1, \sigma_2 \in \Sigma_2$ and therefore $\sigma_1 \wedge \sigma_2 \in \Sigma$ we define an additive measure $\Sigma \xrightarrow{\eta} \breve{K}(\mathscr{H}_1)$ by means of

$$\eta(\sigma_1 \cap \sigma_2) = \mathbf{S}(1, 2; W_2, F_2(\sigma_2))\eta_1(\sigma_1) \quad (4.2.16)$$

which is a preparator of the ensemble

$$R_1[S(\eta_1(\varepsilon_1) \times W_2)S^+], \quad (4.2.17)$$

where ε_1 corresponds to the unit element of Σ_1. This preparator has the following meaning: It is the preparator for systems of type 1 which are obtained from the preparator $\Sigma_1 \xrightarrow{\eta_1} \breve{K}(\mathscr{H}_1)$ with the help of a scattering on systems of type 2 and of a further decomposition according to the responses of the $\sigma \in \Sigma_2$ (where σ is the idealized registration for systems 2).

4.3 Collapse of Wave Packets?

In this section we shall consider a concept which is widely found in the literature of the measurement process in quantum mechanics and has led to many mystical ideas in quantum mechanics—the notion of the collapse (or reduction) of wave packets. Indeed, this notion has lead to apparent problems and to various attempts to solve these apparent problems. The mystification arose from a misunderstanding of J. von Neumann's description of the measurement process presented in his famous book *Mathematical Foundations of Quantum Mechanics* [25]. In that classic work J. von Neumann considered the time dependence of a system 1 under the unitary transformations

U_t as described in X. In addition he considered measurement processes in which a $W_1 \in K(\mathscr{H}_1)$ is transformed by the measurement process into another W'_1. J. von Neumann obtained the transition from W_1 to W'_1 by making certain assumptions about the measurement process which we shall consider in more detail in §5. On the basis of this abstract axiomatic form it is possible to arrive at the misunderstanding mentioned above in which the transition $W_1 \to W'_1$ is not only considered to be a mathematical discontinuity but is also considered to be a physical one. The mystifying notions arise from attributing physical reality to the "jump" at a given time t.

J. von Neumann has shown that the two time dependent processes are essentially different. The fact that they are different is clear on the basis of the representation and the interpretation of the two processes that we have given in VII, VIII, X and in this chapter.

The processes described by U_t describe the "free" system in the sense described in VII, namely, in the sense that the registration apparatus is displaced by a time interval t. The "free" propagation of systems 1 between the preparation and registration apparatus is extended by a time interval of length t.

By a change from W_1 to W'_1 which occurs as the result of the measurement process we mean the change of the ensemble W_1 before the application of the registration procedure to W'_1 after the application of the registration procedure which results from interaction with the registration apparatus. But what change did J. von Neumann intend in his axiomatic formulation?

Disregarding the special form of the J. von Neumann postulate (which we shall consider in detail in §5), the transition from W_1 to W'_1 is nothing other than the process described in §4.2 with the aid of the transpreparators (4.2.3) and (4.2.4) as follows:

$$W_1 \to W'_1 = \lambda\chi(\sigma)W_1 = \lambda \mathbf{S}(1, 2; W_2, F_2(\sigma))W_1, \tag{4.3.1}$$

where λ is a normalization factor, so that $\mathrm{tr}(W'_1) = 1$. Thus W'_1 is obtained by decomposition according to the response of σ.

We shall now describe how it was possible to imagine (within the context of the decompositions (4.3.1)) a collapse of the wave packet by using the gedanken experiment described in §2.1. In this experiment we find a measurement scattering process between the light quanta and the electrons to be measured. Then the light quanta are registered with the aid of a microscope and photographic plate. Suppose that the ensemble of electrons prepared before the scattering W_1 is one of the P_φ where φ has extremely precise momentum (in the object plane), then in the position representation $\langle x_1, x_2 | \varphi\rangle$ is a very broad function in the object plane.

Without examining the implications of "additional concepts" in quantum mechanics, many people consider the "wave packet" $\langle x_1, x_2 | \varphi\rangle$ to represent the state of a single electron. Providing that the scattered light quantum is registered upon the photographic plate, we may (approximately) infer the location of the electron from the position of the blackened silver grain on the photographic plate. Suppose, for example, the position of the measured electron is approximately $x_1^{(0)}, x_2^{(0)}$. The following conjecture may appear to be reasonable: After the measurement not all positions associated with the wave packet $\langle x_1, x_2 | \varphi \rangle$ are possible but only positions in the vicinity of $x_1^{(0)}, x_2^{(0)}$, that is, "the wave packet $\langle x_1, x_2 | \varphi \rangle$ has, as a result of the measurement process, undergone a 'collapse' to a wave packet $\langle x_1, x_2 | \psi \rangle$ which is strongly concentrated in the vicinity of $x_1^{(0)}, x_2^{(0)}$."

The somewhat mystifying mode of description outlined above is a result of the addition of unnecessary preconceived notions to quantum mechanics. This will become more evident when we consider the following modification of the gedanken experiment described in Figure 44. The modification consists of locating the photographic plate in the focal plane instead of in the image plane. From the location of the blackened points on the photographic plate it is possible to infer the propagation direction of the scattered light quantum. If we choose the ensemble W_2 of the light quanta before scattering to be one for which the momentum is precisely fixed, and assume that the absolute value of the momentum of the light quanta is not changed by the scattering, then we may infer the momentum change of the electron from the direction of the scattered light quantum (since the momentum of the electron before the scattering is fixed). In this modified experiment the wave packet $\langle x_1, x_2 | \varphi \rangle$ does not collapse in space but undergoes a sudden transition into another wave packet $\langle x_1, x_2 | \psi \rangle$ which has fixed momentum and therefore has a large spread in position x_1, x_2.

The form of the new wave packet which results from the "collapse" of the wave packet $\langle x_1, x_2 | \varphi \rangle$ depends strongly upon where the photographic plate is placed. In principle it is possible to imagine a microscope which is so large that we have several hours (after the scattering) to decide where to place the photographic plate. Here the blackening of the photographic plate which happens several hours after the scattering will decide the manner in which the "jump" of the wave packet has occurred (several hours before the blackening!). This "decision" is, in part, also a decision of our free will!

It is not surprising that the range of explanations for this imagined collapse of the wave packet is very broad, from the modification of the Schrödinger

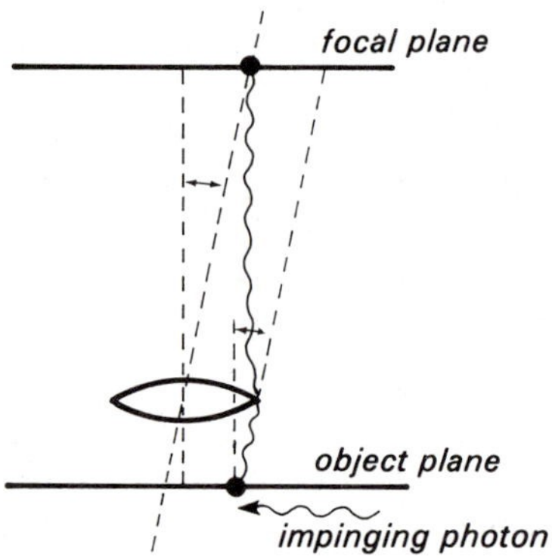

Figure 44

equation to the interpretation of the wave packet $\langle x_1, x_2 | \varphi \rangle$ as a mathematical description of our "knowledge about the electron under consideration". We do not need to examine all these attempts, because, according to the interpretation presented in II, the entire problem of the "collapse of the wave packet" is not a real problem but is a problem which arises from the addition of unnecessary ideas.

According to our formulation of quantum mechanics nothing is discontinuous. In particular, the above gedanken experiment may be described as follows: Electrons 1 of the ensemble $W_1 = P_\varphi$ are scattered by light quanta 2 from ensemble W_2, whereupon after the scattering the ensemble is described by $S(P_\varphi \times W_2)S^+$ where S is the scattering operator. The "jump from $P_\varphi \times W_2$ before the scattering to $S(P_\varphi \times W_2)S^+$ after the scattering is only a computational rule which permits us to avoid the need to provide a detailed description of the behavior of the system during the scattering. The investigations presented in XVI associated with the introduction of the scattering operator S have made this clear. A true sudden "jump" does not occur.

The scattered light quanta 2 can be registered by means of a microscope and photographic plate, the results of which are described by the effects F_2. If Σ_2 is, as described in §2.1, the Boolean ring of surface elements in the image plane, then we may decompose the ensemble $W_1^f = R_1[S(P_\varphi \times W_2)S^+]$ for the electrons with the aid of the preparator $\Sigma_2 \xrightarrow{\eta} \breve{K}(\mathscr{H}_1)$ where

$$\eta(\sigma) = \mathrm{S}(1, 2; W_2, F_2(\sigma)P_\varphi$$

according to which surface elements σ of the photographic plate have been blackened. By choosing a "small" σ then $\eta(\sigma)$ is an ensemble with negligible dispersion of the position.

We may also consider the Boolean ring Σ_2' of the surface elements of the focal plane; then, from the observable $\Sigma_2' \xrightarrow{F_2'} L(\mathscr{H}_2)$ we obtain a preparator $\Sigma_2' \xrightarrow{\eta'} \breve{K}(\mathscr{H}_1)$ where

$$\eta'(\sigma) = \mathrm{S}(1, 2; W_2, F_2'(\sigma))P_\varphi$$

permits another decomposition of the *same* ensemble W_1^f. If $\sigma \in \Sigma_2'$ is a small σ then $\eta'(\sigma)$ is an ensemble which has small dispersion in momentum (providing that P_φ and W_2 have small dispersion in momentum).

The two preparators $\Sigma_2 \xrightarrow{\eta} \breve{K}(\mathscr{H}_1)$ and $\Sigma_2' \xrightarrow{\eta'} \breve{K}(\mathscr{H}_1)$ are, of course, not coexistent! Therefore the corresponding decompositions of W_1^f are not coexistent. The transpreparators

$$\Sigma_2 \xrightarrow{\mathrm{S}(1,\, 2;\, W_2,\, F_2(\cdots))} \Pi, \qquad \Sigma_2' \xrightarrow{\mathrm{S}(1,2;\, W_2,\, F_2'(\cdots))} \Pi$$

are therefore not coexistent. Noncoexistent transpreparators are, however, typical for quantum mechanics, as we have already seen in IV, §§5 and 6.

The process of decomposition by means of preparators, as described in the form of Boolean rings $Q(a)$ which has already been introduced in II will therefore be mystified by such modes of description as the collapse of wave packets. The possibility of two noncoexistent decompositions of the same ensemble W_1^f with the aid of a photographic plate in either the image plane or focal plane is therefore only an example of the model described in §4.2 with the decomposition possibilities given by (4.2.8) to (4.2.10).

Nevertheless, we should not conceal the fact that the fundamental quantum fact: the existence of noncoexistent (indeed, complementary, see IV, D 6.4) preparators raises a peculiar problem, which, psychologically, is difficult to avoid. This problem is known as the "Einstein–Podolski–Rosen Paradox", although it is not a real paradox, that is, it is not a contradiction in quantum mechanics. We shall consider this "paradox" in the following section.

4.4 The Einstein–Podolski–Rosen Paradox

The starting point for our discussion of the so-called Einstein–Podolski–Rosen (EPR) paradox is the discussion in §4.1 (and the somewhat more specialized discussion in §4.2).

For an observable $\Sigma_2 \overset{F_2}{\rightarrow} L(\mathscr{H}_2)$ we obtain a transpreparator

$$\Sigma_2 \overset{\chi}{\rightarrow} \Pi, \tag{4.4.1}$$

where

$$\chi(\sigma) = \mathbf{S}(1, F_2(\sigma)). \tag{4.4.2}$$

$\mathbf{S}(1, F_2)$ is defined by (4.1.5) or (4.1.6). For an arbitrary $W \in K(\mathscr{H}_1 \times \mathscr{H}_2)$ from the above transpreparator we obtain a preparator

$$\Sigma_2 \overset{\eta}{\rightarrow} \breve{K}(\mathscr{H}_1), \tag{4.4.3}$$

where

$$\eta(\sigma) = \chi(\sigma)W \tag{4.4.4}$$

for the ensemble $W_1 = R_1 W$. The preparator (4.4.3) permits coexistent decompositions of W_1, that is, according to a decomposition σ_i of the unit ε of Σ_2 as follows:

$$W_1 = \eta(\varepsilon)W = \sum_i \eta(\sigma_i)W. \tag{4.4.5}$$

Everything depends on the observable $\Sigma_2 \overset{F_2}{\rightarrow} L(\mathscr{H}_2)$. There are now many experiments by which we may measure the observable $\Sigma_2 \overset{F_2}{\rightarrow} L(\mathscr{H}_2)$ for systems of type 2 providing that there is no longer any interaction between systems of type 1 and type 2 for the ensemble W. Examples of such measurements are the measurements after scattering of systems 1 onto systems 2 which we have discussed in §4.2.

Since there is no longer any "interaction" between systems 1 and systems 2, the measurement may no longer change the ensemble of systems 1, that is, it should not make a difference whether systems 2 are measured (and what observables are measured for systems 2), that is, $W_1 = R_1 W$ is not changed by making an additional measurement (such as scattering systems 2 on systems 3—see §6.1). This description is completely consistent with quantum mechanics. However, with the aid of different observables $\Sigma_2 \xrightarrow{F_2} L(\mathscr{H}_2)$ we may obtain different preparators (4.4.3) and therefore obtain different decompositions of the same ensemble W_1. The discussions in §4.2 (4.2.8)–(4.2.10) provide an example.

In order to illustrate the so-called EPR paradox we shall construct examples of two observables $\Sigma_2^{(1)} \xrightarrow{F_2^{(1)}} L(\mathscr{H}_2)$ and $\Sigma_2^{(2)} \xrightarrow{F_2^{(2)}} L(\mathscr{H}_2)$ and we shall choose an ensemble W such that the corresponding preparators $\Sigma_2^{(1)} \xrightarrow{\eta^{(1)}} \breve{K}(\mathscr{H}_1)$ and $\Sigma_2^{(2)} \xrightarrow{\eta^{(2)}} \breve{K}(\mathscr{H}_1)$ of the same ensemble $W_1 = R_1 W$ are complementary in the sense of IV, D 6.4.

We shall now describe a simple example. Let the systems 1 and 2 be elementary systems of spin $\frac{1}{2}$. Let the apparatus denoted by (1, 2) in Figure 43 prepare an ensemble $W = \varphi(a) \in K(\mathscr{H}_1 \times \mathscr{H}_2)$ for which the "total spin" is zero, that is, W satisfies the relation $\operatorname{tr}(WP) = 1$ where P is the following projection operator in $\mathscr{H}_b \times \imath$

$$P = \mathbf{1} \times P_\phi \quad \text{where} \quad \phi = \frac{1}{\sqrt{2}} [u_+(1)u_-(2) - u_+(2)u_-(1)] \in \imath$$

(where $\mathscr{H}_1 = \mathscr{H}_{1b} \times \imath_1$, $\mathscr{H}_2 = \mathscr{H}_{2b} \times \imath_2$, $\mathscr{H}_b = \mathscr{H}_{1b} \times \mathscr{H}_{2b}$, $\imath = \imath_1 \times \imath_2$). W therefore has the form

$$W = W_b \times P_\phi, \tag{4.4.6}$$

where W_b is an operator in $\mathscr{H}_b$.

We may choose the apparatus (1, 2) such that (see Figure 45) the systems 1 and 2 leave in completely different directions and therefore after the preparation are no longer in interaction.

Let the apparatus (3, 2) measure the observable $\Sigma_2 \xrightarrow{F_2} L(\mathscr{H}_2)$. This observable will be the decision observable "the measurement of spin in a given direction $\vec{n}$", that is, we can choose the Boolean ring $\Sigma_2^{(\vec{n})}$ which is generated by the operators P_{u_+} and P_{u_-} in $\imath_2$ where $u_+, u_- \in \imath_2$ are eigenvectors of the spin component of systems 2 in direction $\vec{n}$. $\Sigma_2^{(\vec{n})}$ will therefore consist of the elements $1, P_{u_+}, P_{u_-}, 0$ (as operators in $\imath_2$).

Then, from (4.4.2), (4.4.3) and (4.1.5), after a simple computation, we obtain:

$$\eta(P_{u_+}) = \tfrac{1}{2} W_{1b} \times P_{u_-} \tag{4.4.7}$$

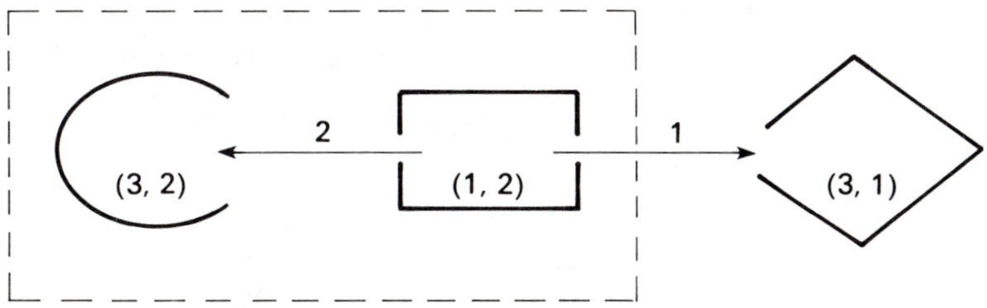

Figure 45

as an operator in $\mathscr{H}_{1b} \times \imath_1$, where W_{1b} denotes the restriction $R_1 W_b$ of W_b in $\mathscr{H}_b = \mathscr{H}_{1b} \times \mathscr{H}_{2b}$. In the same way we obtain

$$\eta(P_{u_-}) = \tfrac{1}{2} W_{1b} \times P_{u_+}. \tag{4.4.8}$$

Therefore, with the aid of the observable $\Sigma_2^{(\vec{n})} \overset{F_2}{\rightarrow} L(\mathscr{H}_2)$ we obtain the decomposition of the ensemble

$$W_1 = R_1 W = \tfrac{1}{2} W_{1b} \times \mathbf{1} \tag{4.4.9}$$

in the form

$$W_1 = \tfrac{1}{2} W_{1b} \times P_{u_+} + \tfrac{1}{2} W_{1b} \times P_{u_-}. \tag{4.4.10}$$

Here we recommend that the reader verify how the decomposition (4.4.9) of the ensemble W_1 (prepared by the apparatus which consists of the apparatuses (1, 2) and (3, 2)—see Figure 45) is achieved in terms of the responses of the apparatus (3, 2). The apparatus (3, 1) serves only to test whether our theory is correct. If, for example, we use (3, 1) to measure spin in the $\vec{n}$ direction on the component $\frac{1}{2} W_{1b} \times P_{u_-}$ we obtain the value $(-\frac{1}{2})$ with certainty since $\mathbf{1} \times P_{u_-}$ has no dispersion in $\frac{1}{2} W_{1b} \times P_{u_-}$. The response $(+\frac{1}{2})$ of the apparatus (3, 2) is therefore strongly correlated with the response $(-\frac{1}{2})$ of the apparatus (3, 2). The fact that it is possible to produce an ensemble $\frac{1}{2} W_{1b} \times P_{u_-}$ in the manner described in Figure 45 with the aid of the composite apparatus (1, 2) and (3, 2) is certainly not alarming. On the other hand, it is alarming to find that we may, without changing the apparatus (1, 2), by only rotating the apparatus (3, 2), obtain another preparator which is complementary to the original!

Choose, for example, first $\vec{n}$ in the direction of axis 1 of the laboratory system, and second, in the direction of axis 2; we then obtain two observables $\Sigma_2^{(1)} \overset{F_2^{(1)}}{\rightarrow} L(\mathscr{H}_2)$ and $\Sigma_2^{(2)} \overset{F_2^{(2)}}{\rightarrow} L(\mathscr{H}_2)$ which lead to two preparators of the form given in (4.4.7) and (4.4.8):

$$\eta^{(1)}(P_{u_+}) = \tfrac{1}{2} W_{1b} \times P_{u_-}, \qquad \eta^{(1)}(P_u) = \tfrac{1}{2} W_{1b} \times P_{u_+} \tag{4.4.11}$$

and

$$\eta^{(2)}(P_{v_+}) = \tfrac{1}{2} W_{1b} \times P_{v_-}, \qquad \eta^{(2)}(P_{v_-}) = \tfrac{1}{2} W_{1b} \times P_{v_+}, \tag{4.4.12}$$

where u_+, u_- are the eigenvectors of spin in the 1-direction and v_+, v_- are the eigenvectors of the spin in the 2-direction.

We now show that from $0 \leq W \leq W_{1b} \times P_{u_-}$ we obtain the following relation $W = W' \times P_{u_-}$. For $0 \leq W \leq W_{1b} \times P_{u_-}$ and $0 \leq W < W_{1b} \times P_{v_-}$ then it follows that $W = W' \times P_{u_-} = W'' \times P_{v_-}$ and therefore $W = 0$ since $\operatorname{tr}((W' \times P_{u_-})(1 \times P_{u_+})) = 0$ and $\operatorname{tr}((W'' \times P_{v_-})(1 \times P_{u_+})) = \operatorname{tr}(W'')$ $\operatorname{tr}(P_{v_-} P_{u_+})) = \operatorname{tr}(W'')|\langle v_- | u_+ \rangle|^2$ and $|\langle v_- | u_+ \rangle|^2 \neq 0$. According to IV, Th. 6.1 the two decompositions

$$W_1 = \tfrac{1}{2} W_{1b} \times \mathbf{1} = \eta^{(1)}(P_{u_+}) + \eta^{(1)}(P_{u_-})$$

and

$$W_1 = \eta^{(2)}(P_{v_+}) + \eta^{(2)}(P_{v_-})$$

are complementary, and, according to IV, D 6.5 the two preparators $\Sigma_2^{(1)} \overset{\eta^{(1)}}{\rightarrow} \breve{K}(\mathscr{H}_1)$ and $\Sigma_2^{(2)} \overset{\eta^{(2)}}{\rightarrow} \breve{K}(\mathscr{H}_1)$ are complementary

Since this is a very illustrative example for two complementary preparators, we will now describe this experiment in detail more precisely with the aid of preparation and registration procedures. Let a denote the preparation procedure associated with the apparatus (1, 2). We therefore have $\varphi(a) = W$ where W is given by (4.4.6). Let $b_0^{(1)}$, $b_0^{(2)}$ denote the registration methods defined by the apparatus (3, 2) for the registration of spin in the 1 and 2 directions, respectively. $\mathscr{R}(b_0^{(1)})$ consists of the sets $b_0^{(1)}, b_+^{(1)}, b_-^{(1)}, \varnothing$ where $\psi(b_0^{(1)}, b_+^{(1)}) = \mathbf{1} \times (\mathbf{1} \times P_{u_+})$ and $\psi(b_0^{(1)}, b_-^{(1)}) = \mathbf{1} \times (\mathbf{1} \times P_{u_-})$ with respect to the representation $\mathscr{H} = \mathscr{H}_b \times (\imath_1 \times \imath_2)$ of the Hilbert space $\mathscr{H}$. A similar situation holds for $\mathscr{R}(b_0^{(2)})$.

According to (4.1.1) there exists a k given by

$$(a, b_0^{(1)}) \overset{k}{\rightarrow} \tilde{a}^{(1)},$$
$$(a, b_+^{(1)}) \overset{k}{\rightarrow} \tilde{a}_+^{(1)},$$
$$(a, b_-^{(1)}) \overset{k}{\rightarrow} \tilde{a}_-^{(1)},$$

where $\mathscr{R}(b_0^{(1)})$ is isomorphically mapped onto $\mathscr{Q}(\tilde{a}^{(1)})$. It follows that

$$\varphi(\tilde{a}^{(1)}) = W_1 = \tfrac{1}{2}W_{b1} \times \mathbf{1} \in K(\mathscr{H}_1),$$
$$\varphi(\tilde{a}_+^{(1)}) = W_{1b} \times P_{u_-},$$
$$\varphi(\tilde{a}_-^{(1)}) = W_{1b} \times P_{u_+}.$$

From $\tilde{a}^{(1)} = \tilde{a}_+^{(1)} \cup a_-^{(1)}$ and $\tilde{a}^{(1)} \cap \tilde{a}_-^{(1)} = \varnothing$ we obtain the following decomposition (see IV, §5)

$$\varphi(\tilde{a}^{(1)}) = \tfrac{1}{2}\varphi(\tilde{a}_+^{(1)}) + \tfrac{1}{2}\varphi(\tilde{a}_-^{(1)}).$$

If we replace (1) by (2) we obtain the analogous formulas by replacing u_+, u_- by v_+, v_-.

Thus we have obtained two different preparation procedures $\tilde{a}^{(1)}$ and $\tilde{a}^{(2)}$ where $\varphi(\tilde{a}^{(1)}) = \varphi(\tilde{a}^{(2)}) = W_1 \cdot \tilde{a}^{(1)}$ is obtained by combining a with $b_0^{(1)}$ and $\tilde{a}^{(2)}$ by combining a with $b_0^{(2)}$.

$\mathscr{Q}(\tilde{a}^{(1)}) \overset{\varphi_{\tilde{a}^{(1)}}}{\rightarrow} \breve{K}(\mathscr{H}_1)$ and $\mathscr{Q}(\tilde{a}^{(2)}) \overset{\varphi_{\tilde{a}^{(2)}}}{\rightarrow} \breve{K}(\mathscr{H}_1)$ where $\varphi_{\tilde{a}^{(1)}}$ and $\varphi_{\tilde{a}^{(2)}}$ are the maps introduced in §5 therefore represent two complementary preparators. Since they are complementary, according to IV, D 6.4 we must obtain

$$\tilde{a}^{(1)} \cap \tilde{a}^{(2)} = \varnothing. \tag{4.4.13}$$

Therefore (4.4.13) is a consequence of the quantum mechanical axioms. Thus we find that this is a nice example of the structures described more generally in IV, D 6.4.

The same can be done for the example described in §4.3 of the decomposition of an ensemble W_1^f of electrons after the scattering with the aid of the registration points on the photographic plate. With the aid of a photographic plate in the image plane we may decompose W_1^f into ensembles which have little dispersion in position. The same ensemble may also be decomposed, with the aid of a photographic plate in the focal plane into ensembles which have little dispersion in momentum. Since, according to the Heisenberg uncertainty relation (see IV (8.3.16)) there are no ensembles which have little dispersion in both position and momentum, we therefore obtain two complementary decompositions of W_1^f without influencing the scattered electrons, namely, by placing the plates in the image and focal planes. For both these preparations of electrons $\tilde{a}^{(1)}$ (the photographic plate in the image plane) and $\tilde{a}^{(2)}$ (the photographic plate in the focal plane) we obtain $\varphi(\tilde{a}^{(1)}) = \varphi(\tilde{a}^{(2)}) = W_1^f$; Despite this, (4.4.13) must be satisfied!

In this way we obtain a contradiction-free explanation of the EPR "problem" in quantum mechanics. We obtain EPR paradox from the EPR problem when we make (apparently) "plausible" conclusions from the observed phenomena. The apparent "plausibility" results when we believe that objective content is being transmitted through space by the microsystems (even though the latter are not "tiny spheres").

The fact that it is possible to decompose ensembles in a complementary way without making use of the interactions with the microsystems of the ensemble can only rest upon the fact (or at least a widely held opinion) that the microsystems are already distinguishable with respect to a "property" that they will either respond or not respond when they are subjected to an effect procedure (b_0, b) if $\psi(b_0, b)$ is a decision effect, that is, if $\psi(b_0, b) \in G$. Otherwise, how would it be possible, with the aid of a particular decomposition procedure, to select, without interaction, those microsystems which respond with certainty to (b_0, b)?

These considerations make it plausible to claim the existence of a set $\mathscr{E}_h$ satisfying the axioms AH 1 and AH 2 which have been introduced in IV, §8.3. Since we have to introduce AH 1 and AH 2 as axioms, it is clear that our "plausible" arguments do not represent deductions made from the EPR phenomena. The mathematical form of a theory cannot be outwitted by pretending that AH 1 and AH 2 are theorems. If we add AH 1 and AH 2 as axioms then we obtain a theory which is contradictory and is in this sense a paradox. The consequence has to be that the so-called "plausibility" arguments (leading to AH 1 and AH 2) are false!

5 Measurements of the First Kind

In the historical development of quantum mechanics a great role has been played by "idealized measurements" which are defined by formal axioms. Such axioms can be found in the first edition of this book (see [26], II, §3). In this edition such axioms are not needed for the fundamentals presented in II and III. However, since they have played such an important role, we shall now consider these "idealized measurements" in order to consider the dubious existence of such measurements. We cannot derive the existence of such "idealized measurements" from the axioms we have used for the development of quantum mechanics.

We now return to the discussion in §§2 and 4. From $\mathbf{T}(1, 2; W_2)$ we obtain a new observable $\Sigma_2 \xrightarrow{F_1} L(\mathscr{H}_1)$ from the observable $\Sigma_2 \xrightarrow{F_2} L(\mathscr{H}_2)$ as follows

$$F_1(\sigma) = \mathbf{T}(1, 2; W_2)F_2(\sigma). \tag{5.1}$$

In addition, we obtain the transpreparator $\Sigma_2 \xrightarrow{\chi} \Pi$ as follows:

$$\chi(\sigma) = \mathbf{S}(1, 2; W_2, F_2(\sigma)). \tag{5.2}$$

We define a measurement of the first kind as a measurement scattering together with an observable $\Sigma_2 \xrightarrow{F_2} L(\mathscr{H}_2)$ such that the relation

$$\mathrm{tr}_1([\chi(\sigma)W_1](\mathbf{1} - F_1(\sigma)) = 0 \tag{5.3a}$$

is satisfied for all $\sigma \in \Sigma_2$ and all $W_1 \in K(\mathscr{H}_1)$. Equation (5.3a) is equivalent to

$$\mathrm{tr}(S(W_1 \times W_2)S^+((\mathbf{1} - F_1(\sigma)) \times F_2(\sigma)) = 0. \tag{5.3b}$$

The requirement (5.3a) for a measurement of the first kind has the following interpretation: After the measurement all those systems of type 1 which are sorted according to the "response of σ" will correspond to an ensemble in which $F_1(\sigma)$ will respond with certainty providing the observable $\Sigma_2 \xrightarrow{F_1} L(\mathscr{H}_1)$ is measured again.

Measurements which are not measurements of the first kind, that is, for which (5.3) is not satisfied for all $\sigma \in \Sigma_2$ are often called measurements of the second kind. If $\Sigma_2 \xrightarrow{F_1} G(\mathscr{H}_1)$, we then speak of measurements of the first kind for a decision observable. Since in the first decades of the development of quantum mechanics physicists have only considered decision observables, in the literature the use of the expression "measurements of the first kind" will, in the terminology used here, always refer to measurements of the first kind for a decision observable.

The condition (5.3a) is identical to the condition IV (5.7) for the preparator $W_1(\sigma) = \chi(\sigma)W_1$ since $F_1(\varepsilon + \sigma) = \mathbf{1} - F_1(\sigma)$. Condition (5.3a) therefore has, as a consequence, the fact that the preparator $\chi(\sigma)W_1$, which is associated with the measurement of $\Sigma_2 \xrightarrow{F_2} L(\mathscr{H}_2)$, for each W_1 is a dispersion-free preparator with respect to the measured observable $\Sigma_2 \xrightarrow{F_1} L(\mathscr{H}_1)$. According to IV, §5 it follows that (5.3a) can be satisfied only if, according to IV (5.12) where $\tilde{E}\mathscr{H}$ is the support of $W_1^f = \chi(\varepsilon)W_1$ the measure $F_1(\sigma)$ has the form

$$F_1(\sigma) = E_1(\sigma) + (\mathbf{1} - \tilde{E})F_1(\sigma)(\mathbf{1} - \tilde{E}), \tag{5.4}$$

where $E_1(\sigma) \in G(\mathscr{H}_1)$ and $E_1(\sigma) \leq \tilde{E}$, and the Boolean subring (generated by the $E_1(\sigma)$ with $\tilde{E}$ as the unit element) is atomic. $\Sigma_2 \xrightarrow{E_1} G(\mathscr{H}_1)$ is a σ-additive measure on Σ_2 with $E_1(\varepsilon) = \tilde{E}$, since $F_1(\varepsilon) = \mathbf{1}$.

In order to evaluate this condition, we shall first modify the Boolean ring Σ_2. It is possible that $F_1(\sigma) = 0$ for $\sigma \neq 0$. Since F_1 is a σ-additive measure and Σ_2 is complete, there exists a uniquely determined σ_s which is the support of the measure $F_1(\sigma)$ (see IV, Th. 2.1.5, where we need only replace $m(\sigma)$ by $\mathrm{tr}_1(W_1F_1(\sigma))$ where W_1 is effective). We therefore obtain $F_1(\sigma_s) = \mathbf{1}$ and $F_1(\sigma) \neq 0$ for $0 \neq \sigma \leq \sigma_s$. Instead of Σ_2 we use the Boolean ring $\tilde{\Sigma}_2$ of all $\sigma \leq \sigma_s$ where σ_s is the unit element of $\tilde{\Sigma}_2$. The observable $\Sigma_2 \xrightarrow{F_1} L(\mathscr{H}_1)$ may therefore be replaced by the effective observable $\tilde{\Sigma}_2 \overset{F_1}{\to} L(\mathscr{H}_1)$. $\tilde{\Sigma}_2 \overset{F_2}{\to} L(\mathscr{H}_2)$ is a generalized observable since $F_2(\sigma_s) \neq \mathbf{1}$. Since $E_1(\sigma) \leq F_1(\sigma)$ we therefore obtain $E(\sigma) = E(\sigma \wedge \sigma_s)$ for all $\sigma \in \Sigma_2$. Therefore we may also consider $E_1(\sigma)$ to be a measure on $\tilde{\Sigma}_2$: $\tilde{\Sigma}_2 \overset{E_1}{\to} G(\mathscr{H}_1)$ where $E_1(\sigma_s) = \tilde{E}$.

We will now show that if W_1 is effective then the projection $\tilde{E}$ will be such that the map $\tilde{\Sigma}_2 \overset{E_1}{\to} G(\mathscr{H}_1)$ is injective, that is, $\tilde{\Sigma}_2$ can be identified with the Boolean ring generated by the $E_1(\sigma)$ (where $\tilde{E}$ is the unit element). Thus $\tilde{\Sigma}_2$ must be atomic.

In order to show that this is the case, choose an effective W_1. The map $\tilde{\Sigma}_2 \overset{E_1}{\to} G(\mathscr{H}_1)$ is injective if and only if it follows that if $\sigma \in \tilde{\Sigma}_2$ and $E_1(\sigma) = 0$ it follows that $\sigma = 0$. Therefore let $E_1(\sigma) = 0$. Then since (5.4) it follows that

$$\begin{aligned} \mathrm{tr}_1(\chi(\sigma)W_1) &= \mathrm{tr}_1(W_1 F_1(\sigma)) \\ &= \mathrm{tr}_1(W_1(\mathbf{1} - \tilde{E})F_1(\sigma)(\mathbf{1} - \tilde{E})) \neq 0 \end{aligned}$$

providing that $(\mathbf{1} - \tilde{E})F_1(\sigma)(\mathbf{1} - \tilde{E}) \neq 0$, since W_1 was assumed to be effective.

On the other hand, since $\tilde{\Sigma}$ is the support of $\chi(\sigma)W_1$, and $E_1(\sigma) = 0$ we obtain

$$\mathrm{tr}_1([\chi(\sigma)W_1]F_1(\sigma)) = \mathrm{tr}_1([\chi(\sigma)W_1]E_1(\sigma)) = 0$$

from which it follows that

$$\mathrm{tr}_1([\chi(\sigma)W_1](\mathbf{1} - F_1(\sigma)) = \mathrm{tr}_1(\chi(\sigma)W_1).$$

Suppose that $\mathrm{tr}_1(\chi(\sigma)W_1) \neq 0$, then (5.3a) would fail, and we must have $(\mathbf{1} - \tilde{E})F_1(\sigma)(\mathbf{1} - \tilde{E}) = 0$, that is, from $E_1(\sigma) = 0$ it follows that $F_1(\sigma) = 0$. Therefore $\sigma \wedge \sigma_s = 0$ and since $\sigma \in \tilde{\Sigma}_2$ it follows that σ must be the null element of $\tilde{\Sigma}_2$.

The observable $\tilde{\Sigma}_2 \overset{F_1}{\to} L(\mathscr{H}_1)$ is therefore atomic, and we may replace $\tilde{\Sigma}_2$ by the Boolean ring generated by $E_1(\sigma)$ with $\tilde{E}$ as the unit element. Therefore, instead of $\tilde{\Sigma}_2$ we need only the atoms σ_i of $\tilde{\Sigma}_2$, that is, pairwise orthogonal sets of E_{1i} which satisfy the following relationship $\sum_i E_{1i} = \tilde{E}$.

For (5.4) we now write (where $F_{1i} = F_1(\sigma_i)$:

$$F_{1i} = E_{1i} + (\mathbf{1} - \tilde{E})F_{1i}(\mathbf{1} - \tilde{E}). \tag{5.5}$$

We will now show that for all $W_1 \in K(\mathscr{H}_1)$ the support of $\chi(\sigma)W_1$ (for all σ) must be smaller than $\tilde{E}\mathscr{H}_1$.

We begin by choosing an effective W_1. Since W_1 is effective, $W_1 = \Sigma_\nu w_\nu P_{\varphi_\nu}$ where $w_\nu \neq 0$ and the φ_ν form a complete orthonormal basis. For $\tilde{E}$ we therefore obtain:

$$\mathrm{tr}(S(W_1 \times W_2)S^+((\mathbf{1} - \tilde{E}) \times \mathbf{1})) = 0. \tag{5.6}$$

Thus, for all φ_ν and for $W_2 = \sum_\mu \lambda_\mu P_{\chi_\mu}$ and $\lambda_\mu \neq 0$ we obtain:

$$\begin{aligned} 0 &= \mathrm{tr}(S(P_{\varphi_\nu} \times P_{\chi_\mu})S^+((\mathbf{1} - \tilde{E}) \times \mathbf{1})) \\ &= \mathrm{tr}(P_{S\varphi_\nu\chi_\mu}((\mathbf{1} - \tilde{E}) \times \mathbf{1})) = \|((\mathbf{1} - \tilde{E}) \times \mathbf{1})S\varphi_\nu\chi_\mu\|^2. \end{aligned}$$

Therefore we obtain

$$[(\mathbf{1} - \tilde{E}) \times \mathbf{1}]S\varphi_\nu \chi_\mu = 0$$

for all φ_ν and χ_μ, that is,

$$S\varphi_\nu \chi_\mu \in \tilde{E}\mathscr{H}_1 \times \mathscr{H}_2. \tag{5.7}$$

Conversely, if (5.7) is satisfied for all $\varphi_\nu \chi_\mu$, then (5.6) holds. From (5.7) it follows that

$$S\varphi\chi_\mu \in \tilde{E}\mathscr{H}_1 \times \mathscr{H}_2 \tag{5.8}$$

for all $\varphi \in \mathscr{H}_1$ and all χ_μ. Thus it follows that

$$\operatorname{tr}(S(W_1 \times W_2)S^+((\mathbf{1} - \tilde{E}) \times \mathbf{1})) = 0 \tag{5.9}$$

for all $W_1 \in K(\mathscr{H}_1)$, that is,

$$\operatorname{tr}_1([\chi(\varepsilon)W_1](\mathbf{1} - \tilde{E})) = 0, \tag{5.10}$$

which means that the support of $\chi(\varepsilon)W_1$ is smaller than $\tilde{E}\mathscr{H}_1$. Therefore, since $\chi(\sigma)W_1 \leq \chi(\varepsilon)W_1$ it follows that

$$\chi(\sigma)W_1 = \tilde{E}[\chi(\sigma)W_1]\tilde{E} \tag{5.11}$$

holds for all σ.

The condition (5.3a), with the aid of (5.5) takes on the form

$$\operatorname{tr}_1([\chi(\sigma_i)W_1](\mathbf{1} - E_{1i})) = 0 \tag{5.12}$$

for all i—this is equivalent to the condition

$$\operatorname{tr}(S(W_1 \times W_2)S^+((1 - E_{1i}) \times F_{2i})) = 0, \tag{5.13}$$

where we have used the simpler notation $F_{2i} = F_2(\sigma_i)$ for the atoms σ_i of $\tilde{\Sigma}_2$.

Since (5.13) holds for all $W_1 \in K(\mathscr{H}_1)$, it follows, as we have seen above, that $S\varphi\chi_\nu$ must, for all $\varphi \in \mathscr{H}_1$ and all χ_ν be eigenvectors having eigenvalue 0 for all operators $(\mathbf{1} - E_{1i}) \times F_{2i}$. What is the subspace of all vectors $\phi \in \mathscr{H}_1 \times \mathscr{H}_2$ for which the condition

$$[(\mathbf{1} - E_{1i}) \times F_{2i}]\phi = 0 \tag{5.14}$$

is satisfied for all i?

ϕ can consist only of components Ψ for which (for each i) either $[(\mathbf{1} - E_{1i}) \times \mathbf{1}]\Psi = 0$ or $(\mathbf{1} \times F_{2i})\Psi = 0$. We note, however that $[(\mathbf{1} - E_{1i}) \times \mathbf{1}]\Psi = 0$ can hold for at most one i because the E_{1i} are pairwise orthogonal. Let $[(\mathbf{1} - E_{1i_0}) \times \mathbf{1}]\Psi = 0$, that is, $(E_{1i_0} \times \mathbf{1})\Psi = \Psi$. Thus $(\mathbf{1} \times F_{2i})\Psi = 0$ for all $i \neq i_0$, that is, only $(\mathbf{1} \times F_{2i_0})\Psi$ can be nonzero. If $\mathscr{H}_{2i_0}$ is the subspace of $\mathscr{H}_2$ for which all F_{2i} for $i \neq i_0$ act as null operators then the space of the ϕ for which (5.14) is satisfied must be equal to

$$\sum_i \oplus E_{1i}\mathscr{H}_1 \times \mathscr{H}_{2i}. \tag{5.15}$$

Since $\mathbf{1} = F_1(\sigma_s) = \sum_i F_{1i}$ it follows that

$$\operatorname{tr}(S(W_1 \times W_2)S^+(\mathbf{1} \times \textstyle\sum_i F_{2i})) = 1. \tag{5.16}$$

This can be the case only if, for all $\varphi \in \mathscr{H}_1$ and all χ_ν the vector $S\varphi\chi_\nu$ is an eigenvector of $(\mathbf{1} \times \sum_i F_{2i})$ with eigenvalue 1. For $S\varphi\chi_\nu = \phi$ and ϕ_i as the components of ϕ in $E_{1i}\mathscr{H}_1 \times \mathscr{H}_{2i}$ from (5.15) it follows that

$$\phi = \sum_i \phi_i = (\mathbf{1} \times \textstyle\sum_i F_{2i})\phi = \displaystyle\sum_i (\mathbf{1} \times F_{2i})\phi_i$$

from which we obtain

$$\phi_i = (\mathbf{1} \times F_{2i})\phi_i.$$

Let $\mathscr{H}_{2i}^{(1)}$ denote the eigenspace of F_{2i} with eigenvalue 1; then from

$$\sum_i F_{2i} = \sum_i F_2(\sigma_i) = F_2(\sigma_s) = \mathbf{1}$$

it follows that $F_{2j}\mathscr{H}_{2i}^{(1)} = 0$ for all $j \neq i$, and we find that $\mathscr{H}_{2i}^{(1)} \subset \mathscr{H}_{2i}$. Therefore the following relationship must be satisfied:

$$S\varphi\chi_\nu \in \sum_i \oplus E_{1i}\mathscr{H}_1 \times \mathscr{H}_{2i}^{(1)} \tag{5.17}$$

for all $\varphi \in \mathscr{H}_1$ and all χ_ν. If $\mathscr{H}_{W_2}$ is the support of W_2 then (5.17) is equivalent to

$$S(\mathscr{H}_1 \times \mathscr{H}_{W_2}) \subset \sum_i \oplus E_{1i}\mathscr{H}_1 \times \mathscr{H}_{2i}^{(1)}. \tag{5.18}$$

Conversely, if (5.18) is satisfied, then it follows that (5.13) and therefore (5.3b) will be satisfied. Certainly we need to guarantee that (5.1) and (5.4) must also be satisfied in addition to (5.5), that is,

$$\begin{aligned} \operatorname{tr}(S(W_1 \times W_2)S^+(\mathbf{1} \times F_{2i})) &= \operatorname{tr}_1(W_1 F_{1i}) \\ &= \operatorname{tr}_1(W_1 E_{1i}) + \operatorname{tr}_1(W_1(\mathbf{1} - \tilde{E})F_{1i}(\mathbf{1} - \tilde{E})) \end{aligned} \tag{5.19}$$

must be satisfied. For $W_1 = P_\varphi$ where $\varphi \in F_{1i}\mathscr{H}_1$ the following condition must also be satisfied:

$$\operatorname{tr}(S(P_\varphi \times W_2))S^+(\mathbf{1} \times F_{2i})) = 1,$$

that is, $(\mathbf{1} \times F_{2i})S\varphi\chi_\nu = S\varphi\chi_\nu$ must be satisfied for all χ_ν. From these considerations we obtain the following condition which is stronger than (5.18):

$$S(E_{1i}\mathscr{H}_1 \times \mathscr{H}_{W_2}) \subset E_{1i}\mathscr{H}_1 \times \mathscr{H}_{2i}^{(1)} \quad \text{for all } i, \tag{5.20}$$

and

$$S((\mathbf{1} - \tilde{E})\mathscr{H}_1 \times \mathscr{H}_{W_2}) \subset \sum_i \oplus E_{1i}\mathscr{H}_1 \times \mathscr{H}_{2i}^{(1)}.$$

From (5.20) it follows that (5.5) holds.

The conditions (5.20) are therefore necessary and sufficient in order to obtain a measurement of the first kind from S, W_2 and $\Sigma_2 \xrightarrow{F_2} L(\mathscr{H}_2)$.

From $S(\mathscr{H}_1 \times \mathscr{H}_2) \subset \mathscr{H}_1 \times (\sum_i \oplus \mathscr{H}_{2i}^{(1)})$ it follows that the only portion of the observable $\Sigma_2 \xrightarrow{F_2} L(\mathscr{H}_2)$ which is essential lies in the subspace $\sum_i \oplus \mathscr{H}_{2i}^{(1)}$. In $\sum_i \oplus \mathscr{H}_{2i}^{(1)}$, however, the observable $\Sigma_2 \xrightarrow{F_2} L(\mathscr{H}_2)$ behaves like a decision observable! We may therefore replace it by the decision observable generated by the projections $E_{2i}^{(1)}$ on $\mathscr{H}_{2i}^{(1)}$! Therefore a measurement of the first kind can only take place if, for all "practical purposes" we measure an atomic decision observable for system 2. Nevertheless $\Sigma_2 \xrightarrow{F_1} L(\mathscr{H}_1)$ need not be a decision observable—it only needs to have the structure described by (5.5).

If, however $\Sigma_2 \xrightarrow{F_1} G(\mathscr{H}_1)$ is a decision observable, that is, the F_{1i} in (5.5) are projection operators, then instead of (5.20) we obtain the following necessary and sufficient condition for a measurement of the first kind:

$$S(F_{1i}\mathscr{H}_1 \times \mathscr{H}_{W_2}) \subset F_{1i}\mathscr{H}_1 \times \mathscr{H}_{2i}^{(1)} \quad \text{for all } i. \tag{5.21}$$

For the case of a decision observable $\Sigma_2 \xrightarrow{F_1} G(\mathscr{H}_1)$ we define an *idealized* measurement of the first kind to be one for which the relations

$$\chi(\sigma_i)W_1 = F_{1i}W_1F_{1i} \tag{5.22}$$

are satisfied for the atoms σ_i of Σ_2 and for all $W_1 \in K(\mathscr{H}_1)$. From (5.22) it follows that (5.3a) holds for all σ.

Here (5.22) is satisfied for all $W \in K_1(\mathscr{H}_1)$ if it is satisfied for all P_φ for which $\varphi \in \mathscr{H}_1$. Since

$$\chi(\sigma_i)P_\varphi = R_1[(\mathbf{1} \times F_{2i}^{1/2})S(P_\varphi \times W_2)S^+(\mathbf{1} \times F_{2i}^{1/2})]$$

it follows from (5.21) that for all $\varphi \in F_{1i}\mathscr{H}_1$

$$(\mathbf{1} \times F_{2i}^{1/2})S(P_\varphi \times W_2)S^+(\mathbf{1} \times F_{2i}^{1/2}) = S(P_\varphi \times W_2)S^+.$$

On the other side, $F_{1i}P_\varphi F_{1i} = P_\varphi$. Therefore, according to (5.22)

$$P_\varphi = R_1 S(P_\varphi \times W_2)S^+$$

for which we conclude that $S(P_\varphi \times W_2)S^+ = P_\varphi \times \tilde{W}_2$ for a suitable $\tilde{W}_2$.

Together with (5.21) it follows that S behaves as an operator of the form

$$S = \sum_i F_{1i} \times A_{2i} \tag{5.23}$$

in the subspace $\mathscr{H}_1 \times \mathscr{H}_{W_2}$ of $\mathscr{H}_1 \times \mathscr{H}_2$ where the A_{2i} are isometric maps $\mathscr{H}_{W_2} \xrightarrow{A_{2i}} \mathscr{H}_{2i}^{(1)}$.

The ideal measurements of the first kind are a mathematical example in which the mappings $\mathbf{T}(1, 2; W_2)$ and $\mathbf{S}(1, 2; W_2)$ can be expressed in a simple form. If (5.23) holds then $\mathbf{T}(1, 2; W_2)$ is given by

$$F_1 = \mathbf{T}(1, 2; W_2)F_2 = \sum_i F_{1i}\,\mathrm{tr}_2(W_2 A_{2i}^+ F_2 A_{2i})$$

and $\mathbf{S}(1, 2; W_2, F_2)$ is given by

$$\mathbf{S}(1, 2; W_2, F_2)W_1 = \sum_i F_{1i}W_1F_{1i}\,\mathrm{tr}_2(W_2 A_{2i}^+ F_2 A_{2i}).$$

The example defined by (2.3.3) and used in §§2.3, 3.2 and 4.2, for the case in which $\chi_\nu = \varphi_\nu$ and the η_ν are pairwise orthogonal is a special case of (5.23), namely

$$S = \sum_\nu P_{\varphi_\nu} \times A_\nu \quad \text{where} \quad A_\nu\psi = \eta_\nu.$$

The proof that the requirements for a measurement of the first kind and for ideal measurements of the first kind do not contradict the fact that the scattering operator is unitary does not mean that such scattering processes are physically possible because the scattering operator cannot be arbitrarily chosen but is a consequence of the interaction Hamiltonian operators. We have not shown that the requirements for measurements of the first kind or for idealized measurements of the first kind do not contradict the interaction description described in VIII. In this connection a series of research papers [27] are of interest because they show that conservation laws (such as the energy conservation law) already frustrate the possibility to satisfy all the conditions for measurements of the first kind, that is, contradict the requirements for measurements of the first kind (except for measurements of the conserved quantities). Despite this situation, in certain limiting cases these conditions can be "approximately" satisfied.

Although there are no measurements of the first kind for sufficiently many observables, this situation does not invalidate the possibility towards making measurements at all and is far from invalidating quantum mechanics although that impression can be obtained from earlier works which claim that the possibility of measurement of the first kind is a crucial issue for quantum mechanics. The special consistency of the use of scattering for the purpose of registration and preparation in quantum mechanics is assured, that is, does not lead to contradictions. The complete consistency of the viewpoint presented in II, III with a theory of the measurement process itself is only provable if we use a theory of macroscopic systems, a subject we shall return to in XVIII, and will be described in detail in [7].

6 The Physical Importance of Scattering Processes Used for Registration and Preparation

In this book the special consistency problem presented in §§2–4 is of particular importance. In fact, we have described physical processes which also play an important role in experiments with microsystems, a topic which is largely neglected in most books on quantum mechanics. The reason for this neglect arises from the fact that most theoretical physicists consider themselves only responsible for providing a theoretical description of the "objective structure of microsystems", and that the problem of experimental physicists is to make measurements of the theoretically predicted results. Thus, applications of the theory described in §§2–4 are, for the most part, only found in books about experimental physics, although they will not generally be expressed in terms of the abstract formulation presented in §§2–4. Actually, it is possible to fill many volumes with such applications.

Since the description of the measurement process in quantum mechanics in terms of "pre-theories" is not possible, the "usual" neglect of a quantum mechanical description of the various experimental measurement procedures (or at least of portions of them) cannot be justified. Unfortunately, because of size limitations, we will not attempt to provide examples of the approximate computation of measurement scattering morphisms, measurement transformation morphisms or transpreparators in this book. In order to illustrate the importance of the theory described in §§2–4 for experiments we shall provide a few simple practical applications.

6.1 Sequences of Measurement Scatterings and Measurement Transformations

The gedanken experiment described in §2.1 shows that several of the processes described in §§2–4 can be carried out consecutively. In the example given in §2.1 first there was a measurement scattering of electrons by light quanta, then there was a measurement transformation of light quanta by the lens, and finally the registration by the photographic plate.

Symbolically it is possible to represent such a sequence by means of Figure 46. In this diagram 1 represents the microsystems which are to be registered. First, these microsystems are scattered by systems 2. Then the systems of type 2 encounter systems of type 3 (which may represent an external field) for the purpose of undergoing a measurement transformation. The systems of type 2 are then registered according to the procedure b_0.

In quantum mechanics this process can be represented in Hilbert space by the scattering operator

$$S = (\mathbf{1}_1 \times S_{23})(S_{12} \times \mathbf{1}_3) \tag{6.1.1}$$

(at least with respect to the ensembles used for systems 2 and 3) where S_{12} is the scattering operator in $\mathscr{H}_1 \times \mathscr{H}_2$ for the scattering of systems 1 on 2 and S_{23} is the scattering operator in $\mathscr{H}_2 \times \mathscr{H}_3$ for systems (2, 3). From

$$\mathrm{tr}_2(W'_2F'_2) = \mathrm{tr}_{23}(S_{23}(W'_2 \times W_3)S_{23}^{+}(F_2 \times \mathbf{1}_3))$$

we obtain the measurement transformation morphism

$$\mathrm{T}(2, 2; W_3)F_2 = F'_2.$$

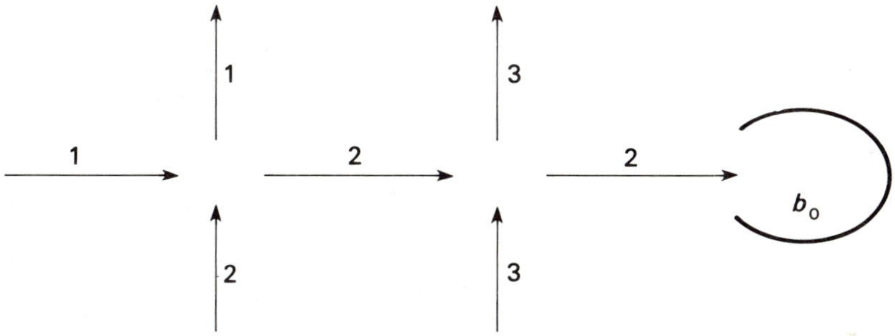

Figure 46

From

$$\mathrm{tr}_1(W_1F_1) = \mathrm{tr}_{12}(S_{12}(W_1 \times W_2)S_{12}^+(\mathbf{1} \times F_2))$$

we obtain the measurement scattering morphism

$$\mathbf{T}(1, 2; W_2)F_2 = F_1.$$

Thus, from the scattering of 1 by the system pairs (2, 3), using the operator (6.1.1), from

$$\mathrm{tr}_1(W_1F_1) = \mathrm{tr}_{123}(S(W_1 \times W_2 \times W_3)S^+(\mathbf{1}_1 \times F_2 \times \mathbf{1}_3))$$

we obtain the measurement scattering morphism

$$\mathbf{T}(1, 2, 3; W_2, W_3)F_2 = F_1.$$

From

$$S(W_1 \times W_2 \times W_3)S^+ = (\mathbf{1}_1 \times S_{23})[S_{12}(W_1 \times W_2)S_{12}^+ \times W_3](\mathbf{1}_1 \times S_{23}^+)$$

we obtain the following relationships in a stepwise manner:

$$\begin{aligned}
&\mathrm{tr}_{123}(S(W_1 \times W_2 \times W_3)S^+(\mathbf{1}_1 \times F_2 \times \mathbf{1}_3))\\
&\quad = \mathrm{tr}_{123}((\mathbf{1}_1 \times S_{23}[S_{12}(W_1 \times W_2)S_{12}^+ \times W_3](\mathbf{1}_1 \times S_{23}^+)(\mathbf{1}_1 \times F_2 \times \mathbf{1}_3))\\
&\quad = \mathrm{tr}_{12}(S_{12}(W_1 \times W_2)S_{12}^+[\mathbf{T}(2, 2; W_3)F_2])\\
&\quad = \mathrm{tr}_1(W_1[\mathbf{T}(1, 2; W_2)\mathbf{T}(2, 2; W_3)F_2]).
\end{aligned}$$

Therefore, in this case we obtain

$$\mathbf{T}(1, 2, 3; W_2, W_3) = \mathbf{T}(1, 2; W_2)\mathbf{T}(2, 2; W_3). \tag{6.1.2}$$

There is no difficulty extending the above results to sequences of arbitrary length, as described by Figure 47. Here we obtain

$$\begin{aligned}
&\mathbf{T}(1, 2, \ldots, n; W_2, \ldots, W_n)\\
&\quad = \mathbf{T}(1, \cdot; W_2)\mathbf{T}(\cdot, \cdot; W_3) \cdots \mathbf{T}(\cdot, \cdot; W_\nu)\mathbf{T}(\cdot, \cdot; W_{\nu+1}) \cdots \mathbf{T}(\cdot, \cdot; W_n).
\end{aligned} \tag{6.1.3}$$

This result has played an important role in the historical development of quantum mechanics. It is possible to separate the chain in Figure 47 anywhere (i.e. between the νth and $(\nu + 1)$th) and consider the remainder as the measurement. Of course the fact of the possibility of separation was not

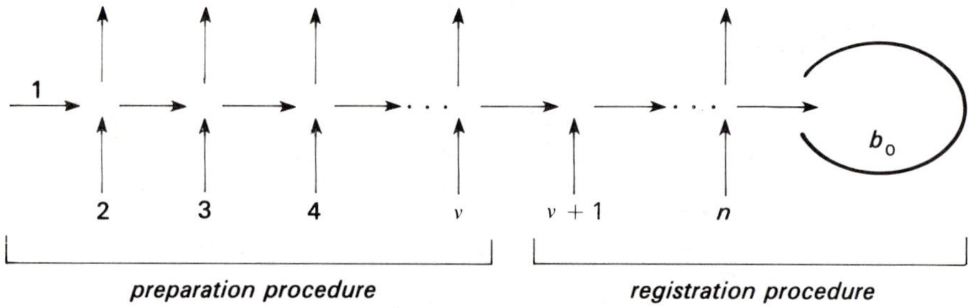

Figure 47

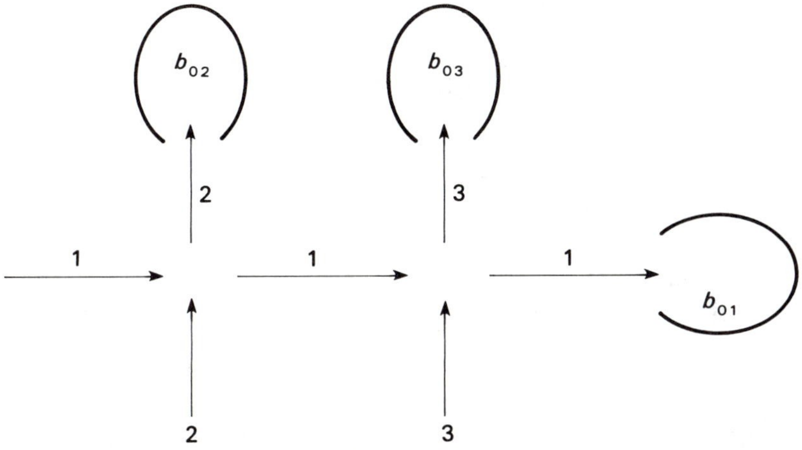

Figure 48

historically proven by equation (6.1.3). From this equation it is trivial to prove that it is possible to divide such a chain arbitrarily into preparation and registration (see Figure 47) and remain consistent with quantum mechanics. Here we do not intend to dwell on the "special consistency" in the remainder of this section. Here we only intend to symbolically represent the experience of experimental physicists by means of pictures of the form of Figure 47.

In addition to the registration apparatus at the end of the chain, other registration apparatuses may be used at other points in the chain, as illustrated by Figure 48. The three apparatuses b_{01}, b_{02}, b_{03} in Figure 48 can be considered to be a registration method b_0 with respect to the systems of type 1 entering from the left. If we identify these three methods by three observables $\Sigma_1 \xrightarrow{\tilde{F}_1} L(\mathscr{H}_1)$, $\Sigma_2 \xrightarrow{\tilde{F}_2} L(\mathscr{H}_2)$, $\Sigma_3 \xrightarrow{\tilde{F}} L(\mathscr{H}_3)$, we then obtain an observable $\Sigma \xrightarrow{F_1} L(\mathscr{H}_1)$ as follows:

$$\begin{aligned} \operatorname{tr}_1(W_1 F_1(\sigma_1 \wedge \sigma_2 \wedge \sigma_3)) = \operatorname{tr}_{123}(S_{13}[S_{12}(W_1 \times W_2)S_{12}^+ \times W_3] \\ \cdot S_{13}^+(\tilde{F}(\sigma_1) \times \tilde{F}_2(\sigma_2) \times \tilde{F}_3(\sigma_3))) \end{aligned}$$

which is defined on the free Boolean ring Σ generated by $\Sigma_1, \Sigma_2, \Sigma_3$ (see AI, §3).

These examples suffice to show the multiplicity of such sequences for the construction of registration methods.

6.2 Physical Importance of Measurement Scattering and Measurement Transformations

We shall now exhibit examples of measurement scattering and measurement transformations. These examples only serve to illustrate the meaning of these well-known experimental procedures under this viewpoint. Furthermore, it is possible to construct more complicated apparatuses from individual known experimental procedures which can be used for experiments.

The use of "external electromagnetic fields" for the purpose of the measurement transformation is extensive. An example is the use of electric and magnetic fields for the deflection of a beam of charged particles for the "determination of momentum" and the measurement of the quantity e/m. We will now illustrate the latter experiment.

We will use the form of the Hamiltonian operator given by VIII (6.31). We will assume that we may produce ensembles which do not have broad position and momentum dispersion and that we are dealing with fields which are not varying rapidly in order that the time dependence of the expectation values of position satisfy Newton's equations of motion for a mass-point:

$$m\ddot{\vec{r}} = e(E + \dot{\vec{r}} \times \vec{B}). \tag{6.2.1}$$

Derivations of this equation from quantum mechanics can be found in many of the quantum mechanics textbooks; see, for example, [6], XI, §1.2.

After the interaction with the field has taken place, it is possible to measure the effect associated with microsystems striking the surface F (see XVI, §6.1). From (6.2.1) it follows that the quantity e/m can be determined from these measurements. A precise description of such an apparatus cannot be given here.

In other well-known apparatuses, fields are used for the determination of the energy. Other examples of the application of fields are the electromagnetic lenses used in electron microscopes. We also draw attention to the Stern–Gerlach experiment in which an inhomogeneous magnetic field is used in order to transform the effect of striking a surface into a measurement of spin. Although this example is relatively simple, its theoretical description in terms of quantum mechanics is already complicated and can only be carried out in an approximate manner (see, for example, [6], XI, §7.2). A comparison of such a theoretical description of the Stern–Gerlach experiment with the description given by the experimentalists (which is also theoretical) will be very informative. It will show the skills required by the experimentalist to construct with minimum possible work a usable (but not better than necessary) approximate theory for the description of the processes in his measurement apparatus. The bending of the beam in the external magnetic field in the Stern–Gerlach experiment is described in a form which is simplified as much as possible in order to exhibit the "connection" of the splitting of the beam with the spin and magnetic moment of the atoms. This "connection" is nothing other than what we have called a measurement transformation morphism.

Scattering by other microsystems will frequently also be used (even in cases in which the complications of the theoretical description are extensive) for the purpose of measurement transformations, for example, in order to measure the "polarization" (spin direction).

It is impossible to enumerate all examples in which measurement scattering is used. We shall only mention a few examples: Scattering of atoms or molecules (systems 1) by electrons (systems 2), and the subsequent registration of the electrons which are interpreted as measurements of the atoms or molecules. An example of historical importance is the Frank–Hertz experiment in which the energy levels of atoms are measured. Here, in addition to the scattering of the electrons, a deceleration field is used as a measurement transformation. Scattering of neutrons (systems 1) by atomic nuclei (systems 2) are used to register neutrons with the aid of effects associated with the atomic nuclei.

Measurement scattering and, in particular, measurement transformations are often used in subtle ways in order to increase the accuracy of experiments, even though the final registration cannot be precisely described. The example given in §2.1 of the microscope is a good illustration. The fact that the process of the sensitization of the silver bromide crystals in a photographic plate is not well understood and the imprecision in the location of the silver grains is unimportant because the theoretically derived resolution of the microscope yields the essential imprecision if the magnification is sufficiently great. Then we may replace the effect of the blackening of the silver bromide crystals by the idealized effect of the incidence of a photon on a surface element $\mathscr{F}$ of the plate.

Despite all the "tricks" in the application of measurement scattering and measurement transformations, the following problem remains for the experimental physicist: the fact that microsystems must finally be registered by a macroscopic registration apparatus, that is, by actual processes on the macroscopic system. The technological possibilities of such macroscopic registration apparatuses are very limited. By the use of measurement scattering processes and measurement transformations it is possible not only to improve the precision of measurements but also to essentially increase the set of experimentally usable registration procedures.

This increase in the possibilities for measurements is not only important from the experimental point of view but it is also indicative of a theoretically important problem, which we cannot ignore in this book because of its relevance for our description of quantum mechanics.

One of the unpleasant aspects of the theory presented in XVI–XVII is that, for certain effect procedures the corresponding effect operators must be "guessed" and then asserted as axioms. Since quantum mechanics does not permit direct measurements (that is, measurements which may be described in terms of pre-theories alone—see I and [8], §10), there was no other way from II to XVI.

There are, however, two decision observables, which are obtained in a purely theoretical way from the physical interpretation of Galileo transformations for elementary systems: Position and Momentum. Clearly the structure of the corresponding apparatuses is not theoretically determined (see IV, §4). It is, however possible to test whether an apparatus is built in such a manner as to satisfy the claims for measurements of position or momentum. In this sense the position and momentum observables of elementary systems can be considered to be observables within the framework of the theory developed in II–X.

As we have seen above we may obtain new registration procedures by the concatenation of measurement scattering processes and measurement transformations. Such new registration procedures may be totally described only by quantum theory (using the interaction Hamiltonians in VIII) if the final registration in the chain is a registration of the position or momentum of an elementary system (or of an impact of such a system on a surface $\mathscr{F}$ described by XVI (6.1.13)). What is the subset $L_r(\mathscr{H}) \subset L(\mathscr{H})$ which can be theoretically achieved by such chains? In terms of the theory there is no need for additional axioms about concrete effect procedures and the corresponding effects $F \in L(\mathscr{H})$, providing that one is satisfied with the effects of this subset $L_r(\mathscr{H})$. We could be more completely satisfied if we could show that the subset $L_r(\mathscr{H})$ so obtained is sufficient to approximate every physical effect of $L(\mathscr{H})$. We are, however, far from a solution of this problem.

6.3 Chains of Transpreparations

The total sequence illustrated in Figure 48 can also be considered to be the preparation procedure for the final ensemble entering into the apparatus b_0. If we use the observables $\Sigma_2 \xrightarrow{\tilde{F}_2} L(\mathscr{H}_2)$, $\Sigma_3 \xrightarrow{\tilde{F}_3} L(\mathscr{H}_3)$ for the description of the registration methods b_{02}, b_{03}, then the sequence illustrated in Figure 48 can be considered to represent a transpreparator $\Sigma_{23} \xrightarrow{\chi} \Pi$, where Σ_{23} is the Boolean ring freely generated by Σ_2, Σ_3 and χ is defined by

$$\chi(\sigma_2 \wedge \sigma_3) = \mathbf{S}(1, 3; W_3, \tilde{F}_3(\sigma_3))\mathbf{S}(1, 2; W_2, \tilde{F}_2(\sigma_2)). \tag{6.3.1}$$

Here it is again trivial to write transpreparators for arbitrary long sequences. In addition, it is possible to pass over from systems of type 1 to systems of another type by scattering, and consider the entire chain to be a preparation process by which systems are obtained from the final link in the chain. In addition, individual elements of the sequence may consist of the use of external fields. In the latter case, the transpreparator takes on the following form:

$$\mathbf{S}W = SWS^+,$$

where S is the scattering operator for the external field.

According to quantum mechanics it is, in principle, possible to compute the transpreparator for complicated sequences. In order, however, to obtain a preparator, it is necessary that the beginning of such a sequence be a macroscopic apparatus. If such is characterized by a preparator $\Sigma_1 \xrightarrow{\eta_1} \breve{K}(\mathscr{H}_1)$, then with the aid of the transpreparator χ defined by (6.3.1) we obtain a preparator $\Sigma \xrightarrow{\eta} \breve{K}(\mathscr{H}_1)$ where Σ is the Boolean ring freely generated by $\Sigma_1, \Sigma_2, \Sigma_3$ and η is given by

$$\eta(\sigma_1 \wedge \sigma_2 \wedge \sigma_3) = \chi(\sigma_2 \wedge \sigma_3)\eta_1(\sigma_1). \tag{6.3.2}$$

If we have a longer sequence, then we can cut the sequence at an arbitrary place, and interpret the "left-hand part" as the "starting preparator"

$$\Sigma_1 \xrightarrow{\eta_1} \breve{K}(\mathscr{H}_1).$$

6.4 The Importance of Transpreparators for the Preparation Process

It is possible to fill volumes with the description of preparation apparatuses which are used for transpreparators and also sequences of transpreparators. Again we must be satisfied by mentioning a number of applications demonstrating the importance of transpreparators for experimental physics.

The great accelerators are probably the best-known examples of preparation apparatuses in which external fields, often in subtle form are used for transpreparations. In other apparatuses scattering of microsystems is often used in order to generate other system types, for example, certain ions, neutrons and short-lived radioactive nuclei.

In §6.2 we have noted that experimental physicists have a knack in choosing suitable measurement scattering processes and measurement transformation processes such that the imprecise theoretical understanding concerning the last registration apparatus will be meaningless. Similarly experimental physicists suceed with the aid of transpreparations, to obtain a theoretically well-defined ensemble even though at the beginning of the sequence we began with a poorly defined ensemble.

Transpreparators such as measurement scattering and measurement transformation morphisms are essential items for the verification of quantum mechanics since in practice we have very few macroscopic apparatuses which either product microsystems or finally register them.

Is it possible (by analogy with the introduction of position and momentum observables) to theoretically define certain initial ensembles from the theory developed in I–X (without the need for the inelegant axioms which were defined, for example, in XI, §2) in order that we obtain the largest possible subset of K by use of the transpreparators? Such an implicit definition of ensembles (such as the implicit definition of position and momentum observables) is not currently available. If we would have a sufficient large subset $L_{ex} \subset L$ of effects such that L_{ex} separates the ensembles (that is, $\mu(w_1, g) = \mu(w_2, g)$ for all $g \in L_{ex}$ implies that $w_1 = w_2$), then it would be possible to test the produced ensembles with the aid of the effects in L_{ex}.

From this viewpoint it is understandable that in most presentations of quantum mechanics the primary emphasis is not in the preparation problem but in the measurement problem because in making measurements it is always possible to test ensembles. This testing can, in many cases, be very simple. Indeed, in many cases it suffices to test an ensemble using a single effect F in order to determine W.

It is clear that mixtures of ensembles are not as well suited for the purpose of experiments as the components of the mixtures (see III, §2). For this reason experimental physicists generally do not use direct mixtures (III, D 2.3) of preparation procedures.

The most desirable ensembles from an experimental physicist's point of view will be those which correspond to the extreme points of the set K (see III, D 6.1 and III, Th. 6.5) and are called "irreducible ensembles" or "pure states". Here we shall avoid the expression "pure state" because this notion is often misunderstood—it gives the impression that all systems of a pure state are "identical". (This notion leads to peculiar difficulties and to problems in logic.) Elements x of a preparation procedure a for which $\varphi(a)$ is an irreducible ensemble can be distinguished by the effects produced by them, that is, by the registration procedures b of which they are elements. Microsystems are not physical objects (see IV, §8) and therefore cannot be separated into "objective identical" categories.

Experimentally, the extreme points of K have the desirable property that they can be tested by means of a single effect. Suppose that we have constructed an effect procedure (b_0, b) for which $\psi(b_0, b) = P_\varphi$ (see III, Th. 6.5), then we may easily test whether a $W \in K$ is equal to P_φ because from $\operatorname{tr}(WP_\varphi) = 1$ it follows that $W = P_\varphi$. In such cases we do not need many effects in order to test W. Therefore, if we can control the registration, we may then easily test whether we have prepared an irreducible ensemble. How then may we prepare such irreducible ensembles?

Clearly with the aid of scattering processes and registration procedures, that is, with well-chosen transpreparators! If we control the registration, then we may seek a scattering process and a particular observable $\Sigma_2 \overset{F_2}{\rightarrow} L(\mathscr{H}_2)$ such that, for certain elements σ_i of Σ_2, the ensembles W_{1i} given by

$$W_{1i} = \mathbf{S}(1, 2; W_2, F_2(\sigma_i))W_1 \tag{6.4.1}$$

are irreducible ensembles. Of course, such a selection will unfortunately not only depend upon the choice of the observable $\Sigma_2 \overset{F_2}{\rightarrow} L(\mathscr{H}_2)$ and the scattering process but also upon the ensembles W_1 and W_2. The advantage of such an application of (6.4.1) can be the fact that the methods for the production of ensembles W_1, W_2 are already known, but that there is no method for the production of the desired irreducible ensemble W_{1i}. For example, it is easy to produce atoms (as systems 1) in the ground state. By scattering with electrons (system 2) it is possible to excite the atoms. By choosing W_2 to be an ensemble which does not have much dispersion in the energy (for example, by accelerating the electrons in an external field), then by measuring the energy of the electrons after the scattering (that is, the observable $\Sigma_2 \overset{F_2}{\rightarrow} L(\mathscr{H}_2)$) we may "determine" into which state the atom is excited, that is, we can select an ensemble W_{1i} of atoms in a particular excited state.

We have presented this example in order to make it clear that the transpreparations are not generally associated with registrations in the sense of measurements of the first kind (§5).

6.5 Unstable States

Unstable states are frequently found in the realm of atomic dimensions. The most obvious example is that of radioactive atomic nuclei. The "excited states" of atoms and molecules described in XI–XV are also unstable states, which undergo transitions by the emission of photons.

There is an extensive literature devoted to unstable states [28]. Since somewhat complex interactions are required in order to investigate unstable states, it is understandable that the study of unstable states has lead to a great variety of approximation methods. We shall not attempt to describe these methods in this volume.

An attentive reader of the literature will, despite the matters of detail, readily become aware of the fact that there are deep conceptual uncertainties associated with the quantum mechanical treatment of unstable states. Indeed, the concept of an "unstable state" has not been precisely formulated. This is precisely the point where we must use the theory presented earlier in order to place this problem in correct perspective.

At first we ask whether "unstable state" refers to the notion of state defined in III, D 1.1. In this volume we have often used the expression "ensemble" instead of "state". Since this problem is indeed concerned with "certain" ensembles, the problem of "unstable ensembles" is a *preparation problem.*

Many of the attempts to solve this problem result from the widespread notion that the observables and states are attributed to the miscosystems, avoiding the fact that the problem of unstable states must be a preparation problem, that is, a problem of producing ensembles with the aid of preparation procedures. Often physicists are unaware that the production of such ensembles is intimately tied up with the preparation process and cannot be separated from the latter. The usual approach is to seek a formulation of quantum mechanics, the concepts of which should be defined only in terms of self-existing properties of microsystems—properties which are detached from all apparatuses. As we have already seen, such a formulation of quantum mechanics is unrealistic. Any attempts in this direction must ultimately fail. In this book we have emphasized the fact that it does not make sense to define observables without taking into account the apparatus. Despite this we have often "guessed" the effect operators without knowing the detailed structure of the apparatus but using only a rough knowledge of the mode of operation of the apparatus. In this way we have, for example, guessed an *approximate* operator for the effect of the impact of a system on a surface (see XVI, §6).

If, after considering the examples from quantum mechanics described in XI–XVI, we remain unaware of the fact that effects such as ensembles for a given procedure are "guessed", then we may easily be deluded into believing that these "guessed" quantities are objective quantities (that is, independent of the procedures used) and that the experimental physicist has to find apparatuses which measure these quantities. If, on the other hand, we are aware, that it is impossible to separate the concepts of an observable and a state from the procedures used, then we will not be under the illusion that it is possible to introduce the notion of an "unstable state" independent of the preparation procedures.

If we consider the experimental situations in which unstable states occur then we discover how easy it is to produce such ensembles. In contrast to this, it appears to be theoretically difficult to specify preparation procedures for unstable states and their corresponding operators $W \in K$. One frequently used possibility for the preparation of unstable states is by "irradiating" microscopic systems, that is, by using scattering processes.

Scattering processes have already been extensively discussed in XVI. Should we have already treated unstable states there? Yes and no. Yes, in the sense that we could have already considered the more complicated structures associated with scattering operators than we have already done. No, because it is not as simple as in the case of the introduction of the "effect of microsystems striking a surface" to discover the operators which accurately describe the experimental process of the transpreparation for unstable states. Intuitively the procedure of irradiation can be roughly described as follows: After the irradiation, some of the miscrosystems are left in unstable states. In more detail, after the scattering process we are left with an ensemble of microsystems, which is generally a mixture which can be decomposed into components, one of which is that of an unstable state. The latter is made evident from the fact that, at a later time, microsystems are emitted, usually obeying an experimental decay law: The number of nondecayed systems is given by $N(t) = N(0)e^{-t/\tau}$; the quantity τ is called the mean lifetime.

In order to provide the principal aspects of this problem we shall consider a simple example, For system 1, which we will "irradiate" with systems of type 2, we shall consider a composite system consisting of two elementary systems, one of which is so massive that we may describe it by means of an external potential field $V(r)$. We shall assume that $V(r)$ has the form as given in Figure 49. We may then describe system 1 as an elementary system moving in the potential field $V(r)$. Systems of type 2 which are used to irradiate may

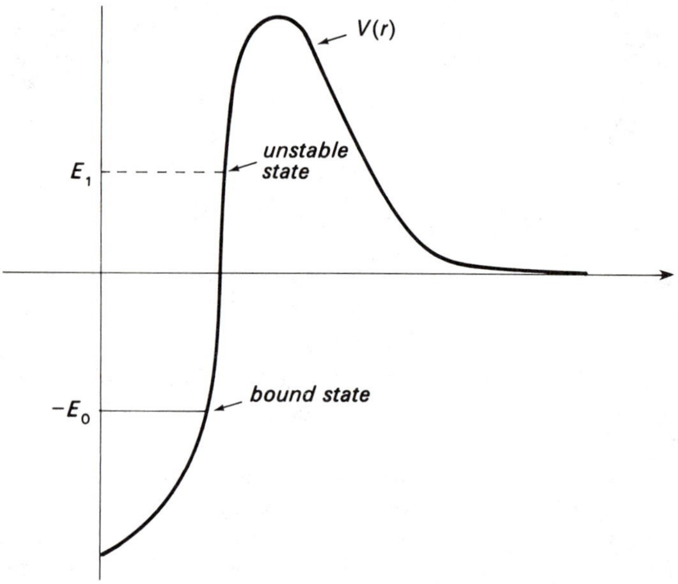

Figure 49

have a repulsive interaction $W(r_{12})$ with system 1, and will have no interaction with the "heavy" system which produces the field $V(r)$. The Hamiltonian will therefore be given by:

$$H = \frac{1}{2m_1}\vec{P}_1^2 + \frac{1}{2m_2}\vec{P}_2^2 + V(r_1) + W(r_{12}) \tag{6.5.1}$$

We shall assume that the operator $H_{10} = (1/2m_1)\vec{P}_1^2 + V(r_1)$ has a single discrete nondegenerate eigenvalue $-E_0$ with corresponding eigenvector $\varphi_0(\vec{r}_1)$ which we shall call the "bound ground state". We shall assume that the initial ensemble has the form $P_{\varphi_0} \times W_2$. Therefore in the Heisenberg picture we have to set $W = \Omega_-(P_{\varphi_0} \times W_2)\Omega_-^+$ (see VI, §5). We assume that the dispersion in energy of W is much smaller than E_0, and that the mean energy is greater than $V_m + E_0$ where V_m is the maximum of $V(r)$. Then

$$W = \Omega_-(P_{\varphi_0} \times W_2)\Omega_-^+$$

describes the "elastic" scattering of system 2 as well as the inelastic scattering. In elastic scattering system 1 remains in the ground state; in an inelastic scattering system 1 will be expelled and can be detected far from the scattering system by a detection surface.

In an inelastic scattering it is possible that a "nearly bound state" of energy E_1 (see Figure 49) exists in which system 1 may be found long after the scattering. This is an example of an unstable state. E_1 would be a bound state if the potential $V(r)$ does not sink below E_1 for large r. Here we speak of the "tunnel effect" in which system 1 "tunnels" through the "mountain" of $V(r)$. A classical mass-point of energy E_1 cannot climb over the mountain, that is, is "imprisoned" by the potential, even in the case in which the potential drops below E_1 for large r.

How is it possible to prepare an ensemble corresponding to the "unstable state" having energy E_1 intuitively described above?

We can detect the systems 1 and 2 after scattering by the effect of striking a spherical surface of radius R. Let T denote the time before which all the systems of type 2 have struck the surface. The experiment shows that systems of type 1 strike the surface after T and that they have nearly the energy E_1, and that the number of system 1 striking the sphere per unit time decreases with a time dependence $e^{-t/\tau}$. In this experiment R must not be too large. The following condition must be satisfied:

$$R \ll \tau \frac{k_1}{m_1}, \tag{6.5.2}$$

where k_1 is defined by $E_1 = k_1^2/2m_1$, that is, k_1/m_1 is the mean velocity of the systems 1.

Experimentally the "unstable state" of systems 1 can be singled as follows: All systems of type 1 are selected where the systems of type 2 have lost more energy than E_0 and have struck the sphere before time T, but none of the systems of type 1 have struck the sphere. This experimental situation characterizes a preparation procedure a for which $\varphi(a) \in K(\mathscr{H}_1)$ determines the so-called "unstable state". How do we determine the operator $\varphi(a)$?

First, on a purely quantum mechanical basis, from the Hamiltonian operator H (6.5.1) it is possible to compute the operation $\mathbf{T}_c$ which maps $P_{\varphi_0} \times W_2$ onto a $W \in \breve{K}(\mathscr{H}_1 \times \mathscr{H}_2)$, which, for large distances r_2 is identical to the outgoing waves for the inelastic scattering.

Using the expressions in XVI, §6.2 for the effect of the systems striking a surface, we may calculate the probability that systems 1 will strike the spherical surface in the time interval t_1 to t_2. For $t > T$ this probability should approximately have the time dependence $e^{-t/\tau}$. $\mathbf{T}_c$ does not, however, determine $\tilde{W} = \varphi(a)$ with the above defined a. We need an additional operation $\tilde{\mathbf{T}}$ which maps $\mathbf{T}_c(P_{\varphi_0} \times W_2)$ onto a $\tilde{W} \in \breve{K}(\mathscr{H}_1)$ and corresponds to selecting the cases in which no system 1 strikes the surface up to time T. The knowledge of the "guessed" operator for the effect of striking of system 1 up to time T is not sufficient in order to determine $\tilde{\mathbf{T}}$. In order to compute $\tilde{\mathbf{T}}$ we need a theory of the interaction of the microsystem with the material of the real sphere, that is, with a macroscopic system. Therefore it is very understandable that we avoid this physical problem by making some idealized postulates about $\tilde{\mathbf{T}}$. We have to ask whether such postulates are realistic. We shall later show how we may, by guessing, obtain completely unrealistic results.

As we have seen above, the problem of obtaining a preparation procedure for "unstable states" clearly shows the following: First, that it is not possible to define unstable states by means of a limiting process as we have done for the case of scattering, because, in the limit $t \to \infty$ nothing remains of the unstable states. Second, there is no unique definition because the unstable states prepared depend precisely on the preparation procedure a that was used. The physical distinction between such ensembles produced by different a's is so minimal, that any definition method for unstable states is admissible (for example, on poles in the complex plane of the scattering amplitude) if it is practical and we obtain the physical same results as for the different experimental preparation methods a.

In the above example $\varphi(a)$ does not, for all practical purposes, depend upon R and T if (6.5.2) is satisfied and R and T are neither too small or too large.

Providing that we realize that there are only approximate methods for the definition of unstable states, we will be able to avoid (as mentioned above) making errors in the guessing of $\tilde{\mathbf{T}}$. Now we wish to demonstrate an error which results in a misunderstanding of quantum mechanics in that measurements are decisions on propositions rather than about processes which must be described in physical terms. In this misunderstanding it is believed that ideal measurements of the first kind are the only "true" measurements, that is, those which correspond to decisions on propositions. Then it is believed that all measurements in physics must refer back to ideal measurements of the first kind. Then it becomes possible to believe that a measurement that a system which has not passed through the spherical surface in the time interval t to $t + \Delta t$ can be idealized to a decision that if the system is in the sphere at time t it will also be there at time $t + \Delta t$. If $E(t)$ is the projection (in the Heisenberg picture) for the decision effect that the position at time t is within the sphere, then the idea that the particle does not pass through the sphere during the time t_1 to t_2 has to be replaced (with $\Delta t = (t_2 - t_1)/n$) by a segment of a subsequent ideal measurement of the first kind that "the position of the particle at all times $t + m\Delta t\,(m = 0, 1, \ldots, n)$ must lie within the sphere". Then the corresponding operation will, according to §5 be given by

$$W \to E(t_2) \cdots E(t_1 + \Delta t)E(t_1)WE(t_1)E(t_1 + \Delta t) \cdot E(t + 2\Delta t) \cdots E(t_2) \tag{6.5.3}$$

We may prove that, in the limits $\Delta t \to 0$ (that is, $n \to \infty$) from (6.5.3) we obtain an operation T_{t_2, t_1} for which

$$\operatorname{tr}(T_{t_2,t_1} W_{t_1}) = \operatorname{tr}(E(t_1) W_{t_1} E(t_1)) \tag{6.5.4}$$

holds. If the system is in the sphere at time t, then it will be there at all other times—it cannot come out. The probability that the system is in the sphere is 1 at all times, although it must, according to experience, decrease according to $e^{-t/\tau}$. If we set $\tilde{\mathbf{T}} = T_{t_2, t_1}$ for sufficiently small t_1 and sufficiently large t_2, then we will be in contradiction with experience. Apparently we have imposed idealized requirements upon $\tilde{\mathbf{T}}$ which are absolutely unrealistic, since these requirements implicitly presuppose that the spherical surface reflect the systems 1 instead of (in actual experiments) absorbing them and detecting them. Unfortunately, this situation is described as a paradox, although it is not one. The fact that an idealized registration defined by means of imposing requirements is very different from an actual registration of the incidence of a system has nothing to do with a paradox but only shows that it is physically unrealistic to describe all experiments in terms of decision effects and of ideal decision measurements of the first kind.

That (6.5.4) holds can be proven exactly [29]. We shall only make (6.5.4) plausible. From

$$E(t + \Delta t) = E(t) + i\Delta t[HE(t) - E(t)H] + \Delta t^2 \cdots$$

it follows that

$$\begin{aligned} &E(t + \Delta t)E(t)WE(t)E(t + \Delta t) \\ &\quad = E(t)WE(t) + i\Delta t[HE(t)WE(t) - E(t)HE(t)WE(t) \\ &\qquad + E(t)WE(t)HE(t) - E(t)WE(t)H] + \Delta t^2 \cdots. \end{aligned}$$

From which we obtain

$$\operatorname{tr}(E(t + \Delta t)E(t)WE(t)E(t + \Delta t)) = \operatorname{tr}(E(t)WE(t)) + \Delta t^2 \cdots.$$

In the limit $n \to \infty$, for n steps, we obtain (6.5.4).

The fact that for a realistic treatment of the preparation of unstable states it is necessary to introduce an absorbing spherical surface can be experimentally verified as follows: Suppose, for example, we have an excited atom in the vicinity of a dielectric surface (a macroscopic system which does not absorb photons), then the lifetime τ, that is, the transition probability for the emission of a photon is changed. An unstable state is always characterized by an absorbing neighborhood, which, must, of course, be taken into account. This makes it "easy" to produce unstable states.

The reader may compare the above considerations to the classical absorber theory of Wheeler–Feynman [30] and its application to quantum mechanics by Süssmann [31].

7 Complex Preparation and Registration Processes

In previous discussions (§§2–6) we described processes for which there is a scattering operator. For more complicated processes the mathematical description is more difficult. For all collision experiments, however, there exists a wave operator Ω_- which maps the Hilbert space of the incoming channels into the Hilbert space of the interacting total system. In place of (2.2.2) we then have the equation

$$W = \Omega_-(W_1 \times W_2)\Omega_-^+, \tag{7.1}$$

where we must pay attention to the fact that in the formula

$$\operatorname{tr}(WF) = \operatorname{tr}(\Omega_-(W_1 \times W_2)\Omega_-^+ F) \tag{7.2}$$

the time dependence of F must be taken in the Heisenberg picture with the entire Hamiltonian operator (including the interaction) (see XVI, §4.9).

By analogy to the discussion in §2.2, it follows that

$$\operatorname{tr}(\Omega_-(W_1 \times W_2)\Omega_-^+ F) = \operatorname{tr}_1(W_1 F_1) \tag{7.3}$$

defines a B-continuous effect morphism $L(\mathscr{H}) \to L(\mathscr{H}_1)$. Since Ω_- is isometric, it is easy to verify that for $W_2 = P_\psi$, this effect morphism is surjective (choose W_1 to be an arbitrary P_φ and set $F = P_{\Omega_-\varphi\psi}$), that is, no information possibilities are lost. Only in the case in which W_2 is not irreducible is it possible to lose information possibilities.

Scattering of the most complicated type can therefore be used for the purpose of registration. Thus, it is not necessary to introduce any new considerations other than those previously introduced in §§2 and 3.

This short discussion was introduced in order to more easily realize the manifold ways in which scattering processes described in quantum mechanical terms may be used in order to obtain new possibilities for registration of effects by using "known" registration and preparation processes, and to invent new preparation possibilities by the use of "known" preparation and registration possibilities.

All the above proves only the "special consistency" of the measurement problem with quantum mechanics. The "general" consistency problem, that the preparation and registration procedures (as presented in II) are, in general, consistent with quantum mechanics, is much more difficult. This problem will be discussed in XVIII, §3, and solved in [7].

Readers interested in problems described in XVII are referred to the book of K. Kraus [42].

CHAPTER XVIII

Quantum Mechanics, Macrophysics and Physical World Views

In closing this book, we shall now recall the route of the development of quantum mechanics in II–XVII in order to come to some judgment concerning the place of quantum mechanics in physics as a whole and to discuss remaining open problems which cannot be examined in this book. We will consider the question whether it is necessary to consider quantum mechanics as a revolution in the world view of physics. No other theory has given rise to more varied and peculiar notions for the meaning of a physical theory. No other theory has produced as much philosophical discussion.

1 Universality of Quantum Mechanics?

A more or less known decisive preliminary decision concerning the domain of validity of quantum mechanics has played an important role in discussions about the philosophical meaning and the physical consequences of quantum mechanics. It is the preliminary decision that quantum mechanics is, at least in its fundamental principles, universal, that is, is valid for all physics, and that all other nonquantum mechanical theories, such as classical, are only approximation theories relative to quantum mechanics.

The belief in the universality of quantum mechanics has its historical origins. Despite this, it is necessary to examine this idea with a critical and sceptical eye.

In the beginning of quantum mechanics after a tentative investigation towards a comprehensive interpretation of quantum mechanics physicists were lead to the usual statistical interpretation based upon the fundamental concepts such as "state" and "observable." We will not attempt to investigate all the more or less precise ideas which have led to these two concepts. In practice everything functions well if we are sufficiently practiced, that is, we have a certain familiarity of these fundamental concepts.

This "familiarity" often lets us forget that quantum mechanics formulated in this way was not a correct $\mathscr{PT}$ (see I and [1]) because it did not give any indication what observables mean and in what state microsystems are to be found. Did quantum mechanics really make no indication about these? Certainly it did—with the aid of the correspondence principles. For quantum mechanics the correspondence principle was not only a means by which we can guess the route to quantum mechanics but was also a part of the quantum theory itself, a fact which was clear to more perceptive physicists (see, for example, [32]). Indeed, at the beginning of quantum mechanics there was nothing left over but to discover the quantum mechanical observables from the classical observables using the correspondence principle. Here it was unconsciously hoped that the classical mechanical observables, the meaning of which is already known, will also determine the meaning of the corresponding quantum mechanical operators (according to the correspondence principle). Physicists hoped that the unknown measurement methods for the quantum mechanical observables will, in principal, be determined by the known measurement methods for the corresponding classical observables.

Many physicists were so accustomed to this situation that they were often unaware that every new operator obtained from the correspondence principle (for a measurement method for an experiment) is more correctly a new axiom, and hence represents an extension of quantum mechanics to a more comprehensive theory and that the correspondence principle does not give a measurement method for these observables. Only the more critical physicists recognized this gap in quantum mechanics. What is then the general validity of quantum mechanics, that is, the universality of quantum mechanics if we always must rely upon the use of the correspondence principle and unknown measurement methods?

Bound up with the ideal that the concepts of "state" and "observable" refer only to "micro-objects," and that measurements consist of the determination of observable properties of these micro-objects, many physicists have not viewed as a defect the need to guess the operators and their ignorance of the measurement methods. The latter were regarded as a purely experimental problem. The physical meaning of, for example, the "position," the "momentum," the "angular momentum," the "energy" did not seem to be questionable. In "practice" the quantum mechanical description appeared to function, indeed, even in macroscopic physics, witness the great success of quantum mechanics in the explanations of the properties of macroscopic systems, particularly for such curious phenomena such as superconductivity and superfluidity. Are these successes not a proof of the universality of quantum mechanics? Should we not assume that all theories of macroscopic systems are limiting cases for the more exact theory of quantum mechanics?

Although these successes may make the idea of universality of quantum mechanics more compelling, it is important to note that these successes are not a proof for the validity of the fundamental principles of a theory. The great success of classical mechanics had led to the notion that "all" of physics can be reduced to complicated mechanical motion, an idea which, for example, Laplace has expressed in his tract on probability theory (see, for example, [6], VI, §4). And, of course, as we now know, physics cannot be reduced to mechanics.

Since the adherents of the universality of quantum mechanics do not justify their position only on the successes of quantum mechanics, they have presented other arguments. Since the author of this book is not an adherent of the universality viewpoint, he would not be a suitable advocate for the universality position. For this reason, we refer interested readers to the original literature (see the extensive bibliography in [33]).

Even if the adherents of the universality of quantum mechanics avoid the problem of elaborating the bounds (limits) of quantum mechanics, they necessarily introduce a new difficulty, namely, the need to answer the following question: How do we obtain a "valid determination" of measurement results? Here again we find a variety of different approaches, which range from attempts to show that probability theory itself gives the valid determination to the introduction of consciousness of the observer or to the so-called "many worlds" interpretation of quantum mechanics. For such attempts we refer readers to the original literature [33] and [34].

We now recall that the formulation of the foundations of quantum mechanics given here began with the notions of preparation and registration procedures, and makes no claim of universality because quantum mechanical systems are not physical objects in the sense of III, §4.1 (see also IV, §8.1), that is, they cannot be objectively described while the selection procedures themselves require an objective mode for their description! Mathematically this is expressed in the fact that the structure underlying the notion of a selection procedure is very similar to that of a Boolean ring while the structure of species pseudoproperties is only an orthocomplemented (not Boolean) lattice (see III, §4 and IV, §8). Here, are we not placed in the position in which the use of selection procedures as our starting point is made doubtful because the apparatuses used for preparation and registration are built of atoms and therefore must be properly considered to be quantum mechanical systems and are therefore not physical objects? Must we therefore switch over to the universality notion for quantum mechanics which we have just rejected and seek another basis for the foundations of quantum mechanics?

2 Macroscopic Systems

We are of the opinion that the foundations of quantum mechanics presented here yields a true picture of reality. Instead of finding it necessary to introduce fantastic ideas in order to answer the question about the validation of measurement results, we must, on the basis of our requirements about a $\mathscr{PT}$ (see I and [8]), examine the following actual physical question: How is it possible, on the one hand, to objectify the macroscopic systems, that is, how we are to describe them as physical objects and, on the other hand, apply quantum mechanics to macrosystems and be successful in explaining their properties (properties in the sense of II, §4.1, not pseudoproperties!)?

In the usual formulation of quantum mechanics we begin with the fundamental concepts of state and observable and note the existence of noncommensurable observables as part of the fundamental structure. This fundamental structure is determined by the lattice of closed subspaces of a Hilbert space, and is often called a quantum logic. In our formulation of quantum mechanics the derived concepts of state and decision observable are defined in terms of the mathematical description of preparation and registration procedures as presented in II. In the usual formulation the description of experiments, that is, of preparation and registration, can only be done in terms of ordinary language. Therefore it is understandable that many physicists consider the notion of "quantum logics" to be more fundamental than the "objective" logic of preparation and registration. Do we have any experimental evidence that there exist noncoexistent observables for macrosystems? On the contrary, is it not the case that the so-called fundamental structure of quantum mechanics, the existence of noncommensurable decision effects has not been experimentally confirmed for macrosystems? Conversely, is it not the case that one of the typical aspects of macroscopic systems is that only coexistent observables can be measured? What, then, are the great successes for the application of quantum mechanics to macrosystems?

First we may question whether the existence of noncommensurable observables has been experimentally proven for the case of microsystems. This proof has already been given in XVII. The most transparent case, although a somewhat idealized gedanken experiment, has been exhibited in XVII, §§2.1 and 4.3—the experiment with the microscope and the photographic plate in the image or focal plane.

If we seek to realize similar experiments for macrosystems we find that there are insurmountable difficulties in order to reach accuracy sufficiently high to prove the noncommensurability of the observables under consideration. Or, expressed differently, for two such noncommensurable decision observables only those approximation observables (which are not decision observables) are realizable (in the sense of IV, §4) which are all coexistent. Therefore there can be no claim that the "validity of quantum logic" has been experimentally proven for the case of macrosystems.

The formulation of quantum mechanics based upon the structure given in II permits us to see more clearly what is common and what is different between microsystems and macrosystems.

It is not known whether there is any one application of statistics in physics which cannot be carried out with the aid of the structure presented in II (see also [8], §§11, 12). These structures can be applied regardless of whether we are dealing with micro- or macrosystems. The distinction between macro- and microsystems becomes evident only when we exceed the axioms introduced in II. We must therefore ask, in a critical way, whether structure axioms have been introduced in III which are probably not applicable to macrosystems.

Some relevant structures have already been discussed in III. There, for example, the set C, is defined differently for micro- and macrosystems (see III, §1). There Axiom AV 4s (in III, §3) is not suitable for the description of macrosystems. Macrosystems, on the contrary, can be described objectively in a state space Z, and in terms of a time dependent trajectory $z(t)$ in Z. All known descriptions of macrosystems have this form, whether they are describing hydrodynamics, the Boltzmann transport equation or Brownian motion. We do not have space to outline or explain these cases. A presentation of such a description can be found in [7] and in [35]; a simplified mathematical introduction can be found in [6], XV.

We shall now attempt to clarify one aspect of this problem: The circumstances under which it is possible to objectify the mode of description of the macrosystems which are used for the preparation and registration of microsystems. First, we note that the measurable attribute for a macrosystem, that is, its state $z \in Z$ at different times t is defined by pre-theories. For example, for hydrodynamics it is assumed that concepts such as position, time, density, flow velocity, temperature, etc. are already understood, and that we know, on the basis of pre-theories, how to measure these quantities. With respect to microsystems we are not (!) in a position to describe the so-called observables in terms of pre-theories, as we have already discussed in I.

On the other hand, macrosystems are also produced for experiment, that is, they are prepared, and their trajectories are measured, that is, registered. Often such a description of the preparation and registration is ignored, in particular, when the behavior of the macrosystems is dynamically determined, that is, when the trajectories $z(t)$ are determined by their value at a given time, for example, by $z(0)$. If the use of statistics is required, then macrosystems can be described in terms of preparation and registration procedures. We note, however, that it is always possible to describe dynamically determined systems in terms of statistics as a special case of the fundamental structures introduced in II. Herein lies the generality of the formulation of statistics introduced in II. Let $M_m, \mathcal{Q}_m, \mathcal{R}_{0m}, \mathcal{R}_m$ be the sets defined (by analogy with II) for the macrosystems under consideration. How do these sets relate to the state space description?

A state space description means that for every system $x \in M_m$ there is a corresponding trajectory $z(t)$. For macrosystems $z(t)$ is not, in general, defined for all values of t, particularly for arbitrary times $t \to -\infty$ before the macrosystem was produced. It is mathematically convenient to assume that $z(t)$ is defined only for $t > 0$ where $t = 0$ is the time at which the macrosystem was prepared. Let Y denote the space of trajectories $z(t)$, where we frequently impose certain continuity requirements upon the elements $z(t)$ of Y (see [7], [35], [6], XV). A description of macrosystems in state space means that, in the theory of these systems, a map $M_m \xrightarrow{f} Y$ is defined.

A structure $\mathscr{E}$ of objective properties is defined on Y as follows: Let Σ be the Boolean ring of subsets of Y (for example, the set of Borel sets if Y has a uniform structure). A property structure on M_m is defined by

$$\mathscr{E} = \{f^{-1}(\sigma) \mid \sigma \in \Sigma\}$$

which, on the basis of the physical meaning of f is a structure of objective properties, that is, between $\mathscr{E}$, $\mathscr{Q}_m$, $\mathscr{R}_{0m}$, $\mathscr{R}_m$ the relations given in III, D 4.1.2 are satisfied. In this case $\mathscr{E}$ is not only "virtual" (see III, D 4.1.1) because $\mathscr{E}$ is distinguished by the physically interpreted structure $M \xrightarrow{f} Y$ and is therefore a set of real properties.

In [7], [35] it is shown how to describe macroscopic systems more precisely in terms of trajectories in state space, that is, how it is possible to define in a mathematically more precise manner that a $b \in \mathscr{R}_m$ is a registration procedure for trajectories $z(t)$ in the state space. For us it is only important that the $b \in \mathscr{R}_m$ does not register the systems better than the $p \in \mathscr{E}$ and that the $p \in \mathscr{E}$ are, so to speak, idealizations of the procedures $b \in \mathscr{R}_m$.

Therefore M_m, $\mathscr{Q}_m$, $\mathscr{R}_{0m}$, $\mathscr{R}_m$, $\mathscr{E}$ represent an objective description of physical objects, namely, the macrosystem under consideration. What, however, can this objective description of macrosystems have to do with a quantum mechanics of macrosystems? What do we mean by a quantum mechanical description of macrosystems?

By a quantum mechanical description of macrosystems we mean that a macrosystem is considered to be composed (according to VIII, §4) of very many atom nuclei and electrons. When we assert that this quantum mechanics of "many-particle systems" is not a closed theory (not a g.G. closed theory in the sense of [8]), but contains many unrealistic aspects, which must be resolved by an extension of the theory, thus we assert nothing other than that the structure introduced in VIII for composite systems is insufficient to describe reality.

From VIII, §2 it follows that the description of composite systems introduced there requires a certain set of registration possibilities, which just give a physical meaning of the representation of the Hilbert space in the form of product spaces. Do the assumed registration possibilities exist in the case in which the system consists of very many elementary systems? Is it possible to measure, for example, the positions and momenta of all these electrons? If it were not possible then the description of composite systems in the form presented in VIII is only an approximation for the case in which the systems are composed of only a few elementary systems. The fact that this form is questionable for the case of relativistic systems has already been mentioned in VIII. It is important that the reader be aware of the limitations of this structure for the description of the composition of elementary systems because some physicists assume that this structure is universal, many obscure conclusions are drawn from this generality. Both macrosystems and relativistic systems indicate that there are limitations of the structure for composite systems described in VIII, a result which is already evident in the physical basis of this structure given in VIII.

There is another very striking argument that the structure described in VIII cannot be universal. The method of product spaces can be extended to *arbitrary* many systems. But there are not arbitrary many systems in the world. There must be a limit of the validity of this structure.

Nevertheless we may formulate quantum mechanics for "many" particles using VIII, that is, for such systems which can also be described in terms of state spaces Z. If we restrict quantum theory to such systems types of many particles we will denote the theory by $\mathscr{PT}_{q\,\exp}$, where the index exp shall indicate that we are considering the formal extrapolation of quantum mechanics to macrosystems, even though we know that $\mathscr{PT}_{q\,\exp}$ has unrealistic consequences. Although $\mathscr{PT}_{q\,\exp}$ is a restriction of the total quantum mechanics we will, for simplicity, use the same letters M, $\mathscr{Q}$, $\mathscr{R}_0$, $\mathscr{R}$ as in II.

We now have two theories for the same domain of facts, one given by $\mathscr{PT}_m$ with M_m, $\mathscr{Q}_m$, $\mathscr{R}_{0m}$, $\mathscr{R}_m$, Z, Y and $\mathscr{E}$, and the other $\mathscr{PT}_{q\,\exp}$ with M, $\mathscr{Q}$, $\mathscr{R}_0$, $\mathscr{R}$. We conjecture that not all of the elements of $\mathscr{Q}$, $\mathscr{R}_0$, $\mathscr{R}$ are realizable, but are in part only invented preparation and registration procedures, invented in order to maintain the structures presented in III–VIII for $\mathscr{PT}_{q\,\exp}$. Therefore it is reasonable that the relation between $\mathscr{PT}_{q\,\exp}$ and $\mathscr{PT}_m$ can be described in the following way: There exists a bijective map $M_m \xrightarrow{i} M$ for which $i\mathscr{Q}_m \subset \mathscr{Q}$, $i\mathscr{R}_{0m} \subset \mathscr{R}_0$, $i\mathscr{R}_m \subset \mathscr{R}$, and the representation of the Galileo group in $\mathscr{PT}_m$ and its representation in $\mathscr{PT}_{q\,\exp}$ are compatible. We shall only examine the last condition for the physically important time translation.

The meaning of the time translation T_τ as a displacement of the registration relative to a preparation in $\mathscr{PT}_{q\,\exp}$ has been described in VII, §1. Here we only need the fact that $b \to T_\tau b$ is a map of $\mathscr{R}$ into $\mathscr{R}$.

In $\mathscr{PT}_m$ a time translation operator T_τ which maps $\mathscr{R}_m$ into $\mathscr{R}_m$ can be defined using the trajectories $z(t) \in Y$ and has the same physical meaning as T_τ in $\mathscr{PT}_{q\,\exp}$: For this purpose we define for $\tau > 0$

$$\tilde{T}_\tau z(t) = z(t + \tau).$$

$T_\tau b$ is then the registration which responds to $z(t)$ in the same manner as does b to $z'(t) = z(t + \tau)$. In particular, for $p \in \mathscr{E}$:

$$T_\tau p = \{z(t) \,|\, z'(t) = z(t + \tau) \in p\}.$$

For macrosystems T_τ is meaningfully defined only for $\tau > 0$, that is, for the semigroup of positive time translations, since $z(t)$ is defined only for $t > 0$. (Here we suggest that the reader examine the combination problem (see III, §1) for the case of macrosystems for negative time displacements!)

For the injection i (as mappings of $\mathscr{Q}_m$ in $\mathscr{Q}$ and $\mathscr{R}_m$ in $\mathscr{R}$) we require that

$$\begin{aligned} \lambda_m(a \cap b_0, a \cap b) &= \lambda((ia) \cap (ib_0), (ia) \cap (ib)), \\ \lambda((ia) \cap (iT_\tau b_0), (ia) \cap (iT_\tau b)) &= \lambda((ia) \cap (T_\tau ib_0), (ia) \cap (T_\tau ib)), \end{aligned} \tag{2.1}$$

where λ_m is the probability function in $\mathscr{PT}_m$.

The mapping T_τ in $\mathscr{P}\mathscr{T}_{q\,\text{exp}}$ is determined by the operator H given in VIII (5.8). Then, for given i, the behavior of the trajectories $z(t)$ is determined by (2.1) and the Hamiltonian operator. This is precisley the essence of what is called "statistical mechanics".

The above requirements (2.1) for the embedding i of the theory $\mathscr{P}\mathscr{T}_m$ into $\mathscr{P}\mathscr{T}_{q\,\text{exp}}$ represent a research program. Here we may only present a few aspects of this program, for a more complete description, see [7] and [35]. We shall only briefly mention a few important aspects.

According to the meaning of $\mathscr{P}\mathscr{T}_m$ there exists, as defined by the pre-theories, a measurement of the state z "at time" t. In $\mathscr{P}\mathscr{T}_{q\,\text{exp}}$ such a measurement "at time" t is not defined, a fact we have shown at the beginning of VII, and again in XVII, and we shall again consider in §4. Therefore a meaningful connection between $\mathscr{P}\mathscr{T}_m$ and $\mathscr{P}\mathscr{T}_{q\,\text{exp}}$ can be made only with the help of T_τ, and not with the help of the concept of measurements "at time" t.

It follows that (2.1) cannot be exactly satisfied for all $\tau \geq 0$. Physically it is sufficient if (2.1) is satisfied (providing that τ is not too large), to such an approximation that the differences between the left- and right-hand sides of (2.1) cannot be tested because it is impossible to perform sufficiently many experiments.

The existence of such a map i is called compatibility condition of the objective mode of description of macrosystems with $\mathscr{P}\mathscr{T}_{q\,\text{exp}}$. The opinion proposed here is that the subsets $i\mathscr{Q}_m$ and $i\mathscr{R}_m$ describe the "physically possible" preparation and registration procedures.

A general theory which determines the map i has not yet been found. In a few special cases (e.g. the Boltzmann distribution function) the map i can be constructed. In practice physicists handle this generally unsolved problem by "guessing" some of the observables which are in the range of i (some of the so-called macro-observables) with the aid of the correspondence principle, physical intuition and luck.

For some it is unacceptable to abandon quantum mechanics as the most comprehensive theory. Proponents of this viewpoint will often seek, within the context of $\mathscr{P}\mathscr{T}_{q\,\text{exp}}$, to deduce the structures of $\mathscr{P}\mathscr{T}_m$. The first step of such a deduction would be the derivation of the sets $i\mathscr{Q}_m$, $i\mathscr{R}_{0m}$, $i\mathscr{R}_m$ alone with the aid of the structure of Hilbert spaces described in VIII and with the aid of the Hamiltonian operator. Different such attempts have taken place, using the aid of limiting processes in the sense of increasing "particle number." But the structures present in $\mathscr{P}\mathscr{T}_{q\,\text{exp}}$ are too weak for such a deduction. We may, for example, derive electrostatics from electrodynamics, and obtain a theory which is valid for a restricted domain. Here, however, we require more—why are only such preparation and registration procedures from $i\mathscr{Q}_m$ and $i\mathscr{R}_m$ realizable, and not those from all of $\mathscr{Q}$ and $\mathscr{R}$? If we take this physical interpretation of $i\mathscr{Q}_m$ and $i\mathscr{R}_m$ seriously then $\mathscr{P}\mathscr{T}_{q\,\text{exp}}$ is not a closed theory. By replacing $\mathscr{Q}$, $\mathscr{R}_0$, $\mathscr{R}$ by the subsets $i\mathscr{Q}_m$, $i\mathscr{R}_{0m}$, $i\mathscr{R}_m$ then $\mathscr{P}\mathscr{T}_{q\,\text{exp}}$ no longer has the form of an axiomatic basis (see I). In particular, the equivalence class decomposition of $i\mathscr{Q}_m$, etc, is totally changed, that is, we obtain new ensembles and effect sets, from which we obtain a total change of the structures set down in III for $\mathscr{P}\mathscr{T}_{q\,\text{exp}}$.

Although the compatibility of $\mathscr{PT}_m$ with $\mathscr{PT}_{q\,\mathrm{exp}}$ could not be proven in general, the present studies of this problem show that this compatibility appears to hold with almost certainty (see [7]).

3 Compatibility of the Measurement Process with Preparation and Registration Procedures

In XVII we have seen that the quantum mechanical description of some of the parts of the measurement process is compatible with the viewpoint presented in II. On the contrary, we cannot, using the methods described in XVII, explain how we may establish measurement results as objective entities. Quantum mechanics by itself is not sufficient. Assuming the compatibility of $\mathscr{PT}_{q\,\mathrm{exp}}$ and $\mathscr{PT}_m$, then, in principle, it is possible to prove that a physical (but not a purely quantum mechanical) description is compatible with the viewpoint presented in II.

Here we shall not provide a detailed analysis of the individual steps (see, for example, [36], [7] and [37]). It suffices, on the contrary, to make clear that the structure in II can also be deduced from the structures of $\mathscr{PT}_m$. We shall not describe these deductions mathematically here; these deductions can be found in [6], XVI or (somewhat more difficult, but with more detail) in [7].

Here it is sufficient to outline the deductions as follows: The structure of experiments with microsystems outlined in I and described mathematically in II is developed from preparation and registration apparatuses, where the latter are macrosystems. Preparation and registration apparatuses combined to an actual experiment also represent a macrosystem which is composed out of two macrosystems. Thus, to experiment with microsystems we are immediately confronted only with macrosystems and their trajectories. The single characteristic of the experiment is that the apparatus is composed of two parts, and that the first part (the preparation apparatus), influences the behavior of the second part (the registration apparatus) but not vice versa. The directedness of this interaction from the preparation part to the registration part may be purely defined with the help of macroscopic description of trajectories (see [6], XVI, §1.3 and [7]).

Thus experiments with microsystems can be described entirely in the macroscopic domain without the "use of microsystems."

If we now consider the structure of this directed interaction we must restrict our considerations (as in the case of any physical theory) to a certain domain of application, that is, to the fundamental domain. Its boundaries can be gradually ascertained by introducing additional axioms (called normative axioms). These axioms can be introduced in such a manner that we may deduce the structure in II and III, where the set M is, at first, nothing other than the set of individual interaction processes (see [7] and [6], XVI). In this way the microsystems appear to be physically real entities which carry the interaction from the preparation part to the registration part. The historical discovery of microsystems can therefore be duplicated in a mathematical form within the context of a theory of macrosystems!

The theory $\mathscr{P}\mathscr{T}_{q\,\text{exp}}$ can be applied to the combined preparation and registration apparatuses as a macrosystem. If the compatibility discussed in §2 between $\mathscr{P}\mathscr{T}_{q\,\text{exp}}$ and $\mathscr{P}\mathscr{T}_m$ is valid then it also applies between the macroscopic description of experiments with microsystems, that is, between the structure introduced in II and III and the theory $\mathscr{P}\mathscr{T}_{q\,\text{exp}}$. Nevertheless there exists an essential difference between the physical description of the subprocesses of preparation and registration in XVII and the description of the entire processes of preparation and registration in $\mathscr{P}\mathscr{T}_m$, where $\mathscr{P}\mathscr{T}_m$ and $\mathscr{P}\mathscr{T}_{q\,\text{exp}}$ are compatible. This difference arises from the fact that for macrosystems we regard $\mathscr{P}\mathscr{T}_m$ rather than $\mathscr{P}\mathscr{T}_{q\,\text{exp}}$ as the more comprehensive theory, and therefore we do not regard all aspects of $\mathscr{P}\mathscr{T}_{q\,\text{exp}}$ as real. For example, the registration process is closed, providing that the trajectories on the registration apparatus have taken place despite the question whether these trajectories are ascertained by an "observer" or not? The formal possibilities in the theory $\mathscr{P}\mathscr{T}_{q\,\text{exp}}$ to measure other observables than those which correspond to the macroscopic observables associated with the apparatus do not possess any reality. These formal properties belong to the unrealistic part of $\mathscr{P}\mathscr{T}_{q\,\text{exp}}$. *The microscopic observable that is measured by the registration apparatus* is solely determined by this apparatus (and not by observers) (for a more precise mathematical formulation of this result see [7]).

We may now illustrate this result using the example discussed in XVII—the example of the microscope. After the scattering of the photon and the passage of the photon through the lens, we may still determine whether the photographic plate lies either in the image or focal plane. If, for example, the photographic plate lies in the image plane, then the observable is fixed; in this case, as an approximation observable for the position of the electron.

A bizarre example of the problem of "fixing" the registration result has been given by Schrödinger—Schrödinger's cat [38]:

> We may construct a bizare case. A cat is imprisoned in a steel chamber, together with the following infernal machine (which is out of reach of the cat): in a geiger counter there is a small amount of radioactive substance, so small, that in the duration of a few hours as few as a single atom decays with equal probability for none. If the counter responds, a relay is activated, and a hammer breaks a vial of cyanide. If the system is observed in an hour, we will say that the cat is living if no atom has decayed. At the first atom decay the cat will be poisoned. The Ψ function of the entire system is then expressed in the form that the cat is in a mixed state of being dead and alive.

In our description of quantum mechanics the result is objective, and does not depend upon whether someone observes the system or not. There are no "measurement possibilities" in the macroscopic domain other than the determination of objective results, whether the cat is dead or alive. The objective completion of the registration is already determined by the action or inaction of the hammer, that is, the effect of the microsystem is made evident in the macrosystem. The objective completion of the registration exists as soon as the action of the microsystem has reached the macroscopic domain. Here the objection is sometimes raised that such a description is inconsistent because it would be necessary to introduce a "discontinuity," at a suitable "size" of the system at which the quantum mechanical description is suddenly transformed into the macroscopic, objective one. We shall return to this objection in §5.

At the end of a quantum mechanical measurement process we do not need to consider such notions as the consciousness of an observer nor "my consciousness" or that of "a friend" (see such discussions in [34]). The "completion" exists, in our description, in the macroscopic realm, and it does not matter who notices the result nor in what form the result is ascertained and "fixed" (for example, in a computer).

4 "Point in Time" of Measurement in Quantum Mechanics?

In VII and in XVII we have shown that, with respect to the registration of microsystems, it is not physically meaningful to attribute a "point in time" at which the measurement occurs. In VII we could not exclude the possibility of introducing additional structure which would permit us to speak of making a measurement at a particular instant of time. In XVII we saw that quantum mechanics makes such a possibility doubtful because measurement scattering processes, like any scattering processes, last a finite time. However, we could imagine that we speak of idealizations if we speak of the instant of a measurement, that is, of the idealization that the scattering processes used are "very short." The duration of such measurement scattering processes is not so much determined by the interaction of the systems 1 and 2 rather than by the ensembles W_1 and W_2: The duration is the time where the W_t (see XVI, §§1 and 2) perceptibly differ from W_t^i and W_t^f.

Even if this duration is very short, the measurement scattering morphism $\mathbf{T}(1, 2; W_2)$ (see XVII (2.2.7)) does not directly depend on the "point in time" of the scattering, since S does not depend on the time where W_t differs from W_t^i and W_t^f (see XVI, §4.4). $\mathbf{T}(1, 2; W_2)$ depends only indirectly from this "time" by the ensemble W_2 (see also the gedanken experiment in XVII, §2.1).

Therefore a definition of a point in time where the measurement occurs is, in general, impossible and very artificial for a few cases. We will now show that such a definition is also not necessary for the interpretation of quantum mechanics.

But what is then the meaning of the time parameter for a time dependent decision observable $A(t)$? In many textbooks we can read that t is the point of time at which the observable is measured. Is it therefore necessary to restrict the measurement of time dependent observables $A(t)$ to measurement processes with "very short" measurement scatterings? Many textbooks seem to answer this question in the affirmative. Nevertheless this restriction to very short measurements is not necessary. Only in the case of approximation of external fields in VIII, §6 it was necessary to presuppose the existence of "short" measurement processes, but short only compared with the variations of the fields.

The opinion that the time parameter in $A(t)$ denotes the time where the observable is measured results from an uncritical transfer from classical mechanics to quantum mechanics.

For macrosystems we have presupposed (see §2) that it is clear from pre-theories what it means to measure the state that a system has at a given time t. We say that the pre-theories define the "direct" measurements of the states at various times t. We presuppose, for instance, for a system of mass points that pre-theories tell us what is meant by position and velocity of the mass points at a time t. Such a pre-theory may be, for example, geometric optics. These classical concepts cannot be transferred to quantum mechanics since the pre-theories for quantum mechanics do not tell us, what is meant by the position and momentum observables $Q(t)$ and $P(t)$. The pre-theories for quantum mechanics define only the preparation and registration procedures and especially the trajectories of the macroscopic preparation and registration apparatuses, that is, the states which the apparatuses have at various times t.

The direct measurements of the states of a classical system at a time t are already defined *without* any use of the dynamics of the classical system, for example, of the dynamics of the system of mass points. But also for classical systems we may use the dynamics to enlarge the possibilities for measurement by "indirect" means (for the concept of indirect measurements see [8], §10). For the system of mass points we may directly measure $\vec{r}^{(i)}(t_1)$ and $\dot{\vec{r}}^{(i)}(t_1)$. Then from the dynamics we obtain the values $\vec{r}^{(i)}(t_2)$, $\dot{\vec{r}}^{(i)}(t_2)$ for another time t_2, that is, also for a time t_2 where direct measurement will no longer be possible. For example, we may indirectly measure the positions of the planets for many years B.C.

In quantum mechanics *all* observables can only be indirectly measured on the basis of the directly measured trajectories of the apparatuses. Therefore it is impossible to define a time dependent observable $A(t)$ in the same way as in classical mechanics, that is, by a direct measurement of A at a time t. To see this more clearly, we reflect on the following two time dependent observables $A(t)$ and $B(t)$ where $B(t) = A(t_1 + t)$ for a fixed t_1. Both $A(t)$ and $B(t)$ satisfy the relations

$$A(t) = e^{iHt}A(0)e^{-iHt},$$

$$B(t) = e^{iHt}B(0)e^{-iHt}.$$

For which of these two observables $A(t)$ or $B(t)$ is t the so-called time of measurement? It is just not possible to distinguish $A(t)$ and $B(t)$ by the (nonexistent) criterion that, for example, the time t in $A(t)$ is the "point in time" for the measurement of $A(t)$ while the point in time of the measurement of $B(t)$ is $t_1 + t$.

Is there no possibility at all of distinguishing between $A(t)$ and $B(t)$? There is another possibility: The behavior of the observables under Galileo transformations. This behavior is a good *intrinsic* possibility in quantum mechanics which we have utilized in the definition of $Q(t)$ in VII, §4. In general $A(0)$ and $B(0)$ do not have the same behavior under Galileo transformations.

The fact that there is no possibility in $\mathscr{PT}_{q\,\mathrm{exp}}$ to define the values of the observable "at a time t" does not contradict the compatibility between $\mathscr{PT}_m$ and $\mathscr{PT}_{q\,\mathrm{exp}}$. The macro-observables which correspond to the trajectories are not defined in $\mathscr{PT}_{q\,\mathrm{exp}}$ alone but are defined in $\mathscr{PT}_{q\,\mathrm{exp}}$ by means of the imbedding map i. The macro-observables for the states of the macrosystems at time t are defined in this way.

5 Relationships Between Different Theories and Quantum Mechanics

Since we do not consider quantum mechanics to be a universal theory, it is important to survey the relationships between the different theories and their domains of validity (fundamental domains—see I). We shall use the general analysis from [8], §8 (see also [6], III, §7). Familiarity with the analysis presented there is not necessary for an understanding of this section. Our discussion will be based upon the following diagram, which we shall provide a detailed explanation later:

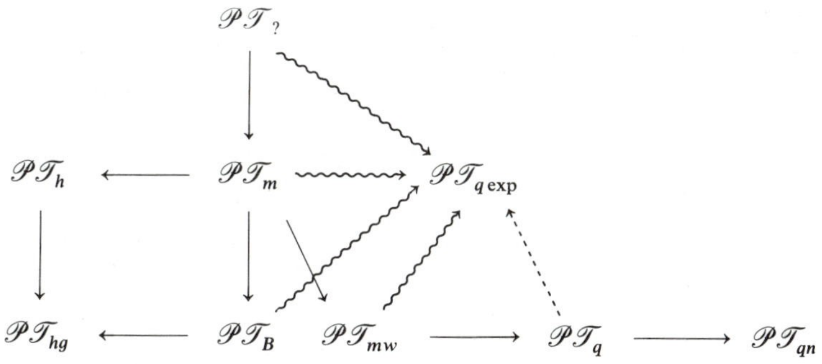

Here we do not claim completeness. We only intend to give an example of the network associated with physical theories. The following will also serve to help to correctly describe the place of quantum mechanics $\mathscr{PT}_q$ in physics. Here $\mathscr{PT}_q$ will denote the theory which we have presented in II–XVII. In the diagram there are also new theories which we have not described which, except for knowledge of some of their general structures, are at present only partially understood. In the diagram these are denoted by $\mathscr{PT}_?$ and $\mathscr{PT}_m$.

In §2 we have briefly described the meaning of the theories $\mathscr{PT}_?$ and $\mathscr{PT}_m$: $\mathscr{PT}_?$ is the desired most comprehensive theory of macrosystems, which includes the mutual interactions of macrosystems. The expression "macro" characterizes the domain of application of the theory. Here we must be careful not to include cosmology, because such an inclusion would introduce totally new problems with respect to preparation and registration (see §6). We shall restrict $\mathcal{PT}_?$ to a domain, which we shall call the laboratory domain. Then in the above diagram $\mathcal{PT}_?$ will be the most comprehensive theory. It may, however (at least among the theories presented here), not apply to high energy interactions in order that it not include elementary particle theory.

We may already imagine some of the structure of $\mathcal{PT}_m$. It should describe macrosystems in terms of trajectories in a state space, as described in §2. $\mathcal{PT}_m$ would then be a restriction of the more comprehensive theory $\mathcal{PT}_?$. The arrow in the diagram from $\mathcal{PT}_?$ to $\mathcal{PT}_m$ represents this restriction.

Many specializations, that is, restrictions of the partially understood theory $\mathcal{PT}_m$ are familiar to physicists. We shall mention only a few of these restrictions $\mathcal{PT}_h$, $\mathcal{PT}_{hg}$, $\mathcal{PT}_B$. Here $\mathcal{PT}_h$, $\mathcal{PT}_B$ and $\mathcal{PT}_{hg}$ are the theories of hydrodynamics as described by the Navier–Stokes equations, the Boltzmann transport equation, and the theory of hydrodynamics as applied to the special case of gases. The arrows again illustrate the restrictions, for example, indicating that $\mathcal{PT}_{hg}$ may be obtained from $\mathcal{PT}_B$ by restriction.

Of particular interest here is the part of the above diagram which lies to the right of the vertical line from $\mathcal{PT}_?$ to $\mathcal{PT}_m$.

$\mathcal{PT}_{mw}$ may be the restriction of the theory $\mathcal{PT}_m$ which is appropriate for the description of directed interaction processes between two macrosystems (see §3). The theory $\mathcal{PT}_q$ can be obtained by a restriction of the theory $\mathcal{PT}_{mw}$. The restriction consists of those directed interactions for which the axioms from III–XVII are usable. In brief, we say that $\mathcal{PT}_q$ describes the interaction by means of "microsystems." $\mathcal{PT}_{qn}$ may be the theory which is obtained from $\mathcal{PT}_q$ if we begin by using the sets K of ensembles and the sets L of effects as fundamental sets and "forget" that K and L are obtained from preparation and registration procedures. $\mathcal{PT}_{qn}$ is therefore the "normal" version of quantum mechanics. It is customary to formulate this "normal" version by beginning with the sets $\partial_e K$ and $\partial_e L$ of extreme points of K and L where the elements of $\partial_e K$ are represented by the normed vectors of a Hilbert space $\mathcal{H}$ and are called states. The elements of $\partial_e L$ are the projection operators and are frequently called yes–no observables or propositions.

Even though we are only certain that $\mathcal{PT}_q$ is a theory of microsystems, using the structure presented in VIII we often extrapolate $\mathcal{PT}_q$ to "many-particle systems," that is, to macrosystems. This tendency of physicists to extrapolate the fundamental domain of a theory as far as possible, even when such an extension may be questionable, is completely legitimate in the evolution of physics. Only by such investigation is it possible to discover the limitations of a theory. If we extend the fundamental domain of $\mathcal{PT}_q$ in this manner to systems composed of arbitrarily many microsystems, then by restriction of this theory we may obtain the theory $\mathcal{PT}_{q\,\mathrm{exp}}$ which is only applicable to many-particle systems. The dashed line in the above diagram refers to such a restriction, in order to indicate that the validity of all the axioms from III and XVIII for many particle systems (that is, macrosystems) is highly questionable.

The wavy lines in the above diagram indicate embeddings. In §2 we have described the embedding of $\mathscr{PT}_m$ into $\mathscr{PT}_{q\,\exp}$ with the aid of the embedding map i. The arrow from $\mathscr{PT}_?$ to $\mathscr{PT}_{q\,\exp}$ means that the embedding of $\mathscr{PT}_m$ in $\mathscr{PT}_{q\,\exp}$ is perhaps only a special case of the more general embedding of $\mathscr{PT}_?$ into $\mathscr{PT}_{q\,\exp}$. For the discussion here it is important to note that we consider $\mathscr{PT}_m$ to be a more comprehensive theory than $\mathscr{PT}_{q\,\exp}$ and that we believe that the only part of $\mathscr{PT}_{q\,\exp}$ which reflects reality is the part which is the "image" of $\mathscr{PT}_m$ under the embedding.

Why then do we actually consider the embedding of $\mathscr{PT}_?$ and $\mathscr{PT}_m$ into $\mathscr{PT}_{q\,\exp}$? That is why the theories $\mathscr{PT}_?$ and $\mathscr{PT}_m$ are not completely known. For the case of $\mathscr{PT}_m$ only certain partial structures are known, which can, with the aid of the map i, be completed by adding additional structures. The known Hamiltonian operator in $\mathscr{PT}_{q\,\exp}$ (although we have been very critical of its introduction in VIII) makes it possible to make precise statements concerning the dynamics of trajectories in $\mathscr{PT}_m$. Thus, we can, for example, obtain the structure of $\mathscr{PT}_B$ from the partially known structure of $\mathscr{PT}_m$ and the embedding i, if we restrict our considerations in $\mathscr{PT}_m$ and in $\mathscr{PT}_{q\,\exp}$ to "attenuated" gases. In [6], XV, §10.2 we have outlined how it is possible (in the above sense) to derive $\mathscr{PT}_B$ (and the interaction cross section used in $\mathscr{PT}_B$)—in the approximation that $\mathscr{PT}_{q\,\exp}$ is approximated by a classical picture. A more precise description of such a derivation of $\mathscr{PT}_B$ can be found in [39]. It is in this sense the wavy arrow from $\mathscr{PT}_B$ to $\mathscr{PT}_{q\,\exp}$ is to be understood to be a specialization of the embedding of $\mathscr{PT}_m$ in $\mathscr{PT}_{q\,\exp}$.

Similarly the wavy arrow from $\mathscr{PT}_{mw}$ to $\mathscr{PT}_{q\,\exp}$ is a specialized case of the embedding of $\mathscr{PT}_m$ into $\mathscr{PT}_{q\,\exp}$. This embedding of $\mathscr{PT}_{mw}$ into $\mathscr{PT}_{q\,\exp}$ guarantees the compatibility of quantum mechanics with the introduction of the preparation and registration structures presented in II, as we have shown in §3.

A realistic development of $\mathscr{PT}_{mw}$, including its embedding in $\mathscr{PT}_{q\,\exp}$ would supply a solution to the following problem which was frequently mentioned in §3—how the theory $\mathscr{PT}_q$ must change as we increase the number of particles in a composite system in order to obtain the theory $\mathscr{PT}_m$ in the case of "many" particles.

Here we do not use the term "change" in the mathematical sense of continuity. The axioms of II and III cannot be changed in a continuous manner. This is, however, not a new phenomena in physics. The comparison between a more comprehensive to a less comprehensive theory cannot always be carried out in terms of continuous mathematical processes. Thus, for example, the Newtonian space time theory and the Einstein space time theory of special relativity are both "correct," that is, are both useful theories; Einstein's theory is more comprehensive than Newton's, and both appear to mathematically contradict the other (see, for example, [6], IX, §8).

If in $\mathscr{PT}_{mw}$ we consider interactions which are transmitted by physical systems (as carriers of the interaction—see II, §4.4) where the number of elementary systems comprising these carriers become larger and larger, then we must assume that the assumptions made in II, III, VIII become "less and less realistic." In $\mathscr{PT}_{mw}$ this phenomenon must be present because $\mathscr{PT}_{mw}$ must not describe any unrealistic interactions. Thus, for example, the embedding i of $\mathscr{PT}_m$ into $\mathscr{PT}_{q\,\mathrm{exp}}$ specialized to the embedding of $\mathscr{PT}_{mw}$ in $\mathscr{PT}_{q\,\mathrm{exp}}$ must show that for larger molecules (as carriers of the interaction between the preparation and registration parts) the axioms set down in II, III, VIII can no longer be valid, that is, that there must be a more comprehensive theory $\mathscr{PT}_w$ of interaction carriers than $\mathscr{PT}_q$. Here, for systems consisting of few "particles" $\mathscr{PT}_q$ and $\mathscr{PT}_w$ should be in close agreement. For interaction carriers consisting of large numbers of elementary systems $\mathscr{PT}_w$ should be consistent with $\mathscr{PT}_m$.

When we assert that there must be a more comprehensive theory $\mathscr{PT}_w$ (which we may obtain from $\mathscr{PT}_m$, providing that $\mathscr{PT}_m$ is already known) we should also expect to find noticeable differences in the case of systems composed of some elementary systems. Have such differences already been observed?

They have not, because the interest has been in making comparisons between actual experiments and the theory $\mathscr{PT}_q$, and the question as to what is possible in experiments has not, in general, been made. For the most part the consequences of the structure of the Hamiltonian operator have been tested rather than the experimental consequences of the axioms in III.

We now seek to determine the fundamental domain of $\mathscr{PT}_q$. For elementary systems, including coupling with external fields (in the sense of VIII, §6) $\mathscr{PT}_q$ appears to describe the possibilities of preparation and registration very well. The description of the possibilities to prepare composite systems in "bound" states is excellent providing that the eigenvalues of H are not "too close"—where by "too close" we mean that the transition frequency is of the order of days or years. Thus it is questionable whether, for large molecules (which have both right- and left-handed forms) the eigenstates of H can be prepared or whether H can be measured as an observable since the right- and left-handed forms are not eigenstates of H. The eigenstates of H are, on the contrary, superpositions of right- and left-handed forms with transition frequencies which, for large molecules, are very small, with the consequence that there are no transitions between right- and left-handed forms.

It is questionable whether all ensembles for unbounded states, that is, for scattering states can be prepared. We are hard pressed if we inquire about the experimental possibility to prepare ensembles W' which are obtained from X (4.15) from W where W is an ensemble prepared in the usual way from known scattering experiments (see the remarks in X, §4). It seems impossible to prepare $W' = CWC^{-1}$ if W is an unstable state (XVII, §6.5).

We deceive if we believe that although it may be extremely difficult to experimentally realize many preparations and registrations that they may yet be possible "in principle." Can they in fact be realizable "in principle" given sufficient financial resources? It is precisely here that the more comprehensive theory $\mathscr{PT}_w$ makes actual physical assertions about the possibilities of realizing such preparations and registrations.

It is precisely here that the embedding of $\mathscr{PT}_m$ in $\mathscr{PT}_{q\,\mathrm{exp}}$, that is, the specification of the embedding map i together with the Hamiltonian operator

H from $\mathscr{PT}_{q\,\exp}$ are sufficient to enable us to determine $\mathscr{PT}_w$. If the domain of realistic preparation and registration procedures are known for macrosystems, then they can be found for other not-so-large systems. A jump from $\mathscr{PT}_q$ to $\mathscr{PT}_m$ or $\mathscr{PT}_?$ will therefore not exist. On the contrary there must be a theory $\mathscr{PT}_w$ which is more comprehensive than $\mathscr{PT}_q$ which describes the preparation and registration possibilities more realistically than does $\mathscr{PT}_q$; for systems composed of "fewer particles" $\mathscr{PT}_w$ and $\mathscr{PT}_q$ will ((approximately) make the same assertions about these possibilities.

From this conception of $\mathscr{PT}_w$ let us return to the considerations of measurement scattering in XVII, §2. Then it will be clear that the following description is unsatisfactory: In a sequence of measurement scattering processes in which every step is described by $\mathscr{PT}_q$, the final step of a scattering on a macrosystem is a sudden jump which fixes the measurement results. On the contrary, the registration possibilities become more restricted the larger the systems which are used to scatter the microsystems. Thus, in the case of measurement scattering experiments with large molecules measurement results can be fixed in the molecules themselves, even though additional physical processes may be required to make the fixed result accessible to us. The "development" of a silver bromide grain sensitized by a photon resulting in a silver grain and its observation by a microscope is an example of such a physical process which makes the objective property "sensitized" accessible to us.

With the consideration of the fundamental domain of $\mathscr{PT}_q$ and the discussion of the possibilities for a more comprehensive theory $\mathscr{PT}_w$, we have achieved what a physicist would wish for the formulation of a theory and the determination of its bounds. In II to XVII we have developed $\mathscr{PT}_q$ and illustrated it by means of examples. Here in XVIII we have shown the limits of application of $\mathscr{PT}_q$. The fact that sharp boundaries cannot be drawn is typical for any $\mathscr{PT}$. The similarities of a mathematical picture $\mathscr{MT}$ in a $\mathscr{PT}$ with reality are not exact in the fundamental domain, and become less and less valid as they leave the domain.

We have now concluded our description of $\mathscr{PT}_q$. Since some physicists have used quantum mechanics to justify certain ideas, we will now pose the question as whether these ideas can be justified by means of the theory $\mathscr{PT}_q$ developed here.

6 Quantum Mechanics and Cosmology

In cosmology we seek to obtain general statements about the universe as a whole. Here we find typical problems concerning the physical reality of statements about the universe (see, for example, [8], X, §10). In this section we shall only be concerned with the relationship between quantum mechanics and cosmology.

In §2 we have given physical arguments for the fact that $\mathscr{PT}_{q\,\exp}$ cannot be the most comprehensive theory for macrosystems. An application of quantum mechanics to the universe as a whole is clearly not possible in the context of the formulation of quantum mechanics presented in II–XVII. In this formulation the notion of statistics presented in II plays a central role. Statistics is a theory concerned with frequencies of occurrence associated with selection procedures. Such frequencies of events can only occur *in* the universe. The universe is, as a whole, unitary. Therefore it is not meaningful to make statements of frequency about the universe itself because, in principle, there are not different universes, but only different parts of the same universe. Certainly we could imagine universes different from the one in which we live. Such a universe could not have any physical interaction with our universe otherwise both would be parts of a more comprehensive universe. Those who disagree with the analysis presented in §2 which shows that $\mathscr{P}\mathscr{T}_{q\,\exp}$ can only approximately describe macrosystems must seek elsewhere for the limits of applicability of $\mathscr{P}\mathscr{T}_{q\,\exp}$ because the universe, as a whole, cannot belong to the domain of application of $\mathscr{P}\mathscr{T}_{q\,\exp}$.

Certainly there are attempts which seek to apply quantum mechanics to the entire universe [34]. What can we make of these attempts? If we examine these attempts we find that the meaning of quantum mechanics is different than the theory denoted by $\mathscr{P}\mathscr{T}_q$ and developed in II–XVII. Expressions like "state," "probability," "possibility," etc. will have a different meaning than that used in the theory presented here. Expressions such as "state," "probability," "possibility" are associated with certain philosophical ideas which cannot be subjected to physical verification. We do not wish to prohibit the practice of adding philosophical ideas to a physical theory, but think that it is reasonable to do so *after* a theory has a physical interpretation. We know that many physicists think that every interpretation presupposes philosophical ideas. One of the reasons for writing this book was to show that it is possible to interpret quantum mechanics without such philosophical presuppositions, that is, to interpret it only on the basis of pre-theories.

If, for example, a vector ψ from a Hilbert space $\mathscr{H}$ is said to represent the "state of the universe," it would not make sense physically because no one knows what the "state of the universe" should mean as a physical concept. If by state we man what is defined in III, D 1.1, then the universe "has" no state ψ because there cannot be an ensemble of universes. The preparation procedures in $\mathscr{P}\mathscr{T}_{q\,\exp}$ will, in the application of $\mathscr{P}\mathscr{T}_{q\,\exp}$ to the universe, lead to absurdity because there is no single preparation procedure for universes, but there is only a single universe.

On the other hand, no one is forbidden to propose a mathematical theory $\mathscr{M}\mathscr{T}$ and to add "philosophical" ideas to such a theory. Since an $\mathscr{M}\mathscr{T}$ cannot in itself have a physical interpretation, it is necessary therefore to supply $\mathscr{M}\mathscr{T}$ by the introduction of mapping principles ($\mathscr{M}\mathscr{A}\mathscr{P}$) (see I) in order to obtain a $\mathscr{P}\mathscr{T}$. If we use the mathematical theory of Hilbert spaces for $\mathscr{M}\mathscr{T}$ in an attempt to apply it to cosmology, then it is necessary to provide a different interpretation than that provided here for quantum mechanics in II and III. Physically what is important is not the species of structure upon which $\mathscr{M}\mathscr{T}$ is formally constructed, but what physical interpretation is attributed to the elements of $\mathscr{M}\mathscr{T}$ (see I and [8], §5). Only if $\mathscr{M}\mathscr{T}$ is an axiomatic basis can we easily determine what structure aspects actually represent reality. The formal species of structure of Hilbert space cannot be obtained from an axiomatic basis (as obtained in II and III) *if* we seek to describe cosmology as a $\mathscr{PT}$. If we examine attempts to formulate a "quantum theory of the universe" then we shall readily see that philosophical grounds lead us to introduce such notions as "state of the universe" as a concept already understood before the theory, that is, *a priori*. Then it is possible to arbitrarily argue without coming to a conclusion whether it is meaningful to formulate a "quantum mechanics of the universe." The exponents of such a theory will take the *a priori* ideas (which cannot be defined on the basis of pre-theories) as the sole approach to understanding physics. Others, such as the author of this book cannot attribute any physical meaning to such ideas and consider them to be meaningless word games. It is possible only to agree on the portion of the interpretation of a physical theory, the portion which was denoted by $\mathscr{MAP}$ in I (see also [8] and [6], III), namely, the portion which connects experiments with the mathematical picture $\mathscr{MT}$.

For the case of a "physical" cosmology the problem of mapping principles $\mathscr{MAP}$ is in no way simpler than it was in the case of quantum mechanics. Cosmological statements about time–space structure cannot be described in terms of pre-theories based upon experiment. On the contrary, pre-theories only describe phenomena which can be described "in the laboratory," which, in turn, is influenced by the universe. For microsystems such experiments in a laboratory are carried out in a "shielded" environment; for a theory of the universe experiments are required which are essentially influenced by the universe. In the case of quantum mechanics we seek to draw conclusions about the structure of the much smaller microsystems from processes involving "macrosystems" in the laboratory domain. In cosmology we seek to draw conclusions about the structure of the "much larger" universe from processes in the laboratory domain. The problem of mapping principles for cosmology is an interesting one, but one which, for space limitations, cannot be discussed further in this volume.

7 Quantum Mechanics and Physical World Views

It is often said that special relativity, general relativity and quantum mechanics have lead to "revolutions" in the world view of physics. Particularly, in connection with quantum mechanics there have been the most extensive different philosophies introduced. We refer readers who are interested in the various philosophical discussions about quantum mechanics to [40] in which there are thorough critical examinations of the various philosophical points of view and extensive bibliographies.

The fact that quantum mechanics has been used as the basis for the most varied philosophical ideas clearly shows that not the physical theory noted quantum mechanics has philosophical implications but the reverse—that fundamental philosophical considerations have lead to different ideas about quantum mechanics. Of course, the formulation of quantum mechanics presented here rests upon a fundamental physical outlook which is described in detail in I and in [8] and will now briefly be discussed.

If we consider quantum mechanics to be a theory in the sense described in this book, then we cannot speak of a revolution in the physical world view. Instead, we will find that quantum mechanics has led to a critical reexamination of what physics is about. In this way quantum mechanics has made it necessary to abandon preconceived ideas which were believed to be essential for physics. Such ideas were: physics as the analysis of causal connection, physics as an explanation of the past, physics as a way to predict the deterministic course of the world. What remains is physics as a method to discover the structures of reality, and to use these structures, a method solely based on objective facts and on the well-defined language of craftsmen. This view is described in detail in [8] and [6], III, and was briefly outlined in I.

If we consider a physical theory as a method to discover the structures of reality, and representing these structures by the mathematical structures in $\mathcal{MT}$ (see I or, for more detail, see [8] or [6], III), then there will be no "revolutions" in physics. For example, Newtonian physics and gravitational theory is as useful now as before despite the introduction of general relativity (see [6], IX and X). The astronauts travel between the earth and the moon according to Newton's laws.

The fact that in this analysis of the structure of reality it has been necessary to introduce new structures for the case of the universe as well as for the microworld which surpasses that required for the description of processes in the laboratory could only be puzzling to those who believe (on some philosophical ground) in the homogeneity of the structure of nature relative to scaling in size.

One particular effect of quantum mechanics is particularly important: the need for sober judgment in physics and to reduce physical theory to what is real. Thus quantum mechanics forms a catalyst for a new and fruitful reflection about problems in the philosophy of science. Thus the discussions about quantum mechanics presented here are leading towards an examination of the "fundamantal structure of a physical theory" as presented in [8].

Thus the reduction of prejudices about physics resulting from a reexamination of quantum mechanics make it possible to discuss other aspects of reality in a more meaningful way than before, for example, the problem of "consciousness" and the problem of free will. In [6], XVII these problems are briefly discussed.

APPENDIX V

Groups and Their Representations

In the space available here it will not be possible to develop the theory of group representations in the generality used in VI–VIII. Here we shall only attempt to provide the most important aspects of representation theory which are necessary for practical applications in the theory of atomic and molecular structure. It is hoped that structures used in VI–VIII can be understood without the proofs. The proofs of the more general theorems can be found in the specialized literature [41].

1 Groups

A set $\mathscr{G}$ is said to be a group if there is a structure defined on $\mathscr{G}$ in terms of a map $\mathscr{G} \times \mathscr{G} \to \mathscr{G}$ (which for $a, b, c \in \mathscr{G}$ we will denote by $(a, b) \to ab = c$) which satisfies the following axioms:

(1) $(ab)c = a(bc)$;
(2) there exists an $e \in \mathscr{G}$ such that $ea = a$ for every $a \in \mathscr{G}$;
(3) to each $a \in \mathscr{G}$ there exists an element $b \in \mathscr{G}$ such that $ba = e$.

From (1) to (3) for $x = ab$ it follows that $bx = bab = eb = b$. Since, according to (3) there exists a c such that $cb = e$ it follows that $cbx = c(bx) = cb = e$ and $cbx = (cb)x = ex = x$ and we therefore obtain $x = e$, that is, $ab = e$. Thus it follows that $ae = a(ba) = (ab)a = ea = a$.

Suppose that there exists a b' for which $b'a = e$. Then it follows that $b' = b'e = b'ab = eb = b$, that is, the element b in (3) is uniquely determined by a. We therefore write $b = a^{-1}$. We obtain $a^{-1}a = aa^{-1} = e$ and $(a^{-1})^{-1} = a$.

Similarly it follows that the element e in (2) is uniquely defined since from $e'a = a$ for all $a \in \mathscr{G}$ it follows that $e' = e'e = e$.

If, in addition to (1), (2), (3), the following condition

(4) $ab = ba$

is satisfied for all $a, b \in \mathscr{G}$ then $\mathscr{G}$ is called an Abelian group.

A subset $\mathscr{H} \subset \mathscr{G}$ is said to be a subgroup if the group operations in $\mathscr{G}$ satisfy the relation $\mathscr{H} \times \mathscr{H} \to \mathscr{H}$ and if the relations (1), (2), (3) are satisfied. In order that $\mathscr{H}$ be a subgroup it is necessary and sufficient that $a, b \in \mathscr{H}$ implies $ab^{-1} \in \mathscr{H}$.

Proof. For $a \in G$ it follows that $aa^{-1} = e \in \mathscr{G}$; from $e \in \mathscr{G}$ and $a \in \mathscr{G}$ it follows that $ea^{-1} = a^{-1} \in \mathscr{G}$; from $a, b \in \mathscr{G}$ it follows that $ab^{-1} \in \mathscr{G}$ and therefore $a(b^{-1})^{-1} = ab \in \mathscr{G}$.

All powers a^m (where $a^0 = e$, $a^{-n} = (a^{-1})^n$) of the element a form a subgroup. A group for which all the elements are powers of a single element is said to be cyclic, and is therefore Abelian. Either $a^n \neq a^m$ for $m \neq n$ or there exists a $k > 0$ for which $a^k = e$ because from $a^n = a^m$ it follows that $a^{n-m} = a^{m-n} = e$. If v is the smallest positive integer for which $a^v = e$, then the elements $a^0 = e, a^1, \ldots, a^{v-1}$ are all different and form the entire cyclic group because, for $0 \leq r < v$ it follows that $a^m = a^{qv+r} = a^r$. If $\mathscr{H}$ is a subgroup of this cyclic group and $s > 0$ is the smallest exponent with a^s in $\mathscr{H}$ then for $a^m \in \mathscr{H}$ it follows that $a^m = a^{qs+r} = (a^s)^q a^r$ and therefore $a^r \in \mathscr{H}$; since $0 \leq r < s$ it follows that $r = 0$. Therefore in $\mathscr{H}$ we find only elements of the form $(a^s)^q$. Since $a^n = e$ in $\mathscr{H}$, we must have $n = p \cdot s$, that is s must be a divisor of n. For each divisor s of n the $(a^s)^q$ form a subgroup.

2 Cosets and Invariant Subgroups

Let $\mathscr{H}$ be a subset of $\mathscr{G}$ (not necessarily a subgroup). Let $a\mathscr{H}$ denote the subset of $\mathscr{G}$ which is obtained by multiplying the elements of $\mathscr{H}$ on the left by a. If $\mathscr{H}$ is a subgroup, then we say that $a\mathscr{H}$ (where $a \in \mathscr{G}$) is a left coset and $\mathscr{H}a$ is a right coset. If $a \in \mathscr{H}$ then $a\mathscr{H} = \mathscr{H}$ and conversely since $ae = a \in \mathscr{H}$. In addition $a\mathscr{H} = b\mathscr{H}$ if and only if $a^{-1}b \in \mathscr{H}$ since if $a^{-1}b \in \mathscr{H}$ then $b\mathscr{H} = aa^{-1}b\mathscr{H} = a(a^{-1}b\mathscr{H}) = a\mathscr{H}$; if $a\mathscr{H} = b\mathscr{H}$ then $a^{-1}b\mathscr{H} = \mathscr{H}$ from which it follows that $a^{-1}b \in \mathscr{H}$. Different cosets have no common elements; if $a\mathscr{H}$ and $b\mathscr{H}$ had a common element $ac_1 = bc_2$ then $a^{-1}b = c_1c_2^{-1} \in \mathscr{H}$ and the cosets would be identical. Each a belongs to a left coset namely $a\mathscr{H}$. Each coset has the same cardinality since $ac \leftrightarrow bc$ (for $c \in \mathscr{H}$) is a bijective map of $a\mathscr{H}$ onto bh. The entire group is the union of its cosets. If j is the number of different cosets (the "index" of $\mathscr{H}$), N the cardinality of G and n the cardinality of $\mathscr{H}$, then $N = jn$.

If the left cosets are also right cosets then $a\mathscr{H} = \mathscr{H}a$ because $\mathscr{H}a$ is the right coset which contains a. Then we have $\mathscr{H} = a\mathscr{H}a^{-1}$, and $\mathscr{H}$ is called an invariant subgroup.

3 Isomorphisms and Homomorphisms

If $\mathscr{G}$ and $\mathscr{G}'$ are two groups and if there exists a surjective map $\mathscr{G} \to \mathscr{G}'$ such that $ab \to a'b'$, then $\mathscr{G}$ is said to be homomorphic to $\mathscr{G}'$, and we write $\mathscr{G} \sim \mathscr{G}'$. From $ea = a$ it follows that $e'a' = a'$ from which we conclude that e' is the identity in $\mathscr{G}'$. From $a^{-1}a = e$ it follows that $(a^{-1})'a' = e'$ from which we conclude that $(a^{-1})' = (a')^{-1}$. If $\mathscr{G} \xrightarrow{f} \mathscr{G}'$ is a (not necessarily surjective) map satisfying $ab \to a'b'$ then $f(\mathscr{G})$ is a subgroup of $\mathscr{G}'$ and $\mathscr{G} \xrightarrow{f} f(\mathscr{G})$ is a homomorphism. If the map $\mathscr{G} \to \mathscr{G}'$ is bijective, then $\mathscr{G}$ is isomorphic to $\mathscr{G}'$ which we denote by $\mathscr{G} \cong \mathscr{G}'$.

If $\mathscr{G} \sim \mathscr{G}'$ and if $\mathscr{N}$ is the subset of $\mathscr{G}$ which is mapped upon the unit element e' then $\mathscr{N}$ is an invariant subgroup in $\mathscr{G}$.

Proof. If $a, b \in \mathscr{N}$ then $ab^{-1} = e'(e') = e'$ from which we conclude that $ab^{-1} \in \mathscr{N}$ and $\mathscr{N}$ is a subgroup. If d is an arbitrary element of $\mathscr{G}$ then $da \to d'e' = d'$, that is, $d\mathscr{N} \to d'$. Conversely, if $x \to d'$ then $d^{-1}x \to (d')^{-1}d' = e'$ from which we conclude that $d^{-1}x \in \mathscr{N}$ and that $x \in d\mathscr{N}$. The residue classes $d\mathscr{N}$ is precisely the set of elements which are mapped upon d'. In the same way we may show that $\mathscr{N}d$ is the set of elements which are mapped upon d'. Therefore $d\mathscr{N} = \mathscr{N}d$.

The set of cosets of an invariant subgroup forms a group with the operation $d_1\mathscr{N}d_2\mathscr{N}$ because $d_1\mathscr{N}d_2\mathscr{N} = d_1d_2\mathscr{N}\mathscr{N} = d_1d_2\mathscr{N}$. $\mathscr{N}d\mathscr{N} = d\mathscr{N}$ (therefore $\mathscr{N}$ is the unit element), $d^{-1}\mathscr{N}d\mathscr{N} = \mathscr{N}$. This group is called the factor group $\mathscr{G}/\mathscr{N}$. From $\mathscr{G} \sim \mathscr{G}'$ it therefore follows that $\mathscr{G}' \cong \mathscr{G}/\mathscr{N}$ where $\mathscr{N}$ is the subgroup of $\mathscr{G}$ (kernel) which is mapped upon e'. On the other hand, $\mathscr{G} \sim \mathscr{G}/\mathscr{N}$ on the basis of the map $a \to a\mathscr{N}$. For a homomorphic map $\mathscr{G} \to \mathscr{G}'$ the diagram

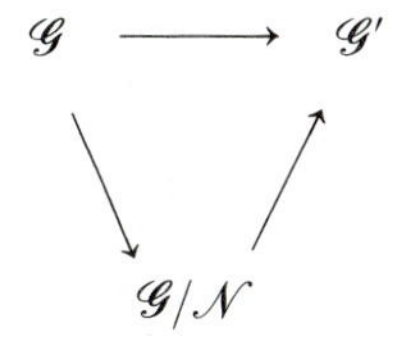

is commutative.

4 Isomorphism Theorem

Let $\mathscr{G} \sim \bar{\mathscr{G}}$. Then $\bar{\mathscr{G}} \cong \mathscr{G}/\mathscr{N}$ where $\mathscr{N}$ is the set of elements of $\mathscr{G}$ which are mapped upon $\bar{e}$, the unit element of $\bar{\mathscr{G}}$. Let $\mathscr{H}$ be a subgroup of $\mathscr{G}$ which is mapped by the homomorphism (of $\mathscr{G}$ onto $\bar{\mathscr{G}}$) onto $\bar{\mathscr{H}}$. $\bar{\mathscr{H}}$ is then clearly a subgroup of $\bar{\mathscr{G}}$. Let $\mathscr{K}$ denote the set of all elements of $\mathscr{G}$ which are mapped onto $\bar{\mathscr{H}}$. Then $\mathscr{H} \subset \mathscr{K}$. If $a \in \mathscr{H}$ then $\mathscr{K}$ is the ensemble of all $a\mathscr{N}$ where $a \in \mathscr{H}$, that is, $\mathscr{K} = \mathscr{H}\mathscr{N}$. From this it follows that $\mathscr{K}$ is a subgroup of $\mathscr{G}$ since if an_1 and bn_2 ($a, b \in \mathscr{H}$ and $n_1, n_2 \in \mathscr{N}$) are two arbitrary elements of $\mathscr{K}$ then $an_1(bn_2)^{-1} = an_1n_2^{-1}b^{-1} = an_3b^{-1}$ (where $n_3 = n_1n_2^{-1} \in \mathscr{N}$). Since $\mathscr{N}b^{-1} = b^{-1}\mathscr{N}$ it follows that $an_3b^{-1} = ab^{-1}n_4 \in \mathscr{K}$ because $ab^{-1} \in \mathscr{H}$ since $\mathscr{H}$ is a subgroup. $\mathscr{K}$ is therefore homomorphically mapped upon $\bar{\mathscr{H}}$. The set of elements which are mapped onto $\bar{e}$ is $\mathscr{N}$ because $\mathscr{N}$ is a subset of $\mathscr{K}$. Therefore $\bar{\mathscr{H}} \cong \mathscr{K}/\mathscr{N} = \mathscr{H}\mathscr{N}/\mathscr{N}$. From the homomorphism of $\mathscr{H}$ onto $\bar{\mathscr{H}}$ it follows that $\bar{\mathscr{H}} \cong \mathscr{H}/\mathscr{H} \cap \mathscr{N}$ since $\mathscr{H} \cap \mathscr{N}$ is the set of elements of $\mathscr{H}$ which are mapped onto $\bar{e}$.

Therefore we obtain

$$\mathscr{H}\mathscr{N}/\mathscr{N} \cong \mathscr{H}/\mathscr{H} \cap \mathscr{N}. \tag{4.1}$$

Let $\mathscr{G}$ be mapped homomorphically onto $\bar{\mathscr{G}}$ and $\bar{\mathscr{G}}$ be homomorphically mapped onto $\bar{\mathscr{G}}/\bar{\mathscr{M}}$ where $\bar{\mathscr{M}}$ is an invariant subgroup of $\bar{\mathscr{G}}$. In this way $\mathscr{G}$ is homomorphically mapped upon $\bar{\mathscr{G}}/\bar{\mathscr{M}}$ and we therefore find that $\bar{\mathscr{G}}/\bar{\mathscr{M}} \cong \mathscr{G}/\mathscr{M}$ where $\mathscr{M}$ is the set of elements which are mapped upon the unit element of $\bar{\mathscr{G}}/\bar{\mathscr{M}}$—precisely the same elements which are mapped by the first map of $\mathscr{G}$ onto $\bar{\mathscr{G}}$ to elements of $\bar{\mathscr{M}}$. If $\mathscr{N}$ is the invariant subgroup for which the map of $\mathscr{G}$ onto $\bar{\mathscr{G}}$ to the unit element of $\mathscr{G}$, then we find that $\bar{\mathscr{G}} \cong \mathscr{G}/\mathscr{N}$. For this homomorphism of $\mathscr{G}$ onto $\bar{\mathscr{G}}$ we find that $\mathscr{M}$ (which contains $\mathscr{N}$) is mapped onto $\bar{\mathscr{M}}$, and we therefore obtain $\bar{\mathscr{M}} \cong \mathscr{M}/\mathscr{N}$. Thus, from $\bar{\mathscr{G}}/\bar{\mathscr{M}} \cong \mathscr{G}/\mathscr{M}$ it follows that

$$\mathscr{G}/\mathscr{M} \cong (\mathscr{G}/\mathscr{N})/(\mathscr{M}/\mathscr{N}). \tag{4.2}$$

5 Direct Products

A group $\mathscr{G}$ is said to be a direct product of two subsets $\mathscr{N}_1$ and $\mathscr{N}_2$

$$\mathscr{G} = \mathscr{N}_1 \times \mathscr{N}_2$$

if and only if:

(1) $\mathscr{G} = \mathscr{N}_1\mathscr{N}_2$.
(2) $\mathscr{N}_1$ and $\mathscr{N}_2$ are invariant subgroups in $\mathscr{G}$.
(3) $\mathscr{N}_1 \cap \mathscr{N}_2 = \{e\}$.

This is equivalent to the following:

(a) Each of the elements g of $\mathscr{G}$ can be written in the form $g = n_1n_2$ where $n_1 \in \mathscr{N}_1$, $n_2 \in \mathscr{N}_2$.
(b) The elements of $\mathscr{N}_1$ commute with all the elements of $\mathscr{N}_2$.
(c) n_1, n_2 are uniquely determined by the relation $g = n_1n_2$.

PROOF. (1), (2), (3) imply (a), (b), (c).

(a) is an immediate consequence of (1). In addition $n_1n_2n_1^{-1} \in \mathscr{N}_2$ because $\mathscr{N}_2$ is an invariant subgroup. Therefore $n_1n_2n_1^{-1}n_2^{-1} \in \mathscr{N}_2$. $n_2n_1^{-1}n_2^{-1}$ is, however, an element of $\mathscr{N}_1$, and the same is true for $n_1n_2n_1^{-1}n_2^{-1}$. According to (3) it follows that

$n_1n_2n_1^{-1}n_2^{-1} = e$, from which it follows that (b) holds. If $n_1n_2 = n_1'n_2'$ then from $(n_1')^{-1}n_1 = n_2'n_2^{-1}$ is an element of both $\mathcal{N}_1$ and $\mathcal{N}_2$ and is therefore equal to e, from which we have proven (c).

(a), (b), (c) imply (1), (2), (3):

(1) follows directly from (a). Since n_2 commutes with the elements of $\mathcal{N}_1$, it follows that $g\mathcal{N}_1g^{-1} = n_1n_2\mathcal{N}_1n_2^{-1}n_1^{-1} = n_1\mathcal{N}_1n_1^{-1} = \mathcal{N}_1$, from which we have proven (2). If m is an element of both $\mathcal{N}_1$ and $\mathcal{N}_2$, then it can be represented in two ways in the form n_1n_2: $m = me = em$. From uniqueness (c) it follows that $m = e$.

The generalization of this result to several factors

$$\mathcal{G} = \mathcal{N}_1 \times \mathcal{N}_2 \times \cdots \times \mathcal{N}_r$$

is immediate: $g = n_1n_2 \cdots n_r$ where the n_ν all commute and are unique.

For several independently given groups $\mathcal{G}_1, \mathcal{G}_2, \ldots, \mathcal{G}_r$ we may define a new group with elements $g = g_1g_2 \cdots g_r$ where $gg' = (g_1g_1')(g_2g_2') \cdots (g_rg_r')$. It is easy to show that $\mathcal{G} = \mathcal{G}_1 \times \mathcal{G}_2 \times \cdots \times \mathcal{G}_r$.

If $\mathcal{G} = \mathcal{N}_1 \times \mathcal{N}_2 \times \cdots \times \mathcal{N}_r$ and $\mathcal{A}_\rho = \mathcal{N}_1 \times \mathcal{N}_2 \cdots \times \mathcal{N}_{\rho-1} \times \mathcal{N}_{\rho+1} \times \cdots \times \mathcal{N}_r$, then $\mathcal{G} = \mathcal{N}_\rho \times \mathcal{A}_\rho$ and $\mathcal{N}_\rho \cap \mathcal{A}_\rho = \{e\}$. Therefore, according to the isomorphism theorem, $\mathcal{G}/\mathcal{N}_\rho \cong \mathcal{A}_\rho$ and $\mathcal{G}/\mathcal{A}_\rho \cong \mathcal{N}_\rho$.

6 Representations of Groups

Let $\mathcal{G}$ be a group and let a map Ω of $\mathcal{G}$ into the set of linear operators for a vector space $\mathcal{X}$ be defined such that, for $a \in \mathcal{G}$ and $a \to \Omega(a)$ the following equation

$$\Omega(a)\Omega(b) = \Omega(ab)$$

is satisfied, then it follows that the vector space $\mathcal{X}$ can be written as a direct sum $\mathcal{X} = \imath + \imath_0$ where $\imath = \Omega(e)\mathcal{X}$ and $\imath_0 = (\mathbf{1} - \Omega(e))\mathcal{X}$. For $u \in \imath$ and for $v \in \imath_0$ it follows $\Omega(e)u = u$ and $\Omega(a)v = 0$ for all $a \in G$. If $\mathcal{X}$ is a Hilbert space and $\Omega(a)$ is a unitary operator it directly follows that $\Omega(e) = \mathbf{1}$. If $\Omega(e) = \mathbf{1}$ then it follows that $\Omega(a^{-1}) = \Omega(a)^{-1}$.

For a fixed representation, for $u \in \mathcal{X}$ we shall write au instead of $\Omega(a)u$ by means of which $\mathcal{G}$ is a set of linear operators of $\mathcal{X}$. We then say that $\mathcal{G}$ is an operator domain of $\mathcal{X}$ (see AIV, §14).

If $\mathcal{X}$ is a finite-dimensional Hilbert space and if $\mathcal{G}$ is a unitary representation (that is, all the $\Omega(a)$ are unitary) then since $\Omega(a^{-1}) = \Omega(a)^+$, the operator domain $\mathcal{G}$ is completely reducible (see AIV, §14). If $\mathcal{X}$ is finite dimensional (but not necessarily a Hilbert space) and if $\mathcal{G}$ is a finite group, then the operator domain $\mathcal{G}$ is completely reducible. This can be shown by introducing an inner product in such a way that the operators in $\mathcal{G}$ are unitary. If $u_1, \ldots, u_n$ is a basis for $\mathcal{X}$, then we define a function $F(u, v)$ as follows:

$$F(u, v) = \sum_{\nu=1}^{n} \bar{\alpha}_\nu\beta_\nu \quad \text{for} \quad u = \sum_{\nu=1}^{n} u_\nu\alpha_\nu \quad \text{and} \quad v = \sum_{\nu=1}^{n} u_\nu\beta_\nu.$$

Let the elements of the group (and therefore the operators of $\mathscr{X}$) be denoted by $a, b, c, \ldots$. $\sum_a$ or $\sum_b$ shall denote that we are to sum over all finitely many group elements. We define the inner product as follows:

$$\langle u, v\rangle = \sum_a F(au, av).$$

If b is a different element, then, for $c = ab$ we obtain

$$\langle bu, bv\rangle = \sum_a F(abu, abv) = \sum_c F(cu, cv) = \langle u, v\rangle$$

because, as we run through all a, $c = ab$ (for fixed b!) runs through all the elements of the group exactly once because the equation $c = ab$ has an inverse $a = cb^{-1}$. Thus all group elements b are unitary with respect to the above defined inner product.

In §10.6 we shall find that each unitary representation of a finite group or a compact group is completely reducible in an infinite-dimensional Hilbert space, and we may apply the theorems of AIV, §14 to $\mathscr{G}$ (instead of M).

7 The Irreducible Representations of a Finite Group

In order to find all possible irreducible representations of a finite group $\mathscr{G}$ we shall consider the group algebra or group ring $\mathscr{R}_{\mathscr{G}}$. Its elements are defined by the set of all (formal) sums $\sum_a \beta_a a$ where the β_a are arbitrary complex numbers. The sum of two ring elements is defined by $(\sum_a \beta_a a) + (\sum_a \gamma_a a) = \sum_a (\beta_a + \gamma_a)a$; the product of two elements is defined by $(\sum_a \beta_a a)(\sum_b \gamma_b b) = (\sum_{a,b} \beta_a \gamma_b ab)$. It is easy to verify that the usual rules of addition and multiplication for ring elements are satisfied. The unit elements e of $\mathscr{G}$ is also the unit elements of the ring.

With respect to addition $\mathscr{R}_{\mathscr{G}}$ is therefore an h-dimensional vector space (h = order of the group $\mathscr{G}$) with the operator domain $\mathscr{G}$ where the product of a with a ring element $t = \sum_b \beta_b b$ is defined as follows:

$$at = \sum_b \beta_b ab. \tag{7.1}$$

This representation of $\mathscr{G}$ in $\mathscr{R}_{\mathscr{G}}$ is called the "regular" representation. It is, according to §6, completely reducible.

An invariant subspace is a subspace of $\mathscr{R}_{\mathscr{G}}$ which has the following property: if the subspace contains t, then it contains all at ($a \in \mathscr{G}$), and therefore contains all $(\sum_a \beta_a a)t = st$ where $s = \sum_a \beta_a a$.

Therefore the invariant subspaces are those subsets ℓ of $\mathscr{R}_{\mathscr{G}}$ which contain the difference $s - t$ for $s, t \in \ell$ and the product st for $t \in \ell$ and arbitrary s in $\mathscr{R}_{\mathscr{G}}$. Such subsets of a ring are called left ideals; if a subset contains ts instead, it is called a right ideal. If both, then it is a two-sided (dual) ideal. $\mathscr{R}_{\mathscr{G}}$ is therefore a direct sum of irreducible left ideals,

$$\mathscr{R}_{\mathscr{G}} = \ell_{11} + \ell_{12} + \cdots + \ell_{1\sigma} + \cdots + \ell_{q\rho} \tag{7.2}$$

where the $\ell_{\nu\mu}$ with the same first index are isomorphic.

We now have the following important theorem:

Every irreducible representation of $\mathscr{G}$ is equivalent to a representation of $\mathscr{G}$ in one of the ℓ's. In other terms, every irreducible vector space $(\mathscr{G})\mathscr{R}$ where $\mathscr{G}$ is the operator domain is isomorphic to one of the ℓ's.

We note that every representation of $\mathscr{G}$ can be extended to a representation of $\mathscr{R}_{\mathscr{G}}$ defined in $\mathscr{R}$ where: $(\sum_a \beta_a a)u = \sum_a \beta_a(au)$. The decomposition of a vector space $\mathscr{R}$ into irreducible components is not changed as a result of this extension of the operator domain.

Let $(\mathscr{R}_{\mathscr{G}})\mathscr{R}$ be an irreducible representation space of $\mathscr{R}_{\mathscr{G}}$. Let v be an arbitrary element of $\mathscr{R}$. We shall now consider the following map of $\mathscr{R}_{\mathscr{G}}$ onto $\mathscr{R}$ or a subset of $\mathscr{R}$: $s \to sv$ (where $s \in \mathscr{R}_{\mathscr{G}}$). This map is a homomorphism (here $\mathscr{R}_{\mathscr{G}}$ is considered to be a vector space with the operator domain $\mathscr{R}_{\mathscr{G}}$) because, for $t \in \mathscr{R}_{\mathscr{G}}$:

$$ts \to (ts)v = t(sv).$$

Thus $\mathscr{R}$ must be isomorphic to a direct sum of some of the ℓ (see AIV, §14). Since $\mathscr{R}$ should be irreducible, it follows that $\mathscr{R}$ is isomorphic to one of the $\ell_{\nu\mu}$.

If, in the decomposition (7.2) we collect the classes of isomorphic $\ell_{\nu\mu}$ by one subspace t_ν, the decomposition of $\mathscr{R}_{\mathscr{G}}$ can be written in the form

$$\mathscr{R}_{\mathscr{G}} = t_1 + t_2 + \cdots + t_q \tag{7.3}$$

with uniquely determined t_ν (see AIV, §14).

We shall now consider the homomorphic maps of $\mathscr{R}_{\mathscr{G}}$ onto itself. Let T denote one of these. Suppose that the unit element is mapped upon t:

$$Te = t. \tag{7.4}$$

Since T is a homomorphism, it commutes with the elements s of $\mathscr{R}_{\mathscr{G}}$ as operators; it follows that:

$$Ts = Tse = sTe = st. \tag{7.5}$$

Each homomorphism T therefore corresponds to a ring element t for which $Ts = st$. Conversely, the map $s \to st$ is, for arbitrary t, a homomorphism since.

$$rs \to (rs)t = r(st).$$

If we carry out the two homomorphic maps T_1, T_2 successively, it follows that $T_1T_2s = st_2t_1$, that is, T_1T_2 corresponds to t_2t_1. The map $T_1 + T_2$ clearly corresponds to $t_1 + t_2$.

The homomorphic maps of a complete reducible vector space are, from AIV, §14, already known to us. The ring $\mathscr{R}_{\mathscr{G}}$ is therefore "oppositely" isomorphic to the ring of homomorphic maps of $\mathscr{R}_{\mathscr{G}}$ onto itself, "opposite" in the sense that T_1T_2 corresponds to the product t_2t_1.

Let $E^{(1)}_{\alpha\beta}$ denote one of the isomorphic maps of $\ell_{1\alpha}$ onto $\ell_{1\beta}$ (where $E^{(1)}_{\alpha\beta}s = 0$ for all $s \in \ell_{\nu\mu}$ with $\nu \neq 1$ or $\mu \neq \alpha$), then these maps correspond to elements $e^{(1)}_{\alpha\beta}$ of $\mathcal{R}_{\mathcal{G}}$. Now, according to AIV, §14, $T = \sum_{\alpha\beta\nu} \tau^{(\nu)}_{\alpha\beta} E^{(\nu)}_{\alpha\beta}$ and therefore every t can be written in the form $t = \sum_{\alpha\beta\nu} \tau^{(\nu)}_{\alpha\beta} e^{(\nu)}_{\alpha\beta}$. From $E^{(1)}_{\alpha\beta} E^{(1)}_{\gamma\alpha} = E^{(1)}_{\gamma\beta}$ it follows that $e^{(1)}_{\gamma\alpha} e^{(1)}_{\alpha\beta} = e^{(1)}_{\gamma\beta}$. Similarly, in general we obtain:

$$e^{(\nu)}_{\alpha\beta} e^{(\mu)}_{\gamma\delta} = \delta_{\nu\mu} \delta_{\beta\gamma} e^{(\nu)}_{\alpha\delta}. \tag{7.6}$$

In the $e^{(\nu)}_{\alpha\beta}$ we have found the decomposition of $\mathcal{R}_{\mathcal{G}}$ into irreducible left ideals: $e^{(1)}_{11}, e^{(1)}_{21}, e^{(1)}_{31}, \ldots$ form the basis (the $e^{(\nu)}_{\rho\sigma}$ are linear independent because the $E^{(\nu)}_{\beta\alpha}$ are) of an irreducible left ideal ℓ_{11} because

$$te^{(1)}_{\rho 1} = \sum_{\alpha\beta\nu} \tau^{(\nu)}_{\alpha\beta} e^{(\nu)}_{\alpha\beta} e^{(1)}_{\rho 1} = \sum_{\alpha} \tau^{(1)}_{\alpha\rho} e^{(1)}_{\alpha 1}. \tag{7.7}$$

Therefore t will be represented in this left ideal by this basis in terms of the matrix $\tau^{(1)}_{\alpha\rho}$. The dimension n_1 of ℓ_1 is therefore equal to the number of the $\ell_{1\alpha}$, that is, the number of left ideals isomorphic to ℓ_{11}. The left ideals $\ell_{1\rho} = (e^{(1)}_{1\rho}, e^{(1)}_{2\rho}, \ldots)$ are, however, isomorphic to ℓ_{11}. We obtain $\ell_{11} + \ell_{12} + \cdots = \ell_1$, which therefore has the basis $e^{(1)}_{\alpha\beta}$. The ℓ_ν are therefore rings of dimension n^2_ν. In particular, we obtain $e = \sum_{\rho,\nu} e^{(\nu)}_{\rho\rho}$.

The number of different irreducible representations of a group is therefore equal to the number of the ℓ_ν. If the representation generated by the ring ℓ_ν is of degree n_ν then the dimension of ℓ_ν is equal to n^2_ν, and the dimension of the ring $\mathcal{R}_{\mathcal{G}}$ is equal to $\sum_\nu n^2_\nu$. Since the dimension of $\mathcal{R}_{\mathcal{G}}$ is also equal to the order h of the group, we obtain the theorem of Burnside: $h = \sum_\nu n^2_\nu$. If we consider a special element a of the group and its decomposition

$$\sum_{\rho\sigma\nu} g^{(\nu)}_{\rho\sigma}(a) e^{(\nu)}_{\rho\sigma},$$

then the $g^{(\nu)}_{\rho\sigma}$ form, as a function of a, that is, as a function on the group manifold, a complete linear independent system. It is only necessary to demonstrate the linear independence because the completeness follows from the fact that there are exactly as many functions as group elements. Suppose that $\sum_{\rho\sigma\nu} \lambda^{(\nu)}_{\rho\sigma} g^{(\nu)}_{\rho\sigma}(a) = 0$ for all a. Then since $t = \sum_a \beta_a a = \sum_{\rho\sigma\nu} \tau^{(\nu)}_{\rho\sigma} e^{(\nu)}_{\rho\sigma}$ it follows that $\tau^{(\nu)}_{\rho\sigma} = \sum_a \beta_a g^{(\nu)}_{\rho\sigma}(a)$ and we therefore obtain $\sum_{\rho\sigma\nu} \lambda^{(\nu)}_{\rho\sigma} \tau^{(\nu)}_{\rho\sigma} = 0$ for all possible values of $\tau^{(\nu)}_{\rho\sigma}$. This is only possible if all the $\lambda^{(\nu)}_{\rho\sigma}$ are equal to 0.

The center $\mathcal{Z}$ of a ring is the set of all elements of the ring which commute with all the elements of the ring. $\mathcal{Z}$ is itself a ring. Let $z = \sum_a \beta_a a$ belong to the center $\mathcal{Z}$ of the ring $\mathcal{R}_{\mathcal{G}}$. Then for all elements of the ring $t \sum_a \beta_a a = \sum_a \beta_a at$. This condition is satisfied if it is satisfied for all the elements of the group, that is, $b \sum_a \beta_a a = \sum_a \beta_a ab$ and therefore $\sum_a \beta_a bab^{-1} = \sum_a \beta_a a$. This is the case if and only if $\beta_a = \beta_{bab^{-1}}$ for all b. An element c from $\mathcal{G}$ is said to be conjugate to d if there exists an element b such that $c = bdb^{-1}$, from which it follows that d is also conjugate to c. If c is conjugate to d and if d is conjugate to f then c is also conjugate to f. The mutually conjugate elements form separate equivalence classes. Thus, for an element z of the center of $\mathcal{R}_{\mathcal{G}}$ the β_a must be identical for the elements of the same class. If $k = \sum'_a a$, where the sum is taken over all the a in a given class, we can therefore write $z = \sum_k \gamma_k k$, where this sum is to be taken over all classes. For arbitrary γ_k this sum is also an element of the center. The dimension of the center is therefore equal to the number of classes in $\mathscr{G}$.

On the other hand, z can be expanded as follows: $z = \sum_{\rho\sigma\nu} \xi^{(\nu)}_{\rho\sigma} e^{(\nu)}_{\rho\sigma}$. In order that z commutes with all the elements of $\mathscr{R}_{\mathscr{G}}$ it suffices to show that z commutes with all the $e^{(\mu)}_{\alpha\beta}$. Here we directly see that z must have the form

$$z = \sum_{\nu} \xi^{(\nu)} e^{(\nu)} \quad \text{where} \quad e^{(\nu)} = \sum_{\rho} e^{(\nu)}_{\rho\rho}. \tag{7.8}$$

Therefore the number of irreducible representations is equal to the number of classes of conjugate elements.

8 Orthogonality Relations for the Elements of Irreducible Representation Matrices

In the previous section we have seen that the $g^{(\nu)}_{\rho\sigma}(a)$ form a complete linearly independent system of functions on the group. In addition, they satisfy certain orthogonality relations. To simplify notation we shall write $G^{(\nu)}(a) = (g^{(\nu)}_{\rho\sigma}(a))$. If $G^{(\nu)}(a)$ is the matrix of the νth irreducible representation in a vector space $\imath^{(\nu)}$ (for example, $\ell_{\nu 1}$ in the previous section) and $G^{(\mu)}(a)$ is the matrix for the μth representation in $\imath^{(\mu)}$ then the matrix product

$$T = \textstyle\sum_a G^{(\nu)}(a) C G^{(\mu)}(a^{-1})$$

with an arbitrary matrix C of $n^{(\nu)}$ rows and $n^{(\mu)}$ columns is a homomorphism of $\imath^{(\mu)}$ onto $\imath^{(\nu)}$ since

$$\begin{aligned} bT &= G^{(\nu)}(b) \sum_a G^{(\nu)}(a) C G^{(\mu)}(a^{-1}) \\ &= \sum_a G^{(\nu)}(ba) C G^{(\mu)}(a^{-1}) \\ &= \sum_c G^{(\nu)}(c) C G^{(\mu)}(c^{-1} b) = \sum_c G^{(\nu)}(c) C G^{(\mu)}(c^{-1}) G^{(\mu)}(b) \\ &= Tb. \end{aligned}$$

Thus (see AIV, §14) T must be the null operator for $\nu \neq \mu$; for $\nu = \mu$, T must be a multiple of the unit matrix E, that is, $T = \delta_{\nu\mu} \beta^{(\nu)} E$. Therefore we obtain

$$\sum_{a,\lambda,\rho} g^{(\nu)}_{\kappa\lambda}(a) c_{\lambda\rho} g^{(\mu)}_{\rho\tau}(a^{-1}) = \delta_{\nu\mu} \beta^{(\nu)} \delta_{\kappa\tau}. \tag{8.1}$$

Since the $c_{\lambda\rho}$ are completely arbitrary, it follows that

$$\sum_a g^{(\nu)}_{\kappa\lambda}(a) g^{(\mu)}_{\rho\tau}(a^{-1}) = \delta_{\nu\mu} \delta_{\kappa\tau} \beta^{(\nu)}_{\lambda\rho}. \tag{8.2}$$

For $\nu = \mu$ we set $\kappa = \tau$ and sum over τ. Since

$$\sum_{\tau} g^{(\nu)}_{\tau\lambda}(a) g^{(\nu)}_{\rho\tau}(a^{-1}) = \delta_{\rho\lambda}$$

it therefore follows that

$$\sum_a \delta_{\rho\lambda} = \beta^{(\nu)}_{\lambda\rho} \sum_{\kappa} \delta_{\kappa\kappa}, \quad \text{that is,} \quad h\delta_{\rho\lambda} = \beta^{(\nu)}_{\lambda\rho} n^{(\nu)},$$

where h is the order of the group and $n^{(\nu)}$ is the degree of the representation. Thus we obtain

$$\frac{1}{h}\sum_a g_{\kappa\lambda}^{(\nu)}(a)g_{\rho\tau}^{(\mu)}(a^{-1}) = \delta_{\nu\mu}\delta_{\rho\lambda}\delta_{\kappa\tau}\frac{1}{n^{(\nu)}}. \tag{8.3}$$

In this sense the functions $g_{\rho\sigma}^{(\nu)}(a)$ are mutually orthogonal. If the representations are unitary, then we obtain

$$g_{\rho\tau}^{(\mu)}(a^{-1}) = \overline{g_{\tau\rho}(a)}$$

and we therefore obtain

$$\frac{1}{h}\sum_a \overline{g_{\tau\rho}^{(\mu)}(a)}g_{\kappa\lambda}^{(\nu)}(a) = \delta_{\nu\mu}\delta_{\tau\kappa}\delta_{\rho\lambda}\frac{1}{n^{(\nu)}}. \tag{8.4}$$

If $\mathscr{X}$ is an arbitrary representation space, if v is a vector from $\mathscr{X}$, then for fixed λ and ν the $u_\tau^{(\nu)} = (n^{(\nu)}/h)\sum_a g_{\lambda\tau}^{(\nu)}(a^{-1})av$ span an invariant subspace $\imath$ of $\mathscr{X}$ since

$$\begin{aligned} bu_\tau^{(\nu)} &= \frac{n^{(\nu)}}{h}\sum_a g_{\lambda\tau}^{(\nu)}(a^{-1})bav \\ &= \frac{n^{(\nu)}}{h}\sum_c g_{\lambda\tau}^{(\nu)}(c^{-1}b)cv \\ &= \frac{n^{(\nu)}}{h}\sum_{c\rho} g_{\lambda\rho}^{(\nu)}(c^{-1})cvg_{\rho\tau}^{(\nu)}(b) = \sum_\rho u_\rho^{(\nu)}g_{\rho\tau}^{(\nu)}(b). \end{aligned}$$

$\imath$ is therefore either null or the representation in $\imath$ is equivalent to $G^{(\nu)}(a)$.

If $\mathscr{X}$ is finite dimensional, then the representation in $\mathscr{X}$ can be reduced with the aid of the $g_{\lambda\tau}^{(\nu)}(a)$ as follows: Beginning with a basis $\{v_\nu\}$ of $\mathscr{X}$, construct, for each v_ν, in the manner described above, the subspace $\imath_\nu$. Then $\mathscr{X} = \sum_i \imath_{\nu_i}$ where we eliminate null spaces and only include multiple occurring spaces $\imath_\nu$ once in the sum (since the $\imath_\nu$ are irreducible, we either have $\imath_\nu \cap \imath_\mu = \{0\}$ or $\imath_\nu \cap \imath_\mu = \imath_\nu$).

If we only wish to know the number of times a certain irreducible representation occurs, then the character of the representation is often the most suitable way. The character of a representation is a function defined on the group as follows

$$\chi(a) = \operatorname{tr}(G(a)), \tag{8.5}$$

where $G(a)$ is the matrix for the representation. Since the trace is independent of the choice of basis, χ is, the same function for all isomorphic representation spaces.

Since $\chi(a) = \chi(bab^{-1})$, χ is a function of the class of conjugate elements. Let $\chi^{(\nu)}(a)$ denote the characters of the irreducible representations. We shall now show that the characters form a complete system of class functions. Since there are exactly as many characters as there are classes, it is only necessary to prove their linear independence. Since $\sum_{\nu\rho\sigma}\lambda_{\rho\sigma}^{(\nu)}g_{\rho\sigma}^{(\nu)}(a) = 0$ is possible only for the case $\lambda_{\rho\sigma}^{(\nu)} = 0$. It follows if we set $\lambda_{\rho\sigma}^{(\nu)} = \lambda^{(\nu)}\delta_{\rho\sigma}$, that

$\sum_{\nu\rho\sigma} \lambda_{\rho\sigma}^{(\nu)} g_{\rho\sigma}^{(\nu)}(a) = \sum_{\nu} \lambda^{(\nu)} \chi^{(\nu)}(a) = 0$ can only be satisfied if all the $\lambda^{(\nu)} = 0$. The character χ can be uniquely expressed in terms of the $\chi^{(\nu)}$ as follows:

$$\chi(a) = \sum_{\nu} m_{(\nu)} \chi^{(\nu)}(a) \tag{8.6}$$

It is easy to show that $\chi(a)$ contains a particular $\chi^{(\nu)}(a)$ as often as the νth irreducible representation occurs, that is, $m_{(\nu)}$ times. The $m_{(\nu)}$ are therefore integers.

If in (8.4) we set $\kappa = \lambda$ and $\rho = \tau$, and we sum over κ and ρ, then we would obtain

$$\frac{1}{h} \sum_{a} \chi^{(\nu)}(a) \chi^{(\mu)}(a^{-1}) = \delta_{\mu\nu}$$

and, since, for finite groups, all representations are unitarily equivalent, we may choose the representations to be unitary, and obtain

$$\frac{1}{h} \sum_{a} \chi^{(\nu)}(a) \overline{\chi^{(\mu)}(a)} = \delta_{\mu\nu}. \tag{8.7}$$

Thus, from (8.7) and by multiplication with $\overline{\chi^{(\mu)}(a)}$ and summing over a we obtain

$$m_{(\nu)} = \frac{1}{h} \sum_{a} \overline{\chi^{(\nu)}(a)} \chi(a). \tag{8.8}$$

9 Representations of the Symmetric Group

The group of $f!$ permutations of f things is called the symmetric group **S**. If ℓ_1 is a left ideal of the group ring $\mathscr{R}_{\mathrm{S}}$ then it is possible to decompose $\mathscr{R}_{\mathrm{S}}$ as follows: $\mathscr{R}_{\mathrm{S}} = \ell_1 + \ell_2$. For the unit element e we therefore obtain $e = e_1 + e_2$. Multiplying on the left by e_1 we obtain $e_1 = e_1^2 + e_1 e_2$ and therefore $e_1^2 = e_1$ and $e_1 e_2 = 0$. e_1 is called an idempotent element. $\mathscr{R}_{\mathrm{S}} e_1$ is identical to ℓ_1 since $\mathscr{R}_{\mathrm{S}} e_1$ is a left ideal and must be contained in ℓ_1; for all elements a of ℓ_1 it follows from $e = e_1 + e_2$, on the other hand, that $a = ae_1 + ae_2 = ae_1$. In order to find the irreducible representation of **S**, we shall only seek those idempotent e_ν for which the left ideal defined by $\mathscr{R}_{\mathrm{S}} e_\nu$ is irreducible.

We shall now consider the permutations p to be operators which permute the order of f things $a_1, a_2, \ldots, a_f$:

$$p a_1 a_2 \cdots a_f = a_{\alpha_1} a_{\alpha_2} \cdots a_{\alpha_f}.$$

For

$$q a_1 a_2 \cdots a_f = a_{\beta_1} a_{\beta_2} \cdots a_{\beta_f}$$

we obtain

$$p q a_1 a_2 \cdots a_f = a_{\beta_{\alpha_1}} a_{\beta_{\alpha_2}} \cdots a_{\beta_{\alpha_f}}.$$

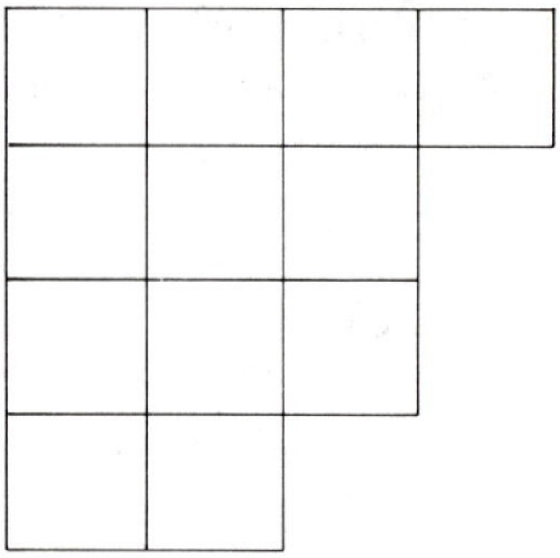

Figure 50

We construct a tableau with k rows with $n_1, n_2, \ldots, n_k$ places ($n_1 \geq n_2 \geq n_3 \geq \cdots \geq n_k$, $\sum_\nu n_\nu = f$), as is shown in Figure 50 for the case in which $f = 12$, $k = 4$, $n_1 = 4$, $n_2 = 3$, $n_3 = 3$, $n_4 = 2$. We shall represent such a tableau by T_n where n is an abbreviation for $n_1, n_2, \ldots, n_k$. Let m_1 denote the number of places in the first column, m_2 for the second, etc. Then $m_1 \geq m_2 \geq \cdots \geq m_k$, $\sum_\mu m_\mu = f$ and $m_1 = k$.

If we arrange the f things a_i into the places in the tableau, we then obtain a scheme. A particular sequence $a_1 \cdots a_n$ of the a_i may be called the initial sequence. If we arrange the a_i into the tableau T_n in its initial sequence beginning with the first row, then the second, etc., we obtain a special scheme which we shall denote by S_n. We shall denote the result of applying a permutation p to the a_i in the scheme S_n by pS_n. For a given tableau T_n there are $f\,!$ different schemes pS_n.

For each tableau the permutations r which only permute the elements of each row of S_n in itself are uniquely determined; the same is also true for the permutations s which permute only the elements in the same column. There are therefore $\prod_\nu (n_\nu)!$ permutations r and $\prod_\mu (m_\mu)!$ permutations of the form s.

The $(-1)^p$ are equal to $+1$ for an even and -1 for an odd permutation p. (A permutation p is said to be even or odd depending on whether the permutation of the u_ν results in the expression $\prod_{\nu<\mu} (u_\nu - u_\mu)$ changing sign or not.) For the elements

$$I_n = \sum_r r \quad \text{and} \quad K_n = \sum_s (-1)^s s \tag{9.1}$$

of the group ring it follows that

$$rI_n = I_n r = I_n; \qquad (-1)^s s K_n = K_n (-1)^s s = K_n. \tag{9.2}$$

Thus it follows that $I_n^2 = \prod_\nu (n_\nu)!\, I_n$ and $K_n^2 = \prod_\mu (m_\mu)!\, K_n$. Therefore I_n and K_n will be, with suitably chosen factors, idempotent elements. $\mathscr{R}_S I_n$ and $\mathscr{R}_S K_n$ are therefore left ideals in $\mathscr{R}_S$. They are, in general, not irreducible. We will now show that for a suitable actor γ_n

$$e_n = \gamma_n I_n K_n \tag{9.3}$$

is an idempotent element and the left ideals $\mathscr{R}_S\, e_n$ are irreducible, and, for different n (that is, for different tableaux) are inequivalent. For this purpose we shall now prove the following combinatorial lemma: If pS_n and $p'S_{n'}$ are two schema for which $n \geq n'$ (that is $n_1 > n_1'$ or $n_1 = n_1'$ and $n_2 > n_2'$, or ...) such that in pS_n two of the a_i never are found in the same row which occur in $p'S_{n'}$ in the same column, then $n = n'$ and there exist two permutations r, s such that $pS_n = rsp'S_n$, that is, $rsp' = p$.

PROOF. Assume that $n_1 \geq n_1'$. Since the a_i of the first row in pS_n must be placed into different columns in $p'S_{n'}$ $T_{n'}$ must have at least n_1 columns. It follows that $n_1 = n_1'$. There exists a permutation s_1' such that in the schema $s_1'p'S_{n'}$ all of the a_i of the first row of pS_n must be in the first column of $s_1'p'S_{n'}$. Since all elements a_i of the second row of pS_n must be in different columns of $s_1'p'S_{n'}$, it follows that $n_2' \geq n_2$, from which, together with $n \geq n'$ it follows that $n_2' = n_2$. By a suitable permutation s_2' it is possible that all the elements of the second row of pS_n will also be in the second row of $s_2's_1'p'S_{n'}$. By a repetitive argument, it follows that $n = n'$ and that there is a sequence of permutations s_ν' such that in $s_{m_1}'s_{m_1-1}' \cdots s_2's_1'p'S_{n'}$ the elements a_i appear in the same rows as in pS_n. Therefore there exists a permutation r such that $pS = rsp'S$ where $s = s_{m_1}' \cdots s_1'$.

If p is an arbitrary permutation, then for $n > n'$ there exists in S_n a pair of elements in the same row which occupies places of the same column in $p^{-1}S_{n'}$. If t is the permutation which exchanges these two a_i in S_n and t' is the corresponding permutation in $p^{-1}S_{n'}$, then $t' = p^{-1}tp$. Thus, from (9.2) and since $t^2 = 1$ it follows that

$$I_n'pK_{n'} = I_n pp^{-1}tpt'K_{n'} = I_n tpt'K_{n'} = -I_n pK_{n'}$$

from which it follows that $I_n pK_{n'} = 0$.

Since this is true for each p, it is true for each element of $\mathscr{R}_S$, that is,

$$I_n \mathscr{R}_S K_{n'} = 0 \quad \text{for } n > n'.$$

For $L_n = I_n K_n$ ($L_n \neq 0$ because $\sum_{r,s}(-1)^s rs$ contains only one term with the unit element and the factor 1, namely, the case in which $r = s = e$) it follows that $rL_n = L_n = L_n s(-1)^s$. Conversely, if

$$ra = a = as(-1)^s \tag{9.4}$$

is satisfied for a ring element a it follows that

$$a = \lambda L_n.$$

To prove this, we write $a = \sum_p p\alpha(p)$. From $ras(-1)^s = a$ it follows that $\sum_p rps\alpha(p)(-1)^s = \sum_p p\alpha(p)$. Therefore, if we set $p = e$ in the sum on the left and if we set $p = rs$ in the sum on the right we obtain $\alpha(rs) = (-1)^s\alpha(e)$, that is, all coefficients $\alpha(p)$ where p may be written as the product of two permutations r and s, are equal to $(-1)^s\alpha(e)$. If we could show that all the remaining $\alpha(p)$ vanish, then we would obtain $a = \alpha(e)L_n$. If p cannot be written in the form rs, then all the rsS_n are different from S_n. According to the combinatorial lemma, in pS_n there are two elements a_i in the same row which occur in the same column in S_n. If t is the permutation which exchanges these two elements in pS_n, then $p^{-1}tp$ does the same in S_n. Thus t is a permutation of type r and $t' = p^{-1}tp$ is one of type s, so that from

$$\sum_q rqs\alpha(q)(-1)^s = \sum_q q\alpha(q),$$

where we set $r = t$ and $s = p^{-1}tp$ it follows that

$$-\sum_q tqp^{-1}tp\alpha(q) = \sum_q q\alpha(q).$$

Thus for the special case in which $q = p$ we obtain

$$\alpha(p) = -\alpha(p), \quad \text{that is,} \quad \alpha(p) = 0$$

which concludes our proof. Since the element $a = I_n b K_n$ has the property (9.4) for an arbitrary element b of $\mathscr{R}_S$, it therefore follows that

$$I_n b K_n = \lambda_b I_n K_n = \lambda_b L_n$$

and, in particular,

$$L_n^2 = I_n(K_n I_n)K_n = \beta_n L_n.$$

For each b we therefore obtain $L_n b L_n = \mu_b L_n \cdot \ell_n = \mathscr{R}_S L_n$ is an irreducible left ideal otherwise there would be an irreducible left ideal $\ell' \subset \ell_n$, and for an element b' from ℓ' it would follow that $L_n b' = \rho L_n$. Since for b' it follows that all $\alpha b'$ are in ℓ', either we have $L_n \ell' = 0$ or $L_n \ell' = (L_n)$ where (L_n) is the one-dimensional subspace of $\mathscr{R}_S$ spanned by L_n. From $L_n \ell' = (L_n)$ it follows that $\ell_n = \mathscr{R}_S L_n = \mathscr{R}_S L_n \ell' \subset \ell_n \ell' \subset \ell'$, that is, $\ell' = \ell_n$. From $L_n \ell' = 0$ it follows that $\ell' \ell' = 0$. If we decompose $e = e' + \varepsilon$ according to the decomposition $\mathscr{R}_S = \ell' + \imath$, then, upon multiplication with e' it follows that $e'^2 = e'$ and $e\varepsilon = 0$ and finally $\ell' = \mathscr{R}_S e'$. Since $\ell' \ell' = 0$, we obtain $e'^2 = 0$, that is, $e' = 0$ and therefore $\ell' = 0$.

We find that $L_n^2 \neq 0$ and therefore $\beta_n \neq 0$, otherwise from $L_n^2 = 0$, and from $\ell_n L_n = 0$ it follows that the trace of the map of $\mathscr{R}_S$ onto itself defined by $a \to aL_n$ would be zero, a result which is easily obtained from the choice of a basis for the decomposition $\mathscr{R}_S = \ell_n + \imath$. For the basis of the group elements it follows that the trace is $f!\,L_n(e) = f!$ where $L_n(p)$ is given by $L_n = \sum_p pL_n(p)$. From this result it follows that if g is the dimension of ℓ_n for β_n, we obtain

$$\beta_n = g^{-1} f!$$

since the trace is, according to the first method, equal to $\beta_n g$. g is computed in XIV (8.28). From $\beta_n \neq 0$ it follows that $e_n = \beta_n^{-1} L_n$ is an idempotent element (that is, $e_n^2 = e_n$) of ℓ_n. It remains to show that two left ideals $\ell_n, \ell_{n'}$ correspond to inequivalent representations. Suppose, for example, that $n > n'$, then we obtain

$$I_n \ell_{n'} = I_n \mathscr{R}_S L_{n'} = I_n \mathscr{R}_S I_{n'} K_{n'} \subset I_n \mathscr{R}_S K_{n'} = 0.$$

I_n will therefore be represented in $\ell_{n'}$ by 0. Suppose $\ell_n \cong \ell_{n'}$, then I_n must also be represented in ℓ_n by 0. We note, however, that

$$I_n L_n = I_n I_n K_n = \prod_\nu (n_\nu)!\, I_n K_n = \prod_\nu (n_\nu)!\, L_n \neq 0.$$

We therefore have as many irreducible inequivalent representations as there are different tableaux, that is, as many as there are decompositions of f into a sum of integers $f = \sum_\nu n_\nu$. This number is identical to the number of classes of conjugate elements in **S**. In order to prove this we shall write the permutations in "cycle representation"

$$p = z_{n_1} z_{n_2} \cdots z_{n_k}, \tag{9.5}$$

where, for example, the cycle $z_3 = (2\ 4\ 1)$ yields the following permutation

$$(2\ 4\ 1) a_1 a_2 a_3 a_4 a_5 = a_2 a_4 a_3 a_1 a_5.$$

The z_{n_i} in (9.5) contain only different numbers. Since each such p can be uniquely represented in the form (9.5), and the cycles z_{n_i} in (9.5) commute, the desired result is not difficult to prove. It is easy to see that

$$q z_{n_1} \cdots z_{n_k} q^{-1} = z'_{n_1} z'_{n_2} \cdots z'_{n_k},$$

where the cycles z'_{n_i} have the same "length" n_i as do the z_{n_i}, only the numbers in the cycles are permuted by q. A class of conjugate elements will therefore consist of all permutations for which the cycles are of the same length n_i where $\sum_i n_i = f$.

10 Topological Groups

Topological groups play an important role in physics. The physical meaning of topology has been briefly outlined in VI, §1. We shall only attempt to provide a few important fundamental aspects of topological groups without detailed proof. The most important theorems for physics are proved in [41].

10.1 The Species of Structure: Topological Group

A group $\mathscr{G}$ is called a topological group if a topology (AII, §1) is defined upon $\mathscr{G}$ for which the map $(a, b) \to ab$ defined on $\mathscr{G} \times \mathscr{G} \to \mathscr{G}$ is continuous and the map $a \to a^{-1}$ is continuous. These conditions are equivalent to the single condition—the map $(a, b) \to ab^{-1}$ defined on $\mathscr{G} \times \mathscr{G} \to \mathscr{G}$ is continuous.

It therefore follows that, for fixed a, the map $a \to ax$ is a homeomorphism—that is, a topological isomorphism—of G onto itself. The maps $x \to xa$ and $x \to x^{-1}$ are also homeomorphisms.

Thus it follows that if $\mathscr{V}$ is the neighborhood filter of the unit element, then $a\mathscr{V} = \{aV \mid V \in \mathscr{V})$ and $\mathscr{V}a$ are neighborhood filters of a. The topology of a group is therefore determined by the neighborhood filter of the unit element.

Conversely, if a filter $\mathscr{V}$ is given in $\mathscr{G}$ such that the following conditions are satisfied:

(1) The unit element is contained in each element V of $\mathscr{V}$;
(2) for each $U \in \mathscr{V}$ there exists a $V \in \mathscr{V}$ such that $VV \subset U$;
(3) $U \in \mathscr{V}$ implies that $U^{-1} \in \mathscr{V}$;
(4) for each $a \in G$ and $U \in \mathscr{V}$ it follows that $aUa^{-1} \in \mathscr{V}$

(see TG 1—TG 3 in VI, §1.1) then this filter defines a topology on $\mathscr{G}$ for which $\mathscr{G}$ is a topological group. For a neighborhood filter of an arbitrary element a we need only choose the set of all aV where $V \in \mathscr{V}$.

A neighborhood V of the unit element for which $V^{-1} = V$ is said to be symmetric. For arbitrary V, $V \cup V^{-1}$, $V \cap V^{-1}$, VV^{-1} are symmetric neighborhoods. Thus it easily follows that there exists a set of symmetric neighborhoods of unity which form a fundamental system of neighborhoods.

We shall only consider separated topological groups. The topology is separating if and only if the intersection of all neighborhoods of the unit element is the unit element.

The general theorems about subgroups and invariant subgroups in §§1–4 can be extended as follows: If $\mathscr{H}$ is a subgroup of $\mathscr{G}$, then its closure $\overline{\mathscr{H}}$ is also a subgroup. If $\mathscr{H}$ is an invariant subgroup, then so is $\overline{\mathscr{H}}$. The center of $\mathscr{G}$ is a closed subgroup of $\mathscr{G}$.

A subgroup $\mathscr{H}$ of $\mathscr{G}$ is open if it has at least one interior point, because if $a \in \mathscr{H}$ is an inner point of $\mathscr{H}$ then so is $b = (ba^{-1})a$. An open subgroup $\mathscr{H}$ is also closed: Let $a \in \overline{\mathscr{H}}$ and let V be a symmetric neighborhood of the unit element with $V \subset \mathscr{H}$. Then we therefore obtain $aV \cap \mathscr{H} \neq \phi$. Therefore there exists $ab \in aV \cap \mathscr{H}$. Thus, since V is symmetric, $a \in bV \subset b\mathscr{H} \subset \mathscr{H}$, that is, $a \in \mathscr{H}$.

If V is a symmetric neighborhood of the unit element, then the set of all products $\prod_{\nu=1}^{n} a_\nu$ where $a_\nu \in V$ is a subgroup $\mathscr{H}$ of $\mathscr{G}$. Since the unit element is an interior point of $\mathscr{H}$, $\mathscr{H}$ is both open and closed. Therefore $\mathscr{G} \setminus \mathscr{H}$ is an open set. If $\mathscr{H} \neq \mathscr{G}$ then $\mathscr{G}$ is not connected (see AII, §4). If $\mathscr{G}$ is connected, then $\mathscr{G}$ is generated by any neighborhood V of the unit element, that is, every element a of $\mathscr{G}$ may be represented in the form $a = \prod_{\nu=1}^{n} a_\nu$ with $a_\nu \in V$.

Let $\mathscr{N}$ be the connected component (AII, §4) of the unit element. With $a \in \mathscr{N}$, $a^{-1}\mathscr{N}$ must be a connected subset of $\mathscr{G}$. Since $e \in a^{-1}\mathscr{N}$ it follows that $a^{-1}\mathscr{N} \subset \mathscr{N}$, that is, $\mathscr{N}^{-1}\mathscr{N} \subset \mathscr{N}$, which means that $\mathscr{N}$ is a subgroup. For $b \in \mathscr{G}$ it follows that $b\mathscr{N}b^{-1}$ is connected, and $e \in b\mathscr{N}b^{-1}$, that is, $b\mathscr{N}b^{-1} \subset \mathscr{N}$ and therefore $b\mathscr{N}b^{-1} = \mathscr{N}$ since $x \to bxb^{-1}$ is an automorphism of $\mathscr{G}$. Therefore $\mathscr{N}$ is an invariant subgroup.

For $b \in \mathscr{G}$ it follows that $b\mathscr{N} = \mathscr{N}b^{-1}$ is a connected set which contains b, that is, $b\mathscr{N} \subset \mathscr{K}$ where $\mathscr{K}$ is the connected component of b. For $b \in \mathscr{K}$ it follows that $b^{-1}\mathscr{K} \subset \mathscr{N}$, that is, $\mathscr{K} \subset b\mathscr{N}$ and we therefore find that $\mathscr{K} = b\mathscr{N}$.

Since $\mathscr{N}$ is a connected group, it is generated by all products of the form $\prod_{\nu=1}^{n} a_\nu$ where $a_\nu \in V \cap \mathscr{N}$ where V is an arbitrary neighborhood of e.

10.2 Uniform Structures of Groups

In a topological group $\mathscr{G}$ it is a simple matter to define a uniform structure. If for every neighborhood V of the unit element e we assign a vicinity

$$\{(x, y) \mid yx^{-1} \in V\}$$

we obtain a fundamental system of vicinities, and therefore obtain a uniform structure called the right uniform structure. Similarly, a left uniform structure may be defined by $\{(x, y) \mid x^{-1}y \in V\}$. The maps $x \to ax$ and $x \to xa$ are isomorphisms for both uniform structures. The map $x \to x^{-1}$ is an isomorphism between the left and right structures. In physical applications most groups are such that the left and right unform structures are identical.

If the map $x \to x^{-1}$ transforms Cauchy filters of the right uniform structure into Cauchy filters of the right uniform structure, then $\mathscr{G}$ has a completion $\hat{\mathscr{G}}$ and the group structure can be extended to $\hat{\mathscr{G}}$. The right and the left uniform structures produce the same $\hat{\mathscr{G}}$ (for a proof, see [12]).

If $\mathscr{G}$ is complete with respect to the right uniform structure, then it is also complete with respect to the left uniform structure and vice versa (for proof, see [12]).

If a group $\mathscr{G}$ has a neighborhood V of e which is either right or left complete, then $\mathscr{G}$ is complete. In order to prove this statement, consider a Cauchy filter $\mathscr{F}$ in $\mathscr{G}$. Since $\mathscr{F}$ is a Cauchy filter, there exists an element $M \in \mathscr{F}$ which is small of the order of $\{(x, y) \mid yx^{-1} \in V\}$. For $x_1 \in M$ we therefore obtain $M \subset Vx_1$. Since Vx_1 is complete, the Cauchy filter has a limit point $x_0 \in Vx_1$, which also must be a limit point of $\mathscr{F}$ in $\mathscr{G}$.

Since every compact set is complete, it follows that every locally compact group is complete. For a compact group the right- and left-handed uniform structures must, of course, be identical, namely that of the compact space.

If there exists a denumerable neighborhood basis of the unit element, then $\mathscr{G}$ is metrizable, a result which follows easily from AII, §2.

In applications we only find metrizable and separable groups. It is therefore always possible to introduce a new uniform structure ph of physical imprecision such that $\mathscr{G}$ is precompact. For this purpose we choose a metric $d(x, y)$. We consider all maps $\mathscr{G} \xrightarrow{f_{\nu\mu}^{(i)}} [0, 1]$ where $f_{\nu\mu}^{(1)}(x) = \pi^{-1}tg^{-1}d(xa_\nu, a_\mu)$ and $f_{\nu\mu}^{(2)}(x) = \pi^{-1}tg^{-1}d(a_\nu x, a_\mu)$ where $\{a_\mu\}$ is a countable dense subset in $\mathscr{G}$. The initial uniform structure ph corresponding to the $f_{\nu\mu}^{(i)}$ is then precompact and the maps $x \to ax$, $x \to xa$ are uniformly continuous. For the completion $\mathscr{G}_{ph}$ the group structure can be lost!

10.3 Lie Groups

We shall not present a detailed description of the theory of Lie groups. Readers who are interested in the full mathematical theory are referred to the specialized literature, in particular, to [13]. Instead, we shall develop some of the fundamental concepts which are of considerable importance for quantum mechanics. This importance induced some authors to seek a formulation of quantum mechanics based entirely on purely group theoretical methods (possibly in a mystifying way). This section is for those readers who do not yet have an intensive familiarity with quantum mechanics.

A Lie group $\mathscr{G}$ is a finite-dimensional topological group which has an open neighborhood U_0 of the unit element which can be mapped homeomorphically onto an open set of $\mathbf{R}^p$ in such a way that the group operations are continuously differentiable. This means that for $a \in U_0$ there exists a correspondence $a \leftrightarrow \alpha^1, \ldots, \alpha^p$ (where for simplicity we assume that $e \leftrightarrow 0, 0, \ldots, 0$) such that for $b \leftrightarrow \beta^1, \ldots, \beta^p$, for $c = ab$ and $c \leftrightarrow \gamma^1, \ldots, \gamma^p$, the functions $\gamma^\nu = f_\nu(\alpha^1, \ldots, \alpha^p; \beta^1, \ldots, \beta^p)$ are continuously differentiable. The correspondence $a \leftrightarrow \alpha^1, \ldots, \alpha^p$ is called a chart in the neighborhood U_0 of e. Other charts can be obtained by using continuously differentiable transformations.

From $U_c = cU_0$ we obtain a neighborhood of the element c. Each $b \in U_c$ may then be written as follows: $b = ca$ where $a \in U_0$. From $a \leftrightarrow \alpha^1, \ldots, \alpha^p$ and $b = ca \leftrightarrow a \leftrightarrow \alpha^1, \ldots, \alpha^p$ we obtain a chart $b \leftrightarrow \alpha^1, \ldots, \alpha^p$ for the elements of U_c.

Let $c \in U_d$ and $d \in U_c$. $U_c \cap U_d$ is then an open set. Let $f \in U_c \cap U_d$. Since f is an element of U_c we may have $f \leftrightarrow \gamma^1, \ldots, \gamma^p$ in the chart of U_c; since f is an element of U_d we may have $f \leftrightarrow \delta^1, \ldots \delta^p$ in the chart of U_d. Thus we have $\delta^\nu = g_\nu(\gamma^1, \ldots, \gamma^p)$. We will now show that the g_ν are continuously differentiable. We have $f = ca = db$ where $a \leftrightarrow \gamma^1, \ldots, \gamma^p$ and $b \leftrightarrow \delta^1, \ldots, \delta^p$. It follows that $b = d^{-1}f = d^{-1}ca$. Since $c \in U_d$ there exists an $x \in U_0$ such that $c = dx$. Therefore $b = xa$ from which it follows that the g_ν are continuously differentiable. In this way $\mathscr{G}$ is a differentiable manifold. It is not difficult to show, using similar arguments that the map $(a, b) \to ab$ is continuously differentiable.

By left multiplication each $a \in G$ corresponds to a transformation A: $Ab = ab$. For a chart $a \leftrightarrow \alpha', \ldots, \alpha^p$ in the neighborhood of e we may define infinitesimal transformations I_ν as follows: $I_\nu = (\partial a/\partial \alpha^\nu)_{\alpha=0}$. For an a which is a neighbor of e we write $a - e = \delta a = \sum_\nu I_\nu d\alpha^\nu$.

In this way the I_ν form a linearly independent basis in the cotangent space (as a real vector space) at e. The I_ν are not group elements but are maps $b \to I_\nu b$ of the elements $b \in \mathscr{G}$ in the cotangent space at b defined as follows

$$\sum_\nu I_\nu d\alpha^\nu b = (\delta a)b = ab - b = \delta b.$$

If $a(s)$ is a differentiable curve $[0, 1] \to \mathscr{G}$ where $a(0) = e$ and $a(1) = a$, then it follows that $a(s + ds)a(s)^{-1} - e = \Sigma_\nu I_\nu \delta\omega^\nu$ with the differential form

$$\delta\omega^\nu = \sum_\mu q^\nu_\mu(a(s))\delta\beta^\mu,$$

where the β^μ are coordinates of a chart containing $a(s)$. It follows that

$$da(s) = \sum_\nu a(s)\delta\omega^\nu(s). \tag{10.3.1}$$

The differential equations (10.3.1) together with the initial conditions $a(0) = e$ yields the solution $a(1) = a$. The infinitesimal transformations I_ν determine the entire group $\mathscr{G}$ (as transformations of $\mathscr{G}$ into itself).

If we choose a curve $a(s)$ for which $\delta\omega^\nu(s) = \rho^\nu ds$ where ρ^ν is constant, (10.3.1) becomes

$$\frac{da(s)}{ds} = \left(\sum_\nu I_\nu \beta^\nu\right) a(s)$$

with the solution

$$a(s) = \exp\left(\sum_\nu I_\nu \rho^\nu s\right)$$

and we therefore obtain

$$a = a(1) = \exp\left(\sum_\nu I_\nu \rho^\nu\right). \tag{10.3.2}$$

Therefore there exists at least one neighborhood of e in which it is possible to introduce a chart $a \leftrightarrow \{\rho^\nu\}$ such that (10.3.2) is satisfied. Then for the original chart of the α^ν:$(\partial\rho^\mu/\partial\alpha^\nu)_{\alpha=0} = \delta^\mu_\nu$.

The ρ^μ are called the canonical parameters of the Lie group. With respect to the canonical parameters the group operations are analytic functions of the parameters. In particular, in a Lie group it is always possible to choose charts such that the group operations can be differentiated more than once.

Equation (10.3.2) gives us the map $b \to ab$ if the maps $b \to I_\nu b$ are known for all $b \in \mathscr{G}$. According to §10.1 each element of the subset of $\mathscr{G}$ which is connected to e can be expressed as a product of elements from an arbitrary neighborhood U of the element e. Therefore it suffices to know the map $b \to I_\nu b$ for the elements $b \in U$. Since the $b \in U$ are neighbors of the element e, the following question is relevant: What aspects of the structure of I_ν determine the subset of $\mathscr{G}$ which is connected with e? The theory of Lie algebras answers this question. Here we shall only derive the relevant structure, without proving the existence of Lie groups nor providing a general theory of the structure of Lie algebras (see, for example, [13]). This structure is of particular importance for quantum mechanics.

For an element $a = e + \sum_\nu I_\nu d\alpha^\nu$ which is a neighbor of e it follows that, for any $b \in \mathscr{G}$, the element bab^{-1} is a neighbor of e:

$$bab^{-1} = e + \sum_\mu I_\mu \delta\sigma^\mu,$$

that is:

$$bI_\nu b^{-1} = \sum_\mu I_\mu \sigma^\mu_\nu(b). \tag{10.3.3}$$

For $b = b_1 b_2$ it follows that

$$\begin{aligned}\sum_\mu I_\mu \sigma^\mu_\nu(b_1 b_2) &= b_1(b_2 I_\nu b_2^{-1})b_1^{-1} \\ &= b_1 \sum_\rho I_\rho \sigma^\rho_\nu(b_2) b_1^{-1} = \sum_{\rho\tau} I_\tau \sigma^\tau_\rho(b_1)\sigma^\rho_\nu(b_2).\end{aligned}$$

Thus (10.3.3) is a representation of $\mathcal{G}$ in the vector space spanned by the I_ν. If, in (10.3.3) we choose for b an element which is a neighbor of e,

$$b = e + \sum_\rho I_\rho \, d\beta^\rho,$$

then from (10.3.3) it follows that, since $b^{-1} = e - \sum_\rho I_\rho \, d\beta^\rho$

$$[I_\rho, I_\nu] = I_\rho I_\nu - I_\nu I_\rho = \sum_\mu I_\mu c^\mu_{\nu\rho}, \tag{10.3.4}$$

where

$$c^\mu_{\nu\rho} = \left(\frac{\partial \sigma^\mu_\nu(b)}{\partial \beta^\rho}\right)_{\beta=0}.$$

The structure constants $c^\mu_{\nu\rho}$ are therefore well defined by the group $\mathcal{G}$. From (10.3.4) it follows directly that $c^\mu_{\nu\rho} = -c^\mu_{\rho\nu}$. From the easily proven Jacobi identity

$$[I_\rho, [I_\mu, I_\nu]] + [I_\mu, [I_\nu, I_\rho]] + [I_\nu, [I_\rho, I_\mu]] = 0 \tag{10.3.5}$$

it follows that, for the $c^\mu_{\rho\nu}$

$$\sum_\lambda (c^\eta_{\lambda\rho} c^\lambda_{\nu\mu} + c^\eta_{\lambda\mu} c^\lambda_{\rho\nu} + c^\eta_{\lambda\nu} c^\lambda_{\mu\rho}) = 0. \tag{10.3.6}$$

Given a real vector space defined by the basis I_ν, and introducing a "product" of I_ρ and I_ν as follows

$$[I_\rho, I_\nu] = \sum_\mu I_\mu c^\mu_{\nu\rho}$$

for which $[I_\rho, I_\nu] = -[I_\nu, I_\rho]$ and (10.3.5) holds, then this vector space is called a Lie algebra. Is there a Lie group which corresponds to each Lie algebra? This is indeed the case, a fact which we shall not prove here (see, for example, [13]).

In closing this section we shall consider the following special case: Let $\mathcal{G}$ be a transformation group of a space into itself, that is, there exists a map $\mathcal{G} \times X \to X$ such that $h(a_1, x) = h(a_2, x)$ for all x implies $a_1 = a_2$. Instead of writing $h(a, x)$ we shall use the simpler notation ax. If X is a differentiable manifold and if in $\mathcal{G}$ there is a chart for a neighborhood of e such that the map $\mathcal{G} \times X \to X$ is continuously differentiable, then all of the above considerations about $\mathcal{G} \times \mathcal{G} \to G$ are applicable to $\mathcal{G} \times X \to X$. In particular, the infinitesimal transformations I_ν can be considered as transformations

$$\left(e + \sum_\nu I_\nu \, d\alpha^\nu\right)x = x + \delta x,$$

where

$$\delta x = \sum_\nu I_\nu x \, d\alpha^\nu$$

of x into the cotangent space at the position x. Thus, the transformation ax is, known from (10.3.1), (10.3.2) if the $I_\nu x$ are known.

Conversely, if the transformations I_ν of X into the cotangent space are such that they form a Lie algebra, then the transformations a of X into itself are defined by (10.3.2). It can be shown that these transformations form a Lie group (see [13]).

10.4 Representations of Topological Groups

By a representation of a topological group $\mathscr{G}$ we mean a homomorphic and continuous map of $\mathscr{G}$ into a topological group of linear transformations of a vector space (see, for example, the map in the group $\mathscr{A}$ in VI, §1 or in the group of unitary operators of a Hilbert space).

A homomorphism f of a topological group $\mathscr{G}$ into a topological group $\mathscr{G}'$ is continuous if it is continuous at a point (for example, at the unit element). If f is continuous, then the invariant subgroup $f^{-1}(e')$ is closed in $\mathscr{G}$ since e' is a closed set in $\mathscr{G}'$. If $\mathscr{N}$ is the connected component of e in $\mathscr{G}$ (see §10.1) then $f(\mathscr{N})$ is also connected (AII, §4) and is therefore a subgroup of the connected component $\mathscr{N}'$ of e' in $\mathscr{G}'$. If f is a bijection and a homeomorphism, then f is called an isomorphism of the topological groups $\mathscr{G}$, $\mathscr{G}'$.

Let U be a neighborhood of the unit element $e \in \mathscr{G}$ and let f be a continuous map $U \xrightarrow{f} \mathscr{G}'$ which satisfies the following conditions: From $a, b, c \in U$ and $ab = c$ imply $f(a)f(b) = f(c)$. Then f is called a local homomorphism. If $f(U)$ is a neighborhood of e' in $\mathscr{G}'$ and $U \xrightarrow{f} f(U)$ is bijective and f is a homeomorphism of U onto $f(U)$, then f is called a local isomorphy of $\mathscr{G}$ and $\mathscr{G}'$.

The question remains whether a local homomorphism can be extended as a homomorphism over all of $\mathscr{G}$. Certainly this can be the case only if $\mathscr{G}$ is connected, since in §10.1 we have seen that only the component which is connected with e can be represented as a product of finitely many elements of U.

If $\mathscr{G}$ is connected, then every element $b \in \mathscr{G}$ can be written in the form $b = \prod_{\nu=1}^{n} a_\nu$ where $a_\nu \in U$. Thus, in order that f be a homomorphism, $f(b) = \prod_{\nu=1}^{n} f(a_\nu)$ must be satisfied. This condition can be satisfied only if $\prod_{\nu=1}^{n} a_\nu = \prod_{\mu=1}^{m} a'_\mu$ and $a_\nu, a'_\mu \in U$ imply that $\prod_{\nu=1}^{n} f(a_\nu) = \prod_{\mu=1}^{m} f(a'_\mu)$. This is, however, not generally the case.

We say that $\mathscr{G}$ is *simply* connected if $\mathscr{G}$ is linearly connected (see AII, §4—every connected Lie group is also linearly connected) and every closed path can be continuously shrunk to a point. If $\mathscr{G}$ is simply connected, then the local homomorphic map f can be extended to all of $\mathscr{G}$.

In order to prove this statement, we consider a path $b(s)$ with $0 \leq s \leq 1$ and $b(0) = e$, $b(1) = b$. Since the set $0 \leq s \leq 1$ is compact, to each neighborhood V of e there exists an ε such that $b(s') \in b(s)V$, that is, $b(s)^{-1}b(s') \in V$ for $|s' - s| < \varepsilon$.

If U is a neighborhood of e and $U \xrightarrow{f} \mathscr{G}'$ is a local homomorphism, we choose a symmetric neighborhood V for which $VV \subset U$. To V there exists an ε of the type described above. Thus it follows that there exists an increasing sequence of finitely many s_ν (where $s_0 = 0$, $s_n = 1$) such that $b(s)^{-1}b(s') \in V$ for $s_{\nu-1} < s$, $s' < s_{\nu+1}$. For $c_\nu = b(s_{\nu+1})^{-1}b(s_\nu)$ it follows that $\prod_\nu c_\nu = b$. If we set $f(b) = \prod_\nu f(c_\nu)$ it follows that for every refinement of the partition $\{s_\nu\}$ we obtain the same value $f(b)$. Since two partitions have a common refinement, it follows that $f(b)$ depends only on the path $b(s)$. Similarly, we see that a continuous change of the path $b(s)$ from e to b does not change the value $f(b)$. Since $\mathcal{G}$ was assumed to be simply connected, it follows that the local homomorphism can be continued in one and only one way.

The continuation f is a homomorphism if $f(a)f(b) = f(ab)$. In order to show this we consider a path C_1 of e to a and a path C_2 from e to b. Let C_2 be defined by $b(s)$ where $b(0) = e$, $b(1) = b$. Then there is a path aC_2 from a to ab given by $ab(s)$. $C_1 + aC_2$ is then a path from e to a and then to b. $f(ab)$ can be considered to be a continuation of f on the path $C_1 + aC_2$. By using a sufficiently fine partition of the path, from local homomorphisms it follows that $f(ab) = f(a)f(b)$.

If $f(U)$ is a neighborhood of e' in $\mathcal{G}'$ and if $\mathcal{G}'$ is connected then $f(\mathcal{G}) = \mathcal{G}'$, which follows directly from the fact that every element of $\mathcal{G}'$ may be represented as a product of elements from $f(U)$.

If f is a local isomorphism, then for the extended f: $f(\mathcal{G}) = \mathcal{G}'$. The extended f must not be an isomorphism. If $\mathcal{N}$ is the closed invariant subgroup which is mapped upon e', then $\mathcal{N}$ must be a discrete subset of $\mathcal{G}$, that is, around each $n \in \mathcal{N}$ there exists a neighborhood in $\mathcal{G}$ around n which contains only n as elements of $\mathcal{N}$. Otherwise there would be an element n_1 of $\mathcal{N}$ such that every neighborhood of n_1 would contain an additional element of $\mathcal{N}$ and if U is a neighborhood of e then $n_1 U$ would contain an element $n_2 \neq n_1$ (and U would contain the element $n_1^{-1}n_2 \neq e$) which contradicts the fact that f is a local isomorphism.

$\mathcal{N}$ must be a subset of the center of $\mathcal{G}$. Proof: Since $\mathcal{N}$ is an invariant subgroup, $n \in \mathcal{N}$ implies that $ana^{-1} \in \mathcal{N}$. For fixed n, ana^{-1} is continuous with a. Since $\mathcal{G}$ is connected and $\mathcal{N}$ is discrete, ana^{-1} must, for continuously varying a be constant and therefore equal to $ene^{-1} = n$, that is, a commutes with n. Conversely, if $\mathcal{N}$ is a discrete subgroup of the center of $\mathcal{G}$, $\mathcal{G}/\mathcal{N}$ is a topological group if we choose the sets $U/\mathcal{N}$ as neighborhoods of the unit element for $\mathcal{G}/\mathcal{N}$ where U are the neighborhoods of the unit element in $\mathcal{G}$. Here it is clear that the canonical map $\mathcal{G} \to \mathcal{G}/\mathcal{N}$ is a local isomorphism.

In our consideration of the continuation of local homomorphisms we have assumed that $\mathcal{G}$ is simply connected. If $\mathcal{G}$ is (linear, but) multiply connected, then by the transition to the so-called covering group $\mathcal{G}^*$, we may reduce our results for the continuation of local homomorphisms to the case of simply connected, since $\mathcal{G}^*$ is simply connected.

If C_2 is a path from e to b, then aC_2 is a path from a to ab. For C_1 as a path from e to a then $C_1 + aC_2$ is a path from e to a and then to ab. If C_2' is a path from e to b which can be continuously developed from the path C_2 and if C_1' can be continuously developed from C_1, then $C_1' + aC_2'$ is a path from e to ab which can be continuously developed from $C_1 + aC_2$. If instead of the paths from e to a we consider equivalence classes $\mathcal{C}_1$ of paths

C_1 defined from the equivalence relation $C_1' \sim C_1$, then $C_1 + aC_2$ uniquely defines a class which we shall denote by $\mathscr{C}_1 + a\mathscr{C}_2$. Here $C_1' \sim C_1$ means that C_1' can be continuously developed from C_1.

The covering group $\mathscr{G}^*$ is defined as the set of all pairs $(a, \mathscr{C})$ where $a \in \mathscr{G}$ and $\mathscr{C}$ is a class of paths from e to a (since $\mathscr{G}$ can be multiply connected, to a given a there can be several classes) and multiplication is defined by $(a, \mathscr{C}_1)(b, \mathscr{C}_2) = (ab, \mathscr{C}_1 + a\mathscr{C}_2)$. The unit element of $\mathscr{G}^*$ is equal to $(e, 0)$ where 0 is the class of closed paths C from e to e which can be continuously shrunk to a point. The element $(a^{-1}, \mathscr{C}')$ which is reciprocal to $(a, \mathscr{C})$ is obtained by constructing the path $C' : a^{-1}(s)$ corresponding to the path $C : a(s)$ where $a(0) = e$, $a(1) = a$.

The topology in $\mathscr{G}^*$ is introduced by means of a neighborhood filter of $(e, 0) : V = \{(a, \mathscr{C}) | a \in U$, and there exists a $C \in \mathscr{C}$ with $C \subset U\}$ where U runs through the neighborhoods of e in $\mathscr{G}$. In this way $\mathscr{G}^*$ becomes a topological group. The map $(a, \mathscr{C}) \to a$ of $\mathscr{G}^*$ to $\mathscr{G}$ is a continuous homomorphism. The invariant subgroup which is mapped onto e is $\mathscr{N} = \{(e, \mathscr{C})\}$. $\mathscr{N}$ is called the fundamental group of $\mathscr{G}$.

$\mathscr{G}^*$ is simply connected because a continuous path from $(e, 0)$ to $(a, \mathscr{C})$ is characterized by a path C of e to a where $C \in \mathscr{C}$.

If there exists in $\mathscr{G}$ a simply connected neighborhood of e (which is the case for all Lie groups) then the map $(a, \mathscr{C}) \to a$ is a local isomorphism of $\mathscr{G}^*$ to $\mathscr{G}$. The fundamental group is then a subgroup of the center of $\mathscr{G}^*$.

If $\mathscr{G} \to \mathscr{G}'$ is a local homomorphic map then $\mathscr{G}^* \to \mathscr{G} \to \mathscr{G}'$ is also a local homomorphism. The local homomorphism $\mathscr{G}^* \to \mathscr{G}'$ may be uniquely extended as a homomorphism $\mathscr{G}^* \to \mathscr{G}'$.

The most important example physically is the three-dimensional rotation group which was denoted by $\mathscr{D}_{\mathscr{g}}$ in VII. The topological group $\mathscr{D}_{\mathscr{g}}$ can be characterized in terms of the following three-dimensional parameter space (by which $\mathscr{D}_{\mathscr{g}}$ becomes a Lie group): The parameters $\alpha_1, \alpha_2, \alpha_3$ correspond to a unit vector $\alpha_1/\alpha, \alpha_2/\alpha, \alpha_3/\alpha$ where $\alpha = (\alpha_1^2 + \alpha_2^2 + \alpha_3^2)^{1/2}$; the unit vector corresponds to the axis of rotation and α corresponds to the rotation angle (in the right-handed sense). We require that $0 \leq \alpha \leq \pi$. The points $\alpha_1, \alpha_2, \alpha_3$ are therefore the points of a sphere of radius π (see Figure 51). Two rotations $\alpha_1, \alpha_2, \alpha_3$ (with $\alpha = \pi$) and $-\alpha_1, -\alpha_2, -\alpha_3$ are identical, that is, we must identify diagonally opposite points of the sphere (see Figure 51). Clearly $\mathscr{D}_{\mathscr{g}}$ is compact, but is not simply connected, because in Figure 52 there is a closed path which cannot be shrunk to a point. The covering group

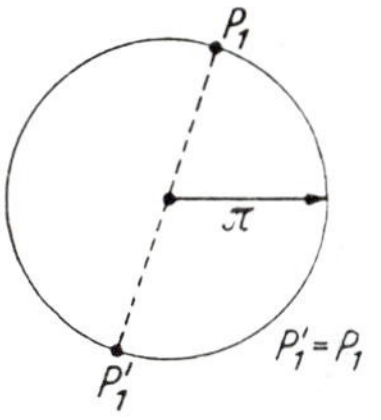

Figure 51

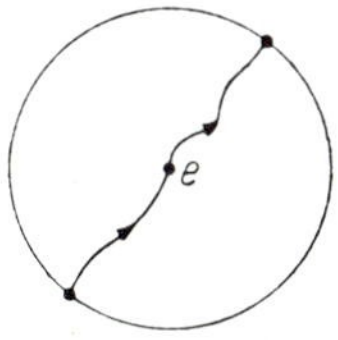

Figure 52

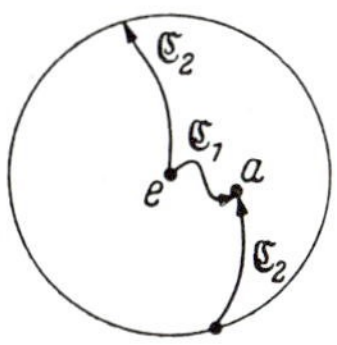

Figure 53

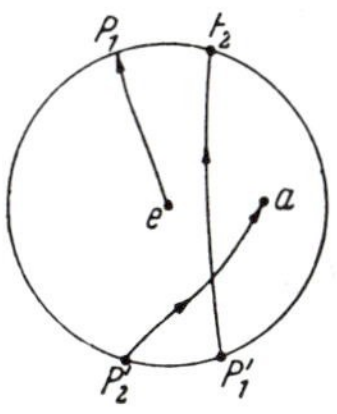

Figure 54

$\mathcal{D}_{\mathcal{G}}^*$ consists (for every $a \in \mathcal{D}_{\mathcal{G}}$) of two elements $(a, \mathcal{C}_1)$, $(a, \mathcal{C}_2)$ where $\mathcal{C}_1$ and $\mathcal{C}_2$ are characterized by the two paths C_1 and C_2 in Figure 53. All other paths from e to a can be continuously transformed into either C_1 or C_2. If, for example, we use two pairs of diametrically opposite points P_1, P' and P_2, P_2' (see Figure 54) then we can move P_2 continuously toward P_1' whereby P_2' is moved towards P_1. The path $P_1'P_2$ can then be ignored and the remaining path can be detached from $P_2' = P_1$ and then formed to C_1. $\mathcal{D}_{\mathcal{G}}$ is isomorphic to $\mathcal{D}_{\mathcal{G}}^*/\mathcal{N}$ where the fundamental group $\mathcal{N}$ consists of the two elements $(e, 0)$ and $(e, \mathcal{C})$ and $\mathcal{C}$ is characterized by the path given in Figure 52. It is easy to show that $(e, \mathcal{C})^2 = (e, 0)$. The fundamental group is cyclic of order 2.

Of particular importance are the above considerations as applied to Lie groups. If in a finite-dimensional vector space a representation of a Lie algebra is given in terms of linear transformations which satisfies the commutation relation (10.3.4), then we obtain a local homomorphism of the Lie group in the form (10.3.2). A representation of the Lie algebra in terms of self-adjoint (not necessarily bounded!) operators in a Hilbert space does not automatically lead to a local homomorphism of the Lie group in terms of unitary operators. Additional conditions for the definition domain must be satisfied (see, for example, [5]).

10.5 Group Rings of Compact Lie Groups

Much of the structure of the representation of finite groups can be extended by analogy to compact groups $\mathcal{G}$. This is a result of the fact that it is possible to define an invariant measure on the Borel subsets of $\mathcal{G}$ and that the measure of $\mathcal{G}$ is finite. Here it is possible to replace sums over finite number of elements of a group by integrals over $\mathcal{G}$.

A left-invariant measure m is defined by the requirement that for every Borel set $\mathcal{T}$ the following relation $m(a\mathcal{T}) = m(\mathcal{T})$ holds. In order to find such measures it is sufficient to determine m for the Borel subsets of a neighborhood U of the unit element, then, for every open set $V \subset U$ the relation $m(aV) = m(V)$ must be satisfied and the aV cover all of G. A measure m for a Lie group can be explicitly given.

We shall use

$$\gamma^\nu = f_\nu(\alpha^1, \ldots, \alpha^p; \beta^1, \ldots, \beta^p)$$

which was given in §10.3 for a chart for the vicinity of the unit element e. A tangential vector d^σ at e will be transformed into the vector δ^ρ at a by the equation

$$\delta^\rho = \sum_\sigma \left(\frac{\partial f_\rho}{\partial \beta^\sigma}\right)_{\beta=0} d^\sigma.$$

Using the functional determinant (Jacobian)

$$\Delta(\alpha_1, \ldots, \alpha_p) = \left|\left(\frac{\partial f_\rho}{\partial \rho^\sigma}\right)_{\beta=0}\right| \tag{10.5.1}$$

permits us to define the following measure m

$$m(\mathcal{T}) = c \int_{\mathcal{T}} \frac{1}{\Delta}\, d\alpha^1 \cdots d\alpha^p = \int_{\mathcal{T}} dm(a) \qquad (10.5.2)$$

which satisfies $m(a\mathcal{T}) = m(\mathcal{T})$. If the group $\mathcal{G}$ is compact, then it is possible to cover $\mathcal{G}$ with a finite number of sets aU where U is an arbitrary neighborhood of e. Thus $m(\mathcal{G})$ is finite. We choose the constant c such that $m(\mathcal{G}) = 1$.

If $\mathcal{G}$ is compact, then the above defined measure is also right invariant. If, for a neighborhood U of e the relation $m(aUa^{-1}) = m(U)$ holds then m is also right invariant because it follows that $m(Ua) = m(a^{-1}Ua) = m(U)$ since m is left invariant. To prove $m(aUa^{-1}) = m(U)$ it suffices to prove that the determinant $A = |\sigma_\nu^\mu(a)| = 1$ where the σ_ν^μ are defined in (10.3.3). According to §10.3 the $\sigma_\nu^\mu(a)$ form a representation of the group. The sequence of elements a^n $(n = 0, 1, 2, \ldots)$ must, in a compact $\mathcal{G}$ have an accumulation point b. Thus there exists a subsequence a^{ν_i} which converges towards b. Thus it follows that $A^{\nu_i} \to B = |\sigma_\nu^\mu(b)|$. Since b^{-1} is represented by the reciprocal matrix, $B \neq 0$ and therefore we find that $A = 1$.

In addition, for m we obtain

$$m(\mathcal{T}) = m(\mathcal{T}^{-1}) \qquad (10.5.3)$$

where $\mathcal{T}^{-1}$ is the set of reciprocal elements of the set $\mathcal{T}$. It suffices to show that this is true for the case in which $\mathcal{T} = aU$ where U is a neighborhood of e. It follows that $\mathcal{T}^{-1} = U^{-1}a^{-1}$. Therefore it is sufficient to show that $m(U) = m(U^{-1})$ for a small neighborhood of e. An element of U then has the parameter $d\alpha^\nu$; its reciprocal has the parameter $-d\alpha^\nu$, from which it follows that $m(U) = m(U^{-1})$.

For an integrable function f over $\mathcal{G}$ we therefore obtain

$$\int_{\mathcal{G}} f(a)\, dm(a) = \int_{\mathcal{G}} f(ab)\, dm(a) = \int_{\mathcal{G}} f(ba)\, dm(a). \qquad (10.5.4)$$

If we reexamine the considerations of §8, it is easy to show that all theorems are applicable providing that we replace $(1/h) \sum_a \cdots$ by $\int \cdots dm(a)$. For the matrix elements $g_{\tau\rho}^{(\nu)}(a)$ of irreducible unitary finite-dimensional representations (where different ν index different representations) the following orthogonality relations hold:

$$\int_{\mathcal{G}} \overline{g_{\tau\rho}^{(\mu)}(a)}\, g_{\kappa\lambda}^{(\nu)}(a)\, dm(a) = \delta_{\mu\nu}\delta_{\tau\kappa}\delta_{\rho\lambda} \frac{1}{n^{(\nu)}} \qquad (10.5.5)$$

and the following equation for the characters hold:

$$\int_{\mathcal{G}} \overline{\chi^{(\mu)}(a)}\, \chi^{(\nu)}(a)\, dm(a) = \delta_{\mu\nu}. \qquad (10.5.6)$$

We have not yet, of course, shown that there are sufficiently many irreducible representations such that the $g_{\tau\rho}^{(\nu)}(a)$ form a complete system of functions over $\mathcal{G}$ and that the $\chi^{(\nu)}(a)$ form a complete system of functions over the classes of $\mathcal{G}$.

That this, is indeed the case we shall show with the help of group rings for compact groups. The measure dm introduced above defines a Hilbert space $\mathscr{L}^2(\mathscr{G}, dm)$ (see the end of AIV, §1). We form an algebra by introducing the following definition of a product: For $f, g \in \mathscr{L}^2(\mathscr{G}, dm)$ we define

$$(f \cdot g)(a) = \int f(ab^{-1})g(b)\, dm(b) \tag{10.5.7}$$

by analogy with the definition for finite groups (§7):

$$\left[\sum_{a'} f(a')a'\right]\left[\sum_{b} g(b)b\right] = \sum_{a',b} f(a')g(b)a'b$$

$$= \sum_a a \sum_b f(ab^{-1})g(b),$$

where we set $a = a'b$. It can be easily proven with the aid of the Schwartz inequality (AIV, §1) that $f \cdot g \in \mathscr{L}^2(\mathscr{G}, dm)$.

We now use the matrix elements $g_{\kappa\lambda}^{(\nu)}(a)$ for the different inequivalent finite-dimensional unitary representations of the group $\mathscr{G}$ which are elements of $\mathscr{L}^2(\mathscr{G}, dm)$. If the $g_{\kappa\lambda}^{(\nu)}(a)$ are continuous, then, since $m(\mathscr{G}) = 1$ we obtain $g_{\kappa\lambda}^{(\nu)}(a) \in \mathscr{L}^2(\mathscr{G}, dm)$. There exists at least one such representation, namely the (one-dimensional) identity representation. According to (10.5.5) the $g_{\kappa\lambda}^{(\nu)}$ form an orthogonal system in $\mathscr{L}^2(\mathscr{G}, dm)$. The fact that this orthogonal system is complete will be proven below. First we shall derive a few relationships which are analogous to those derived in §7.

We define

$$e_{\kappa\lambda}^{(\nu)}(a) = n^{(\nu)} g_{\lambda\kappa}^{(\nu)}(a^{-1}). \tag{10.5.8}$$

Thus, from (10.5.5) we obtain

$$(e_{\kappa\lambda}^{(\nu)} \cdot e_{\tau\rho}^{(\mu)})(a) = n^{(\nu)} n^{(\mu)} \int g_{\lambda\kappa}^{(\nu)}(ba^{-1}) g_{\rho\tau}^{(\mu)}(b^{-1})\, dm(b)$$

$$= n^{(\nu)} n^{(\mu)} \int \sum_\sigma g_{\lambda\sigma}^{(\nu)}(b) g_{\sigma\kappa}^{(\nu)}(a^{-1}) \overline{g_{\tau\rho}^{(\mu)}(b)}\, dm(b)$$

$$= \delta_{\nu\mu}\delta_{\lambda\tau} e_{\kappa\rho}^{(\nu)}(a),$$

that is,

$$e_{\kappa\lambda}^{(\nu)} \cdot e_{\tau\rho}^{(\mu)} = \delta_{\nu\mu}\delta_{\lambda\tau} e_{\kappa\rho}^{(\nu)} \tag{10.5.9}$$

which is identical to the relation (7.6).

The $e_{\kappa\lambda}^{(\nu)}$ (ν and λ fixed) span a finite-dimensional subspace $\ell_\lambda^{(\nu)}$ of $\mathscr{L}^2(\mathscr{G}, dm)$ which is a left ideal because

$$\begin{aligned}(f \cdot e_{\kappa\lambda}^{(\nu)})(a) &= \int f(ab^{-1})e_{\kappa\lambda}^{(\nu)}(b)\, dm(b) \\ &= n^{(\nu)} \int f(ab^{-1})g_{\lambda\kappa}^{(\nu)}(b^{-1})\, dm(b) \\ &= n^{(\nu)} \int f(c)g_{\lambda\kappa}^{(\nu)}(a^{-1}c)\, dm(c) \\ &= n^{(\nu)} \int f(c) \sum_\rho g_{\lambda\rho}^{(\nu)}(a^{-1})g_{\rho\kappa}^{(\nu)}(c)\, dm(c) \\ &= \sum_\rho g_\rho e_{\rho\lambda}^{(\nu)}(a), \quad \text{where} \quad g_\rho = \int f(c)g_{\rho\kappa}^{(\nu)}(c)\, dm(c).\end{aligned}$$

Similarly it can be shown that all the $e_{\kappa\lambda}^{(\nu)}$ (only ν is fixed) span a dual ideal.

In the ring $\mathscr{L}^2(\mathscr{G}, dm)$ the elements of $\mathscr{G}$ are not present. We may, however, define the elements of $\mathscr{G}$ as operators in the Hilbert space $\mathscr{L}^2(\mathscr{G}, dm)$ as follows:

$$(af)(b) = f(a^{-1}b). \tag{10.5.10}$$

It is easy to show that (10.5.10) defines a unitary (infinite-dimensional) representation of $\mathscr{G}$ in $\mathscr{L}^2(\mathscr{G}, dm)$. It follows that

$$\begin{aligned}(ae_{\kappa\lambda}^{(\nu)})(b) &= e_{\kappa\lambda}^{(\nu)}(a^{-1}b) = n^{(\nu)}g_{\lambda\kappa}^{(\nu)}(b^{-1}a) \\ &= \sum_\sigma e_{\sigma\lambda}^{(\nu)}(b)g_{\sigma\kappa}^{(\nu)}(a).\end{aligned}$$

Thus $\ell_\lambda^{(\nu)}$ is an invariant subspace and $\ell_\lambda^{(\nu)}$ contains the νth irreducible representation of $\mathscr{G}$. The $e_{\kappa\lambda}^{(\nu)}$ form an orthogonal system where

$$\int \overline{e_{\kappa\lambda}^{(\nu)}(a)}e_{\rho\sigma}^{(\mu)}(a)\, dm(a) = \delta_{\nu\mu}\,\delta_{\rho\kappa}\,\delta_{\lambda\sigma}n^{(\nu)} \tag{10.5.11}$$

a result which follows directly from (10.5.5) and (10.5.8). If the $e_{\kappa\lambda}^{(\nu)}$ form a complete orthogonal system, $\mathscr{L}^2(\mathscr{G}, dm)$ decomposes in the form

$$\mathscr{L}^2(\mathscr{G}, dm) = \sum_{\nu,\lambda} \oplus\, \ell_\lambda^{(\nu)} = \sum_\nu \oplus\, \ell^{(\nu)}, \tag{10.5.12}$$

where $\ell^{(\nu)} = \sum_\lambda \oplus\, \ell_\lambda^{(\nu)}$. The $\ell^{(\nu)}$ are finite-dimensional and two-sided ideals.

If the $e_{\kappa\lambda}^{(\nu)}$ are not complete, then there exists a vector $r \neq 0$ which is orthogonal to all $e_{\kappa\lambda}^{(\nu)}$. For r we define

$$K(d, b) = \int r(b^{-1}dc)\overline{r(c)}\, dm(c)$$

and we define the operator K:

$$(Kf)(a) = \int K(a, b)f(b)\, dm(b).$$

We may therefore also write $Kf = f \cdot r \cdot \tilde{r}$ where $\tilde{r}(a) = \overline{r(a^{-1})}$. First we show that K is a self-adjoint and compact operator (see AIV, §§4 and 9). The fact that K is self-adjoint follows from

$$K(b, d) = \int r(d^{-1}bc)\overline{r(c)}\, dm(c)$$

$$= \int r(c')\overline{r(b^{-1}\, dc')}\, dm(c') = \overline{K(d, b)}.$$

K is compact, since from

$$|K(d, b)| \leq \int |r(c)|^2\, dm(c)$$

it follows that the following relation

$$\int |K(d, b)|^2\, dm(d)\, dm(b) < \infty$$

holds.

Thus K has only discrete eigenvalues, of which those which are nonzero can only be finitely degenerate. Since $r \neq 0$, from

$$\langle f, Kf \rangle = \int \overline{f(d)} r(b^{-1}\, dc)\overline{r(c)} f(b)\, dm(d)\, dm(b)\, dm(c)$$

$$= \int \overline{f(d)} r(b^{-1}c^{-1})\overline{r(d^{-1}c^{-1})} f(b)\, dm(d)\, dm(b)\, dm(c)$$

$$= \int \left| \int r(b^{-1}c')f(b)\, dm(b) \right|^2 dm(c')$$

it follows that K is not the null operator, that is K must have a nonzero eigenvalue.

From

$$(aKf)(d) = \int K(a^{-1}d, b)f(b)\, dm(b)$$

$$= \int K(d, ab)f(b)\, dm(b)$$

$$= \int K(d, b')f(a^{-1}b')\, dm(b')$$

$$= (Kaf)(d)$$

it follows that K commutes with all elements of $\mathscr{G}$. A finite-dimensional eigenspace $\mathscr{s}$ of K with a nonzero eigenvalue λ is therefore an invariant subspace, that is, a representation space of $\mathscr{G}$. This may be reduced so that in $\mathscr{s}$ there exists an irreducible representation, which we call the νth representation. Let $f_\rho \in \mathscr{s}$ be the basis vectors which corresponds to this νth representation. We obtain $Kf_\rho = \lambda f_\rho$.

We will now show that $f_\rho \in \mathscr{t}^{(\nu)}$ where $\mathscr{t}^{(\nu)}$ is defined above. For $e^{(\nu)} = \sum_\tau e_{\tau\tau}^{(\nu)}$ it follows from $f \in \mathscr{t}^{(\nu)}$ that $e^{(\nu)} \cdot f = f \cdot e^{(\nu)} = f$.

Since $\mathscr{t}^{(\nu)}$ is a two-sided ideal, it follows that for arbitrary g: $e^{(\nu)} \cdot g \in \mathscr{t}^{(\nu)}$ and $g \cdot e^{(\nu)} \in \mathscr{t}^{(\nu)}$. If for a g the relation $e^{(\nu)} \cdot g = g$ it therefore follows that $g \in \mathscr{t}^{(\nu)}$ and $g \cdot e^{(\nu)} = g$.

With $f_\rho \in \mathscr{s}$ defined above it follows that

$$\begin{aligned}
(e^{(\nu)} \cdot f_\rho)(a) &= \int e^{(\nu)}(b) f_\rho(b^{-1}a)\, dm(b) \\
&= \int e^{(\nu)}(b) b f_\rho(a)\, dm(b) \\
&= \int \sum_\sigma e^{(\nu)}(b) f_\sigma(a) g_{\sigma\rho}^{(\nu)}(b)\, dm(b) \\
&= n^{(\nu)} \sum_\sigma f_\sigma(a) \int \sum_\tau \overline{g_{\tau\tau}^{(\nu)}(b)} g_{\sigma\rho}^{(\nu)}(b)\, dm(b) \\
&= f_\rho(a).
\end{aligned}$$

Therefore it follows that $f_\rho \in \mathscr{t}^{(\nu)}$ and also $f_\rho \cdot e^{(\nu)} = f_\rho$. Since $Kf_\rho = f_\rho \cdot r \cdot \tilde{r} = \lambda f_\rho$ we obtain

$$f_\rho \cdot e^{(\nu)} = \frac{1}{\lambda} f_\rho \cdot r \cdot \tilde{r} \cdot e^{(\nu)}.$$

Since r is orthogonal to $\mathscr{t}^{(\nu)}$ it follows that

$$(\tilde{r} \cdot e^{(\nu)})(a) = \int \tilde{r}(b^{-1}) e^{(\nu)}(ba)\, dm(b) = \int \overline{r(b)} e^{(\nu)}(ba)\, dm(b) = 0$$

since $e^{(\nu)}(ba)$ is, for fixed a, an element of the two-sided ideal $\mathscr{t}^{(\nu)}$. Therefore we should obtain $f_\rho = f_\rho \cdot e^{(\nu)} = 0$. Thus it follows that $r = 0$.

For an example of a volume measure on a compact group we shall consider U_n the unitary group in an n-dimensional space $\mathscr{r}$. Each unitary transformation A can be represented in terms of a basis in $\mathscr{r}$ as follows:

$$A = UEU^+, \tag{10.5.13}$$

where E is a diagonal matrix (with the diagonal elements $\varepsilon_\nu = e^{i\alpha_\nu}$) and U is a transformation which transforms the chosen basis of $\mathscr{r}$ (in which E is diagonal) into the basis of the eigenvectors of A. For A we will choose the first n parameters to be the angles α_ν. These parameters have the advantage that two different transformations with the same parameters α_ν belong to the same class.

How many additional parameters do we need in order to determine A? U is, for fixed E, not uniquely determined. Since U transforms the given basis vectors into the eigenvectors of A, that is, into another basis, U is determined by the basis vectors which are determined up to a factor $e^{i\delta}$. For the first basis vector we therefore find that $2(n-1)$ real parameters are necessary. Since the second is orthogonal to the first, only $2(n-2)$ real parameters have to be chosen. For all n vectors we need $n(n-1) = n^2 - n$ parameters. The group U_n is therefore an n^2 parameter group.

From (10.5.13) it follows that

$$dA = dUEU^+ + U\,dEU^+ - UEU^+\,dUU^+.$$

In order to find the volume, we must transport a volume element spanned by vectors dA near A by the transformation A^{-1} to the unit element:

$$\delta A = A^{-1}\,dA = UE^{-1}\delta UEU^{-1} + U\delta EU^{-1} - U\delta UU^{-1},$$

where $\delta U = U^{-1}\,dU$, $\delta E = E^{-1}\,dE$. The volume spanned by the vectors δA must be identical to that spanned by the vectors dA. If we transform the vectors δA into $\delta' A$ by $\delta' A = U^{-1}\delta AU$, then the volumes must be preserved. We have

$$\delta' A = E^{-1}\delta UE - \delta U + \delta E.$$

The matrix elements of $\delta' A$ are therefore given by

$$(\delta' A)_{lk} = (\delta U)_{lk}\left(\frac{\varepsilon_k}{\varepsilon_l} - 1\right) + \delta_{lk}\,i\,d\alpha_k. \tag{10.5.14}$$

If for U we choose the $n(n-1)$ parameter, and the additional n parameters α_ν, then from (10.5.14) it follows that

$$dm(A) = c\prod_{l\neq k}\left(\frac{\varepsilon_k}{\varepsilon_l} - 1\right) d\tilde{m}(A)\,d\alpha_1 \cdots d\alpha_n, \tag{10.5.15}$$

where $d\tilde{m}(A)$ only contains the $n(n-1)$ parameter for U.

Since $\varepsilon_l^{-1} = \bar{\varepsilon}_l$, we obtain

$$\begin{aligned}\prod_{l\neq k}\left(\frac{\varepsilon_k}{\varepsilon_l} - 1\right) &= \prod_{l<k}\left(\frac{\varepsilon_k}{\varepsilon_l} - 1\right)\prod_{k<l}\left(\frac{\varepsilon_k}{\varepsilon_l} - 1\right)\\ &= \prod_{k<l}\left(\frac{\bar{\varepsilon}_k}{\bar{\varepsilon}_l} - 1\right)\prod_{k<l}\left(\frac{\varepsilon_k}{\varepsilon_l} - 1\right)\\ &= \prod_{k<l}\frac{\bar{\varepsilon}_k - \bar{\varepsilon}_l}{\bar{\varepsilon}_l}\prod_{k<l}\frac{\bar{\varepsilon}_k - \varepsilon_l}{\varepsilon_l}\\ &= \prod_{k<l}(\bar{\varepsilon}_k - \bar{\varepsilon}_l)\prod_{k<l}(\varepsilon_k - \varepsilon_l) = |\Delta|^2,\end{aligned}$$

where

$$\Delta = \prod_{k<l} (\varepsilon_l - \varepsilon_k) = \begin{vmatrix} \varepsilon_1^{n-1} & \varepsilon_1^{n-2} & \cdots & \varepsilon_1 & 1 \\ \varepsilon_2^{n-1} & \varepsilon_2^{n-2} & \cdots & \varepsilon_2 & 1 \\ \vdots & & & & \vdots \\ \varepsilon_n^{n-1} & \varepsilon_n^{n-2} & \cdots & \varepsilon_n & 1 \end{vmatrix}.$$

For the integration of the class functions, we may therefore use

$$c|\Delta|^2 \, d\alpha_1 \cdots d\alpha_n \tag{10.5.16}$$

as the volume element.

10.6 Representations in Hilbert Space

A unitary representation of a finite group or a unitary representation of a compact group which is at least measurable in a Hilbert space $\mathscr{H}$ is completely reducible; its irreducible subspaces are finite dimensional. Since the computations for finite groups are identical to those for compact groups providing that sums are replaced by integrals, we shall only prove the above result for compact groups.

We introduce the following operators in $\mathscr{H}$

$$E_{\rho\sigma}^{(\nu)} = \int e_{\rho\sigma}^{(\nu)}(a) U(a) \, dm(a), \tag{10.6.1}$$

where $U(a)$, $a \in \mathscr{G}$ are the unitary representation operators. With the aid of (10.5.9) we obtain:

$$E_{\kappa\lambda}^{(\nu)} E_{\tau\rho}^{(\mu)} = \delta_{\nu\mu} \delta_{\lambda\tau} E_{\kappa\rho}^{(\nu)}. \tag{10.6.2}$$

In particular we obtain:

$$E_{\kappa\kappa}^{(\nu)} E_{\tau\tau}^{(\mu)} = \delta_{\nu\mu} \delta_{\kappa\tau} E_{\tau\tau}^{(\nu)}. \tag{10.6.3}$$

It is easy to show that the $E_{\tau\tau}^{(\nu)}$ are self-adjoint, and are therefore pairwise orthogonal projection operators.

For $e^{(\nu)} = \sum_\tau e_{\tau\tau}^{(\nu)}$ (see §10.5) we therefore find that the operators

$$E^{(\nu)} = \sum_\tau e_{\tau\tau}^{(\nu)}(a) U(a) \, dm(a) = \sum_\tau E_{\tau\tau}^{(\nu)} \tag{10.6.4}$$

are projection operators.

Since the $e_{\kappa\lambda}^{(\nu)}$ form a complete orthogonal system in $\mathscr{L}^2(\mathscr{G}, dm)$, on the basis of (10.5.11), for $h \in \mathscr{L}^2(\mathscr{G}, dm)$ it follows that

$$h = \sum_{\nu\kappa\lambda} \frac{1}{n^{(\nu)}} e_{\kappa\lambda}^{(\nu)} \int \overline{e_{\kappa\lambda}^{(\nu)}(a)} h(a) \, dm(a).$$

Since $\not{t}^{(\nu)}$ (see §10.5) is spanned by the $e_{\kappa\lambda}^{(\nu)}$ (ν fixed) it follows that

$$h^{(\nu)} = \sum_{\kappa\lambda} \frac{1}{n^{(\nu)}} e_{\kappa\lambda}^{(\nu)} \int \overline{e_{\kappa\lambda}^{(\nu)}(a)} h(a)\, dm(a) \in \not{t}^{(\nu)}. \tag{10.6.5}$$

From (10.5.9) and from (10.6.5) it follows that

$$e^{(\mu)} \cdot h^{(\nu)} = \delta_{\nu\mu} h^{(\nu)}.$$

Therefore we obtain

$$h = \sum_{\nu} e^{(\nu)} \cdot h. \tag{10.6.6}$$

For the case in which $h(b) = \langle U(b)f, g\rangle$ it follows that

$$\begin{aligned}\langle U(b)f, g\rangle &= \sum_{\nu} \int e^{(\nu)}(a) \langle U(a^{-1}b)f, g\rangle\, dm(a)\\ &= \sum_{\nu} \int e^{(\nu)}(a) \langle U(b)f, U(a)g\rangle\, dm(a)\\ &= \sum_{\nu} \langle U(b)f, E^{(\nu)}g\rangle.\end{aligned}$$

Since f, g and $U(b)$ are arbitrary, it follows that

$$\sum_{\nu} E^{(\nu)} = \sum_{\tau\nu} E_{\tau\tau}^{(\nu)} = \mathbf{1}. \tag{10.6.7}$$

From (10.6.1) it follows that

$$\begin{aligned}U(b)E_{\rho\sigma}^{(\nu)} &= \int e_{\rho\sigma}^{(\nu)}(a) U(ba)\, dm(a)\\ &= \int e_{\rho\sigma}^{(\nu)}(b^{-1}c) U(c)\, dm(c)\\ &= \int (be_{\rho\sigma}^{(\nu)})(c) U(c)\, dm(c)\\ &= \sum_{\tau} e_{\tau\sigma}^{(\nu)}(c) U(c)\, dm(c) g_{\tau\rho}^{(\nu)}(b),\end{aligned}$$

that is,

$$U(b)E_{\rho\sigma}^{(\nu)} = \sum_{\tau} E_{\tau\sigma}^{(\nu)} g_{\tau\rho}^{(\nu)}(b). \tag{10.6.8}$$

The fact that $(\mathscr{G})\mathscr{H}$ is, in the sense of AIV, §14, completely reducible, can easily proven using (10.6.8). If $\mathscr{T}$ is an invariant subspace, then there exists (using (10.6.7)) for $f \in \mathscr{T}$ a τ for which $E_{\tau\tau}^{(\nu)} f \neq 0$. The $E_{\rho\tau}^{(\nu)} f$ (ν, τ fixed) span a finite-dimensional irreducible subspace of $\mathscr{T}$.

The decomposition of $\mathscr{H}$ in the form AIV (14.7) is obtained from

$$\mathscr{H}_1^{(\alpha)} \times \mathscr{H}_2^{(\alpha)} = E^{(\alpha)}\mathscr{H}. \tag{10.6.9}$$

We may construct the decomposition of $E^{(\alpha)}\mathscr{H}$ in the product form (10.6.9) as follows: Choose an $f \in E^{(\alpha)}\mathscr{H}$ and a τ such that $E^{(\alpha)}_{\tau\tau} f \neq 0$. For $h_1 = E^{(\alpha)}_{\tau\tau} f / \| E^{(\alpha)}_{\tau\tau} f \|$ we introduce the vectors:

$$h_{\rho 1} = E^{(\alpha)}_{\rho\tau} h_1.$$

From (10.5.8) it follows that $E^{(\alpha)+}_{\rho\tau} = E^{(\alpha)}_{\tau\rho}$, and we therefore obtain

$$\langle h_{\rho 1}, h_{\sigma 1} \rangle = \langle h_1, E^{(\alpha)}_{\tau\rho} E^{(\alpha)}_{\sigma\tau} h_1 \rangle = \delta_{\rho\sigma} \langle h_1, E^{(\alpha)}_{\tau\tau} h_1 \rangle = \delta_{\rho\sigma},$$

since $E^{(\alpha)}_{\tau\tau} h = h$. $\mathscr{H}^{(\alpha)}_1$ can be defined in terms of the basis vectors $h_{\rho 1}$. We obtain

$$U(a) h_{\rho 1} = \sum_\sigma h_{\sigma 1} g^{(\alpha)}_{\sigma\rho}(a).$$

We may choose a vector h_2 which is orthogonal to all the $h_{\rho 1}$, and proceed in a manner similar to the treatment of h_1. We obtain in this way the vectors $h_{\rho\nu}$. Therefore there exists a sequence of vectors $h_{\rho\nu}$ for which

$$U(a) h_{\rho\nu} = \sum_\sigma h_{\sigma\nu} g^{(\alpha)}_{\sigma\rho}(a).$$

We now construct the Hilbert space $\mathscr{H}^{(\alpha)}_2$ with a basis u_ν and fix the following isomorphism

$$h_{\rho\nu} \leftrightarrow h_{\rho\nu} u_\nu.$$

According to AIV (14.8) the operators $U(a)$ in $\mathscr{H}$ have the form

$$U(a) = \sum_\alpha E^{(\alpha)}(U^{(\alpha)}(a) \times \mathbf{1}) E^{(\alpha)}, \tag{10.6.10}$$

where the $U^{(\alpha)}(a)$ are operators of the αth irreducible representation. All operators which commute with the representation have, according to AIV (14.9) the following form:

$$B = \sum_\alpha E^{(\alpha)}(\mathbf{1} \times B^{(\alpha)}) E^{(\alpha)}. \tag{10.6.11}$$

10.7 Representations up to a Factor

For quantum mechanics unitary representations up to a factor are of special interest because they correspond to representations in the group $\mathscr{A}$ of $\mathscr{B}$-continuous effect isomorphisms (see VI).

The fact that a unitary representation (up to a factor of a compact group) is completely reducible can easily be proven using the procedure given in §10.5: If $\mathscr{T}$ is an invariant subspace, then there exists a $f \in \mathscr{T}$ such that not all of the

$$\int \chi(b) U(b)\, dm(b) f = 0$$

can be satisfied. For this f the $U(a)f$, $a \in \mathcal{G}$ form an invariant subspace $\mathcal{S}$. In $\mathcal{S}$ we define an operator A as follows:

$$AU(a)f = U(a) \int \chi(b)U(b)\, dm(b) f.$$

The operator $K = A^+A$ is nonzero. K commutes with all the $U(a)$ and is compact. Therefore there exists an invariant finite-dimensional and irreducible subspace in $\mathcal{S}$.

We may therefore restrict our considerations to finite-dimensional unitary irreducible representations. Such a representation may be given by the $U(a)$. Let $|U(a)|$ be the determinant of $U(a)$. It follows that $|U(a)| = e^{i\alpha}$ (α real). By dividing $U(a)$ by one of the nth roots of $e^{i\alpha}$ (where n is the dimension of the representation) we may choose $|U(a)| = 1$. In particular, we may choose $U(e) = 1$. If $|U(a)| = 1$, then each $U(a)$ is uniquely determined up to one of the nth roots of unity. From the continuity of the map VI (3.3.8) we may choose a neighborhood V of the unit element such that $a \to U(a)$ is continuous. From $U(a)U(b) = \omega(a, b)U(a, b)$ it follows that, for $a, b \in V$, the factor $\omega(a, b)$ is continuous. For $b = e$ and $U(e) = 1$ it follows that $\omega(a, e) = 1$. From $|U(a)||U(b)| = \omega(a, b)^n|U(ab)|$ it follows that V can be chosen so small that $\omega(a, b) = 1$ for all $a, b \in V$. The map $a \to U(a)$ is then a local homomorphism of $\mathcal{G}$. According to §10.4 it follows that this map can be extended to a homomorphism, that is, to a representation of the covering group $\mathcal{G}^*$ of $\mathcal{G}$.

Every irreducible unitary representation up to a factor of $\mathcal{G}$ can therefore, by selection of suitable factors for each $U(a)$ be made into an irreducible unitary representation of the covering group $\mathcal{G}^*$: $(a, \mathcal{C}) \to U(a, \mathcal{C})$. Conversely, every irreducible unitary representation of $\mathcal{G}^*$ produces a representation of $\mathcal{G}$ up to a factor as follows: Since we have assumed that the representation $(a, \mathcal{C}) \to U(a, \mathcal{C})$ of $\mathcal{G}^*$ is irreducible, the elements of the fundamental group $\mathcal{N}$ (as a subset of the center of $\mathcal{G}^*$) must be represented by multiples of the unit matrix.

The $U(a, \mathcal{C})$ which represent the elements of a left coset $a\mathcal{N}$ therefore differ only by a factor. If we define a map $\mathcal{G} \to \mathcal{G}^*$ by $a \to (a, \mathcal{C}_k)$ where $(a, \mathcal{C}_k)$ is an arbitrarily chosen element of the coset $a\mathcal{N}$ then from $a \to (a, \mathcal{C}_k) \to U(a, \mathcal{C}_k)$ we obtain a representation of $\mathcal{G}$ up to a factor.

Given a unitary reducible representation up to a factor of $\mathcal{G}$, then we may reduce the representation as follows: $\mathcal{H} = \sum_\nu \oplus \mathscr{h}_\nu$ where the representations in $\mathscr{h}_\nu$ are irreducible. By changing the factors for each $U(a)$ in such a manner that the $U(a)$ may generate in a selected $\mathscr{h}_{\nu 1}$ a local homomorphic representation it follows that the $U(a)$, as operators in all of $\mathcal{H}$, form a local homomorphic representation and therefore form a representation of $\mathcal{G}^*$ in $\mathcal{H}$. Since this representation is both a representation of $\mathcal{G}^*$ as well as a representation up to a factor of $\mathcal{G}$, the elements of the fundamental group $\mathcal{N}$ must be represented by multiples of the unit operator, that is, the representation of the fundamental group $\mathcal{N}$ is equivalent in all the $\mathscr{h}_\nu$. Conversely, by combining such irreducible representations of $\mathcal{G}^*$ which yield equivalent representations of the fundamental group $\mathcal{N}$ to form a reducible representation of $\mathcal{G}^*$, then we obtain a representation of $\mathcal{G}$ up to a factor.

If we apply the above considerations to the three-dimensional rotation group, then from §10.4 it follows that every unitary representation of $\mathcal{D}_{\mathcal{G}}$ up to a factor corresponds to a representation of the covering group $\mathcal{D}_{\mathcal{G}}^*$ in which the element $(e, \mathcal{C})$ where $\mathcal{C}$ is the class of paths of C according to Figure 52 is represented by either 1 or -1. Physically this means that a system type can have only half integer or integer values of spin angular momentum (see, for example, VII, §3).

References

[1] A. Böhm, *The Rigged Hilbert Space and Quantum Mechanics*. Springer Lecture Notes in Physics, vol. 78, 1978.

J. E. Roberts. Rigged Hilbert spaces in quantum mechanics. *Commun. Math. Phys.*, **3**, 98 (1966).

J. P. Antoine. Dirac formalism and symmetry problems in quantum mechanics. I. General Dirac formalism. *J. Math. Phys.*, **10**, 53 (1974).

O. Melsheimer. Rigged Hilbert space formalism as an extended mathematical formalism for quantum systems. I. General theory. *J. Math. Phys.*, **15**, 902 (1974).
O. Melsheimer. Rigged Hilbert space formalism as an extended mathematical formation for quantum systems. II. Transformation theory in non-relativistic quantum mechanics. *J. Math. Phys.*, **15**, 917 (1974).

[2] O. M. Nikodym. *The Mathematical Apparatus for Quantum Theories*. New York–Heidelberg–Berlin: Springer-Verlag, 1966.

[3] R. Werner. *Quantum Harmonic Analysis on Phase Space*. Preprint, Universität Osnabrück, 1983.
R. Werner. Physical uniformities and the state space of nonrelativistic quantum mechanics. *Found Phys.*, **13**, 859 (1983).

[4] M. Reed and B. Simon. *Methods of Modern Mathematical Physics*, Vol. II. New York: Academic Press, 1975.

[5] A. O. Barut and R. Raczka. *Theory of Group Representations and Applications*. Warszawa: PWN-Polish Scientific Publishers, 1977.

[6] G. Ludwig, *Einführung in die Grundlagen der Theoretischen Physik*, Vols. I–IV. Braunschweig: Vieweg, 1974–1979.

[7] G. Ludwig. *An Axiomatic Basis for Quantum Mechanics*, Vols. I–II. New York–Heidelberg–Berlin. Springer-Verlag, 1985.

[8] G. Ludwig. *Die Grundstrukturen einer Physikalischen Theorie*. Berlin–Heidelberg–New York: Springer-Verlag, 1978.

[9] A. Sommerfeld. *Atombau und Spektrallinien*. Braunschweig: Vieweg, 1944.

[10] T. Cato. *Perturbation Theory for Linear Operators*. New York–Heidelberg–Berlin: Springer-Verlag, 1966.

[11] G. Ludwig. Imprecision in physics. In: *Structure and Approximation in Physical Theories*, A. Hartkämper and H.-J. Schmidt (Eds.). New York: Plenum, 1981.

[12] N. Bourbaki. *Topologie General*, Livre III, Chapitre 3. Groupes Topologiques. Paris: Hermann, 1960.
English translation: *Elements of Mathematics: General Topology*, Part I. Reading, Mass.: Addison–Wesley, 1966.

[13] M. Hausner and J. T. Schwartz. *Lie Groups—Lie Algebras*. New York: Gordon & Breach, 1968.

[14] H. Bethe. Quantenmechanik der Ein- und Zwei-Elektronen-probleme. In: *Handbuch der Physik*, Vol. XXIV/1. Berlin–Heidelberg–New York: Springer-Verlag, 1933, p. 530.
M. Kotani, K. Ohno and K. Kayama. Quantum mechanics of electronic structure of simple molecules. In: *Encyclopedia of Physics*, Vol. XXXVII/2, Molecules, II, S. Flügge (Ed.). New York–Heidelberg–Berlin: Springer-Verlag, 1961, p. 17.

[15] H. Bethe. Quantenmechanik der Ein- und Zwei-Elektronen-probleme. In: *Handbuch der Physik*, Vol. XXIV/1. Berlin–Heidelberg–New York: Springer-Verlag, 1933, p. 536.
M. Kotani, K. Ohmo and K. Kayama. Quantum mechanics of electronic structure of simple molecules in: *Encyclopedia of Physics*, Vol. XXXVII/2, Molecules II, Flugge (Ed.). New York–Heidelberg–Berlin: Springer-Verlag, 1961, p. 17.

[16] O. Sinanoglu. *Modern Quantum Chemistry*. New York: Academic Press, 1965.
W. H. Müller and H. F. Schaefer (Eds.). *Modern Theoretical Chemistry*, New York: Plenum, 1976.

[17] G. Herzberg. *Molecular Spectra and Molecular Structure*, 2 vols. Princeton: Van Nostrand, 1966.

[18] W. O. Amrein, J. M. Jauch and K. B. Sinha. *Scattering Theory in Quantum Mechanics*. Reading, Mass.: Benjamin, 1977.
W. O. Amrein, *Non-Relativistic Quantum Dynamics*. Dordrecht: Reidel, 1981.
M. Reed and B. Simon. *Methods of Modern Mathematical Physics*, Vol. III, Scattering Theory. New York: Academic Press, 1979.

[19] R. G. Newton. *Scattering Theory of Waves and Particles*. New York–Heidelberg–Berlin: Springer-Verlag, 1982.
M. Reed and B. Simon. *Methods of Modern Mathematical Physics*, Vol. III, Scattering Theory. New York: Academic Press, 1979.

[20] N. F. Mott and H. S. W. Massey. *The Theory of Atomic Collisions*. Oxford: Clarendon Press, 1965.

[21] H. A. Bethe and E. E. Salpeter, Quantum mechanics of one- and two-electron systems. In: *Encyclopedia of Physics*, Vol. XXXV, Atoms I, S. Flügge (Ed.). New York–Heidelberg–Berlin: Springer-Verlag, 1957, p. 116.

[22] N. F. Mott and H. S. W. Massey. *The Theory of Atomic Collisions*. Oxford: Clarendon Press, 1965.

[23] R. G. Newton. *Scattering Theory of Waves and Particles*. New York–Heidelberg–Berlin: Springer-Verlag, 1982.
A. S. Davydov. *Quantum Mechanics*. Oxford: Pergamon Press, 1965.

[24] B. S. De Witt and N. Graham (Eds.). *The Many-Worlds Interpretation of Quantum Mechanics*. Princeton: Princeton University Press, 1973.
B. d'Espagnat (Ed.). *Foundations of Quantum Mechanics*. New York: Academic Press, 1971.
M. Jammer. *The Philosophy of Quantum Mechanics*. New York: Wiley, 1977.
E. Scheibe. *The Logical Analysis of Quantum Mechanics*. Oxford: Pergamon Press, 1973.
M. Drieschner. *Voraussage–Wahrscheinlichkeit–Objekt*. Springer Lecture Notes in Physics, Vol. 99, 1979.
J. Bub. *The Interpretation of Quantum Mechanics*. Dordrecht: Reidel, 1974.
C. A. Hooker (Ed.). *Contemporary Research in the Foundations and Philosophy of Quantum Theory*. Dordrecht: Reidel, 1973.
C. A. Hooker (Ed.). *The Logico-Algebraic Approach to Quantum Mechanics*, Vol. I, Historical Evolution. Dordrecht: Reidel, 1975.
C. A. Hooker (Ed.). *The Logico-Algebraic Approach to Quantum Mechanics*, Vol. II, Contemporary Consolidation. Dordrecht: Reidel, 1979.
C. A. Hooker (Ed.). *Physical Theory as Logico-Operational Structure*. Dordrecht: Reidel, 1979.
P. Suppes (Ed.). *Logic and Probability in Quantum Mechanics*. Dordrecht: Reidel, 1976.
P. Suppes (Ed.). *Studies in the Foundations of Quantum Mechanics*. Philosophy of Science Association, 1980.
A. R. Marlow (Ed.). *Mathematical Foundations of Quantum Theory*. New York: Academic Press, 1978.
D. G. Holdsworth and C. A. Hooker. A critical survey of quantum logics. In: *Logic in the Twentieth Century*. Milano: Scientia, 1983.

[25] J. von Neumann. *Mathematical Foundations of Quantum Mechanics* (transl. R. T. Beyer). Princeton: Princeton University Press, 1955.

[26] G. Ludwig. *Die Grundagen der Quantenmechanik*. Berlin–Heidelberg–New York: Springer-Verlag, 1954.

[27] K. E. Hellwig. Measuring processes and additive conservation laws. In: *Foundation of Quantum Mechanics*, B. d'Espagnat (Ed.). New York: Academic Press, 1971.

[28] A. Böhm. *Quantum Mechanics*. New York–Heidelberg–Berlin: Springer-Verlag, 1979.

[29] B. Misra and E. C. G. Sudarshan. The Zeno's paradox in quantum theory. *J. Math. Phys.*, **18**, 756–763 (1977).

[30] J. A. Wheeler and R. P. Feynman. Interaction with the absorber as a mechanism of radiation. *Rev. Mod. Phys.*, **17**, 157 (1945).
J. A. Wheeler and R. P. Feynman. Classical electrodynamics in terms of direct interparticle action. *Rev. Mod. Phys.*, **21**, 425 (1949).

[31] G. Süssmann. Die spontane Lichtmission in der unitären Quantenelektrodynamik. Dissertation, Freie Universität Berlin, 1951 (*Z. Phys.*, **131**, 629–662 (1952)).

[32] P. Jordan. *Anschauliche Quantentheorie*. Berlin–Heidelberg–New York: Springer-Verlag, 1936.

[33] M. Drieschner. *Voraussage–Wahrscheinlichkeit–Objekt*. Springer Lecture Notes in Physics, Vol. 99, 1979.

[34] B. S. De Witt and N. Graham (Eds.). *The Many-Worlds Interpretation of Quantum Mechanics*. Princeton: Princeton University Press, 1973.
E. Wigner. The Subject of Our Discussions. In: *Foundation of Quantum Mechanics*, B. d'Espagnat (Ed.). New York: Academic Press, 1971.

[35] G. Ludwig. The connection between the objective description of macrosystems and quantum mechanics of "many particles". In: *Old and New Questions in Physics, Cosmology, Philosophy and Theoretical Biology*, A. van der Merwe (Ed.). New York: Plenum, 1983.

[36] G. Ludwig. Measuring and preparing processes. In: *Foundation of Quantum Mechanics and Ordered Linear Spaces*. Springer Lecture Notes in Physics, Vol. 29, 1974.

[37] G. Ludwig, Quantum theory as a theory of interactions between macroscopic systems which can be described objectively. *Erkenntnis*, **16**, 359–387 (1981).

[38] E. Schrödinger. Die Gegenwärtige Situation in der Quantenmechanik. *Die Naturwiss.*, **23**, 807–812, 823–828, 844–849 (1935).

[39] S. Grossmann. Occupation number representation with localized one-particle functions. *Physica*, **29**, 1373–1392 (1963).
S. Grossmann. Macroscopic time evolution and master equation. *Physica*, **30**, 779–807 (1964).

[40] M. Jammer. *The Philosophy of Quantum Mechanics*. New York: Wiley, 1977.
E. Scheibe. *The Logical Analysis of Quantum Mechanics*. Oxford: Pergamon Press, 1973.

[41] V. S. Varadarajan. *Geometry of Quantum Theory*, Vol. II. New York: Van Nostrand–Reinhold, 1970.

[42] K. Kraus. *States, Effects and Operations*. Springer Lecture Notes in Physics, Vol. 190, 1983.

Index